The Most Successful Way of JEE Preparation

DPP

Daily Practice Problems

JEE Main & Advanced

Electrochemistry & Chemical Kinetics

with Solid State & Solutions

CHEMISTRY Vol-5

Dr. RK GUPTA
Retd. Associate Professor
KGPG College, Simbhawali, GZB

arihant
ARIHANT PRAKASHAN (SERIES), MEERUT

DPP
Daily Practice Problems

arihant

ARIHANT PRAKASHAN (SERIES), MEERUT

Administrative & Production Offices

Corporate Office 'Ramchhaya' 4577/15, Agarwal Road, Darya Ganj, New Delhi -110002
Tele: 011- 47630600, 43518550; Fax: 011- 23280316

Head Office Kalindi, TP Nagar, Meerut (UP) - 250002
Tel: 0121-2401479, 2512970, 4004199; Fax: 0121-2401648

Sales & Support Offices

Agra, Ahmedabad, Bengaluru, Bareilly, Chennai, Delhi, Guwahati, Hyderabad, Jaipur, Jhansi, Kolkata, Lucknow, Meerut, Nagpur & Pune

ISBN : 978-93-13193-44-9

Published by Arihant Publications (I) Ltd.

Production Team

Publishing Managers	Mahendra Singh Rawat & Keshav Mohan	*Inner Designer*	Mazher Chaudhary
Project Head	Sorabh Chaudhary	*Page Layouting*	Sunil Sharma, Hemant
Cover Designer	Shanu Mahad	*Proof Reading*	Salman

For further information about the products from Arihant, log on to www.arihantbooks.com or email to info@arihantbooks.com

/arihantpub /@arihantpub Arihant Publications /arihantpub

ABOUT DPP

Daily Practice Problems

Indian Institutes of Technology better known as IITs have always been the dream destination of each science group students who want to pursue their career as Engineers. These are the institutes which have got very high value not even in India but in the whole world. To get the admission in such institutes one need to face entrance Examinations; JEE Main & JEE Advanced, out of which JEE Main is considered as the preliminary exam & JEE Advanced is the final exam to get into the IITs. Approximately per year 12 lakh students enroll themselves for JEE Main and only the top 2.25 lakh in JEE Main merit can attempt for JEE Advanced, the fact clearly express the degree of toughness of JEE Main & JEE Advanced.

To have success in such an exam one needs to be very planned & systematic in its preparation. A very popular wise saying 'Rome was not built in a day' suggests that to achieve a big target like IITs we need to work continuously. For continuous work, one need to have proper planning that how much work he/she has to do on regular basis, i.e., on daily basis. Each student preparing for JEE has many books but unfortunately none of them gives the students a plan for daily studies, which is the most effective success key for JEE.

Keeping the daily practice needs of the students in mind we have come up with DPP Series (Daily Practice Problems) having approximately 26 books. The primary focus of this series is on achieving success through practice & proper planning. For that we have divided the whole syllabus of JEE Main & JEE Advanced; Physics, Chemistry & Mathematics, into different volumes (around 8 volumes of Mathematics & 9 volumes each of Physics & Chemistry), each volume has Daily Practice Problems Sheets having specific questions on various Topics of the individual chapter of the volume ensuring the complete Practice & Assessment of the topic.

DPP
Daily Practice Problems

In this way we have tried to give you an effective daily planning for JEE keeping in mind the daily hours that a student can devote to his practice, you are just required to practice one sheet of each subject on daily basis. This habit of daily practice will keep you away from exam stress and thoughts like; how much I have prepared/am I preparing properly or not/ do I need to study more on daily basis or what I am doing is ok, etc., which regularly tease your mind and can give you panic at any stage of your preparation.

Some special features that we have tried to incorporate in our Daily Practice Problems (DPP) series are discussed below

1. Micro Level coverage of all the topics of a Chapter/ Unit given in JEE Main & Advanced Syllabus.
2. Each DPP has various types of questions; Subjective Questions, Single/ Multiple Correct Options, Statements type, Comprehension based, Integer Value Answer type, Matching, etc., making you practice & ready for all the question formats of JEE.
3. New & innovative questions in each DPP challenging & improving your problem solving skills.
4. Along with Topical Coverage, Revisal DPPs for JEE Main & JEE Advanced with each Chapter, covering multidisciplinary questions proving comprehensive practice.
5. JEE Main & JEE Advanced Archive (collection of last 13 years' exams questions) with each chapter.
6. Complete Solutions for each DPP.

We hope that DPP series will prove to be very beneficial for the students preparing for JEE, by giving them daily practice in a systematic & planned manner. The suggestions/criticism from the students, teachers & parents about DPP series will be welcomed whole heartedly and we promise to update this series accordingly from time to time.

DPP

Daily Practice Problems

DAILY TASKS...

1 SOLID STATE

2 SOLUTIONS

DPP
Daily Practice Problems

3 ELECTROCHEMISTRY

4 CHEMICAL KINETICS

1. Solid State

DPP
DAILY PRACTICE PROBLEMS

DPP-1 Classification and Characteristics

Only One Option Correct Type

Working Space

Direction *(Q. Nos. 1-12) This section contains 12 multiple choice questions. Each question has four choices (a), (b), (c) and (d), out of which ONLY ONE is correct.*

1 Which of the following conditions favours the existence of a substance in the solid state?

a. Weak cohesive forces
b. High thermal energy
c. Low temperature
d. High temperature

2 Crystalline solids and amorphous solids differ in

a. shape
b. cleavage property
c. heat of fusion
d. All of these

3 A crystalline solid is characterised by following but not by

a. isotropic nature
b. definite characteristic heat of fusion
c. a regular periodically repeated pattern of arrangement of constituent particles in the entire crystal
d. a true solid

4 Consider the following incomplete sentences :

I. When most liquids are cooled, they eventually freeze and form ....
II. Ice is a ....

These are completed by

	I	II
a.	crystalline solid	crystalline solid
b.	supercooled liquid	supercooled liquid
c.	crystalline solid	supercooled liquid
d.	supercooled liquid	crystalline solid

5 Silica (SiO_2) can be crystalline as well as amorphous with following properties:

I. Crystalline.
II. The SiO_4^{2-} tetrahedra are randomly joined giving rise to polymeric chains or three-dimensional sheets.
III. It has high and sharp melting point.

Which of the properties is/are matched with amorphous solids?

a. I and II
b. I and III
c. II and III
d. Only II

6 As it cools, olive oil solidifies and forms a solid over a wide range of temperature. Which term best describes the solid?

a. Ionic
b. Covalent network
c. Metallic
d. Molecular crystal

7 At very low temperature, oxygen (O_2) freezes and forms a crystalline solid. Which term best describes the solid?

a. Covalent network
b. Molecular crystal
c. Metallic
d. Ionic

Working Space

8 Which set forms covalent crystals involving sp^3-hybrid orbitals?

a. C, Si, Ge, Sn b. B, Al, Ga, In c. C, B, N, P d. Li, Na, K, Rb

9 Which of the following characteristics is/are possessed by non-polar molecular solids?

a. Atoms or molecules are held by weak dispersion forces or London forces
b. They are soft and non-conductors of electricity
c. They have low melting point and are usually in liquid or gaseous state at room temperature and pressure
d. All of the above

10 Refractive index of a solid is observed to have the same value along all directions. Solid is said to have

a. isotropy b. anisotropy c. definite shape d. None of these

11 Which of the following statements is not true about amorphous solids?

a. They may become crystalline on heating at a certain temperature
b. When kept for a long time, they may become crystalline
c. They can be moulded by heating
d. They are anisotropic in nature

12 The sharp melting point of crystalline solids compared to amorphous solids is due to

a. same arrangement of constituent particles in different directions
b. different arrangement of constituent particles in different directions
c. a regular arrangement of constituent particles observed over a long distance in the crystal lattice
d. a regular arrangement of constituent particles observed over a short distance in the crystal lattice

One or More than One Options Correct Type

Direction (Q. Nos. 13-19) *This section contains 7 multiple choice questions. Each question has four choices (a), (b), (c) and (d), out of which ONE or MORE THAN ONE are correct.*

13 Of the substances Xe, CH_3Cl and HF,

a. the smallest dipole-dipole forces exist in Xe
b. the largest H-bonding exists in HF
c. the largest dispersion forces exist in Xe
d. the largest dispersion forces exist in CH_3Cl

14 Supercooled liquids or pseudo solids include

a. glass b. rubber and plastics
c. quartz glass d. sodium chloride

15 Select the correct statement(s) about crystalline solids.

a. They are called fluids because of their ability to flow
b. The constituent particles are arranged in regular pattern
c. The constituent particles are at fixed positions and can only oscillate about their mean positions
d. Entropy change, $\Delta S < 0$

16 A solid can be characterised by

a. definite mass, volume and shape b. short intermolecular distances
c. strong intermolecular forces d. rigid and incompressible nature

17 In polar molecular solids,

a. molecules have polar covalent bonds
b. molecules are held by relatively stronger dipole-dipole interactions
c. molecules are held by weak London forces
d. higher melting point as compared to non-polar molecular solids, is observed

Working Space

18 Which of the following characteristics are associated with quartz glass?

a. Definite heat of fusion
b. Supercooled liquid
c. Refractive index same in all the directions
d. A crystalline solid

19 Select the correct statement(s) about graphite and diamond.

a. Both have covalent bonding
b. Both have high melting points
c. Diamond is hard while graphite is soft
d. Diamond is insulator while graphite is conductor

Statement Type

Direction *(Q. Nos. 20-22) This section is based on Statement I and Statement II. Select the correct answer from the codes given below.*

a. Both Statement I and Statement II are correct and Statement II is the correct explanation of Statement I
b. Both Statement I and Statement II are correct but Statement II is not the correct explanation of Statement I
c. Statement I is correct but Statement II is incorrect
d. Statement II is correct but Statement I is incorrect

20 **Statement I** At sufficiently low temperature, constituent particles can still oscillate about their mean positions and the substance exists in solid state.

Statement II At sufficiently low temperature, thermal energy is low and intermolecular forces bring them so close that they cling to one another and occupy fixed position.

21 **Statement I** Amorphous solids are isotropic in nature.

Statement II Physical properties would be same along all directions.

22 **Statement I** A sample of amorphous boric oxide (B_2O_3) could be considered a form of liquid even though it looks like a solid, rigid and shatters, when struck with a hammer.

Statement II It is not crystalline, the B—O network has a random (irregular) arrangement.

Matching List Type

Direction *(Q. Nos. 23 and 24) Choices for the correct combination of elements from Column I and Column II are given as options (a), (b), (c) and (d), out of which one is correct.*

23 Match the nature of bonding in Column I with the solid (having that bonding) in Column II and select the answer from the codes given below.

	Column I		Column II
i.	van der Waals'	p.	Al, Cd
ii.	Ionic	q.	CO_2, H_2
iii.	Metallic	r.	SiO_2, diamond
iv.	Covalent solid	s.	MgO, NaCl

Codes

	i	ii	iii	iv		i	ii	iii	iv
a.	p	q	r	s	b.	q	s	p	r
c.	q	s	r	p	d.	p	r	s	q

Working Space

24 Match the solids given in Column I with their properties given in Column II and select the answer from the codes given below.

	Column I		Column II
i.	Hydrogen bonded molecular solids	p.	Electrical insulators in solid state
ii.	Ionic solids	q.	Highly malleable and ductile
iii.	Metallic solids	r.	Volatile
iv.	Covalent solids	s.	Giant molecules

Codes

	i	ii	iii	iv		i	ii	iii	iv
a.	p	q	r	s	b.	s	p	r	q
c.	r	p	q	s	d.	r	p	s	q

Comprehension Type

***Direction** (Q. Nos. 25 and 26) This section contains a paragraph, which describing theory, experiments, data, etc. Two questions related to the paragraph have been given. Each question has only one correct answer among the four given options (a), (b), (c) and (d).*

Passage

The melting points of RbBr is 955 K while that of NaF is 1260 K.

25 Melting point of NaF is much larger than that of RbBr. It is mainly due to

a. the two crystals are not isomorphous
b. the molar mass of NaF is smaller than that of RbBr
c. the bond in RbBr has more covalent character than the bond in NaF
d. the internuclear distance is greater for RbBr than for NaF

26 Select the correct statement(s).

a. Solubility of RbBr in water is larger than that of NaF
b. Both are covalent compounds
c. Both are insoluble in water
d. Both are coloured compounds

One Integer Value Correct Type

***Direction** (Q. Nos. 27-30) This section contains 4 questions. When worked out will result in an integer from 0 to 9 (both inclusive).*

27 How many of the following are amorphous solids?
Graphite, quartz glass, chrome alum, silicon carbide, quartz crystalline, rubber, plastic.

28 How many of the following are anisotropic in nature?
Graphite, diamond, quartz glass, quartz crystalline, rubber, plastic, sodium chloride.

29 How many of the following are covalent solids?
CO_2, SO_2, SiO_2, MgO, ZnS, SiC, AlN, C (graphite), CaF_2, Mg, C (diamond)

30 Fullerene, an allotrope of the element (X) discovered in 1990 as a constituent of soot with the extraordinary shape of a soccer ball. What is the atomic number of the element (X)?

DPP-2 Unit Cells, Lattices and Bragg's Equation

Only One Option Correct Type

Working Space

Direction (Q. Nos. 1-10) *This section contains 10 multiple choice questions. Each question has four choices (a), (b), (c) and (d), out of which ONLY ONE is correct.*

1 Bravais three-dimensional lattices are of

a. 8 types
b. 9 types
c. 12 types
d. 14 types

2 The lattice site in a pure crystal cannot be occupied by

a. molecule
b. ion
c. electron
d. atom

3 Number of formula units in unit cells of MgO (rock salt), ZnS (zinc blende) and Pt (fcc) respectively are

a. 4, 3, 2
b. 4, 3, 4
c. 4, 4, 4
d. 4, 3, 1

4 A compound is formed by elements *A* and *B* and has cubic structure. *A*-atoms are at the corners and *B*-atoms are at the face centres. Thus, it is

a. AB
b. AB_2
c. AB_3
d. A_3B

5 The mass of a unit cell of NaCl corresponds to

a. 1 Na^+, 6 Cl^-
b. 1 Cl^-, 6 Na^+
c. 1 Na^+, 1 Cl^-
d. 4 Cl^-, 4 Na^+

6 At what angles for the first-order diffraction, spacing between two planes respectively are λ and $\frac{\lambda}{2}$?

a. 0°, 90°
b. 90°, 0°
c. 30°, 90°
d. 90°, 30°

7 A solid has a structure in which W-atoms are located at the cube corners of the unit cell, O-atoms are located at the cube edges and the Na-atoms at the cube centres. Lattice is of the type with formula

a. simple cubic, $NaWO_3$
b. simple cubic, Na_2WO_3
c. tetrahedral, $NaWO_3$
d. body-centred, $NaWO_3$

8 The number of atoms in primitive cubic unit cell, body-centred cubic unit cell and face-centred cubic unit cell are *A*, *B* and *C* respectively. Select the correct values.

	A	*B*	*C*
a.	1	2	3
b.	1	2	4
c.	1	1	3
d.	2	3	4

Working Space

9 For a second order diffraction using X-ray Bragg's equation, for a crystal with separation between two layers as d Å, following graphical study was observed when varying θ at different λ.

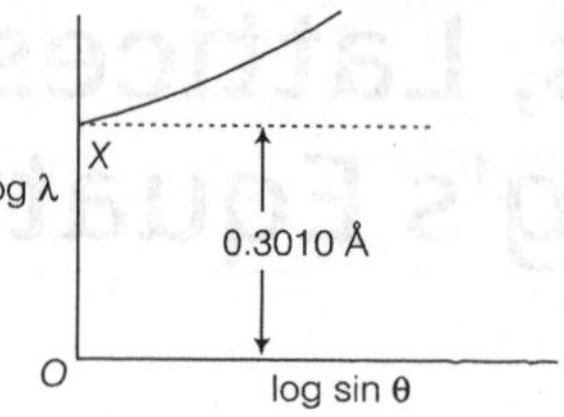

What is the value of d?

a. 0.3010 Å
b. 0.6020 Å
c. 2.0 Å
d. 4.0 Å

10 Match the primitive unit cells given in Column I with their axial angles axial distances given in Column II and select the correct answer from the codes given

	Column I		Column II	
i.	Triclinic	p.	$\alpha = \beta = \gamma = 90°$	$a = b = c$
ii.	Hexagonal	q.	$\alpha = \beta = \gamma = 90°$	$a = b \neq c$
iii.	Cubic	r.	$\alpha \neq \beta \neq \gamma = 90°$	$a \neq b \neq c$
iv.	Tetragonal	s.	$\alpha = \beta = 90°, \gamma = 120°$	$a = b \neq c$

Codes

	i	ii	iii	iv
a.	r	s	p	q
b.	r	s	q	p
c.	s	r	p	q
d.	p	q	r	s

One or More than One Options Correct Type

Direction (Q. Nos. 11-13) *This section contains 3 multiple choice questions. Each question has four choices (a), (b), (c) and (d), out of which ONE or MORE THAN ONE are correct.*

11 In which of the following primitive cells axial distances or edge lengths (a, b, c) are different?

a. Orthorhombic
b. Tetragonal
c. Monoclinic
d. Triclinic

12 In which of the following primitive cells axial distances or edge lengths (a, b, c) are equal?

a. Cubic
b. Hexagonal
c. Orthorhombic
d. Rhombohedral

13 A crystal lattice is characterised by which following facts?

a. Each point in the lattice is called lattice point or lattice site
b. Each point in it is a constituent particle which can be an atom, anion or a molecule
c. Unit cell is the smallest portion of a crystal lattice
d. Unit cell of a crystal lattice is designated by edge lengths a, b, c and axial angles α, β, γ

Comprehension Type

Direction (Q. Nos. 14 and 15) *This section contains a paragraph, which describing theory, experiments, data, etc. Two questions related to the paragraph have been given. Each question has only one correct answer among the four given options (a), (b), (c) and (d).*

Passage

Perovskite (*A*), a mineral containing calcium (o), oxygen (●) and titanium (◎) crystallises in the following cubic unit cell:

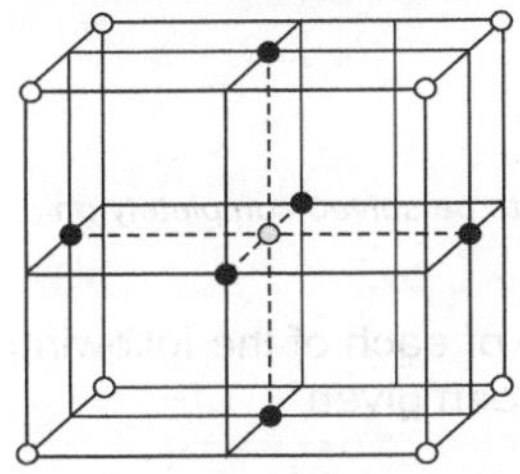

14 What is the formula of *A*?

a. $CaTiO_2$ b. $CaTiO_3$
c. Ca_2TiO_3 d. $Ca(TiO_2)_2$

15 What is the oxidation number of titanium in the given lattice?

a. +2 b. +3
c. +4 d. −4

One Integer Value Correct Type

Direction (Q. Nos. 16-20) *This section contains 5 questions. When worked out will result in an integer from 0 to 9 (both inclusive).*

16 A unit cell is characterised by how many parameters?

17 How many types of primitive unit cells are there?

18 How many of the following primitive unit cells have all the axial angles equal but axial distances different?

Cubic, tetragonal, orthorhombic, hexagonal, trigonal, monoclinic, triclinic.

19 A second order Bragg's diffraction of X-rays from a set of parallel planes separated by 1.155 Å occurs at an angle 60° using wavelength of *x*Å. What is the value of *x*?

20 Number of unit cells in 240 g of element *X* (atomic mass 40) which crystallises in bcc pattern in $y\,N_0$. What is the value of *y*?

Working Space

DPP-3 Packing in Crystals

Subjective Questions

Working Space

Direction (Q. Nos. 1-3) These questions are subjective in nature, need to be solved completely on notebook.

1 Predict the coordination number of the cation in crystals of each of the following compounds for which radii of cations and anions have been given.

	Compound	Radius	
		Cation	Anion
i.	MgO	0.65 Å	1.40 Å
ii.	MgS	0.65 Å	1.84 Å
iii.	CsCl	1.69 Å	1.81 Å

2 Each of the following compounds crystallises in a structure matching that of NaCl, CsCl, ZnS or CaF_2. From the radius ratio, predict which structure is formed by each.

i. NaF ii. CoO iii. CsBr

iv. BeO v. BaF_2

Given, $r_{Na^+} = 95\,pm$, $r_{Co^{2+}} = 32\,pm$

$r_{Cs^+} = 169\,pm$, $r_{Be^{2+}} = 31\,pm$, $r_{Ba^{2+}} = 135\,pm$

$r_{F^-} = 136\,pm$, $r_{Br^-} = 195\,pm$, $r_{O^{2-}} = 140\,pm$

3 Given radius of $I^- = 219$ pm and internuclear distance of

RbI = 367 pm, KCl = 314 pm

RbBr = 344 pm, NaCl = 279 pm

KBr = 329 pm, NaF = 231 pm

Derive the value of radius of F^-.

Only One Option Correct Type

Direction (Q. Nos. 4-14) This section contains 11 multiple choice questions. Each question has four choices (a), (b), (c) and (d), out of which ONLY ONE is correct.

4 CsCl crystallises in a cube that has Cl^- at each corner and Cs^+ at the centre of the unit cell. If $r_{Cs^+} = 169$ pm and $r_{Cl^-} = 181$ pm, then edge length of the cube is

a. 350 pm b. 175 pm
c. 303 pm d. 404 pm

5 A mineral having the formula AB_2 crystallises in the cubic closest - packed lattice with the *A*-atoms occupying the lattice points. Fraction of the tetrahedral sites occupied by *B*-atoms is

a. 0.52 b. 0.68
c. 0.74 d. 1.00

6 Silver metal crystallises in a cubic closest packed arrangement with edge length 407 pm. Thus, radius of the silver atom is

a. 203.5 pm b. 176.2 pm
c. 143.9 pm d. 407.0 pm

7 What is the formula of a magnetic oxide of cobalt used in recording tapes that crystallises with cobalt atoms occupying one-eighth of the tetrahedral holes and one-half of the octahedral holes in a closest packed array of oxide ions?

a. CoO b. Co_2O_3
c. CoO_2 d. Co_3O_4

8 The total number of tetrahedral voids in the face-centred unit cell is

a. 6 b. 8
c. 10 d. 12

9 Packing efficiency (%) of different types of unit cells is given. Select the correct packing efficiency.

	Simple cubic	bcc	hcp	ccp
a.	52	68	74	74
b.	52	74	68	52
c.	52	68	72	74
d.	52	74	74	68

10 Coordination numbers in a square close packed structure and hexagonal close packed structure respectively, are

a. 2, 6 b. 3, 6
c. 4, 6 d. 6, 4

11 The length of the unit cell edge of a bcc lattice metal is 352 pm. Thus, volume of atoms in one mole of the metal is

a. 8.93 cm^3 b. 7.25 cm^3 c. 1.21cm^3 d. 4.00 cm^3

12 A compound M_pX_q has cubic close packing arrangement of X. Its unit cell structure is shown below. The empirical formula of the compound is

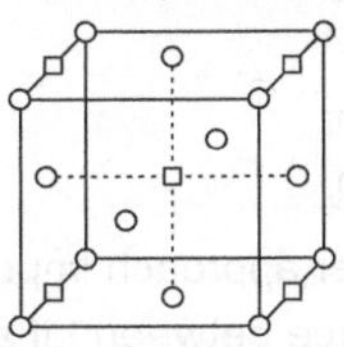

a. MX b. MX_2 c. M_2X d. M_5X_{14} **[IIT JEE 2012]**

13 Radii of A^+ and that of X^- and Y^- have been given as

$$A^+ = 1.00 \text{ pm}$$
$$X^- = 1.00 \text{ pm}$$
$$Y^- = 2.00 \text{ pm}$$

Thus, ratio of volumes of AX and AY unit cells is

a. 0.057 b. 1.0
c. 0.217 c. 17.54

14 In the structure of NaCl given below, ratio r_{Na^+}/r_{Cl^-} is

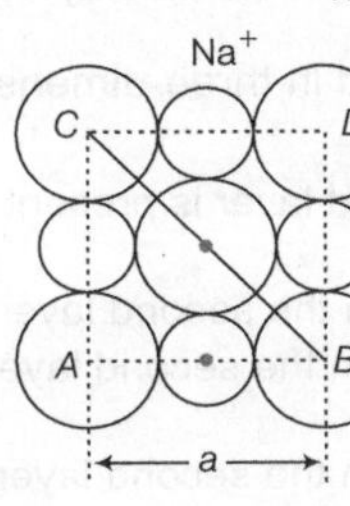

a. 0.414 b. 0.681 c. 0.745 d. 0.520

Working Space

Matching List Type

Direction (*Q. No. 15*) *Choices for the correct combination of elements from Column I and Column II are given as options (a), (b), (c) and (d), out of which one is correct.*

15 Match the crystal given in Column I with the coordination number (of cation and anion) given in Column II and select the answer from the codes given below.

	Column I		Column II
i.	CsBr	p.	4-4
ii.	NaCl	q.	6-6
iii.	ZnS	r.	8-4
iv.	CaF_2	s.	8-8

Codes

	i	ii	iii	iv		i	ii	iii	iv
a.	p	s	r	q	b.	s	q	p	r
c.	s	q	r	p	d.	s	p	q	r

Statement Type

Direction (*Q. No. 16*) *This section is based on Statement I and Statement II. Select the correct answer from the codes given below.*

a. Both Statement I and Statement II are correct and Statement II is the correct explanation of Statement I
b. Both Statement I and Statement II are correct but Statement II is not the correct explanation of Statement I
c. Statement I is correct but Statement II is incorrect
d. Statement II is correct but Statement I is incorrect

16 The metal ion-halide ion distances in several alkali metal halides are given below :

	I^-	Cl^-
Li^+	0.310 nm	0.249 nm
Na^+	0.317 nm	0.279 nm

Statement I Iodide-iodide contact of Li I and NaI makes closer approach impossible.

Statement II There is small difference in internuclear distance between LiI and NaI compared to that between LiCl and NaCl.

One or More than One Options Correct Type

Direction (*Q. Nos. 17-19*) *This section contains 3 multiple choice questions. Each question has four choices (a), (b), (c) and (d), out of which ONE or MORE THAN ONE are correct.*

17 In hexagonal close packing,

a. packing efficiency is 74%
b. the coordination number is 12
c. tetrahedral voids of the second layer are covered by the spheres of the third layer
d. spheres of the fourth layer are exactly aligned with those of the first layer

18 Which of the following are correct about the voids formed in three-dimensional hexagonal close-packed structure?

a. A tetrahedral void is formed when a sphere of the second layer is present above triangular void in the first layer
b. All the triangular voids are not covered by the spheres in the second layer
c. Tetrahedral voids are formed when the triangular voids in the second layer lie above the triangular voids in the first layer
d. Octahedral voids are formed when the triangular voids in the second layer exactly overlap with similar voids in the first layer

19 In which of the following arrangements octahedral voids are formed?

a. hcp
b. bcc
c. Simple cubic
d. fcc

Comprehension Type

Direction *(Q. Nos. 20-23) This section contains 2 paragraphs, each describing theory, experiments, data, etc. Four questions related to the paragraph have been given. Each question has only one correct answer among the four given options (a), (b), (c) and (d).*

Passage I

Potassium crystallises in a bcc lattice as shown

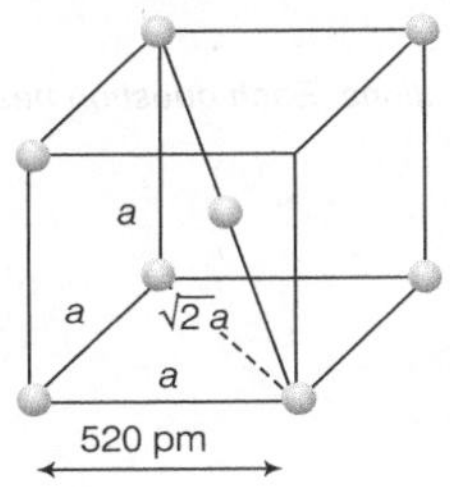

20 Distance between the two nearest neighbours and between next nearest neighbour respectively, is

a. 450 pm, 520 pm
b. 520 pm, 450 pm
c. 450 pm, 450 pm
d. 520 pm, 520 pm

21 Number of neighbours and next - nearest neighbours of K respectively, are

a. 8, 8
b. 6, 6
c. 8, 6
d. 6, 8

Passage II

Metallic gold crystallises in a fcc lattice. The length (a) of the cubic unit cell is 407 pm. Atomic mass of gold is 197 amu.

22 Closest distance between gold atoms is

a. 287.8 pm
b. 407 pm
c. 203.5 pm
d. 235 pm

23 Density of gold is

a. 19.7 g cm^{-3}
b. 19.4 g cm^{-3}
c. 9.7 g cm^{-3}
d. None of these

One Integer Value Correct Type

Direction *(Q. Nos. 24-26) This section contains 3 questions. When worked out will result in an integer from 0 to 9 (both inclusive).*

24 Atoms of an element B form hcp lattice and those of the element A occupy 2/3rd of tetrahedral voids. How many number of atoms of A and B (in total) are present in the hcp lattice?

25 How many octahedral voids are formed in one unit cell of a hcp or ccp lattice?

26 Titanium metal has a density of 4.54 g cm^{-3} and an edge length of 412.6 pm. How many atoms are there in the unit cell? (Ti = 48)

Working Space

DPP-4 Imperfection in Crystals

Only One Option Correct Type

Direction *(Q. Nos. 1-9) This section contains 9 multiple choice questions. Each question has four choices (a), (b), (c) and (d), out of which ONLY ONE is correct.*

Working Space

1 Structure shown in the figure represents

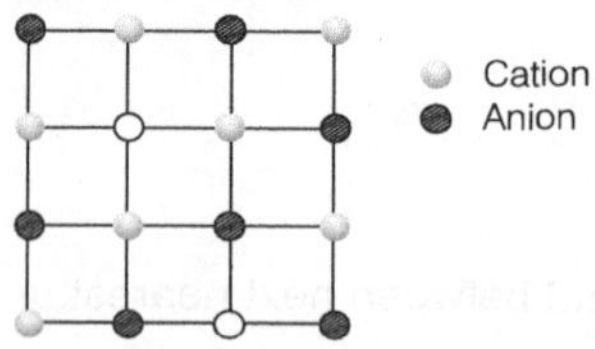

a. Schottky defect
b. Frenkel defect
c. metal excess defect
d. None of these

2 Structure shown in the figure represents

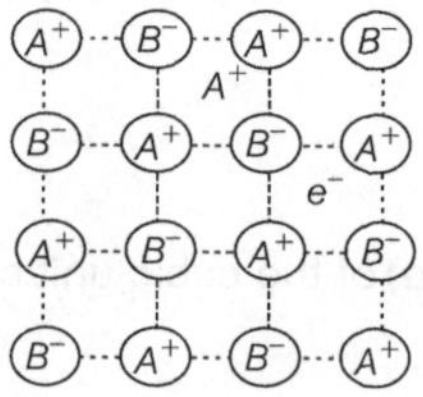

a. metal excess defect because of absence of anion
b. metal excess defect because of excess of cation
c. Frenkel defect
d. Schottky defect

3 Which of the following defects is also known as dislocation defect?

a. Frenkel
b. Schottky
c. Non-stoichiometric
d. Simple interstitial

4 Cations are present in the interstitial sites in

a. Frenkel defect
b. Schottky defect
c. vacancy defect
d. metal deficiency defect

5 Select the correct statement(s) about non-stoichiometric compounds.

a. They are called Berthollide compounds
b. They do not obey the law of constant composition
c. Electrical neutrality is maintained either by having extra electrons in the structure or by changing the charge on some of the metal ions
d. All of the above

6 In a face-centred cubic lattice, atom *A* occupies the corners of the cube and atom *B* occupies the face-centred positions. If one atom of *B* is missing from one of the face-centred points, the formula of the compound is

a. A_2B
b. AB_2
c. A_2B_2
d. A_2B_5

Working Space

7 An element crystallises in fcc lattice having edge length 400 pm. Maximum radius of the atom which can be placed in the interstitial site without distorting the structure is

a. 58.55 pm b. 117 pm c. 166 pm d. 83 pm

8 Experimentally, it was found that a metal oxide has formula $M_{0.98}O$. Metal M is present as M^{2+} and M^{3+} in its oxide. Fraction of the metal which exists as M^{3+} would be **[JEE Main 2013]**

a. 7.01% b. 4.08%
c. 6.05% d. 5.08%

9 If NaCl is doped with 10^{-3} mole % of $SrCl_2$ then, number of cationic vacancies is

a. 10^{-5} mol^{-1} b. 6.02×10^{20} mol^{-1}
c. 6.02×10^{18} mol^{-1} d. 6.02×10^{-18} mol^{-1}

Statement Type

Direction (Q. Nos. 10-13) This section is based on Statement I and Statement II. Select the correct answer from the codes given below.

a. Both Statement I and Statement II are correct and Statement II is the correct explanation of Statement I
b. Both Statement I and Statement II are correct but Statement II is not the correct explanation of Statement I
c. Statement I is correct but Statement II is incorrect
d. Statement II is correct but Statement I is incorrect

10 **Statement I** Zinc oxide is white in colour at room temperature.
Statement II On heating, it loses oxygen and turns yellow.

11 **Statement I** The anionic sites occupied by unpaired electrons are called *F*-centres.
Statement II They impart yellow colour to the crystals of NaCl.

12 **Statement I** Schottky defects involving stoichiometric cation and anion vacancies are likely for compounds and ions of comparable size.
Statement II Schottky defects minimise mechanical distortions and charge displacement.

13 **Statement I** Frenkel defects involving the migration of ions to interstitial sites are more likely for covalent compounds and for large (and highly polarisable) anions.
Statement II Large polarisable anions make it easier to accommodate the displaced cations and minimise repulsive effect.

One or More than One Options Correct Type

Direction (Q. Nos. 14-19) This section contains 6 multiple choice questions. Each question has four choices (a), (b), (c) and (d), out of which ONE or MORE THAN ONE are correct.

14 AgBr(*s*) crystals can exhibit one or more types of defect which are

a. Schottky defect b. Frenkel defect
c. metal excess defect d. metal deficiency defect

15 Select the correct statement(s) about Schottky defect.

a. Equal number of cations and anions are missing
b. It increases density of the substance
c. There is one Schottky defect per 10^{16} ions
d. Cations and anions are of different sizes

16 Which of the following are correct defects in crystals?

a.

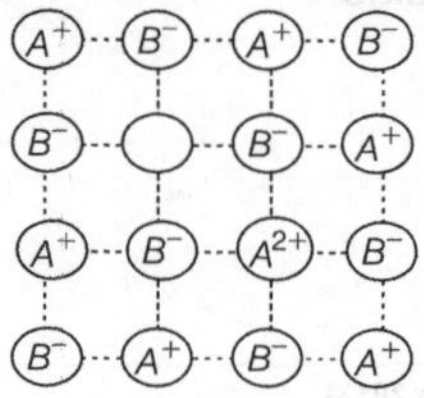

b.

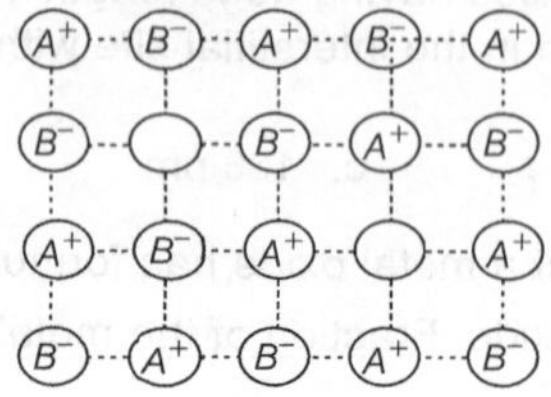

c.

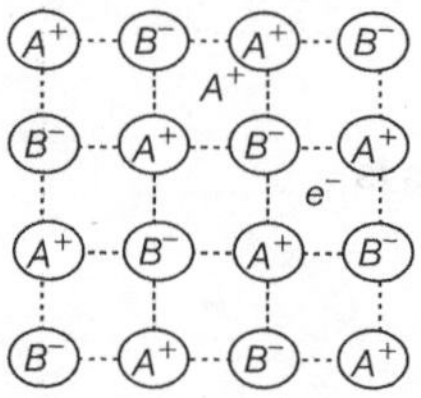

d.

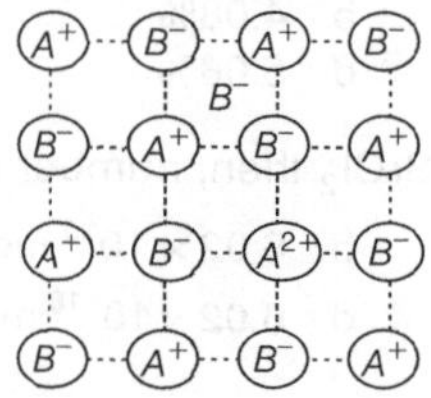

17 Select the correct statements about the given structure.

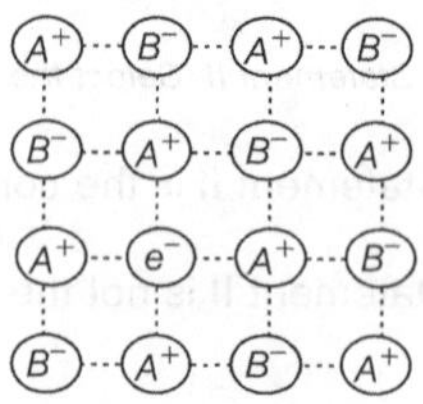

a. This represents metal excess defect and is formed by crystals which would be expected to form Schottky defect
b. Anion sites occupied by electrons are called *F*-centres
c. These are paramagnetic in nature
d. NaCl is yellow and KCl is blue-lilac in colour when heated, probably due to this defect

18 Select the correct statement(s) about following structure.

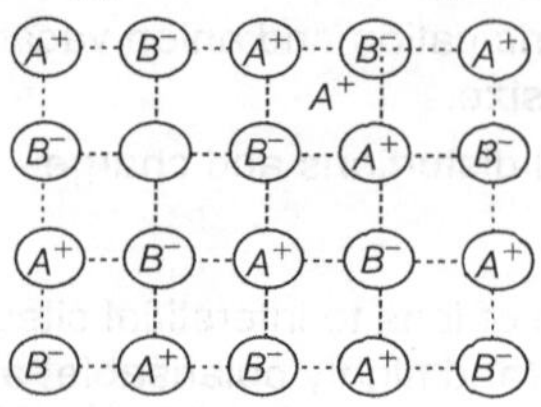

a. It represents Frenkel defect
b. Cation is smaller than anion and there is probability that cations occupy interstitial positions
c. Due to the large difference in size of cation and anion, coordination number is usually low
d. Compounds showing this defect have some covalent character

19 Which of the following statement(s) is/are correct?

a. Vacancy defects result in the decrease in the density of the substance
b. Interstitial defects result in the increase in the density of the substance
c. Impurity defect has no effect on the density of the substance
d. Frenkel defect results in an increase in the density of the substance

Comprehension Type

Direction *(Q. Nos. 20-23) This section contains 2 paragraphs, each describing theory, experiments, data, etc. Four questions related to the paragraphs have been given. Each question has only one correct answer among the four given options (a), (b), (c) and (d).*

Passage I

The crystal *AB* (rock salt structure) has molecular weight 6.02 *y* u, where *y* is an arbitrary unit in u. Also, given minimum distance between cation and anion $= y^{1/3}$ nm and observed density = 20 kg m^{-3}.

20 Theoretical density of *AB* is

a. 10 kg m^{-3} b. 5 kg m^{-3} c. 20 kg m^{-3} d. 7 kg m^{-3}

21 Defect present in the crystal is of the type

a. Schottky b. Frenkel c. metal excess d. metal deficient

Passage II

Titanium (II) oxide has a rock salt structure. X-ray diffraction data shows that the length of one edge of the cubic unit cell for TiO with 1:1 ratio of Ti to O is 418 pm and the density is 4.92 g cm^{-3}. (Atomic mass of Ti = 48).

22 What is the type of defect?

a. Frenkel b. Schottky
c. Both (a) and (b) d. None of these

23 % of vacancies (defects) in the structure is

a. 15.5% b. 84.5% c. 62.0% d. 38.0%

Matching List Type

Direction *(Q. No. 24) Choices for the correct combination of elements from Column I and Column II are given as options (a), (b), (c) and (d), out of which one is correct.*

24 Match the defects given in Column I with the statements given in Column II and select the answer from the codes given below.

	Column I		Column II
i.	Simple vacancy defect	p.	Shown by solids and density of the solid remains the same
ii.	Simple interstitial defect	q.	Shown by non-ionic solids and density of the solid decreases
iii.	Frenkel defect	r.	Shown by ionic solids and density of the solid decreases
iv.	Schottky defect	s.	Shown by non-ionic solids and density of the solid increases

Codes

	i	ii	iii	iv		i	ii	iii	iv
a.	p	q	r	s	b.	r	s	p	q
c.	q	s	p	r	d.	q	s	r	p

One Integer Value Correct Type

Direction *(Q. No. 25) This section contains 1 question. When worked out will result in an integer from 0 to 9 (both inclusive).*

25 Calcium crystallises in a face-centred cubic unit cell with *a* = 0.556 nm and density 1.4848 g/cm^3. Percentage of Schottky defects in this crystal is ... .

DPP-5 Magnetic and Electric Properties of Solids and Semiconductors

Only One Option Correct Type

Working Space

Direction *(Q. Nos. 1-7) This section contains 7 multiple choice questions. Each question has four choices (a), (b), (c) and (d), out of which ONLY ONE is correct.*

1 Consider the following species.

I. $Mg(s)$ II. $TiO(s)$

III. $I_2(s)$ IV. $H_2O(s)$

Which of the above solids is/are electrical conductor?

a. Only I b. Only II

c. II and III d. I and II

2 Which kind of defects are introduced by doping?

a. Dislocation b. Schottky defects

c. Frenkel defects d. Electronic defects

3 Silicon doped with electron-rich impurity forms

a. *p*-type semiconductor b. *n*-type semiconductor

c. intrinsic semiconductor d. insulator

4 Select the correct matching.

a. Ferroelectricity ($\uparrow$) ($\uparrow$) ($\uparrow$) ($\uparrow$) ($\uparrow$)

b. Anti-ferroelectricity ($\uparrow$) ($\downarrow$) ($\uparrow$) ($\downarrow$) ($\uparrow$)

c. Both (a) and (b)

d. None of the above

5 Which is not the correct matching of conductivities?

Species	Conductivities ($ohm^{-1}\ m^{-1}$)
a. Metals	10^4 to 10^7
b. Semiconductors	10^{-6} to 10^4
c. Insulators	10^{-20} to 10^{-10}

d. None of the above (all are correct matching)

6 How many energy levels are present in 3*s* conduction band of a single crystal of sodium weighing 26.8 mg?

a. 7.01×10^{20} b. 7.01×10^{23}

c. 6.02×10^{23} d. 2.10×10^{21}

7 In which of the following oxides conducting or insulating property of is dependent on temperature?

a. TiO b. SiO_2

c. Ti_2O_3 d. MgO

Statement Type

Working Space

Direction *(Q. Nos. 8-12) This section is based on Statement I and Statement II. Select the correct answer from the codes given below.*

a. Both Statement I and Statement II are correct and Statement II is the correct explanation of Statement I
b. Both Statement I and Statement II are correct but Statement II is not the correct explanation of Statement I
c. Statement I is correct but Statement II is incorrect
d. Statement II is correct but Statement I is incorrect

8 **Statement I** Graphite is a good conductor of electricity.

Statement II In graphite, carbon atom is sp^2 hybridised.

9 **Statement I** A ferromagnetic substance becomes a permanent magnet, when it is placed in a magnetic field.

Statement II All the domains get oriented in the direction of magnetic field.

10 **Statement I** In the case of anti-ferromagnetic substances, magnetic moment is zero.

Statement II Domains get oriented in the direction of applied magnetic field.

11 **Statement I** The electrical conductivity of a metal decreases as temperature increases.

Statement II This impedes the flow of electrons, when an electric field is applied.

12 **Statement I** Electrical conductivity of semiconductors increases with rise in temperature.

Statement II There is a small energy gap between conduction band and valence band.

One or More than One Options Correct Type

Direction *(Q. Nos. 13-17) This section contains 5 multiple choice questions. Each question has four choices (a), (b), (c) and (d), out of which ONE or MORE THAN ONE are correct.*

13 Select the correct statement(s) about the movement of electrons and holes in a *p*-type semiconductor when an electric field is applied.

a. Electron will move towards positively charged plate through electronic holes
b. Holes will appear to be moving towards the negatively charged plate
c. Both electrons and holes appears to move towards the positively charged plate
d. Neither electrons nor holes move

14 Select the correct statement(s).

a. When silicon is doped with a group 15 elements (P or As), *n*-type semiconductor is formed
b. When silicon is doped with a group 13 elements (B, Al, Ga), *p*-type semiconductor is formed
c. In *n*-type semiconductor, increase in conductivity is due to the negatively charged electrons
d. In *p*-type semiconductor, electrons move towards the positively charged plate through electronic holes under the influence of electric field

15 Select the correct statement(s).

a. Due to the pairing of electrons, magnetic moment is cancelled
b. Paramagnetic substances are weakly attracted by magnetic field
c. Ferromagnetic substances cannot be magnetised permanently
d. The domains in antiferromagnetic substances are oppositely oriented w.r.t. each other

Working Space

16 Germanium has been doped with impurity as shown in I, II, III, IV. Select the correct statement(s).

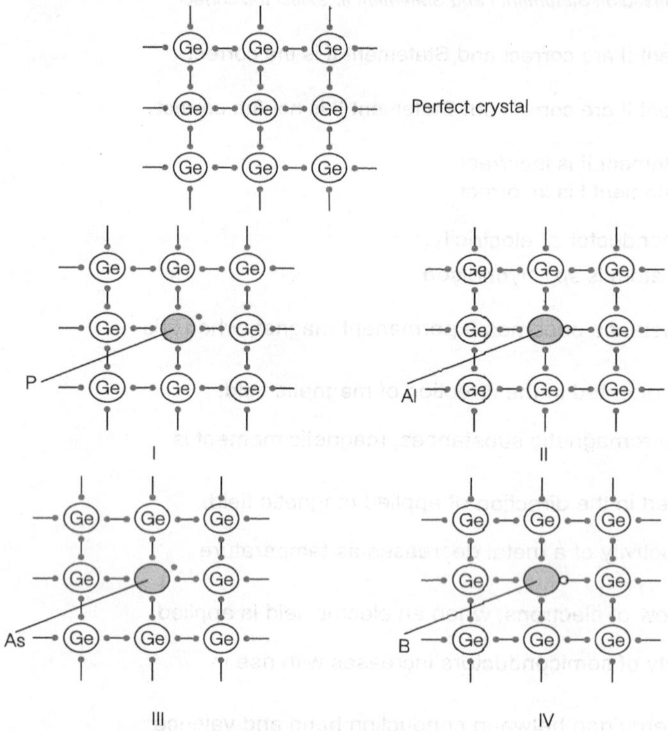

a. *n*-type semiconductor is formed in I and III
b. *p*-type semiconductor is formed in II and IV
c. *n*-type semiconductor is formed in II and IV
d. *p*- type semiconductor is formed in I and III

17 On heating,

a. ferrimagnetic substances lose ferrimagnetism and become paramagnetic
b. ferrimagnetic substances do not lose ferrimagnetism and remain ferrimagnetic
c. a small electric current is produced in polar crystal (pyroelectricity)
d. 6:6 of NaCl coordination changes to 8:8 coordination

Matching List Type

Direction *(Q. No. 18) Choices for the correct combination of elements from Column I and Column II are given as options (a), (b), (c) and (d), out of which one is correct.*

18 Match the items in Column I with the terms in Column II and select the answer from the codes given:

	Column I		Column II
i.	Si doped with As	p.	Electrolytic conductors
ii.	Ge doped with boron	q.	Electronic conductors
iii.	Ag in solid state	r.	*p*-type semiconductor
iv.	NaCl in molten state	s.	*n*-type semiconductor

Codes

	i	ii	iii	iv		i	ii	iii	iv
a.	p	q	s	r	b.	r	q	s	p
c.	s	r	q	p	d.	s	r	p	q

Working Space

Comprehension Type

Direction *(Q. Nos. 19 and 20) This section contains 1 paragraph, which describing theory, experiments, data, etc. Two questions related to the paragraph have been given. Each question has only one correct answer among the four given options (a), (b), (c) and (d).*

Passage

The following pictures show the electrons population of the composite *s-d* bands for three transition metals :

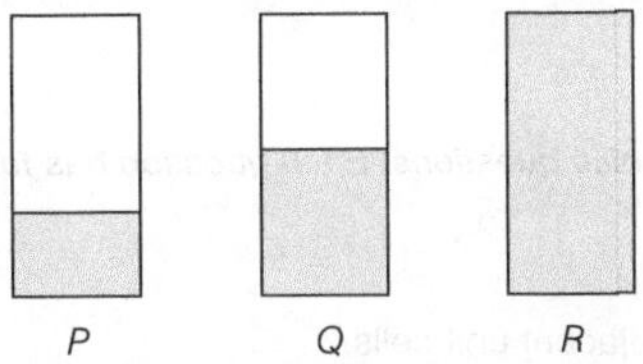

19 Metals with the highest and lowest melting points respectively are

a. *P* and *Q* b. *P* and *R*
c. *Q* and *R* d. *R* and *Q*

20 Hardness of the metals in the increasing order is

a. $P < Q < R$ b. $R < P < Q$
c. $R < Q < P$ d. $Q < R < P$

One Integer Value Correct Type

Direction *(Q. Nos. 21-24) This section contains 4 questions. When worked out will result in an integer from 0 to 9 (both inclusive).*

21 How many of the following species show paramagnetism?

Na^+, O_2, Cu^{2+}, Fe^{2+}, Zn^{2+}, Cr^{3+}, Cu^+, O_2^-, O_2^{2-}

22 Magnetic moment of a compound showing paramagnetism is 54.842×10^{-24} A m^2 (Bohr magneton). How many unpaired electrons are there in the compound?

23 How many of the following are diamagnetic in nature?

TiO_2, VO, V_2O_5, NaCl, C_6H_6, H_2O, CH_3COCH_3, CuCl

24 How many of the following are antiferromagnetic substances?

MnO, MnO_2, Mn_2O, FeO, Fe_2O_3, NiO,
Cr_2O_3, CoO, Co_3O_4, H_2O, NaCl, C_6H_6

SET - 1

Revisal Problems for JEE Main

Only One Option Correct Type

Direction (Q. Nos. 1-18) *This section contains 18 multiple choice questions. Each question has four choices (a), (b), (c) and (d), out of which ONLY ONE is correct.*

Working Space

1 In the cubic close packing, the unit cell has

a. 4 tetrahedral voids each of which is shared by four adjacent unit cells
b. 4 tetrahedral voids within the unit cell
c. 8 tetrahedral voids each of which is shared by four adjacent unit cells
d. 8 tetrahedral voids within the unit cell

2 The unit cell of ammonium chloride is similar to the unit cell of CsCl. Given $r(Cl^-) = 0.181$ nm, then $r(NH_4^+)$ is in the range

a. 0.132 - 0.181 nm
b. 0.075 - 0.132 nm
c. 0.040 - 0.075 nm
d. None of the above

3 M^+ forms $C_{60}M_3$ where, C_{60}^{3-} is fulleride (a superconductor). It forms octahedral holes. If radius of C_{60}^{3-} is 500 pm, then the radius of M^+ is

a. 207 pm
b. 307 pm
c. 400 pm
d. 500 pm

4 The ionic radius of Mn^{2+} ion is 0.080 nm and that of S^{2-} ion is 0.184 nm. Thus, structure of the cubic unit cell of MnS is of the type

a. NaCl
b. ZnS
c. CsCl
d. All of these

5 The density of CaF_2 is 3.180 g cm^{-3} and $a = 0.5451$ nm. Thus, number of Ca^{2+} and F^- ions in the unit cell respectively, are

a. 4 and 4
b. 8 and 8
c. 4 and 8
d. 12 and 8

6 LiCl has NaCl type structure with edge length 514 pm. Assuming that the Li^+ ion is small enough that the chloride ions are in contact, then radius of Cl^- ion is

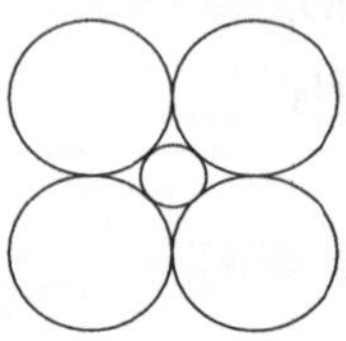

a. 257 pm
b. 182 pm
c. 364 pm
d. 223 pm

7 Angle at which second-order diffraction will occur in an *X*-ray spectrometer when *X*-rays of 200 pm are diffracted by the atoms of a crystal having distance between the two planes as 400 pm is

a. 90°
b. 60°
c. 45°
d. 30°

8 In a closed packed structure of mixed oxides, it is found that lattice has O^{2-} (oxide ion) and one-half of octahedral voids are occupied by trivalent cations (A^{3+}) and one-eighth of tetrahedral voids are occupied by divalent cations (B^{2+}). Thus, mixed oxide is

a. A_2BO_3
b. A_2BO_4
c. AB_2O_4
d. ABO_3

Working Space

9 A compound contains two types of atoms X and Y. Its crystal structure is cubic lattice with X atoms at the corners of the unit cells and Y atoms at the body centres. The simplest formula is

a. XY b. X_2Y
c. XY_2 d. XY_3

10 A crystal is formed by two elements P and Q. The element Q forms cubic close packing and atoms of P occupy one-third of voids. The compound is thus represented as

a. PQ_3 b. P_3Q
c. P_2Q_3 d. P_3Q_2

11 Given, density of NaCl is 2.17×10^3 kg m^{-3}. Thus, edge length of NaCl crystal unit cell is

a. 564 pm b. 282 pm
c. 141 pm d. 56.4 pm

12 Calcium crystallises in a fcc unit cell with $a = 0.556$ nm. It contained 0.1% Frenkel defect and 0.1% Schottky defect. Density with Frenkel defect $= d_f$ and density with Schottky defect $= d_s$, then

a. $d_f = d_s$
b. $d_f > d_s$
c. $d_f < d_s$
d. density is not concerned with defect

13 Unit cell of $(BN)_n$, i.e. boron nitride, is similar to that of graphite except that each layer consists of alternating B and N-atoms in hexagonal rings. It crystallises in a hexagonal unit cell with $a = 0.251$ nm, $z = 2$ for $(BN)_n$.

For graphite, $a = 0.2455$ nm and distance between C-atoms = 0.1415 nm. Thus, distance between B and N-atoms is

a. 0.1415 nm b. 0.2500 nm
c. 0.1450 nm d. 0.1250 nm

14 The following crystallographic data were obtained for a protein.

$$\text{Volume of unit cube} = 1.50 \times 10^{-19}\ \text{cm}^3$$
$$\text{Density} = 1.35\ \text{g cm}^{-3}$$

$z = 4$ and protein fraction = 0.75. Thus, molar mass of protein is

a. 2.3×10^4 g mol^{-1}
b. 3.048×10^4 g mol^{-1}
c. 1.725×10^4 g mol^{-1}
d. None of the above

15 Structure of unit cell is described in the given figure. Fraction of the unit cell is

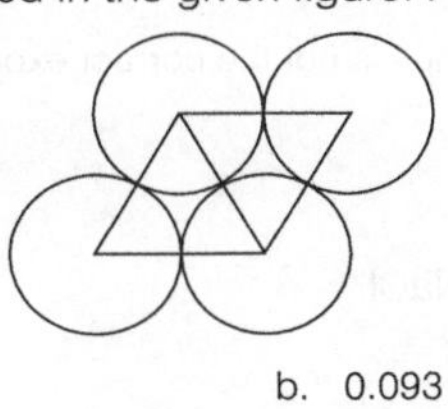

a. 0.907 b. 0.093
c. 0.745 d. 0.521

16 CsCl crystallises in body centred cubic system. If a is its edge length then which of the following expressions is correct? **[JEE Main 2014]**

a. $r_{Cs^+} + r_{Cl^-} = 3a$
b. $r_{Cs^+} + r_{Cl^-} = \dfrac{3a}{2}$
c. $r_{Cs^+} + r_{Cl^-} = \dfrac{\sqrt{3}a}{2}$
d. $r_{Cs^+} + r_{Cl^-} = \sqrt{3}a$

Working Space

17 In the following NaCl type structure :

Salt	Anion-anion distance	Cation-anion distance
MT	2.40 Å	1.70 Å
MQ	1.63 Å	1.15 Å
RT	2.66 Å	1.88 Å
RQ	2.09 Å	1.48 Å

MT is represented as

a. 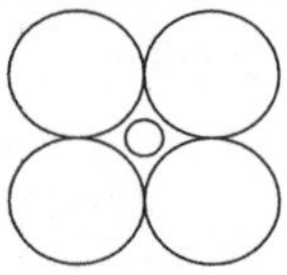

b. 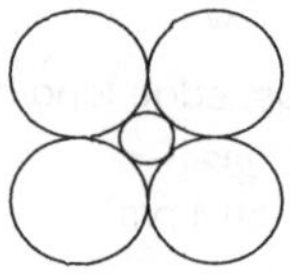

c. 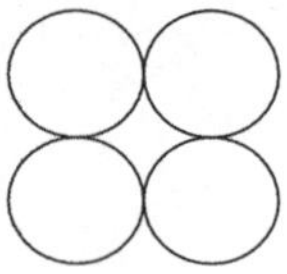

d. 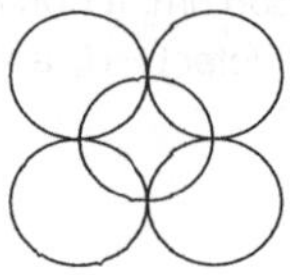

18 For silicon, given

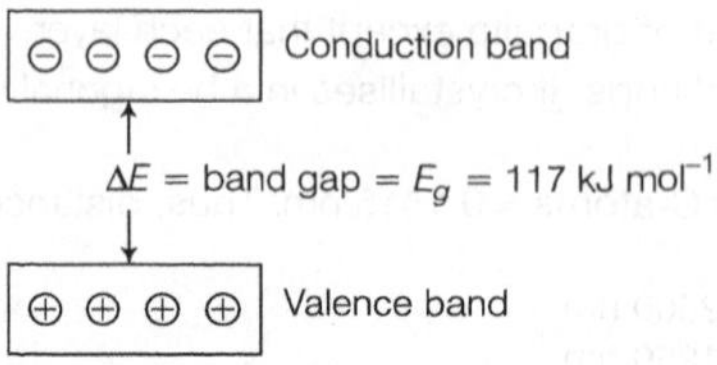

Wavelength of light that can excite an electron from the valence band to the conduction band in silicon is

a. 1022 nm
b. 872 nm
c. 1022 pm
d. 872 pm

Statement Type

***Direction** (Q. Nos. 19 and 20) This section is based on Statement I and Statement II. Select the correct answer from the codes given below.*

a. Both Statement I and Statement II are correct and Statement II is the correct explanation of Statement I
b. Both Statement I and Statement II are correct but Statement II is not the correct explanation of Statement I
c. Statement I is correct but Statement II is incorrect
d. Statement II is correct but Statement I is incorrect

19 **Statement I** NiO is expected to have more defects than NaCl.

Statement II NaCl is an ionic compound.

20 **Statement I** Hexagonal closest-packed structure and a cubic closest-packed structure for a given element would be expected to have the same density.

Statement II The two structures have the same coordination number, hence the same packing fraction (density).

Revisal Problems
for JEE Advanced

Only One Option Correct Type

Working Space

Direction *(Q. Nos. 1-15) This section contains 15 multiple choice questions. Each question has four choices (a), (b), (c) and (d), out of which ONLY ONE is correct.*

1 Gold crystallises in a fcc unit cell. The second order reflection occurs at $\theta = 22.2$.
If wavelength of the light used is 154 pm then density of metallic gold is
(Au = 197, $\sin\theta = 0.378$)

a. 9.68×10^{-3} g cm^{-3}
b. 4.84×10^{-3} g cm^{-3}
c. 15.21×10^{-3} g cm^{-3}
d. 19.36 g cm^{-3}

2 The arrangement of X^- ions around A^+ ion in solid AX is given in the figure (not drawn to scale). If the radius of X^- is 250 pm, the radius of A^+ is **[JEE Advanced 2013]**

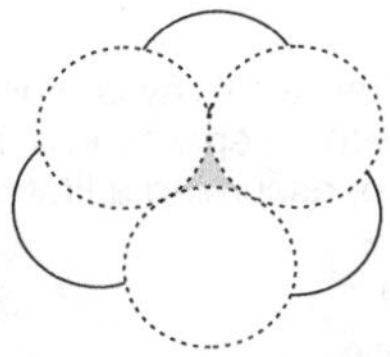

a. 104 pm b. 125 pm
c. 183 pm d. 57 pm

3 Out of V_2O_5 and CoO,

a. V_2O_5 forms *n*-type and CoO forms *p*-type semiconductor
b. V_2O_5 forms *p*-type and CoO forms *n*-type semiconductor
c. Both form *n*-type semiconductor
d. Both form *p*-type semiconductor

4 Shielding factor of Na^+ ($Z = 11$) and F^- ($Z = 9$) is 4.15. Internuclear distance of NaF is 2.31Å.
Thus, the ratio of cation (Na^+) to anion (F^-) is

a. 0.818 b. 0.878
c. 0.752 d. 0.708

5 Atoms of an element are in cubic closest packed structure with radius of atom 141.4 pm.
Thus, edge length of the unit cell is

a. 200 pm b. 300 pm
c. 400 pm d. 100 pm

6 Gold crystallises in a cubic closest packed structure (fcc) and has a density 19.3 g cm^{-3}.
Thus, radius of gold atom is (Au = 197 g mol^{-1})

a. 144 pm b. 288 pm
c. 72 pm d. 408 pm

7 Sodium metal crystallises in body centred cubic lattice with cell edge 429 pm. Thus, radius of sodium atom is

a. 429 pm b. 214.5 pm
c. 186 pm d. 152 pm

Working Space

8 The packing efficiency of the two-dimensional unit cell shown below is **[IIT JEE 2010]**

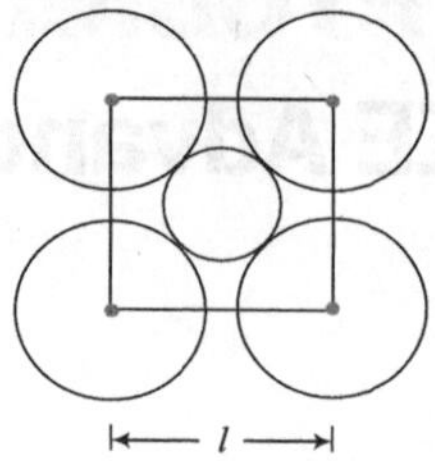

a. 39.27%
b. 68.02%
c. 74.05%
d. 78.54%

9 Radius of the triangular hole shown in the following figure in terms of radius R of the circle is

a. $0.1547\,R$
b. $0.3128\,R$
c. $0.414\,R$
d. $0.2216\,R$

10 When an electron in an excited Mo-atom falls from L to K-shell, an X-ray is emitted. These X-rays are diffracted at an angle of $7.75°$ by planes with a separation of 264 pm. Thus, difference in energy between K-shell and L-shell in Mo, assuming a first order diffraction, is $(\sin 7.75° = 0.135)$

a. 27.86×10^{-16} J
b. 3.59×10^{14} J
c. 1.675×10^{9} J
d. None of these

11 The diffraction of barium with X-radiation of wavelength 229 pm gives a first order diffraction at $30°$. Thus, distance between the two planes is

a. 114.5 pm
b. 229 pm
c. 458 pm
d. 200 pm

12 Consider the following transition of iron at 1200 K.

$$\underset{\text{bcc lattice}}{\text{Fe}(\alpha\text{-form})} \xrightarrow{1200\text{ K}} \underset{\text{fcc lattice}}{\text{Fe}(\gamma\text{-form})}$$

Distance between the nearest neighbours is same in the two forms at the transition temperature 1200 K. Thus, ratio of the densities of the two forms at 1200 K is

a. 1.09
b. 1.00
c. 0.92
d. 2.18

13 The cation/anion radius ratio for a triangular arrangement of anions in which the cation is in contact with the anions (but does not push them apart) as shown in the figure is

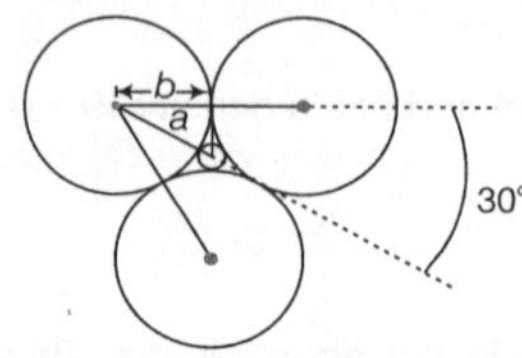

a. 0.866
b. 0.414
c. 0.155
d. 0.500

14 What fraction of the surface of a crystal of Cd at $T = 298$ K consists of vacancies? $\Delta_{sub}H°(\text{Cd}) = 112.01$ kJ mol^{-1}). Energy needed to form a vacancy is approximately 50% of $\Delta_{sub}H°$.

a. 6.02×10^{-23}
b. 6.02×10^{22}
c. 1.5×10^{-10}
d. 1.50×10^{10}

15 For a cubic unit cell, the distance d_{hkl} between planes with Miller indices hkl is given by

Working Space

$$d_{hkl}^2 = \frac{a^2}{h^2 + k^2 + l^2}$$

where, a is the length of the cubic unit cell.

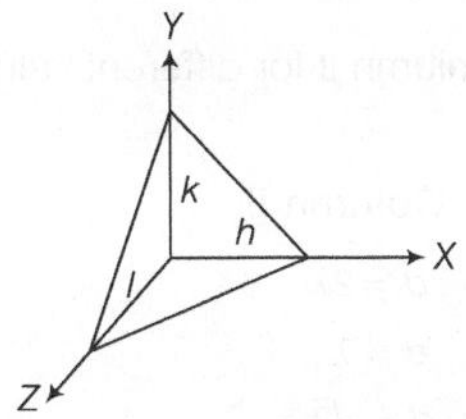

d_{111} spacing for crystalline K is 0.3080 nm. A wavelength of 0.533 nm falls on this plane. The value of sin θ for first order reflection in Bragg's diffraction is

a. 1.00
b. 0.50
c. 0.87
d. 0.71

One or More than One Options Correct Type

***Direction** (Q. Nos. 16-18) This section contains 3 multiple choice questions. Each question has four choices (a), (b), (c) and (d), out of which ONE or MORE THAN ONE are correct.*

16 Boron phosphide (BP) is a semiconductor and a hard, abrasion-resistance material. It crystallises in a zinc blende structure, formed from boron atoms in a fcc lattice and phosphorus atoms in tetrahedral holes. Thus, P-atoms must be

a. $\frac{1}{2}$ of the tetrahedral holes
b. $\frac{1}{8}$ of the octahedral holes
c. $\frac{1}{8}$ of the tetrahedral holes
d. equal to tetrahedral holes

17 In which of the following cases sin θ (with angle of reflection θ) will be equal using X-ray of 400 nm for Ca?

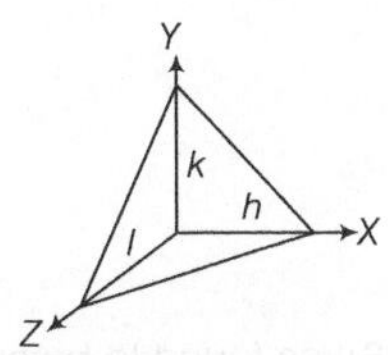

where, d_{hkl} indicates Miller indices on X, Y, Z planes.

	Order of reflection	Plane
a.	1	$d_{111} = 0.321$ nm
b.	2	$d_{222} = 0.161$ nm
c.	1	$d_{222} = 0.161$ nm
d.	2	$d_{111} = 0.321$ nm

18 Consider the following energy diagram.

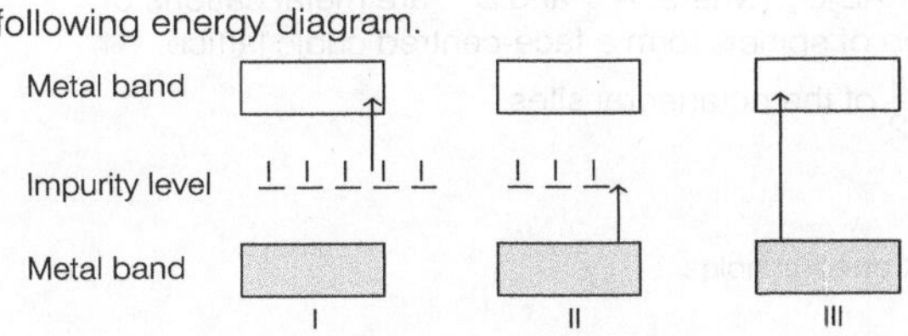

These energy diagrams are respectively, for semiconductors.

	I	II	III
a.	*n*-type	*p*-type	Intrinsic
b.	*n*-type	Intrinsic	*p*-type
c.	*p*-type	Intrinsic	*n*-type
d.	*p*-type	*n*-type	Intrinsic

Working Space

Matching List Type

Direction (Q. No. 19) *Choices for the correct combination of elements from Column I and Column II are given as options (a), (b), (c) and (d), out of which one is correct.*

19 Bragg's equation is given by $n\lambda = 2d \sin \theta$
where, d is edge length.
Match the angle (θ) in Column I with edge length (a) in Column II for different values of n and select the answer from the codes given below.

	Column I		Column II
i.	$\theta = 30°, \; n = 1$	p.	$d = 2\lambda$
ii.	$\theta = 45°, \; n = 2$	q.	$d = \lambda$
iii.	$\theta = 60°, \; n = 3$	r.	$d = \sqrt{2}\,\lambda$
iv.	$\theta = 90°, \; n = 4$	s.	$d = \sqrt{3}\,\lambda$

Codes

	i	ii	iii	iv		i	ii	iii	iv
a.	q	s	p	r	b.	p	s	r	q
c.	q	r	s	p	d.	q	r	p	s

Comprehension Type

Direction (Q. Nos. 20-25) *This section contains 3 paragraphs, each describing theory, experiments, data, etc. Six questions related to the paragraphs have been given. Each question has only one correct answer among the four given options (a), (b), (c) and (d).*

Passage I

Low Energy Electron Diffraction (LEED) can be used to study the structure of solid surfaces. Effective diffraction of the incident beam of electrons is obtained, if the wavelength (λ) of the electrons is similar to the interatomic spacing. Usually, the incident electrons are accelerated by an electrical potential difference of about 150 V.

$$m_e = 9.11 \times 10^{-31} \text{ kg}$$
$$h = 6.6 \times 10^{-34} \text{ Js}$$
$$e = 1.602 \times 10^{-19} \text{ C}$$

20 Wavelength associated with this LEED experiment is

a. 99.7 pm
b. 9.97 nm
c. 99.7 Å
d. 9.97 Å

21 In a LEED experiment on MgO, the lowest diffraction angle θ was found to be such that $\sin \theta = 0.33$ for first-order diffraction on the surface of crystal of MgO. Thus, interionic spacing d is

a. 1.51 Å
b. 151 pm
c. 0.151 nm
d. All of these

Passage II

Spinels are solids with the general formula AB_2O_4 (where, A^{2+} and B^{3+} are metal cations of the same or different metals). The oxide ion of spinels form a face-centred cubic lattice. Cations occupy $\frac{1}{8}$ of tetrahedral sites and $\frac{1}{2}$ of the octahedral sites.

22 In $MgAl_2O_4$,

a. Mg^{2+} ions are in $\frac{1}{8}$ of the eight possible tetrahedral holes
b. Al^{3+} ions are in $\frac{1}{2}$ of the four available octahedral holes
c. Both (a) and (b)
d. None of the above

Working Space

23 In $FeCr_2O_4$,

a. Fe^{2+} ions are in $\frac{1}{8}$ of the eight possible tetrahedral holes and Cr^{3+} ions are in $\frac{1}{2}$ of the four available octahedral holes

b. Fe^{2+} ions are in $\frac{1}{2}$ of the four available octahedral holes and Cr^{3+} ions are in $\frac{1}{8}$ of the eight possible tetrahedral holes

c. Fe^{2+} ions and Cr^{3+} ions both are in $\frac{1}{8}$ of the tetrahedral holes

d. Fe^{2+} ions and Cr^{3+} ions both are in $\frac{1}{2}$ of the octahedral holes

Passage III

Density of crystalline CsCl (168.56 g mol^{-1}) is 4.0 g/cm^3.

24 Volume occupied by a single CsCl ion pair in the crystal is

a. 7.0×10^{-23} cm^3
b. 2.8×10^{-22} cm^3
c. 1.75×10^{-23} cm^3
d. None of these

25 The smallest Cs to Cl internuclear distance in the crystal, assuming each Cs^+ ion to be located in the centre of a cube with Cl^- ion at each corner of the cube is

a. 3.57 Å
b. 1.78 Å
c. 4.12 Å
d. 2.06 Å

One Integer Value Correct Type

Direction *(Q. Nos. 26-33)* ***This section contains 8 questions. When worked out will result in an integer from 0 to 9 (both inclusive).***

26 An element crystallises into a structure which may be described by a cubic type of unit cell having one atom on each corner of the cube and two atoms on one of its diagonals. Volume of the unit cell is 2.4×10^{-23} cm^3 and density is 13.82 g cm^{-3}. Number of atoms present in 200 g of the element is $y\times 10^{23}$. What is the value of y?

27 A newly synthesised copper complex with molecular formula $C_{11}H_{29}N_5B_2Cu$ (molar mass = 316.5 g mol^{-1}) is found to crystallise in the orthorhombic system with following parameters. Thus, the number of molecules of the complex per unit cell is ......

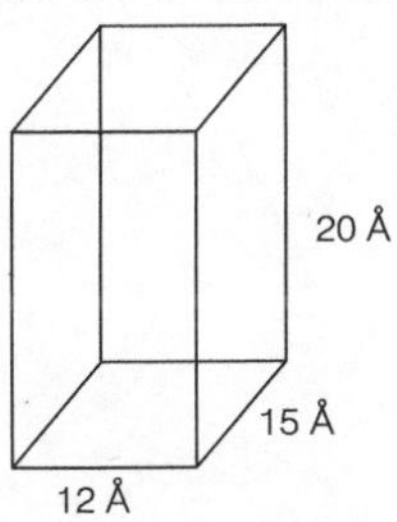

Density = 1.17 g cm^{-3}

28 How many ions are there in the unit cell shown below in the sphalerite structure (zinc sulphide ore)?

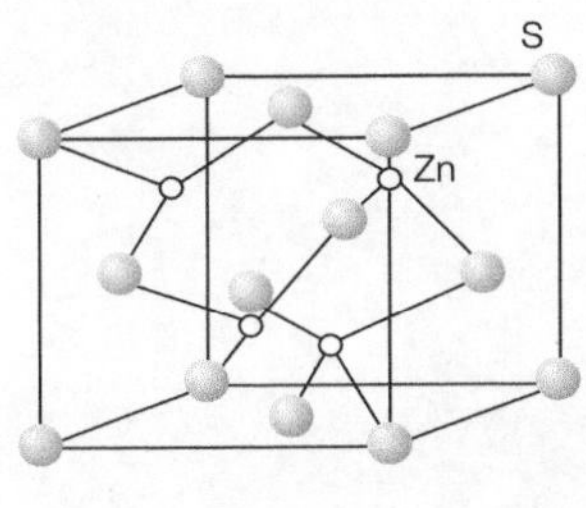

Working Space

29 How many of the following are possible close-packing schemes?

I. *ABCABC* ......, II. *ABAC* ......,

III. *ABABC* ......, IV. *ABCBC* ......,

V. *ABBA* ......, VI. *ABCCAB*

30 There are x primitive unit cells in a crystal. What is the value of x?

31 When nickel (II) oxide is heated in oxygen, some of the cations are oxidised and vacant cation sites are formed according to the equation,

$$4Ni^{2+}(s) + O_2(g) \rightleftharpoons 4Ni^{3+}(s) + 2\,\square_{+} + 2O^{2-}(s)$$

where, $\square_{+}$ denotes a vacant cation site and (s) denotes an ion in solid. Conductivity of $Ni^{3+} \propto (p)^{1/n}$ for small deviation from stoichiometry. What is the value of n?

32 Ice crystallises in a hexagonal lattice. At low temperature at which the structure was determined, the lattice constants were $a = 4.53$ Å and $c = 7.41$ Å. How many H_2O molecules are contained in a unit cell? ($d_{ice} = 0.92$ g/cm^3)

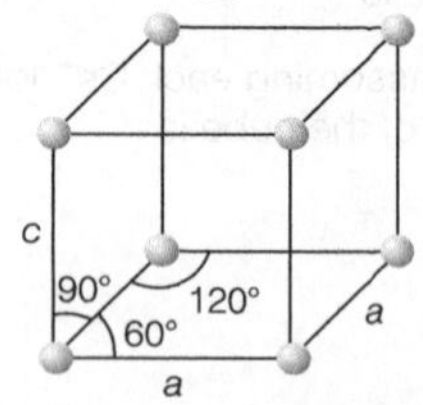

33 Silver (atomic weight, 108 g mol^{-1}) has a density 10.5 g cm^{-3}. The number of silver atoms on a surface area of 10^{-12} m^2 can be expressed in scientific notation as $y \times 10^x$. What is the value of x? **[IIT JEE 2010]**

JEE Main & AIEEE Archive

(Compilation of Last 13 Years' Questions)

1 Sodium metal crystallises in a body centred cubic lattice with a unit cell edge of 4.29 Å. The radius of sodium atom is approximately **(JEE Main 2015)**

a. 0.93 Å b. 1.86 Å
c. 3.22 Å d. 5.72 Å

2 CsCl crystallises in body-centred cubic lattice. If '*a*' its edge length, then which of the following expressions is correct? **(JEE Main 2014)**

a. $r_{Cs^+} + r_{Cl^-} = 3a$
b. $r_{Cs^+} + r_{Cl^-} = \frac{3a}{2}$
c. $r_{Cs^+} + r_{Cl^-} = \frac{\sqrt{3}}{2}a$
d. $r_{Cs^+} + r_{Cl^-} = \sqrt{3}a$

3 Which of the following exists as covalent crystals in the solid state? **(JEE Main 2013)**

a. Iodine b. Silicon
c. Sulphur d. Phosphorus

4 Experimentally, it was found that a metal oxide has formula $M_{0.98}O$. Metal *M*, present as M^{2+} and M^{3+} in its oxide. Fraction of the metal which exists as M^{3+} would be **(JEE Main 2013)**

a. 7.01% b. 4.08%
c. 6.05% d. 5.08%

5 Lithium forms body-centred cubic structure. The length of the side of its unit cell is 351 pm. Atomic radius of the lithium will be **(AIEEE 2012)**

a. 75 pm b. 300 pm
c. 240 pm d. 152 pm

6 In a face-centred cubic lattice, atom *A* occupies the corner positions and atom *B* occupies the face centred positions. If one atom of *B* is missing from one of the face centred points, the formula of the compound is **(AIEEE 2011)**

a. A_2B b. AB_2
c. A_2B_2 d. A_2B_5

7 Copper crystallises in fcc lattice with a unit cell edge of 361 pm. The radius of copper atom is **(AIEEE 2011)**

a. 181 pm b. 108 pm
c. 128 pm d. 157 pm

8 The edge length of a face-centred cubic cell of an ionic substance is 508 pm. If the radius of the cation is 110 pm, the radius of the anion is **(AIEEE 2010)**

a. 288 pm b. 398 pm
c. 618 pm d. 144 pm

9 Percentage of free space in cubic close packed structure and in body-centred packed structure respectively are **(AIEEE 2010)**

a. 30% and 26% b. 26% and 32%
c. 32% and 48% d. 48% and 26%

10 In a compound, atoms of element *Y* form ccp lattice and those of element *X* occupy 2/3rd of tetrahedral voids. The formula of the compound will be **(AIEEE 2008)**

a. X_4Y_3 b. X_2Y_3
c. X_2Y d. X_3Y_4

11 Total volume of atoms present in a face-centred cubic unit cell of a metal is (*r* is atomic radius) **(AIEEE 2006)**

a. $\frac{20}{3}\pi r^3$ b. $\frac{24}{3}\pi r^3$
c. $\frac{12}{3}\pi r^3$ d. $\frac{16}{3}\pi r^3$

12 An ionic compound has a unit cell consisting of *A* ions at the corners of a cube and *B* ions on the centres of the faces of the cube. The empirical formula for this compound would be **(AIEEE 2005)**

a. A_3B b. AB_3
c. A_2B d. AB

13 What type of crystal defect is indicated in the diagram shown below? **(AIEEE 2004)**

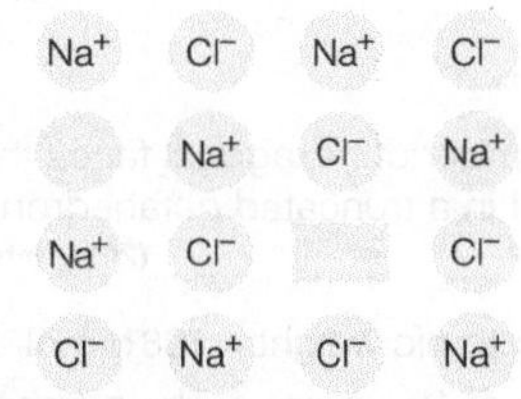

a. Frenkel defect
b. Schottky defect
c. Interstitial defect
d. Frenkel and Schottky defects

14 How many unit cells are present in a cube shaped ideal crystal of NaCl of mass 1.00 g? (Atomic masses of Na = 23, Cl = 35.5) **(AIEEE 2003)**

a. 2.57×10^{21} b. 5.14×10^{21}
c. 1.28×10^{21} d. 1.71×10^{21}

15 Number of atoms in the unit cell of Na (bcc type crystal) and Mg (fcc type crystal) respectively are **(AIEEE 2002)**

a. 4, 4 b. 4, 2
c. 2, 4 d. 1, 1

JEE Advanced & IIT JEE Archive

(Compilation of Last 10 Years' Questions)

1 The arrangement of X^- ions around A^+ ion in solid AX is given in the figure (not drawn to scale). If the radius of X^- is 250 pm, the radius of A^+ is **(2013 Adv., Only One Option Correct Type)**

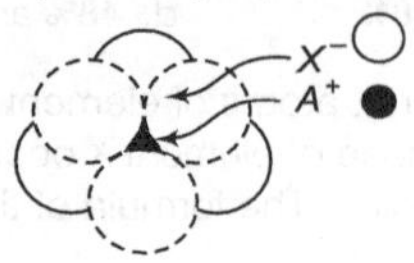

a. 104 pm
b. 125 pm
c. 183 pm
d. 57 pm

2 A compound M_pX_q has cubic close packing (ccp) arrangement of X. Its unit cell structure is shown below. The empirical formula of the compound is **(2012, Only One Option Correct Type)**

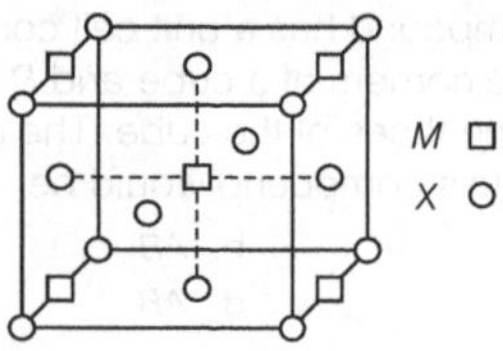

a. MX
b. MX_2
c. M_2X
d. M_5X_{14}

3 The number of hexagonal faces that are present in a truncated octahedron is **(2011, Integer Type)**

4 Silver (atomic weight = 108 g mol^{-1}) has a density of 10.5 g cm^{-3}. The number of silver atoms on a surface of area 10^{-12} m^2 can be expressed in scientific notation as $y \times 10^x$. The value of x is **(2010, Integer Type)**

5 The packing efficiency of the two-dimensional square unit cell shown below is **(2010, Only One Option Correct Type)**

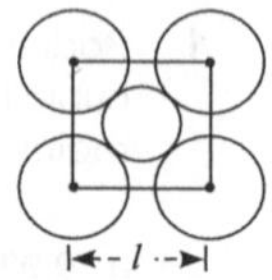

a. 39.27% b. 68.02%
c. 74.05% d. 78.54%

6 Match the crystal system/unit cells mentioned in Column I with their characteristic features mentioned in Column II. **(2007, Matching Type)**

	Column I		Column II
i.	Simple cubic and face-centred cubic	p.	have these cell parameters $a = b = c$ and $\alpha = \beta = \gamma$
ii.	Cubic and rhombohedral	q.	are two crystal systems
iii.	Cubic and tetragonal	r.	have only two crystallographic angles of 90°
iv.	Hexagonal and monoclinic	s.	belong to same crystal system

Codes

	i	ii	iii	iv
a.	p,s	p,q	q	q,r
b.	p	q	r	s
c.	p	q	r	r
d.	q,s	p	r	s

7 The edge length of unit cell of a metal having molecular weight 75 g/mol is 5 Å which crystallises in cubic lattice. If the density is 2 g/cc then, find the radius of metal atom ($N_A = 6 \times 10^{23}$). Give the answer in pm. **(2006, Subjective Type)**

8 An element crystallises in fcc lattice having edge length 400 pm. Calculate the maximum diameter of atom which can be placed in interstitial site without distorting the structure. **(2005, Subjective Type)**

9 Which of the following fcc structure contains cations in alternate tetrahedral voids? **(2005, Only One Option Correct Type)**

a. NaCl b. ZnS c. Na_2O d. CaF_2

2. Solutions

DPP

DAILY PRACTICE PROBLEMS

DPP-1

Different Methods of Expressing Concentration of Solutions

Subjective Questions

Working Space

Direction (Q. No. 1) *This question is subjective in nature, need to be solved completely on notebook.*

1 $Na_2S_2O_3$ solution is 5% by mass of solution. Express the concentration in terms of

i. molarity

ii. molality

iii. mole fraction of solute

The density of solution at 25° C is 1.040 g mL^{-1}.

Only One Option Correct Type

Direction (Q. Nos. 2-24) *This section contains 23 multiple choice questions. Each question has four choices (a), (b), (c) and (d), out of which ONLY ONE is correct.*

2 Which of the following concentration factors is affected by change in temperature?

a. Molarity b. Molality

c. Mole fraction d. Weight fraction

3 6.02×10^{20} molecules of urea are present in 100 mL of its solution. The concentration of urea solution is ($N_0 = 6.02 \times 10^{23}$ mol^{-1}) **[AIEEE 2004]**

a. 0.001 M b. 0.01 M

c. 0.02 M d. 0.10 M

4 Glucose solution is one molal. Glucose present in 1 kg glucose solution is

a. 180 g b. 153 g

c. 820 g d. 1 mole

5 The molality of a urea solution in which 0.0100 g of urea (NH_2CONH_2) is added to 0.300 dm^3 of water at STP is **[AIEEE 2011]**

a. 0.555 molal

b. 5.55×10^{-4} molal

c. 33.3 molal

d. 3.33×10^{-2} molal

6 A solution that is 20% ethanol by volume is found to have a density of 0.977 g/mL. Density of ethanol is 0.789 g/mL. Thus, mass per cent of ethanol solution is

a. 7.89% b. 9.77%

c. 16.15% d. 18.72%

Working Space

7 A solution of NH_4Cl is prepared by dissolving 95 g of NH_4Cl in 200 g of H_2O at 60° C. What is mass per cent when the solution is cooled to 20° C based on this graphical study?

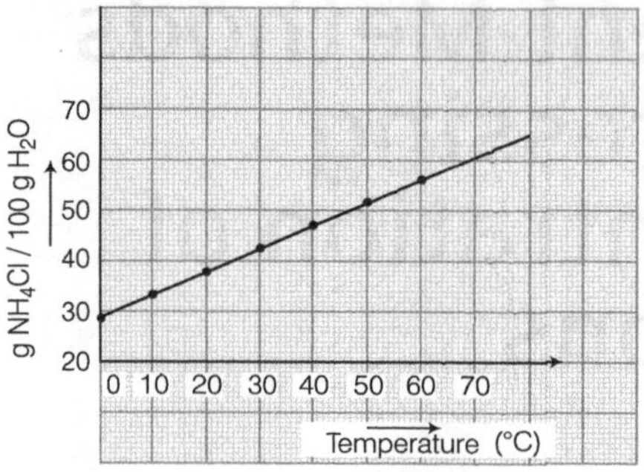

a. 23%
b. 36%
c. 42%
d. 72%

8 Commercial concentrated nitric acid is 15.6 M. To prepare 10 L of 6.0 M nitric acid from it,

a. dilute 3.846 L of 15.6 M to 10.0 L solution
b. add 10.0 L H_2O into 3.846 L of 15.6 M solution
c. add 3.846 L H_2O into 6.154 L of 15.6 M solution
d. None of the above

9 A 5.2 molal aqueous solution of methyl alcohol (CH_3OH) is supplied. What is the mole fraction of methyl alcohol in the solution? **[AIEEE 2011]**

a. 0.100
b. 0.190
c. 0.086
d. 0.050

10 $AgNO_3$ sample is 85% by mass. To prepare 125 mL of 0.05 M $AgNO_3$ solution, $AgNO_3$ sample required is

a. 1.250 g
b. 1.471 g
c. 1.512 g
d. 1.063 g

11 Hardness of a water sample is 200 ppm $CaCO_3$. Thus, molarity of $CaCO_3$ is

a. 2×10^{-3} M
b. 1×10^{-3} M
c. 2×10^{-2} M
d. 2×10^{-4} M

12 100 mL of aqueous solution of 0.01 M $CaCl_2$ is evaporated to dryness when 0.15 g of residue is obtained. Thus, impurity present is

a. 65.86%
b. 35.14%
c. 7.15%
d. None of the above

13 At 25° C, the density of 18 M H_2SO_4 is 1.8 g cm^{-3}. Thus, mass percentage of H_2SO_4 in aqueous solution is

a. 2%
b. 98%
c. 18%
d. 1.8%

14 Laboratory ammonia is 14.8 M $NH_3(aq)$ with a density of 0.8980 g/mL. Mole fraction of NH_3 in this solution is

a. 0.292
b. 0.708
c. 0.148
d. 0.852

Working Space

15 An aqueous solution is 34% H_3PO_4 by mass and has density 1.209 g mL^{-1}. Molarity (I), molality (II) and normality (III) respectively, are

	I	II	III
a.	4.19	5.26	12.87
b.	4.19	5.26	8.38
c.	5.26	4.19	10.52
d.	5.26	4.19	15.78

16 If one assumes the volumes are additive, what is $[NO_3^-]$ in a solution obtained by mixing 275 mL of 0.200 M KNO_3, 325 mL of 0.40 M $Mg(NO_3)_2$ and 400 mL H_2O?

a. 0.185 M
b. 0.600 M
c. 0.315 M
d. 0.630 M

17 How many grams of water would you add to 1.38 moles of CH_3OH in 1 kg water to reduce the molality to 1.00 molal $CH_3OH(aq)$?

a. 180 g
b. 380 g
c. 1380 g
d. 1000 g

18 Three solutions have been provided.

I. 3 g HCl in 1 kg H_2O.
II. 2 L HCl(*g*) and 1 L H_2O at room temperature.
III. Aqueous solution of HCl with $\chi_{HCl} = 1.5 \times 10^{-3}$.

Solutions containing same number of moles of HCl is/are

a. I and II
b. I and III
c. II and III
d. I, II and III

19 Which of the following aqueous solutions has the highest concentration of Na^+?

a. 0.208 M Na_2SO_4
b. A solution containing 1.05 g NaCl/100 mL
c. A solution having 14.7 mg Na^+/mL
d. All have equal $[Na^+]$

20 pH of $Ba(OH)_2$ solution is 12. Number of millimoles present in 100 mL of $Ba(OH)_2$ solution is

a. 0.50
b. 1
c. 0.05
d. 0.10

21 We have

I. 25 mL of 1 M NaOH
II. 10 mL of 0.50 M NaCl

On mixing the two solutions, molar concentrations of Na^+, OH^- and Cl^- respectively, are

	$[Na^+]$	$[OH^-]$	$[Cl^-]$
a.	1.50 M	1.00 M	0.50 M
b.	0.85 M	0.85 M	0.85 M
c.	0.85 M	0.71 M	0.14 M
d.	0.85 M	0.71 M	0.71 M

Working Space

22 A '100 proof' solution of ethanol in water consists of 50.0 mL of $C_2H_5OH(l)$ and 50.0 mL of $H_2O(l)$, mixed at 16°C. Given,

Density of H_2O = 1g mL^{-1}

Density of C_2H_5OH = 0.7939 g mL^{-1}

Density mixture = 0.9344 g mL^{-1}

Volume of the solution is

a. 100 mL
b. 105 mL
c. 96 mL
d. 98 mL

23 How many grams of MgI_2 must be added to 250 mL of 0.0876 M KI to produce a solution with $[I^-]$ = 0.10 M?

a. 0.862 g
b. 0.431 g
c. 0.716 g
d. 0.162 g

24 An aqueous solution is 6% methanol, CH_3OH, by mass with d = 0.988 g mL^{-1}. Thus, molarity of CH_3OH in this solution is

a. 1.88 M
b. 1.90 M
c. 1.80 M
d. 1.85 M

Matching List Type

Direction *(Q. No. 25) Choices for the correct combination of elements from Column I and Column II are given as options (a), (b), (c) and (d), out of which one is correct.*

25 An ethanol-water solution is prepared by dissolving 10.00 mL of ethanol (C_2H_5OH, d = 0.789 g mL^{-1}) in a sufficient volume of water to produce 100 mL of solution with a density, d = 0.982 g mL^{-1}. Match the concentration term in Column I with its value in Column II and select the answer from the codes.

	Column I		Column II
i.	Volume % ethanol	p.	8.03
ii.	Mass % ethanol	q.	3.29
iii.	Mass/Volume % ethanol	r.	1.89
iv.	Mole % ethanol	s.	10.0
v.	Molarity	t.	7.89
vi.	Molality	u.	1.71

Codes

	i	ii	iii	iv	v	vi
a.	p	q	r	t	s	u
b.	s	p	t	q	u	r
c.	s	p	t	q	r	u
d.	q	t	r	u	p	s

One or More than One Options Correct Type

Direction *(Q. Nos. 26 and 27) This section contains 2 multiple choice questions. Each question has four choices (a), (b), (c) and (d), out of which ONE or MORE THAN ONE are correct.*

26 1 kg of aqueous solution has 0.40 kg NaOH. Thus, this solution is

a. 66.67% by weight of water
b. 40.00% by weight of aqueous solution
c. 16.67 molal
d. 16.67 molar

Working Space

27 Mole fraction of ethanol in ethanol-water solution is 0.25. Thus, this solution is

a. 46% by mass of solution
b. 85.2% by mass of water solvent
c. 54% by mass of solution
d. 46% by mass of solvent

Comprehension Type

Direction *(Q. Nos. 28 and 29) This section contains a paragraph, which describing theory, experiments, data, etc. Two questions related to the paragraph have been given. Each question has only one correct answer among the four given options (a), (b), (c) and (d).*

Passage

A handbook gives the concentration of water in a 2.772 M aqueous solution of NaOH as 998.0 g L^{-1}.

28 Molality of solution is

a. 2.772 molal
b. 2.778 molal
c. 3.012 molal
d. 2.172 molal

29 Mole fraction of NaOH in the solution is

a. 2.772×10^{-2}
b. 2.778×10^{-3}
c. 4.766×10^{-2}
d. 4.212×10^{-2}

One Integer Value Correct Type

Direction *(Q. No. 30) This section contains 1 question. When worked out will result in an integer from 0 to 9 (both inclusive).*

30 A certain brine solution has 3.87% NaCl by mass with density 1.036 g mL^{-1}. How many millilitre of this solution should be evaporated to obtain 0.32 g of NaCl?

DPP-2 Henry's Law

Subjective Questions

Direction (Q. No. 1) *This question is subjective in nature, need to be solved completely on notebook.*

Working Space

1 The average human with a body weight of 70 kg has a blood volume of 5.00 L. Henry's law constant $K_H(N_2)$ for solubility in water = 9.04×10^4 bar at 298 K. Density of blood and water = 1.00 kg L^{-1}. Calculate the number of moles of N_2 absorbed in this amount of blood in air of composition 80% N_2 at sea level where the pressure is

i. 1 bar
ii. 50 bar

Only One Option Correct Type

Direction (Q. Nos. 2-12) *This section contains 11 multiple choice questions. Each question has four choices (a), (b), (c) and (d), out of which ONLY ONE is correct.*

2 An unopened soda has an aqueous concentration of CO_2 at 25°C equal to 0.0506 molal. Thus, pressure of CO_2 gas in the can is (K_H = 0.034 mol/kg bar)

a. 0.671 bar
b. 1.49 bar
c. 1.20 bar
d. 1.71 bar

3 $CO(g)$ is dissolved in H_2O at 25°C and 0.010 atm. Henry's law constant for this system is 5.80×10^4 atm. Thus, mole fraction of $CO(g)$ is

a. 1.72×10^{-7}
b. 8.28×10^{-7}
c. 0.99
d. 0.01

4 H_2S gas is used in qualitative analysis of inorganic cations. Its solubility in water at STP is 0.195 mol kg^{-1}. Thus, Henry's law constant (in atm $molal^{-1}$) for H_2S is

a. 0.195
b. 5.128
c. 3.897×10^3
d. 2.565×10^{-4}

5 Variation of K_H (Henry's law constant) with temperature T is shown by following graphs I - IV.

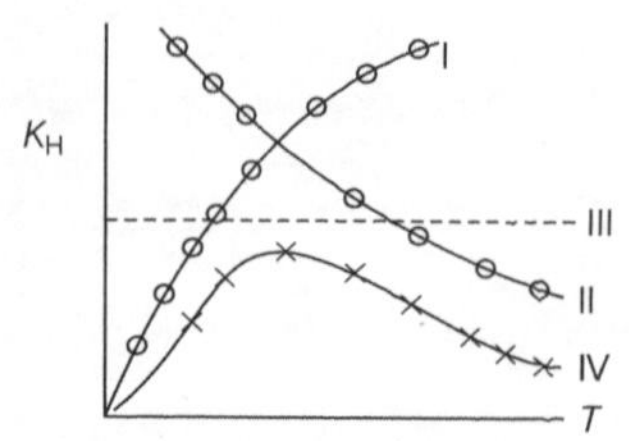

Correct representation is

a. I
b. II
c. III
d. IV

Working Space

6 Concentration of CO_2 (in mole fraction) in fat when partial pressure of CO_2 is 55 kPa at 25° C, is (Henry's law constant of $CO_2 = 8.6 \times 10^4$ torr)

a. 1.6×10^{-4} b. 4.8×10^{-3}

c. 8.6×10^{-4} d. 2.4×10^{-3}

7 What is the concentration of O_2 in a freshwater stream in equilibrium with air at 25° C and 1.0 bar? Given, K_H (Henry's law constant) of $O_2 = 1.3 \times 10^{-3}$ mol/kg bar at 25° C.

a. 8.736×10^{-3} g/kg b. 4.16×10^{-2} g/kg

c. 24.04 g/kg d. 114.5 g/kg

8 Following data has been given for CO_2 for the concentration in H_2O.

Temperature	Henry's constant	Pressure
273 K	728 atm	0.10 atm
333 K	3410 atm	p_2

If solution of CO_2 in H_2O is heated from 273 to 333 K, pressure (p_2) needed to keep CO_2 in the solution is

a. 0.108 atm b. 0.805 atm c. 0.212 atm d. 0.468 atm

9 Relation between the volume of gas (2) that dissolves in a fixed volume of solvent(1) and the partial pressure of gas (2) is (n_t = total moles, K_1 and K_2 are Henry's constants)

a. $n_t RT / K_2$ b. $n_2 RT / K_2 p_2$

c. $n_1 RT / n_2$ d. None of these

10 At 0° C and an O_2 pressure of 1.00 atm, the aqueous solubility of $O_2(g)$ is 48.9 mL O_2 per litre. Thus, molarity of O_2 in a saturated water solution when the O_2 is under the normal partial pressure of 0.2095 atm, is

a. 4.57×10^{-4} M b. 2.18×10^{-3} M

c. 4.57×10^{-2} M d. 2.18×10^{-5} M

11 Henry's law constant for N_2 at 293 K is 76.48 kbar. N_2 exerts a partial pressure of 0.987 bar. If N_2 gas is bubbled through water at 293 K, then number of millimoles of N_2 that will dissolve in 1 L of water is

a. 0.0716 b. 1.29×10^{-5} c. 1.29×10^{-2} d. 5.55

12 A handbook lists the solubility of carbon monoxide in water at 0° C and 1 atm pressure as 0.0354 mL CO per mL of H_2O. What should be the pressure of CO(g) above the solution to obtain 0.010 M CO solution?

a. 0.633 atm b. 6.33 atm c. 1 atm d. 0.01 atm

One or More than One Options Correct Type

Direction *(Q. Nos. 13 and 14) This section contains 2 multiple choice questions. Each question has four choices (a), (b), (c) and (d), out of which ONE or MORE THAN ONE are correct.*

13 Select the correct statement(s).

a. Gases which form ions in the solvent are soluble in that solvent

b. Gases which are liquefied easily are more soluble in water than the other gases which are not liquefied easily

c. Higher the value of K_H at a given pressure, lower is the solubility of the gas in the liquid

d. Higher the value of K_H at a given pressure, larger the solubility of the gas in the liquid

Working Space

14 Select the correct statement(s).

a. K_H (Henry's law constant) is a function of nature of gases
b. Higher the value of K_H at a given pressure, the lower is the solubility of the gas in the liquid
c. K_H increases with increase in temperature
d. K_H increases with increase in partial pressure

Statement Type

Direction (Q. Nos. 15-18) *This section is based on Statement I and Statement II. Select the correct answer from the codes given below.*

a. Both Statement I and Statement II are correct and Statement II is the correct explanation of Statement I
b. Both Statement I and Statement II are correct but Statement II is not the correct explanation of Statement I
c. Statement I is correct but Statement II is incorrect
d. Statement II is correct but Statement I is incorrect

15 **Statement I** Solubility of gases in liquids decreases with rise in temperature.

Statement II $CO_2(g) + H_2O(l) \longrightarrow H_2CO_3(aq); \Delta H = -ve$

16 **Statement I** People living at high altitudes have symptoms of a condition known as anoxia.

Statement II At high altitudes, the partial pressure of oxygen is less than that at the ground level.

17 **Statement I** Soft drinks and soda water (carbonated water) bottles are sealed under high pressure.

Statement II Solubility of CO_2 increases under high pressure.

18 **Statement I** Scuba (instrument used for breathing) divers must cope with high concentration of dissolved gases while breathing at high pressure under water.

Statement II Formation of bubbles in blood blocks capillaries and creates bends which are painful and dangerous to life.

Matching List Type

Direction (Q. No. 19) *Choices for the correct combination of elements from Column I and Column II are given as options (a), (b), (c) and (d), out of which one is correct.*

19 The major components of air are O_2 and N_2 with approximate proportion of 21% and 79% by volume at 298 K. The water is in equilibrium with air at a pressure of 10 atm. At 298 K, $K_H(N_2) = 6.51 \times 10^7$ mm, $K_H(O_2) = 3.30 \times 10^7$ mm

Match the parameters in Column I with their values in Column II and select the answer from the codes given below.

	Column I		Column II
i.	Mole fraction of N_2	p.	4.84×10^{-5}
ii.	Mole fraction of O_2	q.	2.69×10^{-3}
iii.	Moles of N_2 in 1 kg H_2O	r.	9.22×10^{-5}
iv.	Moles of O_2 in 1 kg H_2O	s.	5.12×10^{-3}

Codes

	i	ii	iii	iv		i	ii	iii	iv
a.	p	q	s	r	b.	p	r	s	q
c.	r	p	q	s	d.	r	p	s	q

Working Space

Comprehension Type

Direction (Q. Nos. 20 and 21) This section contains a paragraph, which describing theory, experiments, data, etc. Two questions related to the paragraph have been given. Each question has only one correct answer among the four given options (a), (b), (c) and (d).

Passage

Given figure represents Henry's law at a given temperature for the solubility of HCl gas in cyclohexane (in terms of mole fraction) at different pressures.

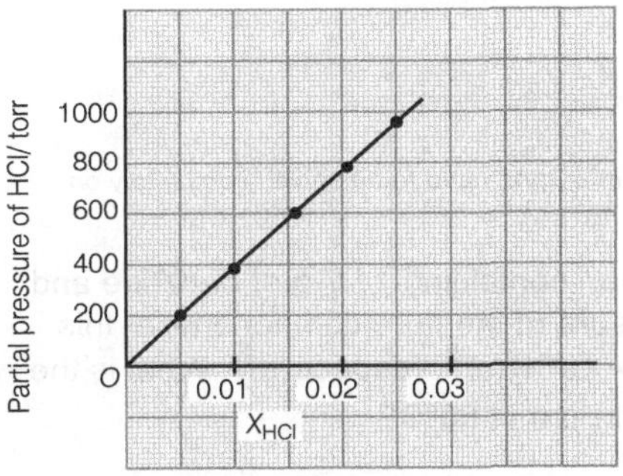

20 Henry's law constant for HCl gas is (in torr)

a. 4×10^4 b. 2×10^4

c. 1×10^4 d. 0.01×10^4

21 What is the ratio of solubility at 400 torr and 960 torr?

a. 2.5 b. 0.4

c. 25 d. 4

One Integer Value Correct Type

Direction (Q. Nos. 22-24) This section contains 3 questions. When worked out will result in an integer from 0 to 9 (both inclusive).

22 In a given aqueous solution of NH_3 at 70°C, $\chi_{NH_3} = 0.03$. Vapour pressure of pure $H_2O(l)$ at 70°C is 266.0 torr. What is the difference between the actual vapour pressure of H_2O in solution and the given vapour pressure of H_2O? (Express value in torr)

23 What is the ratio of the solubilities of N_2 and O_2 in H_2O at 100°C based on the following data?

	Pressure	Henry's constant
N_2	0.80 atm	12.0×10^4 atm
O_2	0.20 atm	6.0×10^4 atm

24 The concentration of O_2 in water required to support aerobic aquatic life is about 3.856 mg dm^{-3}. Minimum partial pressure (in kPa) of oxygen in the atmosphere that can achieve this concentration is
(K_H, Henry's law constant of O_2 = 74.68 kPa m^3 mol^{-1})

DPP-3 Colligative Properties-I (Lowering of Vapour Pressure)

Subjective Questions

Direction (Q. Nos. 1-3) These questions are subjective in nature, need to be solved completely on notebook.

Working Space

1 Calculate the vapour pressure over a solution containing 11.7 g of benzene and 4.6 g of toluene at 50°C, if the vapour pressure of the pure components at this temperature are 3.6×10^4 Nm^{-2} and 1.12×10^4 Nm^{-2}, respectively. What is the composition of the vapour which is in equilibrium at 50°C?

2 6 g of urea is dissolved in 90 g water at 25°C. What is the vapour pressure of the solution if the vapour pressure of water is 40 mm of Hg?

3 At 90°C, the vapour pressure of toluene is 400 torr and that of cresol is 150 torr. What is the composition of the liquid mixture that boils at 90°C when the pressure is 0.50 atm? What is the composition of the vapour produced?

Only One Option Correct Type

Direction (Q. Nos. 4-15) This section contains 12 multiple choice questions. Each question has four choices (a), (b), (c) and (d), out of which ONLY ONE is correct.

4 The vapour pressure of pure liquid solvent is 0.80 atm. When a non-volatile solute *B* is added to the solvent, its vapour pressure drops to 0.60 atm. Thus, mole fraction of the component *B* is

a. 0.50 b. 0.25 c. 0.45 d. 0.75

5 Vapour pressure of pure water at 298 K is 23.8 torr. Thus, vapour pressure of a solution containing 6 g urea and 68.4 g sucrose in 10 moles of water is

a. 23.0 torr
b. 19.2 torr
c. 23.1 torr
d. 21.0 torr

6 Vapour pressure of nitrobenzene (molar mass = $123\,g\,mol^{-1}$) is 3.6 kNm^{-2} and that of water (molar mass = $18\,g\,mol^{-1}$) is 977 $kN\,m^{-2}$. They form immiscible mixture at the given temperature. Thus, percentage of nitrobenzene in the vapour phase is

a. 20.12%
b. 40.28%
c. 79.88%
d. 59.72%

7 The vapour pressure of *n*-hexane at 350 K is 840 torr and that of cyclohexane is 600 torr. Mole fraction of hexane in the mixture that boils at 350 K and 1 atm pressure assuming ideal behaviour is

a. 0.75 b. 0.50 c. 0.80 d. 0.67

8 At 298 K, the vapour pressure of pure water is 23.76 mm Hg and that of an aqueous solution is 22.98 mm Hg. Thus, molality of solution is (assume dilute solution)

a. 0.033 b. 0.67 c. 1.83 d. 5.55

Working Space

9 A binary liquid solution is prepared by mixing *n*-heptane and ethanol. Which one of the following statements is/are correct regarding the behaviour of the solution?
a. The solution formed is an ideal solution **[AIEEE 2009]**
b. The solution is non-ideal showing positive deviation from Raoult's law
c. The solution is non-ideal showing negative deviation from Raoult's law
d. *n*-heptane shows positive deviation while ethanol shows negative deviation from Raoult's law

10 18 g of glucose ($C_6H_{12}O_6$) is added to 178.2 g of water. The vapour pressure of water for this aqueous solution at 100°C is
a. 759.00 torr
b. 7.60 torr
c. 76.00 torr
d. 752.40 torr

11 Which of the following liquid pairs shows a positive deviation from Raoult's law?
a. Water + hydrochloric acid
b. Benzene + methanol
c. Water + nitric acid
d. Acetone + chloroform

12 A solution of two liquids boils at a temperature more than the boiling point of either of them. Hence, the binary solution shows
a. negative deviation from Raoult's law
b. positive deviation from Raoult's law
c. no deviation from Raoult's law
d. positive or negative deviation from Raoult's law depending upon the composition

13 Lowering of vapour pressure of an aqueous solution of a non-volatile non-electrolyte 1 molal aqueous solution at 100°C is
a. 14.12 torr
b. 312 torr
c. 13.44 torr
d. 352 torr

14 The solubility of urea in methanol is 17 g urea/100 mL methanol. Density of methanol is 0.792 g/mL at 293 K at which vapour pressure of methanol is 96.0 mm of Hg. Thus, vapour pressure of saturated solution of urea in methanol is
a. 86.11 mm Hg
b. 90.27 mm Hg
c. 88.03 mm Hg
d. 80.12 mm Hg

15 Which combination of following terms is matched correctly?
I. Vapour pressure
II. Intermolecular forces
III. Latent heat of vaporisation

	I	II	III
a.	High	Weak	Small
b.	High	Strong	Large
c.	Low	Weak	Large
d.	Low	Strong	Small

One or More than One Options Correct Type

Direction *(Q. Nos. 16-18) This section contains 3 multiple choice questions. Each question has four choices (a), (b), (c) and (d), out of which ONE or MORE THAN ONE are correct.*

16 Benzene and naphthalene form an ideal solution at room temperature. For this process, the true statement(s) is/are **[JEE Advanced 2013]**
a. $\Delta G > 0$
b. $(\Delta S)_{system} > 0$
c. $(\Delta S)_{surroundings} = 0$
d. $\Delta H = 0$

Working Space

17 Total vapour pressure of mixture of 1 mole of a volatile component A ($p_A^\circ = 100$ mm Hg) and 3 moles of volatile component B ($p_B^\circ = 60$ mm Hg) is 75 mm Hg. For such case,

a. there is positive deviation from Raoult's law
b. there is negative deviation from Raoult's law
c. boiling point is lowered
d. force of attraction between A and B is smaller than that between A and A, and A and B

18 At 35°C, the vapour pressure of carbon disulphide (CS_2) is 512 mm Hg and that of acetone (CH_3COCH_3) is 344 mm Hg. A solution of CS_2 and CH_3COCH_3 in which mole fraction of CS_2 is 0.25 has a total vapour pressure of 600 mm Hg. Which of the following statement(s) is/are correct?

a. A mixture of 100 mL of CH_3COCH_3 and 100 mL of CS_2 has a volume of 200 mL
b. When CH_3COCH_3 and CS_2 are mixed at 35°C, heat must be absorbed in order to produce a solution at 35°C
c. Process of mixing is exothermic
d. Entropy of mixing is zero

Comprehension Type

Direction (Q. Nos. 19-22) This section contains 2 paragraphs, each describing theory, experiments, data, etc. Four questions related to the paragraphs have been given. Each question has only one correct answer among the four given options (a), (b), (c) and (d).

Passage I

At 300 K, vapour pressure of pure liquid $A = 575$ torr and that of pure liquid $B = 390$ torr

In binary mixture of liquids A and B with ideal behaviour, mole fraction of A in vapour phase $= 0.35$

19 Binary mixture contains

a. 27 mole% A
b. 73 mole% A
c. 50 mole% A
d. 47 mole% A

20 Total vapour pressure of the binary mixture is

a. 440 torr
b. 455 torr
c. 140 torr
d. 483 torr

Passage II

At 40°C, the vapour pressure (in torr) of methanol-ethanol solution is represented by $p_{\text{total}} = 119\chi_m + 135$

where, χ_m is the mole fraction of methanol.

21 Difference of vapour pressures of volatile liquids (methanol-ethanol) in pure state at 40°C is

a. 135 torr
b. 119 torr
c. 254 torr
d. 16 torr

22 What is the mole fraction of methanol in vapour phase in a binary mixture with mole fraction of ethanol 0.50?

a. 0.50
b. 0.35
c. 0.65
d. 0.25

Working Space

Matching List Type

Direction *(Q. No. 23) Choices for the correct combination of elements from Column I and Column II are given as options (a), (b), (c) and (d), out of which one is correct.*

23

Volatile component	Vapour pressure at 350 K	Molar mass
A	24 kPa	60 g mol^{-1}
B	12 kPa	40 g mol^{-1}

100 g of binary mixture of A and B contains 60% of A. Mixture is subjected to the following operation.

Original mixture I $\xrightarrow{\text{Distil}}$ II containing vapours of A and B

$\xrightarrow{\text{Condense}}$ III (liquid) $\xrightarrow{\text{Distil}}$ IV containing vapours of A and B

Match the content of Column I with their values in Column II and select the answer from the codes given below.

	Column I (Mole fraction of A)		Column II
i.	Mixture I	p.	0.25
ii.	Mixture II	q.	0.50
iii.	Mixture III	r.	0.67
iv.	Mixture IV	s.	0.80

Codes

	i	ii	iii	iv		i	ii	iii	iv
a.	p	q	q	s	b.	p	r	s	q
c.	q	r	r	s	d.	s	q	p	p

One Integer Value Correct Type

Direction *(Q. Nos. 24-26) This section contains 3 questions. When worked out will result in an integer from 0 to 9 (both inclusive).*

24 Out of the following pairs of volatile solvents, how many form azeotropic mixture?

	A		B
i.	CCl_4	+	$CHCl_3$
ii.	C_6H_{14}	+	C_7H_{16}
iii.	C_2H_5Br	+	C_2H_5Cl
iv.	C_6H_5Cl	+	C_6H_5Br
v.	C_2H_5OH	+	CH_3COCH_3
vi.	$CHCl_3$	+	CH_3COCH_3
vii.	C_6H_6	+	$C_6H_5CH_3$
viii.	*o*-nitrophenol	+	benzene

25 Number of moles of sucrose that must be added to 360 g of water to lower the vapour pressure of water by 1.19 mm Hg is ....
(Vapour pressure of pure water at the given temperature is 25 mm Hg.)

26 Two liquids A and B form an ideal solution at 50°C. Binary ideal mixture of 1 mole A and two moles B has a vapour pressure of 250 torr and when 1 mole of A is further added to this mixture, vapour pressure increases by 50 torr. What is the ratio of vapour pressure of A to B in pure state?

DPP-4 Colligative Properties-II (Depression in Freezing Point)

Only One Option Correct Type

Working Space

Direction *(Q. Nos. 1-11) This section contains 11 multiple choice questions. Each question has four choices (a), (b), (c) and (d), out of which ONLY ONE is correct.*

1 Addition of 0.643 g of a compound to 50 mL of benzene (density = 0.879 g mL^{-1}) lowers the freezing point from 5.51° C to 5.03° C (K_f of benzene = 5.12 K mol^{-1} kg). Molar mass of the compound is

a. 140 g mol^{-1} b. 137 g mol^{-1}
c. 166 g mol^{-1} d. 156 g mol^{-1}

2 The coolant usually contains a solution of antifreeze prepared by mixing equal volumes of ethylene glycol $C_2H_4(OH)_2$ and water. The density of ethylene glycol is 1.113 g cm^{-3}. Thus, the freezing point of the mixture is (K_f of H_2O = 1.86 K mol^{-1} kg)

a. − 33.4° C b. + 33.4° C
c. − 7.4° C d. + 7.4° C

3 A 5% solution (by mass) of cane sugar in water has freezing point of 272.88 K. Thus, freezing point of 5% glucose in water is (freezing point of pure water is 273.15 K)

a. 272.64 K b. 273.66 K
c. 271.34 K d. 270.12 K

4 30 mL of CH_3OH (d = 0.7980 g cm^{-3}) and 70 mL of H_2O (d = 0.9984 g cm^{-3}) are mixed at 25° C to form a solution of density 0.9575 g cm^{-3}. Thus, depression in freezing point is (K_f for H_2O = 1.86 K mol^{-1} kg)

a. 0.20° b. 19.91° c. 0.73° d. 2°

5 How much ethyl alcohol (C_2H_5OH) must be added to 1 L of water so that solution will not freeze at − 4° F? (K_f of H_2O = 1.86° mol^{-1} kg)

a. 122 g b. 512 g
c. 370 g d. 495 g

6 CH_3OH is 10% by weight of solution of CH_3OH in water (K_f of H_2O = 1.86° mol^{-1} kg). Thus, freezing point of the solution is

a. 266.5 K b. 267.2 K
c. 261.5 K d. 264.8 K

7 The amount of ice that will separate out on cooling a solution containing 50 g of ethylene glycol in 200 g water to − 9.3° C is (K_f of H_2O = 1.86° mol^{-1} kg)

a. 38.7 g b. 161.3 g c. 140.0 g d. 152.7 g

8 A solution of a non-volatile solute in water freezes at − 0.30° C. The vapour pressure of pure water at 298 K is 23.51 torr. For water, K_f = 1.86 K mol^{-1} kg. Thus, vapour pressure of the solution (in torr)

a. 23.44 b. 23.00 c. 20.41 d. 22.47

Working Space

9 A solution is prepared by condensing 4 L of a gas measured at 27°C and 748 mm Hg pressure into 58 g benzene (K_f of benzene = 5.12 K mol^{-1} kg). Thus, depression in freezing point is

a. 5.12° b. 6.16° c. 14.12° d. 7.18°

10 What freezing point depression would be predicted for 0.1 molal solution of benzoic acid in benzene if latent heat of fusion is 30.00 cal g^{-1} at 280 K (freezing point) for benzene? (assume no change in molecular state)

a. 0.52° b. 5.2° c. 0.0052° d. 0.052°

11 Based on the given graph, molar mass of solute X (non-electrolyte) is

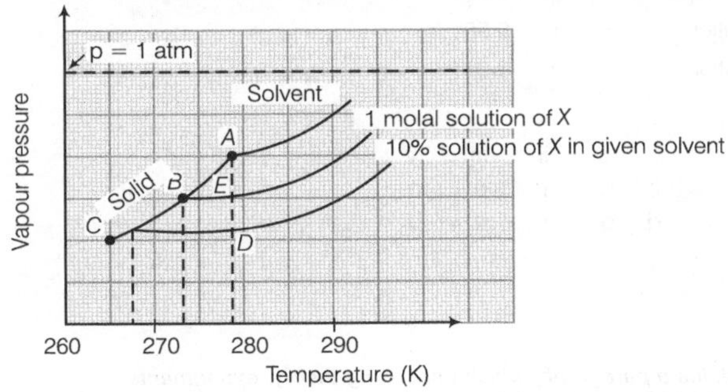

a. 54.5 g mol^{-1} b. 109.0 g mol^{-1} c. 120.0 g mol^{-1} d. 77.2 g mol^{-1}

Statement Type

Direction (Q. No. 12) *This section is based on Statement I and Statement II. Select the correct answer from the codes given below.*

a. Both Statement I and Statement II are correct and Statement II is the correct explanation of Statement I
b. Both Statement I and Statement II are correct but Statement II is not the correct explanation of Statement I
c. Statement I is correct but Statement II is incorrect
d. Statement II is correct but Statement I is incorrect

12 **Statement I** Freezing point of a solution is smaller than that of the solvent.

Statement II At freezing point, solid and liquid exist in equilibrium and $T(\text{fp}) = \dfrac{\Delta H}{\Delta S}$.

One or More than One Options Correct Type

Direction (Q. Nos. 13 and 14) *This section contains 2 multiple choice questions. Each question has four choices (a), (b), (c) and (d) out of which ONE or MORE THAN ONE are correct.*

13 In the depression of freezing point experiment, it is found that the

a. vapour pressure of the solution is less than that of pure solvent
b. vapour pressure of the solution is more than that of pure solvent
c. only solute molecules solidify at the freezing point
d. only solvent molecules solidify at the freezing point

14 A solution containing 3.5 g of solute X is 50.0 g of water has a volume of 52.5 mL and a freezing point of – 0.86°C. Thus, (K_f of H_2O = 1.86° mol^{-1} kg)

a. molality of the solution is 0.46 mol kg^{-1}
b. molarity of the solution is 0.44 mol L^{-1}
c. molar mass of solute X is 152 g mol^{-1}
d. mole fraction of solute X is 0.0082

Working Space

Matching List Type

Direction *(Q. No. 15) Choices for the correct combination of elements from Column I and Column II are given as options (a), (b), (c) and (d) out of which one is correct.*

15 Given, K_f of $H_2O = 1.86°$ mol^{-1} kg.

Match the solutions given in Column I with their depression in freezing points given in Column II and select the answer from the codes given below.

	Column I		Column II
i.	6.0% urea solution	p.	3.72°
ii.	9.6% methanol solution	q.	1.86°
iii.	17.1% sucrose solution	r.	5.58°
iv.	36.0% glucose solution	s.	0.93°

Codes

	i	ii	iii	iv		i	ii	iii	iv
a.	p	q	s	r	b.	p	r	s	q
c.	q	r	p	s	d.	q	r	s	p

Comprehension Type

Direction *(Q. Nos. 16 and 17) This section contains a paragraph, which describing theory, experiments data, etc. Two questions related to the paragraph have been given. Each question has only one correct answer among the given options (a), (b), (c) and (d).*

Passage

1 kg of 1 molal aqueous solution of sucrose is cooled and maintained at $-3.534°C$. $K_f(H_2O) = 1.86$ K mol^{-1} kg

16 Relative lowering of vapour pressure is

a. 0.020
b. 55.56
c. 56.55
d. 0.018

17 How much ice will separate under the given conditions?

a. 745 g
b. 392.0 g
c. 353.0 g
d. 658.0 g

One Integer Value Correct Type

Direction *(Q. Nos. 18-20) This section contains 3 questions. When worked out will result in an integer from 0 to 9 (both inclusive).*

18 At 100°C and 1 atm, following equilibrium is set up

$$\underset{\text{Density } 1.0 \text{ g cm}^{-3}}{H_2O(l)} \rightleftharpoons \underset{\substack{\text{Steam} \\ 0.0006 \text{ g cm}^{-3}}}{H_2O(g)}$$

Volume occupied by water mole. in 10 L steam is x cm^3. What is the value of x?

19 Molecular formula of white phosphorus in benzene is $(P)_x$. What is the value of x based on following experiment?

Freezing point of benzene = 5.45°C K_f(benzene) = 5.12 K mol^{-1} kg

Freezing point of solution containing 0.675 g of white phosphorus in 50.0 g of benzene = 4.89°C.

20 One molal aqueous solution of urea freezes at −1.86°C. Aqueous solution of $C_nH_{2n}O_n$ 36% by mass of water freezes at −3.72°C. What is the value of n?

DPP-5 Colligative Properties-III (Elevation in Boiling Point)

Only One Option Correct Type

Direction *(Q. Nos. 1-13) This section contains 13 multiple choice questions. Each question has four choices (a), (b), (c) and (d) out of which ONLY ONE is correct.*

Working Space

1 Elevation in boiling point of an aqueous solution of a non-electrolyte solute is 1.04°. What is the depression in freezing point of this solution?

$$K_b(H_2O) = 0.52° \text{ mol}^{-1} \text{ kg}$$
$$K_f(H_2O) = 1.86° \text{ mol}^{-1} \text{ kg}$$

a. 1.86° b. 0.52° c. 3.72° d. 0.93°

2 12.8 g of naphthalene ($C_{10}H_8$) is dissolved in 92 g toluene (C_7H_8). Boiling point of this solution is

(boiling point of toluene = 110°C and mole fraction boiling point elevation constant of toluene is 36.1 K)

a. 149.2°C b. 113.3°C c. 116.7°C d. 111.1°C

3 Elevation in boiling point of a molar (1 M) glucose solution ($d = 1.2 \text{ g mL}^{-1}$) is

a. $0.98\, K_b$ b. K_b c. $1.20 K_b$ d. $1.02 K_b$

4 For a dilute solution containing 2.5 g of a non-volatile non-electrolyte solute in 100 g water, the elevation in boiling point at 1 atm pressure is 2° C. Assuming concentration of solute is much lower than the concentration of solvent, the vapour pressure (mm of Hg) of the solution is (Take, $K_b = 0.76 \text{ K mol}^{-1} \text{ kg}$).

a. 724 b. 740 c. 736 d. 718 **[IIT JEE 2012]**

5 Relative lowering of vapour pressure of a non-volatile solute X in water is 0.05. If $K_b(H_2O) = 0.52° \text{ mol}^{-1} \text{ kg}^{-1}$.

(ΔT_b) = elevation in boiling point assuming a dilute solution

$(\Delta T_b)'$ = elevation in boiling point assuming a solution of moderate concentration

then, $(\Delta T_b) - (\Delta T_b)' =$

a. Zero b. 0.08° c. – 0.08° d. 0.16°

6 6% aqueous solution of urea (by mass of solvent) has elevation in boiling point x_1. 18% aqueous solution of glucose (by mass of solvent) has elevation in boiling point x_2. If these two solutions are mixed then, elevation in boiling point is x_3. Then,

a. $x_1 + x_2 = x_3$ b. $3x_1 + x_2 = x_3$ c. $x_1 = x_2 = x_3$ d. $x_1 < x_2 < x_3$

7 AB is a sparingly soluble salt in aqueous solution with $K_{sp} = 1 \times 10^{-8}$. If $K_b(H_2O) = 0.52° \text{ mol}^{-1}$ kg then elevation in boiling point of saturated solution is

a. 5.2×10^{-5} K b. 5.2×10^{-9}K c. 5.2×10^{-8} K d. 5.2×10^{-4} K

8 Elevation in boiling point of a solution of non-electrolyte in non-polar solvent (toluene) is 0.3°. Mole fraction elevation constant of non-polar solvent is 36.1 K. Thus, molality (mol kg^{-1}) of the solution is

a. 8.31×10^{-3} b. 0.091 c. 10.98 d. 1.27

9 Which of the given aqueous solution has maximum boiling point?
(K_b of $H_2O = 0.52°\,mol^{-1}$ kg), T_b (H_2O) = 373 K)

a. 18% glucose solution by weight of solution
b. 34.2% sucrose solution by weight of solution
c. 6% urea solution by weight of solution
d. All have equal boiling points

10 X and Y both are soluble in given solvent. Molar mass of the polymer Y is

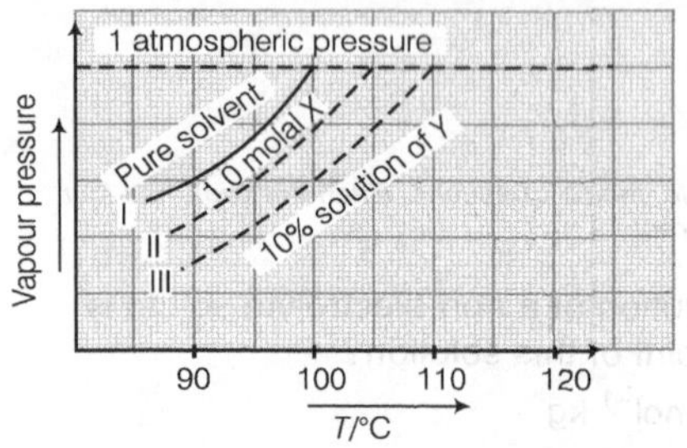

a. 50 g mol^{-1} b. 250 g mol^{-1} c. 110 g mol^{-1} d. 200 g mol^{-1}

11 Given, $H_2O(l) \rightleftharpoons H_2O(g)$ at 373 K, $\Delta H° = 9.72$ kcal mol^{-1}
Thus, boiling point of 0.1 molal sucrose solution is

a. 373.52 K b. 373.052 K c. 374.04 K d. 374.52 K

12 Glucose solution is 20% by weight of solution. Its boiling point is ....
Given 1 molal urea solution boils at 100.52° C and water boils at 1 atm pressure.

a. 100.52° C b. 102.17° C c. 97.83° C d. 100.72° C

13 Latent heat of fusion and latent heat of vaporisation of a solvent are in the ratio of 1: 6 and freezing point and boiling point (in Kelvin scale) are in the ratio of 2 :3. Thus, ratio of elevation in boiling point to depression in freezing point of 1 molal solution is

a. 1/2 b. 1/4 c. 3/4 d. 3/8

One or More than One Options Correct Type

Direction *(Q. Nos. 14 and 15) This section contains 2 multiple choice questions. Each question has four choices (a), (b), (c) and (d), out of which ONE or MORE THAN ONE are correct.*

14 Out of the following, colligative properties are

a. osmotic pressure b. boiling point
c. depression in freezing point d. relative lowering of vapour pressure

15 10% aqueous solution of a non-electrolyte solute of molar mass (m) has elevation in boiling point $x°$, then

a. graphically

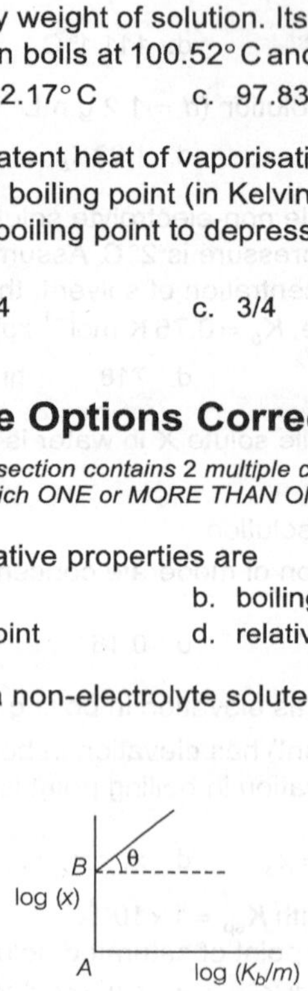

b. $AB = 2$ and $\tan\theta = 1$
c. x is dependent on (K_b/m)
d. $\log\left(\frac{K_b}{m}\right) \rightarrow 0$, when boiling point of the solution is twice the boiling point of water

Working Space

Comprehension Type

Direction *(Q. Nos. 16 and 17) This section contains a paragraph, which describing theory, experiments, data, etc. Two questions related to the paragraph have been given. Each question has only one correct answer among the four given options (a), (b), (c) and (d).*

Passage

Heavy water boils at 101.42° C and its molal elevation constant (K_b) is 10% higher than that of pure water. Elevation in 0.1 molal aqueous urea solution is 0.052°.

16 What is the molal elevation constant (in K mol^{-1} kg) of heavy water?

a. 0.572 b. 0.0572
c. 0.692 d. 0.0692

17 What is the latent heat of vaporisation of heavy water?

a. 0.49 kcal mol^{-1} b. 9.8 kcal mol^{-1}
c. 0.0245 kcal mol^{-1} d. None of these

Matching List Type

Direction *(Q. No. 18) Choices for the correct combination of elements from Column I and Column II are given as options (a), (b), (c) and (d) out of which one is correct.*

18 One molal aqueous sucrose solution boils at 100.52° C and water boils at 100° C. 5% aqueous solution (by mass of solvent) of non-electrolyte AB_2 boils at 100.26° C, 9% aqueous solution (by mass of solvent) of non-electrolyte A_2B boils at 100.52° C. Match the molar masses given in Column II with the species in Column I and select the answer from the codes given below.

	Column I		Column II
i.	Molar mass of AB_2	p.	10
ii.	Molar mass of A_2B	q.	40
iii.	Molar mass of A	r.	90
iv.	Molar mass of B	s.	100

Codes

	i	ii	iii	iv
a.	p	s	r	q
b.	q	r	p	s
c.	s	r	q	p
d.	p	q	s	r

One Integer Value Correct Type

Direction *(Q. Nos. 19 and 20) This section contains 2 questions. When worked out will result in an integer from 0 to 9 (both inclusive).*

19 A solution containing 3.975 g of sulphur in 100 g of CS_2 boils at 319.67 K. The boiling point of pure CS_2 is 319.30 K.

$$K_b(CS_2) = 2.406° \text{ mol}^{-1} \text{ kg}$$

If molecular state of sulphur is $(S)_x$ then what is the value of x?

20 An aqueous solution containing 28.4 g of non-volatile compound having the stoichiometric composition $C_nH_{2n}O_n$ in 100 g water boils at 101.24° C at 1 atm pressure. $K_f(H_2O)$ = 0.52° mol^{-1} kg and boiling point of H_2O = 373 K. What is the value of n?

DPP-6 Colligative Properties-IV (Osmotic Pressure)

Only One Option Correct Type

Working Space

Direction *(Q. Nos. 1-17) This section contains 17 multiple choice questions. Each question has four choices (a), (b), (c) and (d), out of which ONLY ONE is correct.*

1 Osmosis is involved in one or more processes.

I. Excretion of urine
II. Reverse osmosis
III. Evaporation
IV. Interchange of nutrients and waste products between tissue cells and their surroundings.

Select the correct processes.

a. I and III
b. I, II and III
c. I, II and IV
d. All of these

2 Osmotic pressure of insulin solution at 298 K is found to be 0.0072 atm at 298 K. Hence, height of water column due to this pressure is
(density of mercury = 13.6 g mL^{-1})

a. 0.76 cm
b. 0.70 cm
c. 7.44 cm
d. 76 cm

3 Which has maximum osmotic pressure at temperature T K?

a. 100 mL of 1.0 M urea solution
b. 300 mL of 1.0 M glucose solution
c. Mixture of 100 mL of 1.0 M urea solution and 300 mL of 1.0 M glucose solution
d. All of the above are isotonic solutions

4 There is formation of Prussian blue when Fe^{3+} reacts with $[Fe(CN)_6]^{4-}$. Two solutions as shown are separated by a semipermeable membrane AB. Due to osmosis, there is

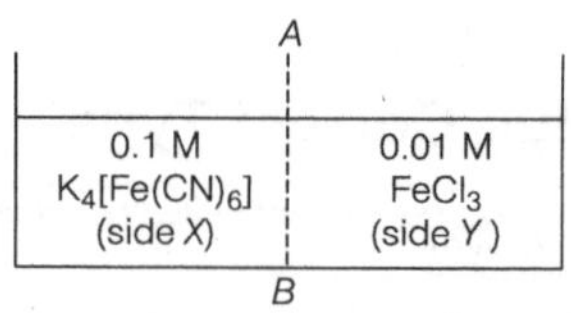

a. blue colour formation in side X
b. blue colour formation in side Y
c. blue colour formation in both the sides X and Y
d. no blue colour formation

5 Three solutions of following in water are taken exerting osmotic pressure under isothermal conditions as given:

Solution	Osmotic pressure
I. 10 g glucose in 200 mL solution	p_1
II. 10 g urea in 100 mL solution	p_2
III. 10 g sucrose in 50 mL solution	p_3

Thus,

a. $p_1 > p_2 > p_3$
b. $p_1 < p_2 < p_3$
c. $p_2 > p_1 > p_3$
d. $p_2 > p_3 > p_1$

Working Space

6 What will be the ratio of the masses of formalin (HCHO) and glucose ($C_6H_{12}O_6$) contained in equal volumes of solutions having the same osmotic pressure at the given temperature?

a. 1 : 1
b. 1 : 2
c. 1 : 3
d. 1 : 6

7 A solution containing 0.4 g of a non-volatile solute in 0.1 dm^3 of the solution exerts an osmotic pressure of 66.66 kPa at 300 K. Thus, molar mass of the solute is

a. 150 g mol^{-1}
b. 300 g mol^{-1}
c. 175 g mol^{-1}
d. 168 g mol^{-1}

8 For a dilute aqueous solution, density is 1.00 g cm^{-3} and g (acceleration due to gravity) is 10.0 ms^{-2}. Then, osmotic pressure set up at highest of 0.17 m is

a. 0.0168 atm
b. 1.7×10^3 Pa
c. Both (a) and (b)
d. None of these

9 Sea water is approximately 0.70 M in NaCl. It can be desalinated by reverse osmosis. For 50% recovery, additional osmotic pressure required (on sea water side) at 298 K is

a. 30 atm
b. 45 atm
c. 68 atm
d. 90 atm

10 A 5% solution of cane sugar (molar mass = 342 g mol^{-1}) is isotonic with 1% of a solution of an unknown non-electrolyte solute at 300 K. The molar mass of unknown solute is (in g mol^{-1}) **[AIEEE 2011]**

a. 136.2
b. 171.2
c. 68.4
d. 34.2

11 5.25% solution of a substance is isotonic with a 21.5% solution of urea (NH_2CONH_2) by weight of solution in the same solvent at 300 K. If the densities of both the solutions are assumed to be equal to 1.0 g cm^{-3} then, molar mass of the substance will be

a. 90 g mol^{-1}
b. 115 g mol^{-1}
c. 105 g mol^{-1}
d. 246 g mol^{-1}

12 Freezing point of a biological fluid is $-0.56°$C in aqueous solution. $K_f(H_2O) = 1.86°$ mol^{-1} kg. Thus, its osmotic pressure at 310 K is (assume molarity = molality)

a. 0.0766 atm
b. 7.66 atm
c. 0.766 atm
d. 0.914 atm

13 100 mL of a solution containing 0.4 g urea (NH_2CONH_2) and 0.9 g of another substance (X) shows an osmotic pressure of 2.85 atm at 25° C. Molar mass of the substance (X) is

a. 180 g mol^{-1}
b. 90 g mol^{-1}
c. 60 g mol^{-1}
d. 120 g mol^{-1}

14 A solution containing 1 g of glucose and 1 g of sucrose in 1 L solution is found to be isotonic with a solution containing 2 g of another unknown solute (X) in 1 L solution at 300 K. Thus, molar mass of the solute X is

a. 180 gmol^{-1}
b. 342 gmol^{-1}
c. 261 gmol^{-1}
d. 236 gmol^{-1}

15 The vapour pressure of pure benzene at 353 K is 10.0268×10^4 Nm^{-2} and that of a solution containing non-volatile solute in benzene is 9.89600×10^4 Nm^{-2}. (d of benzene = 8.149×10^2 g dm^{-3}). Thus, osmotic pressure of the solution is

a. 3.94×10^5 Nm^{-2} b. 4.00×10^5 Nm^{-2}
c. 3.94×10^6 Nm^{-2} d. 4.00×10^6 Nm^{-2}

16 Blood plasma has following composition,

$$Mg^{2+} = 2.0 \text{ milliequivalent } L^{-1}$$

$$PO_4^{3-} = 3.0 \text{ milliequivalent } L^{-1}$$

and other ions. What is the osmotic pressure set up at 300 K due to Mg^{2+} and PO_4^{3-} ions?

a. 0.04926 atm b. 0.2463 atm c. 24.63 atm d. 2.463 atm

17 Osmotic pressure of 3% aqueous solution of a non-volatile, non-electrolyte solute A at 25° C is 3.632 atm. Thus, its vapour pressure is ($p^\circ_{H_2O}$ at 25° C is 23.7 mm) (assume dilute solution)

a. 23.64 mm b. 22.15 mm c. 23.42 mm d. 22.76 mm

Statement Type

Direction (Q. No. 18) *This section is based on Statement I and Statement II. Select the correct answer from the codes given below.*

a. Both Statement I and Statement II are correct and Statement II is the correct explanation of Statement I
b. Both Statement I and Statement II are correct but Statement II is not the correct explanation of Statement I
c. Statement I is correct but Statement II is incorrect
d. Statement II is correct but Statement I is incorrect

18 **Statement I** Osmotic pressure of 34.2% sucrose solution at 300 K is 30.0 atm.

Statement II $\pi = MRT$ is valid for dilute solution.

Matching List Type

Direction (Q. No. 19) *Choices for the correct combination of elements from Column I and Column II are given as options (a), (b), (c) and (d), out of which one is correct.*

19 Match the terms in Column I with the corresponding definition in Column II and select the answer from the codes given below.

	Column I		Column II
i.	Haemolysis	p.	Refers to two or more solutions of same osmolarity.
ii.	Genation	q.	An ionic compound which absorbs water molecules from the atmosphere.
iii.	Hygroscopic	r.	The swelling and bursting of a red blood cell when placed in a solution of lower osmolarity than the cell itself.
iv.	Hypertonic	s.	The shrinking of a red blood cell when placed in a solution of greater osmolarity than the cell itself.
v.	Isotonic	t.	Refers to a solution having a greater osmolarity than 0.30 osmol-the osmolarity of normal blood cells.

Codes

	i	ii	iii	iv	v
a.	p	q	r	s	t
b.	t	p	q	r	s
c.	s	t	p	q	r
d.	r	s	q	t	p

Working Space

Comprehension Type

Direction *(Q. Nos. 20 and 21) This section contains a paragraph, which describing theory, experiments, data, etc. Two questions related to the paragraph have been given. Each question has only one correct answer among the four given options (a), (b), (c) and (d).*

Passage

Consider the following arrangement in which a solution containing 20 g of haemoglobin in 1 L of the solution is placed in the right compartment and pure water is placed in the left compartment. At equilibrium, the height of the water in the right column is 77.8 mm in excess of the height of the solution in the left column at 298 K.

20 Osmotic pressure of the solution is

a. 763 Nm^{-2} b. 7.63 Nm^{-2}
c. 0.763 Nm^{-2} d. 76.3 Nm^{-2}

21 Molar mass of haemoglobin is

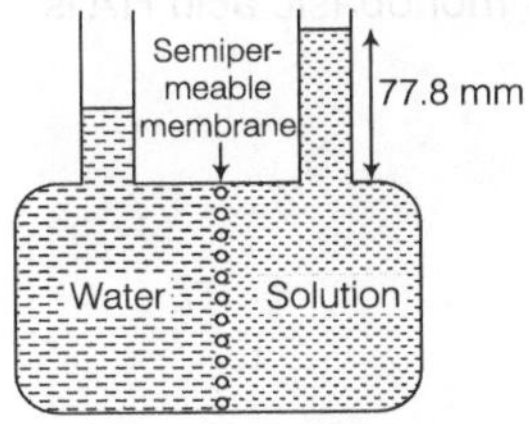

a. 6.49 kg mol^{-1} b. 64.92 kg mol^{-1}
c. 0.649 kg mol^{-1} d. 0.0649 kg mol^{-1}

One Integer Value Correct Type

Direction *(Q. Nos. 22-24) This section contains 3 questions. When worked out will result in an integer from 0 to 9 (both inclusive).*

22 6% aqueous solution of non-electrolyte $C_nH_{2n}O_n$ at 27°C has osmotic pressure of 24.63 atm. Total number of atoms in one molecule is ... .

23 1 molal urea aqueous solution freezes at –1.86°C and 9.00% (w/w) aqueous solution of a non-electrolyte *A* freezes at –0.93°C. Osmotic pressure of 4.41% (w/v) aqueous solution at 298 K is—atmosphere.

24 Variation of π/C vs C (C is the concentration of solute in g dm^{-3}) is given

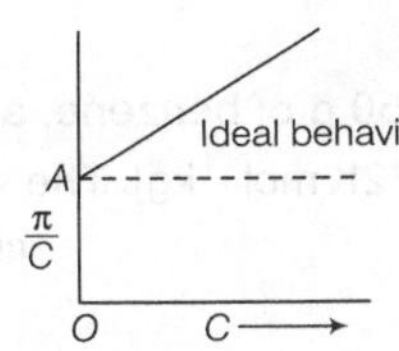

At 300 K for a solute of molar mass 308 gm^{-1}, $OA = x \times 10^{-y}$ in scientific notation. What is the value of x? (R = 0.0821 L atm mol^{-1} K^{-1})

DPP-7 van't Hoff Factor

Subjective Questions

Direction (Q. Nos. 1and 2) *These questions are subjective in nature, need to be solved completely on notebook.*

Working Space

1 What is the osmotic pressure of 5.85% NaCl aqueous solution at 300 K when it is 50% ionised?

2 Depression in freezing point of decimolal aqueous monobasic acid HA is 0.18786°. $K_f(H_2O) = 1.86° \text{ mol}^{-1}$ kg. Calculate

i. i (van't Hoff factor)

ii. α (degree of ionisation)

iii. pH

iv. K_a (ionisation constant)

Only One Option Correct Type

Direction (Q. Nos. 3-9) *This section contains 7 multiple choice questions. Each question has four choices (a), (b), (c) and (d), out of which ONLY ONE is correct.*

3 The degree of dissociation (α) of weak electrolyte A_xB_y is related to van't Hoff factor (i) by the expression **[AIEEE 2011]**

a. $\alpha = \dfrac{i-1}{(x+y-1)}$ b. $\alpha = \dfrac{i-1}{x+y+1}$

c. $\alpha = \dfrac{x+y-1}{i-1}$ d. $\alpha = \dfrac{x+y+1}{i-1}$

4 0.004 M Na_2SO_4 aqueous solution is isotonic with 0.01 M glucose solution at 300 K. Thus, degree of dissociation of Na_2SO_4 is

a. 75% b. 50%

c. 25% d. 85%

5 When 20 g of naphthoic acid $(C_{11}H_8O_2)$ is dissolved in 50 g of benzene, a freezing point depression of 2 K is observed. [K_f (benzene) = 1.72K mol^{-1} kg]. The van't Hoff factor (i) is **[IIT JEE 2007]**

a. 0.5 b. 1

c. 2 d. 3

6 An aqueous solution of a solute AB has boiling point of 101.08° C and freezes at −1.80° C. AB is found to be 100% ionised at boiling point. If $K_b/K_f = 0.3$, then AB

a. is 100% ionised at the freezing point of solution

b. behaves as non-electrolyte at the freezing point of solution

c. forms dimer at the freezing point of solution

d. forms trimer at the freezing point of solution

Working Space

7 Osmolarity of 0.02 M potassium ferrocyanide solution at 300 K is (assume solute is 100% ionised)

a. 0.10 osmol
b. 0.20 osmol
c. 0.02 osmol
d. 0.004 osmol

8 *AB* and *CD* are two non-volatile solutes dissolved in same solvent, showing following behaviour

$$AB \rightleftharpoons A^+ + B^-$$

AB is ionised in solvent, α = degree of ionisation

$$CD \rightleftharpoons \frac{1}{2}(CD)_2$$

CD is dimerised in solvent, β = degree of dimerisation

$$\Delta T_b(AB) = 2\Delta T_b\,(CD)$$

If $\alpha = 0.80$, then β is

a. 0.50
b. 0.40
c. 0.80
d. 0.20

9 Which has the highest boiling point?

a. 0.1 m Na_2SO_4
b. 0.1 m $C_6H_{12}O_6$ (glucose)
c. 0.1 m $MgCl_2$
d. 0.1 m $Al(NO_3)_3$

One or More than One Options Correct Type

Direction (Q. Nos. 10-12) *This section contains 3 multiple choice questions. Each question has four choices (a), (b), (c) and (d), out of which ONE or MORE THAN ONE are correct.*

10 Following solutions have been provided at temperature *T* K.

I. 1M aqueous glucose solution.
II. 1M aqueous sodium chloride solution.
III. 1M aqueous ammonium phosphate solution.
IV. 1M benzoic acid in benzene.

Select correct statements for the above solutions.

a. All are isotonic solutions
b. III is hypertonic of I, II and IV
c. IV is hypotonic of I, II and III
d. II is hypotonic of III but hypertonic of I and IV

11 Which of the following statements are correct?

a. Raoult's law states that the vapour pressure of a component over a solution is directly proportional to its mole fraction
b. The osmotic pressure (π) of a solution is given by the equation $\pi = 2Rt$ of 1 M NaBr solution at temperature t° C
c. The correct order of osmotic pressure of 0.01 M aqueous solution of each compound is sucrose < CH_3COOH < KCl < K_2SO_4
d. Two solutions of urea of 1 mol kg^{-1} in water and in methanol will have the same depression in freezing point

12 In which case, van't Hoff factor (i) remains unchanged?

a. $PtCl_4$ reacts with KCl
b. *aq*. $ZnCl_2$ reacts with *aq*. NH_3 (excess)
c. *aq*. $FeCl_3$ reacts with *aq*. $K_4[Fe(CN)_6]$
d. $HgCl_2$ reacts with excess of KI

Matching List Type

Working Space

***Direction** (Q. Nos. 13 and 14) Choices for the correct combination of elements from Column I and Column II are given as options (a), (b), (c) and (d), out of which one is correct.*

13 Match the different solutions in Column I with their ΔT_b in Column II and select the answer from the codes given below.

	Column I		Column II
i.	1 molal urea solution	p.	K_b
ii.	1 molal KCl solution with 100% ionisation	q.	$1.5\,K_b$
iii.	1 molal K_2SO_4 solution with 75% ionisation	r.	$2.0\,K_b$
iv.	1 molal $K_4[Fe(CN)_6]$ solution with 50% ionisation	s.	$2.5\,K_b$
v.	1 molal iron (III) sulphate (VI) solution with 25% ionisation	t.	$3.0\,K_b$

Codes

	i	ii	iii	iv	v
a.	p	r	s	t	r
b.	p	q	s	t	r
c.	p	r	t	s	q
d.	r	s	t	p	q

14 In a 0.2 molal aqueous solution of a weak acid (HX), depression in freezing point is 0.383° K_f is 1.86° mol^{-1} kg. Assume molarity equal to molality.

Match the parameters in Column I with their values in Column II and select the answer from the codes given below.

	Column I		Column II
i.	van't Hoff factor (*i*)	p.	0.03
ii.	Degree of ionisation	q.	2.22
iii.	pK_a of HX	r.	1.03
iv.	pH of HX	s.	3.73

Codes

	i	ii	iii	iv
a.	p	q	r	s
b.	s	q	p	r
c.	p	s	r	q
d.	r	p	s	q

Comprehension Type

***Direction** (Q. Nos. 15-20) This section contains 3 paragraphs, each describing theory, experiments, data, etc. Six questions related to the paragraphs have been given. Each question has only one correct answer among the four given options (a), (b), (c) and (d).*

Passage I

HgI_2 is insoluble in water. This solute is added into KI solution.

15 In the above case,

a. freezing point is lowered
b. boiling point is elevated
c. freezing point is elevated
d. boiling point and freezing point remain unchanged

16 van't Hoff factor in this example

a. increases by 1 unit
b. decreases by 2 units
c. decreases by 1 unit
d. increases by 2 units

Working Space

Passage II

A solution containing 10 g of a dibasic acid in 1000 g of water freezes at $-0.15°C$. 10 mL of the acid is neutralised by 12 mL of 0.1 N NaOH solution. [$K_f(H_2O) = 1.86° \text{ mol}^{-1} \text{ kg}$]

17 van't Hoff factor (i) of the dibasic acid is

a. 1.34 b. 2.00
c. 2.72 d. 3.00

18 pH of the solution of the acid is

a. 2.00 b. 1.70
c. 3.40 d. 4.20

Passage III

A solution containing 0.684 g of cane sugar ($C_{12}H_{22}O_{11}$) in 100 g water freezes at $-0.037°C$. A solution containing 0.585 g of NaCl in 100 g water freezes at $-0.342°C$.

19 Apparent molecular weight of NaCl is

a. 58.5 g mol^{-1} b. 31.6 g mol^{-1}
c. 29.25 g mol^{-1} d. 117.0 g mol^{-1}

20 van't Hoff factor (i) is

a. 2.0 b. 1.5
c. 1.85 d. 0.85

One Integer Value Correct Type

***Direction** (Q. Nos. 21-25) This section contains 5 questions. When worked out will result in an integer from 0 to 9 (both inclusive).*

21 Relative decrease in vapour pressure of an aqueous NaCl solution is 0.167. Thus, number of moles of NaCl present in 180 g of H_2O is ... .

22 Aqueous solution of barium phosphate which is 100% ionised has $\Delta T_f / K_f$ as 0.40. Hence, given solution is $x \times 10^{-2}$ molal. What is the value of x?

23 0.002 molal aqueous solution of an ionic compound $[Co(NH_3)_5(NO_2)]Cl$ freezes at $-0.00732°C$. [$K_f(H_2O) = 1.86° \text{ mol}^{-1} \text{ kg}$]. How many moles of ions does 1.0 mole of the salt produce on being dissolved in water?

24 Solute X (non-electrolyte, non-volatile) is dimerised in solvent A [$K_f(A) = 1.86° \text{ mol}^{-1} \text{ kg}$] and trimerised in same amount of solvent B [$K_f(B) = 2.72° \text{ mol}^{-1} \text{ kg}$]. Depression in freezing point due to X in solvent (A) is α and depression in freezing point due to X in solvent (B) is β. Then, sum of the ratio of α and β is ....

25 A certain mass of a substance when dissolved in 100 g of benzene lowers its freezing point by 1.28°. The same mass of solute dissolved in 100 g of water lowers its freezing point by 1.40°. Substance has normal molar mass in benzene but completely ionised in water, calculate the number of ions from one unit of substance.

$$K_f(H_2O) = 1.86° \text{ mol}^{-1} \text{ kg}$$
$$K_f(\text{benzene}) = 5.12 \text{ mol}^{-1} \text{ kg}$$

SET - 1

Revisal Problems for JEE Main

Only One Option Correct Type

Working Space

Direction *(Q. Nos. 1-22) This section contains 22 multiple choice questions. Each question has four choices (a), (b), (c) and (d), out of which ONLY ONE is correct.*

1 The partial pressure of ethane over a solution containing 6.56×10^{-3} g of ethane is 1 bar. If the solution contains 5.00×10^{-2} g ethane, then partial pressure is

a. 1.312 bar
b. 1.0 bar
c. 0.762 bar
d. 0.912 bar

2 Concentrated HNO_3 is 63% by mass in aqueous solution and having density 1.5 g mL^{-1}. Molarity of a solution obtained by diluting 10 mL of this HNO_3 to 100 mL solution is

a. 1.50 M
b. 15.0 M
c. 0.15 M
d. None of these

3 Two elements *A* and *B* form compounds AB_2 and AB_4 (both are non-volatile and non-electrolyte) when dissolved in 20.0 g benzene, 1 g of AB_2 lowers the freezing point of benzene by 2.3 K, whereas 1.0 g of AB_4 lowers it by 1.3 K [K_f (benzene) = 5.5 K mol^{-1} kg]. Thus, atomic masses of *A* and *B* respectively, are

a. 120, 212
b. 96, 28
c. 212, 120
d. 28, 46

4 Vapour pressure of pure water at 298 K is 23.8 mm. Thus, vapour pressure of 20% glucose solution (weight/volume of solution) is (density of solution is 0.9971 g/cc)

a. 23.332 mm
b. 23.169 mm
c. 23.334 mm
d. 23.217 mm

5 Pure and dry N_2 gas was bubbled through a solution of 2.25 g of a non-volatile solute *A* in 150 g benzene and through pure benzene.

Loss in weight of solution = 2.154 g and loss in weight of benzene = 0.160 g

Thus, molar mass of solute is

a. 158 g mol^{-1}
b. 170 g mol^{-1}
c. 189 g mol^{-1}
d. 127 g mol^{-1}

6 Mass of methyl alcohol (CH_3OH) which when dissolved in 100 g of water would just prevent formation of ice at $-9.3°$ C is [$K_f(H_2O) = 1.860°$ mol^{-1} kg]

a. 20.0 g
b. 17.20 g
c. 9.3 g
d. 16.00 g

7 At 27° C, an aqueous solution of a carbohydrate containing 18 g per 100 mL of the solution is found to be isotonic with aqueous solution containing 2.925 g NaCl per 100 mL of the solution. Thus, molar mass of the carbohydrate is (assume NaCl is completely ionised)

a. 360.0 g mol^{-1}
b. 180.0 g mol^{-1}
c. 90.0 g mol^{-1}
d. 58.5 g mol^{-1}

8 Mass percentage of the resultant solution obtained by mixing 300 g of 25% solution and 400 g of 40% solution by mass is

a. 37.50%
b. 33.57%
c. 15.0%
d. 32.50%

Working Space

9 1 g mixture of Na_2CO_3 and $NaHCO_3$ each having in 1 : 1 molar ratio required x mL of 0.1 M HCl for complete reaction. What is the value of x?

a. 158 mL
b. 125 mL
c. 147 mL
d. 191 mL

10 The partial molar volumes of propanone (CH_3COCH_3) and butanone ($CH_3COCH_2CH_3$) in a mixture in which mole fraction of propanone is 0.4 are 75 cm^3 mol^{-1} and 80 cm^3 mol^{-1}. Thus, the total volume of solution of total mass 1 kg is

a. 1175 cm^3
b. 1000 cm^3
c. 972 cm^3
d. 1200 cm^3

11 The vapour pressure of an aqueous solution of $CaCl_2$ at 25°C is 0.02970 atm. The vapour pressure of water at the same temperature is 0.03126 atm. Thus, depression in freezing point of the solution is [$K_f(H_2O)$ = 1.86° mol^{-1} kg]

a. 5.17°
b. 1.72°
c. 10.33°
d. 0.86°

12 Correct order of boiling point of the following aqueous solution is (assume molarity = molality)

I. 0.01 M Na_2SO_4,
II. 0.01 M Na_3PO_4,
III. 0.01 M NaBr,
IV. 0.01 M glucose solution

a. equal
b. IV < III < II < I
c. IV < III < I < II
d. II < I < III < IV

13 If sodium sulphate is considered to be completely dissociated into cations and anions in aqueous solution, the change in freezing point of water (ΔT_f), when 0.01 mole of sodium sulphate is dissolved in 1 kg of water is [K_f (H_2O) = 1.86° mol^{-1} kg] **[AIEEE 2010]**

a. 0.0372°
b. 0.0558°
c. 0.0744°
d. 0.0186°

14 122 g of benzoic acid is dissolved in 1000 g of benzene at 10°C at which vapour pressure of benzene is 66.6 cm of Hg. If benzoic acid is completely dimerised in benzene, then vapour pressure of the solution of the benzoic acid in benzene is

a. 66.20 cm of Hg
b. 66.00 cm of Hg
c. 64.10 cm of Hg
d. 65.92 cm of Hg

15 Osmotic pressure of decimolar potassium ferrocyanide solution which is 40% ionised at 300 K is

a. 2.50 bar
b. 3.00 bar
c. 6.5 bar
d. 4.50 bar

16 How many milliliters of glycerol ($C_3H_8O_3$) of density 1.26 g mL^{-1} must be added per kilogram of water to produce a solution with 4.85 mol % of $C_3H_8O_3$?

a. 207 mL
b. 115 mL
c. 215 mL
d. 122 mL

17 A beaker containing 0.010 mol of $C_{12}H_{22}O_{11}$ in 100 g H_2O and a beaker containing 0.020 mol of $C_{12}H_{22}O_{11}$ in 100 g H_2O are placed in a chamber and allowed to equilibrate. Thus, mole fraction of $C_{12}H_{22}O_{11}$ in both solutions is

a. 0.00269, 0.00269
b. 0.00269, 0.00172
c. 0.00172, 0.00269
d. 0.00172, 0.00172

18 The vapour pressure of pure benzene at a certain temperature is 640 mm Hg. A non-volatile, non-electrolyte solid weighing 2.175 g is added to 39.0 g of benzene. The vapour pressure of the solution is 600 mm Hg. Molecular weight of the non-electrolyte is

a. 65.25 g mol^{-1}
b. 60.08 g mol^{-1}
c. 72.08 g mol^{-1}
d. 92.05 g mol^{-1}

Working Space

19 For a dilute solution containing 2.5 g of a non-volatile, non-electrolyte solute in 100 g of water, the elevation in boiling point at 1 atm is 2°C. Assuming concentration of the solute is much lower than that of solvent, vapour pressure (mm of Hg) of the solution is ($K_b = 0.76$ K kg mol^{-1})

a. 724
b. 740
c. 736
d. 718

20 Henry's law constant for CO_2 in water is 1.67×10^8 Pa at 298 K. CO_2 gas is packed in 500 mL H_2O at 298 K under a pressure of 2.5 atm. Thus, quantity of CO_2 in 500 mL H_2O is

a. 0.00152 mol
b. 0.171 mol
c. 0.012 mol
d. 0.042 mol

21 Consider the separate solutions of

0.500 M $C_2H_5OH\ (aq)$
0.100 M $Mg_3(PO_4)_2(aq)$
0.250 M $KBr(aq)$
0.125 M $Na_3PO_4(aq)$

each at 25°C.

Which statement is true about these solutions assuming all salts to be strong electrolytes?

a. They all have same osmotic pressure **[JEE Main 2014]**
b. 0.100 M $Mg_3(PO_4)_2\ (aq)$ has the highest osmotic pressure
c. 0.125 M $Na_3PO_4\ (aq)$ has the highest osmotic pressure
d. 0.500 M $C_2H_5OH\ (aq)$ has the highest osmotic pressure

22 For a weak monobasic acid, $pK_a = 4$ then van't Hoff factor for 0.01 M acid is

a. 1.01
b. 1.02
c. 1.10
d. 1.20

Statement Type

***Direction** (Q. Nos. 23 and 24) This section is based on Statement I and Statement II. Select the correct answer from the codes given below.*

a. Both Statement I and Statement II are correct and Statement II is the correct explanation of Statement I
b. Both Statement I and Statement II are correct but Statement II is not the correct explanation of Statement I
c. Statement I is correct but Statement II is incorrect
d. Statement II is correct but Statement I is incorrect

23 **Statement I** 6% urea solution and 6% acetic acid solution in water are equimolar.
Statement II Equimolar solutions are isotonic.

24 **Statement I** 2 g of benzoic acid (C_6H_5COOH) in 25.0 g of benzene lowered the freezing point of benzene by 1.62 K. K_f (benzene) = 4.86 K kg mol^{-1}.
Statement II Depression in freezing point is a colligative property.

Revisal Problems
for JEE Advanced

Only One Option Correct Type

Working Space

Direction (Q. Nos. 1-17) *This section contains 17 multiple choice questions. Each question has four choices (a), (b), (c) and (d), out of which ONLY ONE is correct.*

1 Mole fraction of non-electrolyte (A) in aqueous solution is 0.07. One molal urea solution freezes at $-1.86°$. Thus, depression in freezing point of the non-electrolyte A is

a. 0.26° b. 1.86°
c. 0.13° d. 7.78°

2 A complex of potassium, iron and cyanide is 100% ionised at 1 m (molal). If the elevation in boiling point is 2.08°, then complex is [K_b (H_2O) = 0.52° mol^{-1} kg]

a. $K_4[Fe(CN)_6]$
b. $K_3[Fe(CN)_6]$
c. $K\,Fe[Fe(CN)_6]$
d. None of the above

3 Select the correct statement(s).

a. Boiling point of 1 molal NaCl solution is twice that of 1 molal sucrose solution
b. Boiling point elevation of 1 molal glucose solution is half that of 1 molal KCl solution
c. Boiling point is a colligative property
d. All of the above

4 What mass of sucrose, $C_{12}H_{22}O_{11}$, should you dissolve in 100.0 g water to obtain a solution in which mole fraction of $C_{12}H_{22}O_{11}$ is 0.124?

a. 124 g b. 269 g
c. 342 g d. 171 g

5 The freezing point (in °C) of a solution containing 0.1 g of $K_3[Fe(CN)_6]$ in 100 g water in which it is 80% ionised is [K_f (H_2O) = 1.86° mol^{-1} kg]

a. -2.3×10^{-2} b. -5.7×10^{-2}
c. -5.7×10^{-3} d. -1.92×10^{-2}

6 Which represents correct difference?

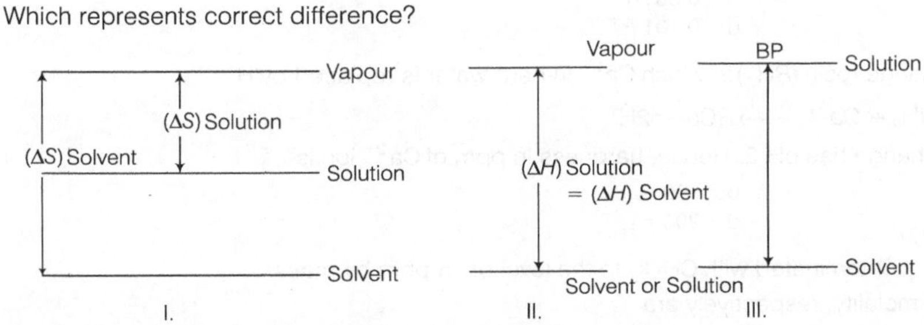

a. I, II, III b. I, III
c. II, III d. I, II

7 60 g of urea is dissolved in 1100 g of solution in water. To keep $\Delta T_f / K_f$ as 1 mol kg^{-1}, water separated in the form of ice is

a. 40 g b. 60 g
c. 100 g d. 200 g

Working Space

8 A solution contains 0.1 mole of acetamide in 1 L of glacial acetic acid. When the solution is cooled, the first crystal that appeared at the freezing point contains the molecule(s) of

a. acetamide
b. acetic acid
c. Both (a) and (b)
d. either acetamide or acetic acid depending on condition of the experiment

9 Two moles of the complex $[(Cu(NH_3)_3Cl]Cl$ are dissolved in 3 moles of H_2O. Relative decrease in vapour pressure was found to be 0.50. If this solution is treated with *aq*. $AgNO_3$ solution, then there is formation of

a. 1.0 mole AgCl
b. 0.25 mole AgCl
c. 0.40 mole AgCl
d. 2.0 moles AgCl

10 Total volume of a mixture of 46 g ethanol (C_2H_5OH) and 72 g water (H_2O) based on the given figure is

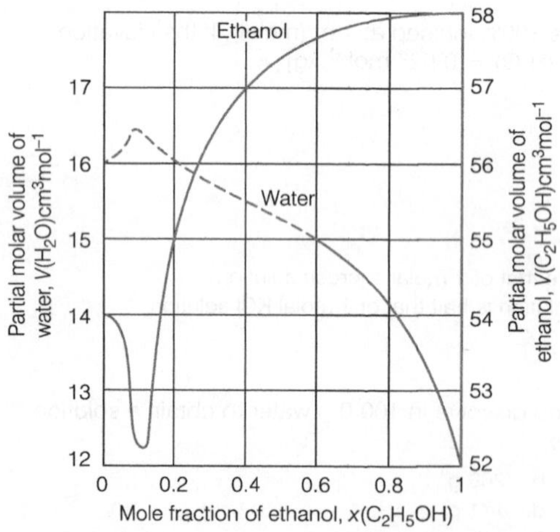

a. 69.0 cm^3
b. 22.2 cm^3
c. 111.0 cm^3
d. 234.0 cm^3

11 pH of 0.1 M dibasic acid is found to be 3. Hence, its osmotic pressure at temperature *T* K is

a. 0.01 *RT*
b. 0.03 *RT*
c. 0.025 *RT*
d. 0.101 *RT*

12 Water is purified by ion-exchange resin (RH_2) in which Ca^{2+} in hard water is replaced by H^+

$$RH_2 + Ca^{2+} \longrightarrow RCa + 2H^+$$

Water coming out of ion-exchange has pH 2. Hence, hardness in ppm of Ca^{2+} ion is

a. 50
b. 100
c. 125
d. 200

13 A sample of drinking water is contaminated with $CHCl_3$ to the level of 15 ppm by mass. Thus, mass percentage and molality, respectively are

	Mass%	Molality
a.	15×10^{-4}	1.255×10^{-4}
b.	15×10^{-2}	1.502×10^{-4}
c.	15×10^{-6}	1.255×10^{-5}
d.	15×10^{-3}	1.255×10^{-6}

Working Space

14 Based on the given diagram, which of the following statements regarding the solutions of two miscible volatile liquids A and B are correct?

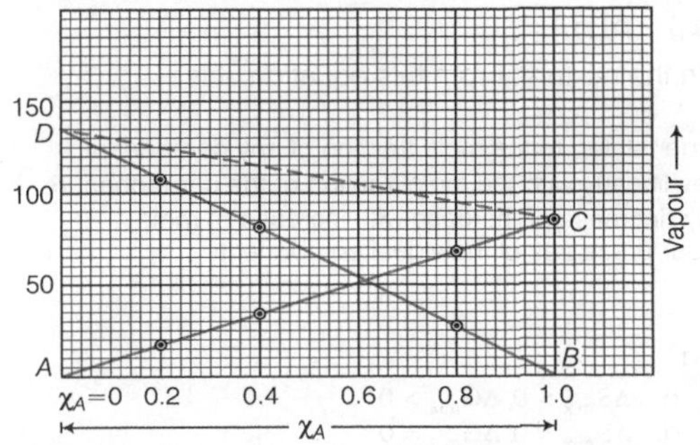

I. Plots AC and BD show that Raoult's law is obeyed for the solution in which A is solute and B is solvent and as well as for that in which B is solute and A is solvent.
II. Plots CD shows that Dalton's law of partial pressure is observed for the binary solution of components A and B.
III. In pure state, vapour pressures of volatile liquid differ by 50 units.
IV. Vapour pressure of a mixture containing 40 mole % of B is about 105 units.

Select the correct answer using the codes given below.

a. All except IV
b. All except I
c. All except III
d. All of these

15 25 mL of an aqueous solution of KCl was found to require 20 mL of 1.0 M $AgNO_3$ solution when titrated using K_2CrO_4 as an indicator. Thus, depression in freezing point of KCl solution with 100% ionisation will be
[$K_f(H_2O) = 2.0°\ mol^{-1}\ kg$, molarity = molality]

a. 5.0°
b. 3.2°
c. 1.6°
d. 0.8°

16 A 0.001 molal solution of $[Pt(NH_3)_4Cl_4]$ in water had a depression in freezing point of 0.0054°. $K_f(H_2O) = 1.86°\ mol^{-1}\ kg$ (assume 100% ionisation). Thus, complex is

a. $[Pt(NH_3)_3Cl_3]\ Cl \cdot NH_3$
b. $[Pt(NH_3)_4Cl_3]Cl$
c. $[Pt(NH_3)_4Cl_2]Cl_2$
d. $[Pt(NH_3)_4Cl_4]$

17 The complete combustion of 2.40 g of a compound of carbon, hydrogen and oxygen yielded 5.46 g of CO_2 and 2.23 g of H_2O. When 8.69 g of the compound was dissolved in 281 g of water, the freezing point depression was found to be 0.97°C. Thus, molecular formula of the organic compound is [$K_f(H_2O) = 1.86°\ mol^{-1}\ kg$]

a. $C_2H_4O_2$
b. C_2H_6O
c. $C_6H_{12}O_6$
d. C_3H_6O

One or More than One Options Correct Type

Direction *(Q. Nos. 18-21) This section contains 4 multiple choice questions. Each question has four choices (a), (b), (c) and (d), out of which ONE or MORE THAN ONE are correct.*

18 You have been provided with 10.00% by mass solution of ethanol in water.

At 15°C, density = 0.9831 g mL^{-1}
At 25°C, density = 0.9804 g mL^{-1}

Select the correct statement(s).

a. Molarity at 15°C is larger than that at 25°C
b. Molarity at 25°C is larger than that at 15°C
c. Molality at both temperatures are equal
d. Molality at 25°C is larger than that at 15°C

19 One molal aqueous glucose solution has density 1.180 g mL^{-1}. Also, $K_f / K_b = 3$. Thus,

a. $\Delta T_b / \Delta T_f = 3$
b. osmotic pressure at 300 K is 24.63 atm
c. osmotic pressure at 300 K is 29.06 atm
d. $\Delta T_b / \Delta T_f = 0.33$

Working Space

20 For a mixture of binary components A and B with vapour pressures p_A° and p_B° in pure state, total vapour pressure is

$$p_{\text{total}} = p_A^\circ + (p_B^\circ - p_A^\circ)\chi_B$$

where, χ_B is mole fraction of component B in the mixture. From this equation, it is concluded that

a. total vapour pressure over the solution can be related to the mole fraction of any component
b. total vapour pressure over the solution varies linearly with the mole fraction of the component B
c. total vapour pressures over the solution may decrease or increase with increase of mole fraction of component A depending on vapour pressures of individual components
d. $p_{\text{total}} = p_B^\circ + (p_A^\circ - p_B^\circ)\chi_A$

21 If liquids A and B form an ideal solution, then

a. $\Delta H_{\text{mix}} = 0, \Delta V_{\text{mix}} = 0$
b. $\Delta S_{\text{mix}} > 0, \Delta G_{\text{mix}} > 0$
c. $\Delta S_{\text{mix}} > 0, \Delta G_{\text{mix}} < 0$
d. $\Delta S_{\text{mix}} = 0, \Delta G_{\text{mix}} = 0$

Comprehension Type

Direction *(Q. Nos. 22-27) This section contains 3 paragraphs, each describing theory, experiments, data, etc. Six questions related to the paragraphs have been given. Each question has only one correct answer among the four given options (a), (b), (c) and (d).*

Passage I

Water boils at 100°C at 1 atm pressure. Latent heat of vaporisation $(\Delta H) = 10$ kcal mol^{-1}.

22 What is the vapour pressure at 40°C?

a. 76.0 torr
b. 58.19 torr
c. 112.7 torr
d. 102.20 torr

23 Lowering of vapour pressure of 0.2 molal urea solution at 40°C is

a. 0.2087 torr
b. 0.2912 torr
c. 1.2000 torr
d. 0.6920 torr

Passage II

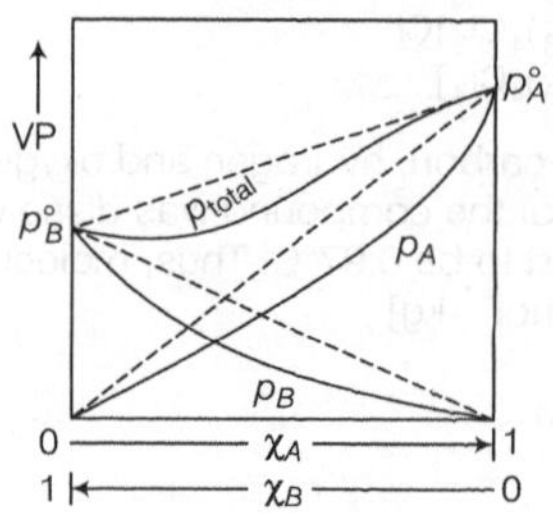

Vapour pressure diagram for real solutions of two liquids A and B that exhibit a negative deviation from Raoult's law. The vapour pressure of both A and B are less than predicted by Raoult's law. The dashed lines represent the plots for ideal solutions.

24 Solution containing components A and B shows this type of deviation from ideal behaviour when

	Attraction $A \cdots B$	ΔH_{mix}	ΔV_{mix}	Boiling point
a.	Larger than average of $A \cdots A$, $B \cdots B$ attraction	+ ve	+ve	Larger than expected
b.	As in (a)	– ve	–ve	As in (a)
c.	Smaller than average of $A \cdots A$, $B \cdots B$ attraction	+ ve	+ve	Smaller than expected
d.	As in (c)	+ve	+ve	As in (c)

25 This type of deviation is also expected in the following mixture.

a. Ethanol and cyclohexane
b. Ethylbromide and ethylchloride
c. Benzonitrile and ethylcyanide
d. Diethylether and chloroform

Working Space

Passage III

Vapour pressure of binary solution containing 6 g of non-volatile solute in 180 g water is 20.0 torr at 298 K. If one mole of water is further added, vapour pressure increases by 0.02 torr.

26 Molar mass of the solute is

a. 18.0 g mol^{-1} b. 60.0 g mol^{-1}
c. 36.0 g mol^{-1} d. 54.0 g mol^{-1}

27 Vapour pressure of H_2O at 298 K is

a. 20.22 torr b. 20.24 torr
c. 20.00 torr d. 19.78 torr

One Integer Value Correct Type

***Direction** (Q. Nos. 28-31) This section contains 4 questions. When worked out will result in an integer from 0 to 9 (both inclusive).*

28 Molarity of Na^+ ions in lake water is 4.58 mol L^{-1}. Density of lake water is 1.25 g mL^{-1}. Thus, molality of Na^+ ion is ...... molal.

29 The so-called ozone layer in the earth's stratosphere has an average total pressure of 10 mm with partial pressure of ozone in the layer 1.2×10^{-5} mm. What is the concentration of ozone (in ppm) assuming that air is a mixture of N_2 and O_2 is 4 : 1 molar ratio.

30 The vapour pressure of a solvent is decreased by 10 mm of Hg when a non-volatile solute was added to the solvent. The mole fraction of the solute in the solution is 0.2. How many times the vapour pressure is decreased when mole fraction of solvent is decreased by 25%?

31 The freezing point depression of a 1.00×10^{-3} molal aqueous solution of $K_x[Fe(CN)_6]$ is 7.10×10^{-3} K. If K_f $(H_2O) = 1.86$ K kg mol^{-1}, what is the value of x?

Matching List Type

***Direction** (Q. No. 32) Choices for the correct combination of elements from Column I and Column II are given as options (a), (b), (c) and (d), out of which one is correct.*

32 17.1 g of sucrose $(C_{12}H_{22}O_{11})$ is dissolved in 1000 g of water.

$$p^\circ (H_2O) = 26.50 \text{ mm}$$
$$K_f (H_2O) = 1.86 \text{ mol}^{-1} \text{ kg}$$
$$K_b (H_2O) = 0.52^\circ \text{ mol}^{-1} \text{ kg}$$

Match the colligative properties in Column I with their values in Column II and select the answers from the codes given below.

	Column I		Column II
i.	Δp (in mm)	p.	1.2315
ii.	ΔT_f (in K)	q.	0.026
iii.	ΔT_b (in K)	r.	0.093
iv.	π (in atm) at 300 K	s.	0.024

Codes

	i	ii	iii	iv		i	ii	iii	iv
a.	p	q	r	s	b.	r	s	p	q
c.	p	r	s	q	d.	s	r	q	p

JEE Main & AIEEE Archive

(Compilation of Last 13 Years' Questions)

1 The vapour pressure of acetone at 20°C is 185 torr. When 1.2 g of a non-volatile substance was dissolved in 100 g of acetone at 20°C, its vapour pressure was 183 torr. The molar mass (g mol^{-1}) of the substance is **(JEE Main 2015)**

a. 488 b. 32
c. 64 d. 128

2 Consider the separate solution of 0.500 M C_2H_5OH (*aq*), 0.100 M $Mg_3(PO_4)_2(aq)$, 0.250 M KBr (*aq*) and 0.125 M $Na_3PO_4(aq)$ at 25° C. Which statement is true about these solutions, assuming all salts to be strong electrolytes? **(JEE Main 2014)**

a. They all have same osmotic pressure
b. 0.100 M $Mg_3(PO_4)_2$ (*aq*) has the highest osmotic pressure
c. 0.125 M $Na_3PO_4(aq)$ has the highest osmotic pressure
d. 0.5000 M C_2H_5OH (*aq*) has the highest osmotic pressure

3 K_f for water is 1.86 K kg mol^{-1}. If your automobile radiator holds 1.0 kg of water, then how many grams of ethylene glycol ($C_2H_6O_2$) must you add to get the freezing point of the solution lowered to –2.8°C? **(AIEEE 2012)**

a. 72 g b. 93 g
c. 39 g d. 27 g

4 A 5.2 molal aqueous solution of methyl alcohol, CH_3OH, is supplied. What is the mole fraction of methyl alcohol in the solution ? **(AIEEE 2011)**

a. 0.100 b. 0.190 c. 0.086 d. 0.050

5 Ethylene glycol is used as an antifreeze in a cold climate. Mass of ethylene glycol which should be added to 4 kg of water to prevent it from freezing at – 6° C will be (K_f for water = 1.86 K kg mol^{-1} and molar mass of ethylene glycol = 62 g mol^{-1}) **(AIEEE 2011)**

a. 804.32 g b. 204.30 g c. 400.00 g d. 304.60 g

6 The degree of dissociation (α) of a weak electrolyte, A_xB_y is related to van't Hoff factor (*i*) by the expression **(AIEEE 2011)**

a. $\alpha = \frac{i-1}{(x+y-1)}$ b. $\alpha = \frac{i-1}{x+y+1}$

c. $\alpha = \frac{x+y-1}{i-1}$ d. $\alpha = \frac{x+y+1}{i-1}$

7 A 5% solution of cane sugar (molar mass = 342) is isotonic with 1% of a solution of an unknown solute at 300 K. The molar mass of unknown solute in g/mol is **(AIEEE 2011)**

a. 136.2 b. 171.2 c. 68.4 d. 34.2

8 The molality of a urea solution in which 0.0100 g of urea, $[(NH_2)_2CO]$ is added to 0.3000 dm^3 of water at STP is **(AIEEE 2011)**

a. 0.555 m b. 5.55×10^{-4} m
c. 33.3 m d. 3.33×10^{-2} m

9 If sodium sulphate is considered to be completely dissociated into cations and anions in aqueous solution, the change in freezing point of water (ΔT_f), when 0.01 mole of sodium sulphate is dissolved in 1 kg of water, is (K_f = 1.86 K kg mol^{-1}) **(AIEEE 2010)**

a. 0.0372 K b. 0.0558 K
c. 0.0744 K d. 0.0186 K

10 On mixing, heptane and octane form an ideal solution. At 373 K, the vapour pressures of the two liquid components (heptane and octane) are 105 kPa and 45 kPa respectively. Vapour pressure of the solution obtained by mixing 25 g of heptane and 35 g of octane will be (molar mass of heptane = 100 g mol^{-1} and of octane = 114 g mol^{-1}). **(AIEEE 2010)**

a. 72.0 kPa b. 75.0 kPa
c. 52.5 kPa d. 45.0 kPa

11 Two liquids *X* and *Y* form an ideal solution at 300 K, vapour pressure of the solution containing 1 mole of *X* and 3 moles of *Y* is 550 mm Hg. At the same temperature, if 1 mole of *Y* is further added to this solution, vapour pressure of the solution increases by 10 mm Hg. Vapour pressure (in mm Hg) of *X* and *Y* in their pure states will be, respectively **(AIEEE 2009)**

a. 200 and 300 b. 300 and 400
c. 400 and 600 d. 500 and 600

12 A binary liquid solution is prepared by mixing *n*-heptane and ethanol. Which one of the following statements is correct regarding the behaviour of the solution? **(AIEEE 2009)**

a. The solution formed is an ideal solution
b. The solution is non-ideal, showing positive deviation from Raoult's law
c. The solution is non-ideal, showing negative deviation from Raoult's law
d. *n*-heptane shows positive deviation while ethanol show negative deviation from Raoult's law

13 At 80°C, the vapour pressure of pure liquid '*A*' is 520 mm Hg and that of pure liquid '*B*' is 1000 mm Hg. If a mixture solution of '*A*' and '*B*' boils at 80°C and 1 atm pressure, the amount of '*A*' in the mixture is (1 atm = 760 mm Hg) **(AIEEE 2008)**

a. 52 mole per cent b. 34 mole per cent
c. 48 mole per cent d. 50 mole per cent

14 The vapour pressure of water at 20°C is 17.5 mm Hg. If 18 g of glucose ($C_6H_{12}O_6$) is added to 178.2 g of water at 20°C, the vapour pressure of the resulting solution will be **(AIEEE 2008)**

a. 17.675 mm Hg b. 15.750 mm Hg
c. 16.500 mm Hg d. 17.325 mm Hg

15 The density (in g mL^{-1}) of a 3.60 M sulphuric acid solution that is 29% H_2SO_4 (molar mass = 98 g mol^{-1}) by mass will be **(AIEEE 2007)**

a. 1.64 b. 1.88
c. 1.22 d. 1.45

16 A mixture of ethyl alcohol and propyl alcohol has a vapour pressure of 290 mm at 300 K. The vapour pressure of propyl alcohol is 200 mm. If the mole fraction of ethyl alcohol is 0.6, its vapour pressure (in mm) at the same temperature will be **(AIEEE 2007)**

a. 350 b. 300 c. 700 d. 360

17 A 5.25% solution of a substance is isotonic with a 1.5% solution of urea (molar mass= 60 g mol^{-1}) in the same solvent. If the densities of both the solutions are assumed to be equal to 1.0 g cm^{-3}, molar mass of the substance will be **(AIEEE 2007)**

a. 90.0 g mol^{-1} b. 115.0 g mol^{-1}
c. 105.0 g mol^{-1} d. 210.0 g mol^{-1}

18 Density of a 2.05 M solution of acetic acid in water is 1.02 g / mL. The molality of the solution is **(AIEEE 2006)**

a. 1.14 mol kg^{-1} b. 3.28 mol kg^{-1}
c. 2.28 mol kg^{-1} d. 0.44 mol kg^{-1}

19 18 g of glucose ($C_6H_{12}O_6$) is added to 178.2 g of water. The vapour pressure of water for this aqueous solution at 100°C is **(AIEEE 2006)**

a. 759.00 torr b. 7.60 torr
c. 76.00 torr d. 752.40 torr

20 Two solutions of a substance (non-electrolyte) are mixed in the following manner. 480 mL of 1.5 M first solution + 520 mL of 1.2 M second solution. What is the molarity of the final mixture? **(AIEEE 2005)**

a. 2.70 M b. 1.34 M
c. 1.50 M d. 1.20 M

21 Equimolar solutions in the same solvent have **(AIEEE 2005)**

a. different boiling point and different freezing point
b. same boiling point and same freezing point
c. same freezing point but different boiling point
d. same boiling point but different freezing point

22 Benzene and toluene form nearly ideal solutions. At 20°C, the vapour pressure of benzene is 75 torr and that of toluene is 22 torr. The partial vapour pressure of benzene at 20°C for a solution containing 78 g of benzene and 46 g of toluene in torr is **(AIEEE 2005)**

a. 53.5 b. 37.5
c. 25 d. 50

23 If α is the degree of dissociation of Na_2SO_4, the van't Hoff factor (i) used for calculating the molecular mass is **(AIEEE 2005)**

a. $1-2\alpha$ b. $1+2\alpha$
c. $1-\alpha$ d. $1+\alpha$

24 6.02×10^{20} molecules of urea are present in 100 mL of its solution. The concentration of urea solution is **(AIEEE 2004)**
(Avogadro constant, $N_A = 6.02 \times 10^{23}$ mol^{-1})

a. 0.001 M b. 0.01 M
c. 0.02 M d. 0.1 M

25 To neutralise completely 20 mL of 0.1 M aqueous solution of phosphorous acid (H_3PO_3), the volume of 0.1 M aqueous KOH solution required is **(AIEEE 2004)**

a. 10 mL b. 20 mL c. 40 mL d. 60 mL

26 Which of the following liquid pairs shows a positive deviation from Raoult's law? **(AIEEE 2004)**

a. Water — hydrochloric acid
b. Benzene — methanol
c. Water — nitric acid
d. Acetone — chloroform

27 Which one of the following aqueous solutions will exhibit highest boiling point? **(AIEEE 2004)**

a. 0.01 M Na_2SO_4 b. 0.01 M KNO_3
c. 0.015 M urea d. 0.015 M glucose

28 Which one of the following statements is false? **(AIEEE 2004)**

a. Raoult's law states that the vapour pressure of a component over a solution is proportional to its mole fraction
b. The osmotic pressure (π) of a solution is given by the equation $\pi = MRT$, where M is the molarity of the solution
c. The correct order of osmotic pressure for 0.01 M aqueous solution of each compound is $BaCl_2 > KCl > CH_3COOH >$ sucrose
d. Two sucrose solutions of same molality prepared in different solvents will have the same freezing point depression

29 If liquid A and B form an ideal solution, the **(AIEEE 2003)**

a. enthalpy of mixing is zero
b. entropy of mixing is zero
c. free energy of mixing is zero
d. free energy as well as the entropy of mixing are each zero

30 In a 0.2 molal aqueous solution of a weak acid HX, the degree of ionisation is 0.3. Taking K_f for water as 1.85, the freezing point of the solution will be nearest to **(AIEEE 2003)**

a. -0.480°C b. -0.360°C
c. -0.260°C d. $+0.480$°C

31 Which of the following concentration factor is affected by change in temperature? **(AIEEE 2002)**

a. Molarity b. Molality
c. Mole fraction d. Weight fraction

JEE Advanced & IIT JEE Archive

(Compilation of Last 10 Years' Questions)

1 For a dilute solution containing 2.5 g of a non-volatile non-electrolyte solute in 100 g of water, the elevation in boiling point at 1 atm pressure is 2°C. Assuming concentration of solute is much lower than the concentration of solvent, the vapour pressure (mm of Hg) of the solution is (take $K_b = 0.76$ K kg mol^{-1}). **(2012, Only One Option Correct Type)**

a. 724 b. 740
c. 736 d. 718

2 The freezing point (in° C) of solution containing 0.1 g of $K_3[Fe(CN)_6]$ (mol. wt. 329) in 100 g of water ($K_f = 1.86$ K kg mol^{-1}) is **(2011, Only One Option Correct Type)**

a. -2.3×10^{-2} b. -5.7×10^{-2}
c. -5.7×10^{-3} d. -1.2×10^{-2}

3 The Henry's law constant for the solubility of N_2 gas in water at 298 K is 1.0×10^5 atm. The mole fraction of N_2 in air is 0.8. The number of moles of N_2 from air dissolved in 10 moles of water of 298 K and 5 atm pressure is **(2009, Only One Option Correct Type)**

a. 4.0×10^{-4} b. 4.0×10^{-5}
c. 5.0×10^{-4} d. 4.0×10^{-6}

Passage for Q. Nos. (4-6)

Properties such as boiling point, freezing point and vapour pressure of a pure solvent change when solute molecules are added to get homogeneous solution. These are called colligative properties.

Applications of colligative properties are very useful in day-to-day life. One of its examples is the use of ethylene glycol and water mixture as anti-freezing liquid in the radiator of automobiles.

A solution M is prepared by mixing ethanol and water. The mole fraction of ethanol in the mixture is 0.9.

Given, freezing point depression constant of water

$(K_f^{water}) = 1.86$ K kg mol^{-1}

Freezing point depression constant of ethanol

$(K_f^{ethanol}) = 2.0$ K kg mol^{-1}

Boiling point elevation constant of water

$(K_b^{water}) = 0.52$ K kg mol^{-1}

Boiling point elevation constant of ethanol

$(K_b^{ethanol}) = 1.2$ K kg mol^{-1}

Standard freezing point of water = 273 K
Standard freezing point of ethanol = 155.7 K
Standard boiling point of water = 373 K
Standard boiling point of ethanol = 351.5 K
Vapour pressure of pure water = 32.8 mm Hg
Vapour pressure of pure ethanol = 40 mm Hg
Molecular weight of water = 18 g mol^{-1}
Molecular weight of ethanol = 46 g mol^{-1}

In answering the following questions, consider the solutions to be ideal dilute solutions and solutes to be non-volatile and non-dissociative. **(2008, Comprehension Type)**

4 The freezing point of the solution M is

a. 268.7 K b. 268.5 K
c. 234.2 K d. 155.9 K

5 The vapour pressure of the solution M is

a. 39.3 mm Hg
b. 36.0 mm Hg
c. 29.5 mm Hg
d. 28.8 mm Hg

6 Water is added to the solution M such that the mole fraction of water in the solution becomes 0.9. The boiling point of this solution is

a. 380.4 K b. 376.2 K
c. 375.5 K d. 354.7 K

7 When 20 g of naphthoic acid ($C_{11}H_8O_2$) is dissolved in 50 g of benzene ($K_f = 1.72$ K kg mol^{-1}), a freezing point depression of 2 K is observed. The van't Hoff factor (i) is **(2007, Only One Option Correct Type)**

a. 0.5 b. 1
c. 2 d. 3

8 75.2 g of C_6H_5OH (phenol) is dissolved in 800 g solvent of $K_f = 14$. If the depression in freezing point is 7 K, then find the percentage of phenol that dimerises. **(2006, Subjective Question)**

9 The elevation in boiling point, when 13.44 g of freshly prepared $CuCl_2$ are added to one kilogram of water, is (Some useful data, $K_b = 0.52$ kg K mol^{-1}, molecular weight of $CuCl_2 = 134.4$ g). **(2005, Only One Option Correct Type)**

a. 0.05 b. 0.1
c. 0.16 d. 0.21

3. Electrochemistry

DPP
DAILY PRACTICE PROBLEMS

DPP-1 Conduction

Subjective Questions

Working Space

Direction (Q. Nos. 1-5) *These questions are subjective in nature, need to be solved completely on notebook.*

1 Resistance of a decimolar KCl solution between two platinum electrodes 0.020 m apart and 0.0006 m^2 in area was found to be 40 Ω.

Calculate

i. Cell constant.

ii. Conductance.

iii. Specific conductance (conductivity).

2 Specific conductance of 0.02 M KCl solution is 0.2768 $\Omega^{-1}\,m^{-1}$ and it has a resistance of 82.4 Ω at 25°C. What is the specific conductance of 0.01 M K_2SO_4 solution at the same temperature if it has a resistance of 300 Ω?

3 A 0.05 M KOH solution offered a resistance of 31.6 Ω in a conductivity cell of constant 0.3792cm^{-1} at 298 K. What is

i. molar conductance?

ii. equivalent conductance?

4 Specific conductance of 0.0025 M K_2SO_4 solution is 0.075 $S\,m^{-1}$ ($\Omega^{-1}\,m^{-1}$).

Calculate

i. molar conductance.

ii. equivalent conductance.

5 Conductivity of 0.00241 M acetic acid is 7.896×10^{-5} $S\,cm^{-1}$. Molar conductance of acetic acid is 390.5$S\,cm^2\,mol^{-1}$ at infinite dilution. Calculate the dissociation constant of acetic acid.

Only One Option Correct Type

Direction (Q. Nos. 6-24) *This section contains 19 multiple choice questions. Each question has four choices (a), (b), (c) and (d), out of which ONLY ONE is correct.*

6 The correct expression in SI system relating the equivalent conductance (Λ_C), specific conductance (κ) and equivalent concentration (C) is (Given κ in $S\,cm^{-1}$, C in equivalent dm^{-3})

a. $\Lambda_C = \dfrac{\kappa}{C}$

b. $\Lambda_C = \dfrac{100\,\kappa}{C}$

c. $\Lambda_C = \dfrac{\kappa \times 10^{-3}}{C}$

d. $\Lambda_C = \dfrac{\kappa \times 10^{-6}}{C}$

7 Conductivity (κ) of 0.01 M NaCl solution is 0.00145 $S\,cm^{-1}$. What happens to the conductivity if extra 100 mL of H_2O be added to the above solution?

a. Increases

b. Decreases

c. Remains unchanged

d. First increases and then decreases

Working Space

8 An increase in equivalent conductance of a strong electrolyte with dilution is mainly due to

a. increase in number of ions
b. increase in ionic mobility of ions
c. 100% ionisation of electrolyte at normal dilution
d. increase in both, i.e. number of ions and ionic mobility of ions

9 Specific conductance of 0.01 N KCl solution is $x\,\text{S cm}^{-1}$ having conductance y S. Thus, specific conductance of 0.01N NaCl having conductance zS is (in S cm^{-1})

a. $\frac{zx}{y}$ b. $\frac{yx}{z}$ c. $\frac{z}{yx}$ d. None of these

10 In terms of Λ_m (molar conductance), Λ_m° (molar conductance of very dilute solution) ionisation constant (K_a) of weak acid is

a. $K_a = \frac{C\Lambda_m^\circ}{(\Lambda_m^\circ - \Lambda_m^\circ)}$ b. $K_a = \frac{C\Lambda_m^2}{\Lambda_m^\circ(\Lambda_m^\circ - \Lambda_m)}$

c. $K_a = \frac{C(\Lambda_m^\circ)^2}{\Lambda_m^\circ(\Lambda_m^\circ - \Lambda_m)}$ d. None of these

11 Given,
limiting ionic conductance of
H^+ ion : $\lambda_{H^+}^\infty = 350\,\text{S cm}^2\,\text{equi}^{-1}$ and of SO_4^{2-} ion : $\lambda_{SO_4^{2-}}^\infty = 80\,\text{S cm}^2\,\text{equi}^{-1}$
Thus, for H_2SO_4, limiting values of molar conductance and equivalent conductances are

	Molar conductance ($\text{S cm}^2\,\text{mol}^{-1}$)	Equivalent conductance ($\text{S cm}^2\,\text{equi}^{-1}$)
a.	430	430
b.	860	430
c.	215	430
d.	430	860

12 500 mL of an aqueous solution contains 0.1 mole of KCl. If its specific conductance is $x\,\text{S cm}^{-1}$, its molar conductance will be (in $\text{S cm}^2\,\text{mol}^{-1}$)

a. $10x$ b. $200x$ c. x d. $5000x$

13 Resistivity of a material is equal to resistance when cell constant is

a. 1cm^{-1} b. 0.1cm^{-1} c. 10 cm^{-1} d. 100cm^{-1}

14 Cell constant is maximum in case of a

a. wire of length 100 cm and area 100cm^2
b. wire of length 10 cm and area 10cm^2
c. one centimetre cube of a material
d. equal in all cases

15 Which quantity is temperature independent?

a. Resistivity b. Conductivity
c. Cell constant d. Cell EMF

16 0.1 M H_2SO_4 solution is diluted to 0.01 M H_2SO_4. Hence, its molar conductivity will be

a. 100 times b. $\frac{1}{10}$th times
c. 10 times d. 1000 times

Working Space

17 Conductivity (Siemen's S) is directly proportional to the area of the vessel and the concentration of the solution in it, and is inversely proportional to the length of the vessel, then constant of proportionality is expressed in

a. $S\,m\,mol^{-1}$ b. $S^2\,m^2\,mol^{-1}$
c. $S\,m^2\,mol^{-1}$ d. $S^2\,m^2\,mol$

18 Given, $\Lambda_\infty (S\,cm^2\,mol^{-1})$ for different electrolytes

Electrolyte	$\Lambda_\infty (S\,cm^2\,mol^{-1})$
KCl	149.9
KNO_3	145.0
HCl	426.2
CH_3COONa	91.0
NaCl	126.5

Thus, Λ_∞ of CH_3COOH is

a. 516.5 b. 420.3
c. 172.9 d. 390.7

19 The equivalent conductance of two strong electrolytes at infinite dilution in H_2O (where ions move freely through a solution) at 25 °C are given below :

$$\Lambda^\circ_{CH_3COONa} = 91.0\,S\,cm^2\,equiv^{-1}$$

$$\Lambda^\circ_{HCl} = 426.2\,S\,cm^2\,equiv^{-1}$$

What additional information/quantity are needed to calculate Λ° of an aqueous solution of acetic acid?

a. Λ° of NaCl
b. Λ° of CH_3COOK
c. the limiting equivalent conductance of H^+
d. Λ° of $CH_2ClCOOH$

20 Resistance of 0.2 M solution of an electrolyte is 50 Ω. The specific conductance of the solution is 1.3 $S\,m^{-1}$. If resistance of the 0.4 M solution of the same electrolyte is 260 Ω, its molar conductivity is **[AIEEE 2011]**

a. $6250\,S\,m^2\,mol^{-1}$ b. $6.25 \times 10^{-4}\,S\,m^2\,mol^{-1}$
c. $62.5 \times 10^{-4}\,S\,m^2\,mol^{-1}$ d. $62.5\,S\,m^2\,mol^{-1}$

21 The equivalent conductance of NaCl at concentration C and at infinite dilution are λ_c and λ_∞ respectively. The correct relationship between λ_c and λ_∞ is given as (where constant B is positive).

a. $\lambda_C = \lambda_\infty + (B)C$ b. $\lambda_C = \lambda_\infty - (B)C$
c. $\lambda_C = \lambda_\infty - (B)\sqrt{C}$ d. $\lambda_C = \lambda_\infty + B\sqrt{C}$

22 At 298 K, given specific conductance of saturated

AgCl solution = $3.41 \times 10^{-6}\,\Omega^{-1}\,cm^{-1}$

and that of water used = $1.60 \times 10^{-6}\,\Omega^{-1}\,cm^{-1}$

Equivalent conductance of saturated AgCl solution = $138.3\,\Omega^{-1}\,cm^2\,equiv^{-1}$

Thus, solubility product (K_{sp}) of AgCl is

a. 1.81×10^{-6} b. 1.31×10^{-5}
c. 1.72×10^{-10} d. 1.72×10^{-8}

Working Space

23 $AgNO_3$ (*aq*) was added to an *aq.* KCl solution gradually and the conductivity of the solution was measured. The plot of conductance (Λ_C) *vs* the volume of $AgNO_3$ is **[IIT JEE 2011]**

a. 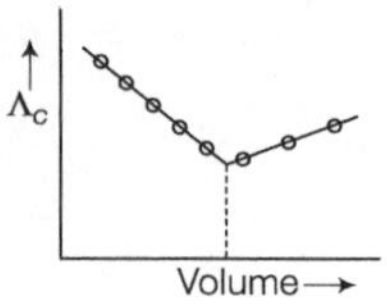

b. 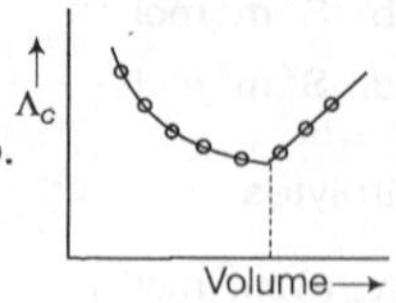

c. 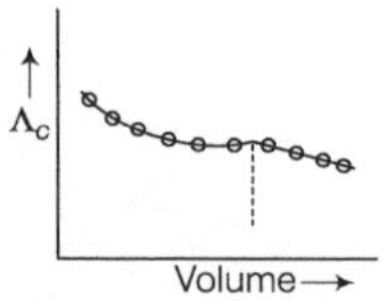

d. 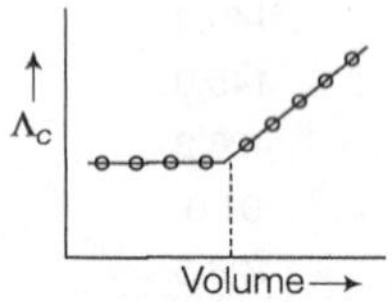

24 Resistance of 0.2 M solution of an electrolyte is $50\,\Omega$. The specific conductance of this solution is $1.4\ \text{S m}^{-1}$. The resistance 0.5 M solution of the same electrolyte is 280Ω. The molar conductivity of 0.5 solution of the electrolyte in $\text{S m}^2\ \text{mol}^{-1}$ is **[JEE Main 2014]**

a. 5×10^{-4} b. 5×10^{-3}
c. 5×10^{3} d. 5×10^{2}

One or More than One Options Correct Type

Direction *(Q. Nos. 25-27) This section contains 3 multiple choice questions. Each question has four choices (a), (b), (c) and (d), out of which ONE or MORE THAN ONE are correct.*

25 Equivalent conductance of 0.01 N $CaCl_2$ solution is $126.36\ \text{S cm}^2\ \text{equiv}^{-1}$. This is equal to

a. $126.36 \times 10^{-4}\ \text{S m}^2\ \text{equiv}^{-1}$ b. $126.36 \times 10^{-4}\ \text{S m}^2\ \text{mol}^{-1}$
c. $252.72 \times 10^{-4}\ \text{S m}^2\ \text{mol}^{-1}$ d. $63.18 \times 10^{-4}\ \text{S m}^2\ \text{mol}^{-1}$

26 Degree of dissociation of very pure water is 1.9×10^{-9}. Also limiting molar conductance of $\lambda_0(H^+) = 350\ \text{S cm}^2\ \text{equiv}^{-1}$
$\lambda_0(OH^+) = 200\ \text{S cm}^2\ \text{equiv}^{-1}$
Thus, equivalent conductance of water is

a. $1.045 \times 10^{-6}\ \text{S m}^2\ \text{equiv}^{-1}$
b. $1.045 \times 10^{-8}\ \text{S m}^2\ \text{equiv}^{-1}$
c. $1.045 \times 10^{-10}\ \text{S m}^2\ \text{equiv}^{-1}$
d. $1.045 \times 10^{-10}\ \text{S m}^2\ \text{mol}^{-1}$

27 Select the correct statement(s).

a. Λ_m^∞ (molar conductance at infinite dilution) of strong electrolytes as well as weak electrolytes can be obtained by Kohlrausch's law. $\Lambda_m^\infty = x_+ \lambda_+^\infty + y_- \lambda_-^\infty$
b. Λ_m^∞ of strong as well as weak electrolytes can be obtained by extrapolation of graph between Λ_m and $\sqrt{C}$ to $\sqrt{C} \longrightarrow 0$
c. Λ_{AgCl}^∞ can be obtained by known value of $\Lambda_{AgNO_3}^\infty$, Λ_{HCl}^∞ and $\Lambda_{HNO_3}^\infty$
d. $\Lambda_{BaSO_4}^\infty = \Lambda_{BaCl_2}^\infty + \Lambda_{H_2SO_4}^\infty - 2\Lambda_{HCl}^\infty$

Working Space

Matching List Type

Direction (Q. Nos. 28 and 29) Choices for the correct combination of elements from Column I and Column II are given as options (a), (b), (c) and (d) out of which one is correct.

28 An aqueous solution of X is added slowly to an aqueous solution of Y as shown in Column I. The variation in conductivity of these reactions is given in Column II. Match Column I with Column II and select the correct answer using the codes given below the lists. **[JEE Advanced 2011]**

	Column I		Column II
i.	$\underset{X}{(C_2H_5)_3N} + \underset{Y}{CH_3COOH}$	p.	Conductivity decreases and then increases.
ii.	$\underset{X}{KI(0.1\ M)} + \underset{Y}{AgNO_3(0.01\ M)}$	q.	Conductivity decreases and then does not change.
iii.	$\underset{X}{CH_3COOH} + \underset{Y}{KOH}$	r.	Conductivity increases and then does not change.
iv.	$\underset{X}{NaOH} + \underset{Y}{HI}$	s.	Conductivity does not change much and then increases.

Codes

	i	ii	iii	iv		i	ii	iii	iv
a.	r	s	q	p	b.	s	r	q	p
c.	q	r	s	p	d.	p	s	r	q

29 Limiting equivalent conductance (in $S\,cm^2\,equiv^{-1}$) of a weak monobasic acid (HA) at infinite dilution is 100 and that of its 0.01 M solution is 5 at 298 K. Match the parameters given in Column I with their values in Column II and select the correct answer from the codes given.

	Column I		Column II
i.	x (degree of ionisation)	p.	3.30
ii.	$[A^-]$	q.	4.58
iii.	pH	r.	0.05
iv.	pK_a	s.	5×10^{-4}

Codes

	i	ii	iii	iv		i	ii	iii	iv
a.	p	q	r	s	b.	s	r	p	q
c.	s	p	q	r	d.	r	s	p	q

Comprehension Type

Direction (Q. Nos. 30-33) This section contains 2 paragraphs, each describing theory, experiments, data, etc. Four questions related to the paragraphs have been given. Each question has only one correct answer among the four given options (a), (b), (c) and (d).

Passage I

Consider the following solutions of an electrolyte

Molarity	0.1 M	0.2 M
Resistance	$100\,\Omega$	$520\,\Omega$
Conductivity	$1.30\ Sm^{-1}$	–

30 Conductivity of 0.2 M solution is

a. $2.60\ Sm^{-1}$ b. $0.25\ Sm^{-1}$ c. $4.00\ Sm^{-1}$ d. $0.50\ Sm^{-1}$

31 Molar conductivity of 0.2 M solution is

a. $1.25 \times 10^{-3}\ Sm^2\,mol^{-1}$ b. $2.50 \times 10^{-3}\ Sm^2\,mol^{-1}$

c. $1.25 \times 10^{-5}\ Sm^2\,mol^{-1}$ d. $2.50 \times 10^{-5}\ Sm^2\,mol^{-1}$

Working Space

Passage II

Given molar conductance of 0.001 M NH_4OH solution at 298K = $3.0 \times 10^{-3}\, Sm^2\, mol^{-1}$.

Limiting molar conductance of

$$aq.\, NH_4Cl = 1.50 \times 10^{-2}\, Sm^2\, mol^{-1}$$

$$aq.\, NaCl = 1.26 \times 10^{-2}\, Sm^2\, mol^{-1}$$

and $$aq.\, NaOH = 2.48 \times 10^{-2}\, Sm^2\, mol^{-1}$$

32 Degree of dissociation of NH_4OH at 298 K is

a. 0.11　　b. 0.011

c. 0.89　　d. 0.78

33 pK_b of NH_4OH is

a. 4.70　　b. 4.80

c. 4.72　　d. 4.87

One Integer Value Correct Type

Direction (Q. Nos. 34 and 35) *This section contains 2 questions. When worked out will result in an integer from 0 to 9 (both inclusive).*

34 0.01 M aqueous solution of a dibasic acid is diluted to 0.004 N such that equivalent conductance is x times. What is the value of x?

35 At what serial number reciprocal of resistance is given?

(1) Volt　　(2) Ampere

(3) Coulomb　　(4) Faraday

(5) Conductivity　　(6) Resistivity

(7) Siemen　　(8) Debye

DPP-2 Electrochemical Series

Subjective Questions

Working Space

Direction *(Q. No. 1) This question is subjective in nature, need to be solved completely on notebook.*

1 Given for half-cell

$Cu^{2+}/Cu \quad E^{\circ}_{Cu^{2+}/Cu} = 0.34\text{ V}, Ag^{+}/Ag \quad E^{\circ}_{Ag^{+}/Ag} = 0.80\text{V}$

i. Construct a galvanic cell using these electrodes so that its standard emf is positive.s

ii. For what concentration of Ag^{+} will the emf of the cell at 298 K be zero, if the concentration of Cu^{2+} is 0.01 M?

Only One Option Correct Type

Direction *(Q. Nos. 2-17) This section contains 16 multiple choice questions. Each question has four choices (a), (b), (c) and (d), out of which ONLY ONE is correct.*

2 Cl_2 gas is passed into a solution containing KF, KI and KBr, and $CHCl_3$ is added. There is a colour in $CHCl_3$ (lower) layer. It is due to

a. formation of I_2 (violet)
b. formation of Br_2 (orange)
c. formation of I_2 and Br_2 both
d. formation of F_2 (colourless)

3 Select the correct statement(s) about galvanic cell (Daniell cell) with $E^{\circ}_{cell} = 1.10\text{V}$.

a. 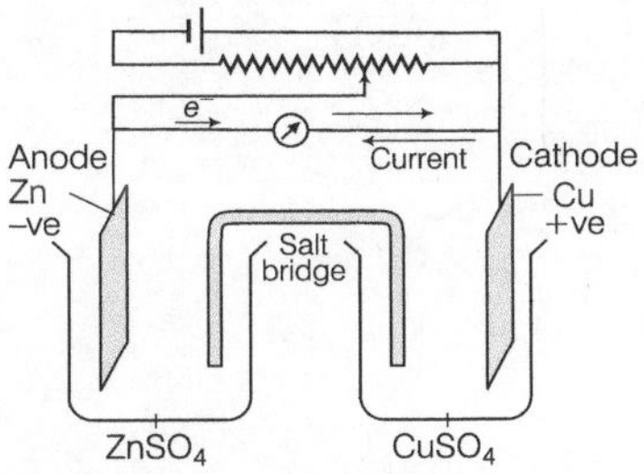

This is observed when $E_{ext} < 1.1\text{V}$, zinc dissolves at anode and copper deposits at cathode

b. 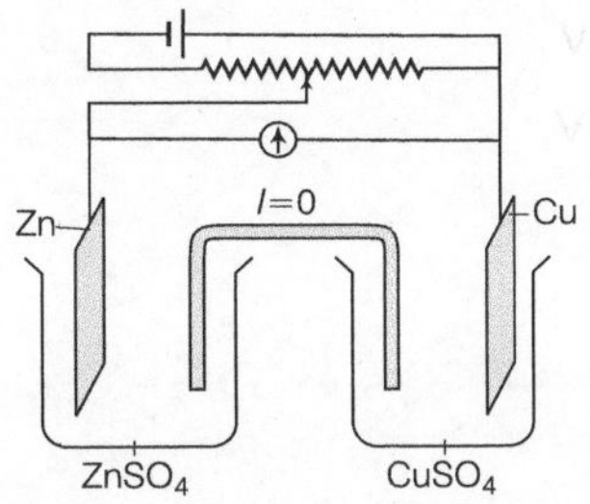

This is observed when $E_{ext} = 1.10\text{ V}$ and no reaction takes place

c. 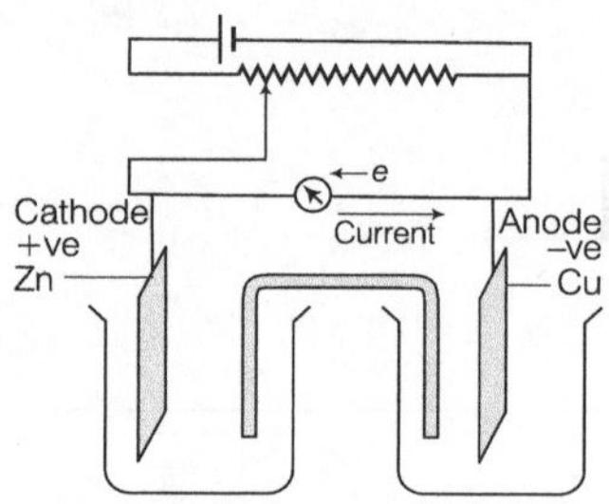

This is observed when $E_{ext} > 1.1V$ zinc is deposited at the zinc electrode and copper dissolves at the copper electrode

d. All of the above

4 Given,

$E^\circ_{Ag^+/Ag} = 0.80\,V$

$E^\circ_{Mg^{2+}/Mg} = -2.37\,V$

$E^\circ_{Cu^{2+}/Cu} = +0.34\,V$

$E^\circ_{Hg^{2+}/Hg} = 0.79\,V$

Which of the following statement(s) is/are correct?

a. $AgNO_3$ can be stored in a copper vessel
b. $Cu(NO_3)_2$ can be stored in a magnesium vessel
c. $CuCl_2$ can be stored in a silver vessel
d. $HgCl_2$ can be stored in a copper vessel

5 One mole of Ag_2CO_3 is strongly heated in an open vessel. Residue will be

a. 1 mole of Ag_2O
b. 2 moles of Ag
c. 1 mole of Ag_2O and 1 mole of CO_2
d. 2 moles of Ag, 0.5 mole of O_2 and 1 mole of CO_2

6 Zinc, silver and iron plates are dipped in $CuSO_4$ solutions placed in different vessels as shown:

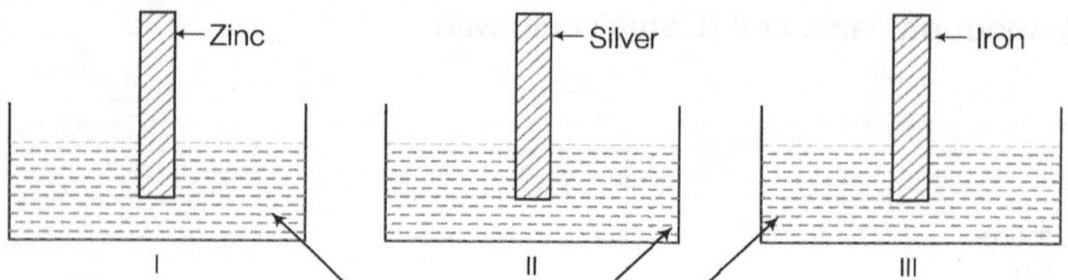

Blue colour of $CuSO_4$ fades in

a. I, II, III
b. I, II
c. II, III
d. I, III

7 For the following half-cell reactions, E° values are:

$$Mn^{2+}(aq) + 2H_2O(l) \longrightarrow MnO_2(s) + 4H^+(aq) + 2e^-, \quad E^\circ = -1.23\,V$$

$$MnO_4^-(aq) + 4H^+(aq) + 3e^- \longrightarrow MnO_2(s) + 2H_2O(l), \quad E^\circ = +1.70\,V$$

Thus,

a. Mn^{2+} reacts with MnO_4^- in acid solution to form MnO_2
b. $Mn(MnO_4)_2$ is stable in acid solution
c. MnO_2 disproportionates to Mn^{2+} and MnO_4^- in acid solution
d. None of the above

Working Space

Working Space

8 Which of the following statement is correct?

a. E_{cell} and $\Delta_r G$ of a cell reaction both are extensive properties
b. E_{cell} and $\Delta_r G$ of a cell reaction both are intensive properties
c. E_{cell} is an intensive property while $\Delta_r G$ of a cell reaction is an extensive property
d. E_{cell} is an extensive property while $\Delta_r G$ of a cell reaction is an intensive property

9 If $Sn^2 + 2e^- \longrightarrow Sn, \quad E^\circ_{Sn^{2+}/Sn} = -0.14\text{ V}$

$Sn^{4+} + 2e^- \longrightarrow Sn^{2+}, \quad E^\circ_{Sn^{4+}/Sn^{2+}} = +0.13\text{ V}$

Then,

a. Sn^{2+} is unstable and disproportionates to Sn^{4+} and Sn
b. Sn^{2+} is stable and disproportionation reaction is not spontaneous
c. Sn^{4+} is easily reduced to Sn
d. None of the above

10 Consider some facts about Standard Hydrogen Electrode (SHE).

I. It is assigned a zero potential at all temperature corresponding to the reaction.

$$H^+(aq) + e^- \longrightarrow \frac{1}{2}H_2(g)$$

half-cell can be represented as $H^+\ (1M)\,|\,Pt(H_2)$
1 bar

II. Platinum electrode coated with platinum black is dipped in acidic solution and H_2 gas is bubbled through it.

III. Concentration of both are oxidised and reduced forms of hydrogen is maintained at unity.

Select the correct facts.

a. I, II, III
b. II, III
c. I, II
d. I, III

11 Given, $E^\circ_{Cr^{3+}/Cr} = -0.74\text{V}, \quad E^\circ_{MnO_4^-/Mn^{2+}} = 1.51\text{V},$

$E^\circ_{Cr_2O_7^{2-}/Cr^{3+}} = 1.33\text{V}, \quad E^\circ_{Cl/Cl^-} = 1.36\text{V}$

Based on the data given above strongest oxidising agent will be **[JEE Main 2013]**

a. Cl_2
b. Cr^{3+}
c. Mn^{2+}
d. MnO_4^-

12 The gas *X* at 1 atm is bubbled through a solution containing a mixture of 1 M Y^- and 1 M Z^- at 298 K. If the standard reduction potential

$$E^\circ_{Z/Z^-} < E^\circ_{Y/Y^-} < E^\circ_{X/X^-}$$

(more positive value)

then,

a. *Y* will oxidise *X* and not *Z*
b. *Y* will oxidise *Z* and not *X*
c. *Y* will oxidise both *X* and *Z*
d. Z^- will reduce both *X* and *Y*

13 The standard reduction potential values of three metallic cations *X*, *Y*, *Z* are 0.52, −3.03 and −1.18 V respectively. The order of reducing power of the corresponding metals is

a. $Y > Z > X$
b. $X > Y > Z$
c. $Z > Y > X$
d. $Z > X > Y$

Working Space

14 A solution containing one mole per litre each of $Cu(NO_3)_2$, $AgNO_3$, $Hg_2(NO_3)_2$ and $Mg(NO_3)_2$ is being electrolysed using inert electrodes. The values of standard reduction potentials are

$$E^\circ_{Ag^+/Ag} = +0.80\,V$$
$$E^\circ_{Hg_2^{2+}/Hg} = +0.79\,V$$
$$E^\circ_{Cu^{2+}/Cu} = +0.34\,V$$
$$E^\circ_{Mg^{2+}/Mg} = -2.37\,V$$

with increasing voltage, the sequence of deposition of metals on the cathode will be

a. Ag, Hg, Cu, Mg
b. Mg, Cu, Hg, Ag
c. Ag, Hg, Cu
d. Cu, Hg, Ag

15 The standard reduction potentials at 298 K for the following half-cell are given

$$Zn^{2+}(aq) + 2e^- \rightleftharpoons Zn(s), \quad E^\circ = -0.76\,V$$
$$Cr^{3+} + 3e^- \rightleftharpoons Cr(s), \quad E^\circ = -0.74\,V$$
$$2H^+ + 2e^- \rightleftharpoons H_2(g), \quad E^\circ = 0.00\,V$$
$$Fe^{3+}(aq) + e^- \rightleftharpoons Fe^{2+}(aq), \quad E^\circ = 0.770\,V$$

Which is the strongest reducing agent?

a. $Zn(s)$
b. $Cr(s)$
c. $H_2(g)$
d. $Fe^{2+}(aq)$

16 The $E^\circ_{M^{3+}/M^{2+}}$ values for Cr, Mn, Fe and Co are -0.41 V, $+1.57$ V, $+0.77$ V and 1.97 V respectively. For which one of these metals the change in oxidation state from +2 to +3 is easiest?

a. Cr
b. Mn
c. Fe
d. Co

17 Given the standard electrode potentials

I. $K^+/K = -2.93\,V$,
II. $Ag^+/Ag = 0.80\,V$,
III. $Hg^{2+}/Hg = 0.79\,V$
IV. $Mg^{2+}/Mg = -2.37\,V$,
V. $Cr^{3+}/Cr = -0.74\,V$

These metals are arranged in the increasing reducing power as

a. I < II < IV < III < V
b. II < III < V < IV < I
c. I < IV < V < III < II
d. None of these

Statement Type

***Direction** (Q. Nos. 18 and 19) This section is based on Statement I and Statement II. Select the correct answer from the codes given below.*

a. Both Statement I and Statement II are correct and Statement II is the correct explanation of Statement I
b. Both Statement I and Statement II are correct but Statement II is not the correct explanation of Statement I
c. Statement I is correct but Statement II is incorrect
d. Statement II is correct but Statement I is incorrect

18 **Statement I** $CuSO_4$ can be stored in a vessel made of zinc.

Statement II $E^\circ_{Zn^{2+}/Zn} = -0.76$ V and $E^\circ_{Cu^{2+}/Cu} = +0.34$ V w.r.t. SHE

Working Space

19 **Statement I** When $AgNO_3$ solution is stirred with a spoon made of copper, solution turns blue.

Statement II In electrochemical series, copper is above silver.

One or More than One Options Correct Type

Direction *(Q. Nos. 20-22) This section contains 3 multiple choice questions. Each question has four choices (a), (b), (c) and (d), out of which ONE or MORE THAN ONE are correct.*

20 The positive value of the standard electrode potential of Cu^{2+}/Cu, ($E^\circ_{Cu^{2+}/Cu} = 0.34$ V) indicates that

a. this redox couple is a stronger reducing agent than H^+/H_2 couple

b. this redox couple is a stronger oxidising agent than H^+/H_2 couple

c. Cu can displace H_2 from acid

d. Cu cannot displace H_2 from acid

21 Consider the following half-reactions :

I. $H^+(aq) + e^- \longrightarrow \frac{1}{2}H_2(g),\quad E^\circ_{cell} = 0.00\text{ V}$

II. $2H_2O(l) \longrightarrow O_2(g) + 4H^+ + 4e^-,\quad E^\circ_{cell} = 1.23\text{ V}$

III. $2SO_4^{2-}(aq) \longrightarrow S_2O_8^{2-}(aq) + 2e^-\quad E^\circ_{cell} = 1.96\text{ V}$

Select the correct statement(s) on the basis of the above data.

a. In dilute sulphuric acid, hydrogen ion will be reduced at cathode

b. In concentrated sulphuric acid, water will be oxidised at anode

c. In dilute sulphuric acid, water will be oxidised at anode

d. In dilute sulphuric acid, sulphate ion is oxidised to $S_2O_8^{2-}$

22 Select the correct statement(s) based on the following half-reaction :

$$Cu^{2+} + 2e^- \longrightarrow Cu,\quad E^\circ = 0.34\text{ V}$$

$$Zn^{2+} + 2e^- \longrightarrow Zn,\quad E^\circ = -0.76\text{ V}$$

$$SO_4^{2-} + 4H^+ + 2e^- \longrightarrow SO_2 + 2H_2O,\quad E^\circ = 1.17\text{ V}$$

a. Copper reacts with dil. H_2SO_4 forming H_2 and SO_2

b. Copper reacts with conc. H_2SO_4 forming SO_2

c. Zinc reacts with conc. H_2SO_4 forming SO_2

d. Zinc reacts with dil. H_2SO_4 forming H_2

Matching List Type

Direction *(Q. No. 23) Choices for the correct combination of elements from Column I and Column II are given as options (a), (b), (c) and (d) out of which one is correct.*

23 Half-cell reactions and E° values are given.

	Reaction	E°
(A)	$O_3(g) + 2H^+ + 2e^- \longrightarrow O_2(g) + H_2O$,	+ 2.075 V
(B)	$O_2(g) + 2H_2O + 4e^- \longrightarrow 4OH^-$,	+ 0.401 V
(C)	$O_2(g) + 4H^+ + 4e^- \longrightarrow 2H_2O$,	+ 1.229 V
(D)	$H_2O_2(aq) + 2H^+ + 2e^- \longrightarrow 2H_2O$,	+ 1.763 V
(E)	$Cr^{3+} + 3e^- \longrightarrow Cr$,	– 0.74 V
(F)	$Cr^{2+} + 2e^- \longrightarrow Cr$,	– 0.91 V
(G)	$2H^+(aq) + 2e^- \longrightarrow H_2(g)$	0.00 V

Working Space

Match the half-cell reaction in Column I with their values in Column II and select the answer from the answer codes at the end.

	Column I		Column II
i.	$O_2(g) + 2H^+ + 2e^- \longrightarrow H_2O_2$	p.	0.695 V
ii.	$2H_2O + 2e^- \longrightarrow H_2 + 2OH^-$	q.	– 0.42 V
iii.	$O_3 + H_2O + 2e^- \longrightarrow O_2 + 2OH^-$	r.	1.245 V
iv.	$Cr^{3+} + e^- \longrightarrow Cr^{2+}$	s.	– 0.83 V

Codes

	i	ii	iii	iv
a.	p	s	r	q
b.	r	q	p	s
c.	s	q	p	r
d.	q	p	s	r

Comprehension Type

Direction *(Q. Nos. 24-29) This section contains 3 paragraphs, each describing theory, experiments, data, etc. Six questions related to the paragraphs have been given. Each question has only one correct answer among the four given options (a), (b), (c) and (d).*

Passage I

Given, $E^\circ_{Cr_2O_7^{2-}/Cr^{3+}} = 1.33\,V$, $E^\circ_{Cl_2/Cl^-} = 1.35\,V$,

$E^\circ_{MnO_4^-/Mn^{2+}} = 1.51\,V$, $E^\circ_{Cr^{3+}/Cr} = -0.74\,V$

24 The most stable oxidised species is

a. Cr^{3+}
b. MnO_4^-
c. $Cr_2O_7^{2-}$
d. Mn^{2+}

25 Reducing power of different species is in the order

a. $Cr^{3+} < Cl^- < Mn^{2+} < Cr$
b. $Mn^{2+} < Cl^- < Cr^{3+} < Cr$
c. $Cr^{3+} < Cl^- < Cr_2O_7^{2-} < MnO_4^-$
d. $Mn^{2+} < Cr^{3+} < Cl^- < Cr$

Passage II

Accidently chewing of a stray fragment of aluminium foil can cause a sharp tooth pain if the aluminium comes in contact with an amalgam filling. This filling, an alloy of silver, tin and mercury acts as the cathode (of a tiny galvanic cell), the aluminium as anode and saliva (of mouth) serves as electrolyte. When the aluminium and the filling come in contact, an electric current passes from aluminium to the filling, which is sensed by a nerve in the teeth.

$$E^\circ_{Al^{3+}/Al} = -1.66\,V$$
$$E^\circ_{O_2, H^+/H_2O} = 1.23\,V$$

26 Net reaction taking place when amalgam is in contact with aluminium foil is

a. $Al + O_2 + OH^- \longrightarrow Al(OH)_4^- + H_2O$
b. $4Al + 3O_2 + 12H^+ \longrightarrow 4Al^{3+} + 6H_2O$
c. $4Al + 3O_2 \longrightarrow 2Al_2O_3$
d. $2H_2 + O_3 \longrightarrow 2H_2O$

27 Standard emf experienced by the person with dental filling is

a. 2.89 V
b. – 2.89 V
c. – 0.93 V
d. + 0.43 V

Working Space

Passage III

When suspected drunk drivers are tested with a breathalyser, the alcohol (ethanol) in the exhaled breath is oxidised to ethanoic acid with an acidic solution of potassium dichromate.

$$3CH_3CH_2OH + 2Cr_2O_7^{2-} + 16H^+ \longrightarrow 3CH_3COOH + 4Cr^{3+} + 11H_2O$$

$$E^\circ_{CH_3COOH/CH_3CH_2OH} = 0.06\,V, \quad E^\circ_{Cr_2O_7^{2-}/Cr^{3+}} = 1.33\,V$$

28 Under standard state, E°_{cell} of the breathalyser would be

a. 1.39 V
b. 1.27 V
c. −1.39 V
d. −1.51 V

29 Under the condition of unit activity of each species, emf recorded in the breathalyser at pH 4 would be

a. 0.31 V
b. −1.01 V
c. 1.64 V
d. 0.95 V

One Integer Value Correct Type

Direction *(Q. No. 30) This section contains 1 question. When worked out will result in an integer from 0 to 9 (both inclusive).*

30 E° values of $M^{n+} + ne^- \longrightarrow M$ are given below

Fe^{2+}/Fe −0.44 V, Mg^{2+}/Mg −2.37 V, $Pt(H_2)/H^+$ 0.00 V,
Al^{3+}/Al −1.66 V, Zn^{2+}/Zn −0.76 V, Cu^{2+}/Cu +0.34 V,
Ag^+/Ag +0.80 V, Ni^{2+}/Ni −0.26 V

How many of the metals can be used to prevent corrosion of Ni to Ni^{2+}?

DPP-3 Reference Half-Cells and Nernst Equations

Subjective Questions

Direction (*Q. Nos. 1-4*) *These questions are subjective in nature need to be solved completely on notebook.*

Working Space

1 Consider the cell at 298 K.

$$Zn|Zn^{2+}(aq)\,(1.0\,M)\,||\,Cu^{2+}(aq)\,(0.1M)\,|Cu$$

Given, $E^{\circ}_{Zn^{2+}/Zn} = -0.76$ V

$E^{\circ}_{Cu^{2+}/Cu} = +0.34$ V

i. Write down the cell reaction.
ii. Calculate the emf of the cell.
iii. If concentrations are exchanged, emf increases or decreases.
iv. Is the cell reaction spontaneous or not?

2 For each of the following combinations of electrodes (A and B) and solution, indicate:

i. the net cell reaction
ii. the directions in which electrons flow spontaneously from (A) to (B) or from (B) to (A)

	A	Solution *A*	*B*	Solution *B*
a.	Cu	1.0 M Cu^{2+}	Fe	1.0 M Fe^{2+}
b.	Pt	1.0 M Sn^{2+}/1.0 M Sn^{4+}	Ag	1.0 M Ag^{+}
c.	Zn	0.10 M Zn^{2+}	Fe	0.001 M Fe^{2+}

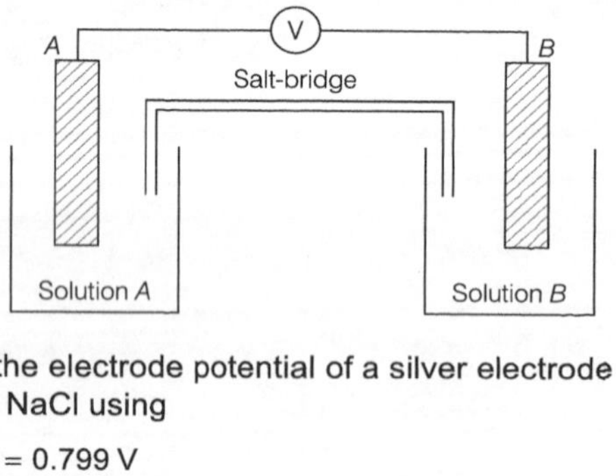

$E^{\circ}_{Ag^{+}/Ag} = 0.80$ V
$E^{\circ}_{Cu^{2+}/Cu} = 0.34$ V
$E^{\circ}_{Fe^{2+}/Fe} = -0.44$ V
$E^{\circ}_{Sn^{4+}/Sn^{2+}} = 0.15$ V

3 Calculate the electrode potential of a silver electrode immersed in a 0.05 M solution of NaCl using

i. $E^{\circ}_{Ag^{+}/Ag} = 0.799$ V

ii. $E^{\circ}_{AgCl/Ag} = 0.222$ V

$[K_{sp}(AgCl) = 1.82 \times 10^{-10}]$

Working Space

4 The standard reduction-potential for the half-cell at 298 K

$NO_3^-(aq) + 2H^+ + e^- \longrightarrow NO_2(g) + H_2O$ is 0.78 V

i. Calculate the reduction-potential in $[H^+] = 8M$.

ii. What will be the reduction-potential of the half-cell in a neutral solution (Assume all the other species to be at unit concentration)?

Only One Option Correct Type

Direction (Q. Nos. 5-22) *This section contains 18 multiple choice questions. Each question has four choices (a), (b), (c) and (d), out of which ONLY ONE is correct.*

5 Which cell will measure the standard electrode potential of zinc electrode?

a. $Pt(s)|H_2(g, 0.1bar)|H^+(aq, 0.1M)||Zn^{2+}(aq, 0.1M)|Zn$

b. $Pt(s)|H_2(g, 1bar)|H^+(aq, 0.1M)||Zn^{2+}(aq, 0.2M)|Zn$

c. $Pt(s)|H_2(g, 1bar)|H^+(aq, 1.0M)||Zn^{2+}(aq, 1.0M)|Zn$

d. $Pt(s)|H_2(g, 1bar)|H^+(aq, 1.0M)||Zn^{2+}(aq, 0.1M)|Zn$

6 For the following cell with hydrogen electrodes at two different pressure p_1 and p_2

$$\underset{p_1}{Pt(H_2)}\,|\,H^+(1M)\,|\,\underset{p_2}{Pt(H_2)}$$

emf is given by

a. $\frac{RT}{F}\log_e\left(\frac{p_1}{p_2}\right)$ b. $\frac{RT}{2F}\log_e\left(\frac{p_1}{p_2}\right)$

c. $\frac{RT}{F}\log_e\left(\frac{p_2}{p_1}\right)$ d. $\frac{RT}{2F}\log_e\left(\frac{p_2}{p_1}\right)$

7 For the cell, $\underset{x\text{ atm}}{Pt(H_2)}\,|\,0.01M\,H^+(aq)\,||\,0.1M\,H^+(aq)\,|\,\underset{y\text{ atm}}{Pt(H_2)}$ $E_{cell} = 0.00V$.

Thus, $\left(\frac{x}{y}\right)$ is

a. 100 b. 10

c. 0.01 d. 0.1

8 A solution of Fe^{2+} is titrated potentiometrically using Ce^{4+} solution.

$$Fe^{2+} \longrightarrow Fe^{3+} + e^-, \qquad E° = -0.77\text{ V}$$

emf of the $Pt\,|\,Fe^{2+}, Fe^{3+}$ pair at 50% and 90% titration of Fe^{2+} are

a. 0.77 V, 0.826 V

b. – 0.826 V, – 0.77 V

c. – 0.77 V, – 0.826 V

d. 0.00 V, 0.00 V

9 Two half-cells are given

$Ag\,|\,AgCl\,|\,KCl\,(0.2\,M)$, $Ag\,|\,AgBr\,|\,KBr\,(0.001M)$,

$K_{sp}(AgCl) = 2.8 \times 10^{-10}$, $K_{sp}(AgBr) = 3.3 \times 10^{-13}$

For a spontaneous cell reaction, cell set up is

a. $Ag|\,AgCl|\,KCl\,(0.2M)\,||\,KBr\,(0.001M)|\,AgBr|\,Ag$

b. $Ag|\,AgBr|\,KBr\,(0.001M)\,||\,KCl\,(0.2\,M)\,|\,AgCl\,|Ag$

c. Both (a) and (b)

d. None of the above

Working Space

10 In the following cell at 298 K, two weak acids (HA) and (HB) with pK_a (HA) = 3 and pK_a (HB) = 5 of equal molarity have been used as shown.

$$\underset{1\,atm}{Pt(H_2)} \mid HA\,(1M) \parallel HB\,(1M) \mid \underset{1\,atm}{Pt\,(H_2)}$$

Thus, emf of the cell is

a. 0.059 V
b. – 0.059 V
c. $\frac{0.0591}{2}$ V
d. $\frac{-0.059\text{ V}}{2}$

11 A concentration cell reversible to anion (Cl^-) is set up $\underset{1\,bar}{Pt\,(Cl_2)} \mid Cl^-\,(aq, C_1) \parallel Cl^-\,(aq, C_2) \mid \underset{1\,bar}{Pt\,(Cl_2)}$ cell reaction is spontaneous, if

a. $C_1 > C_2$
b. $C_1 < C_2$
c. $C_1 = C_2$
d. $C_1 = 0$

12 Which has maximum potential at 298 K (numerical value) for the half-cell reaction?

$$2H^+ + 2e^- \longrightarrow H_2\,(1\,bar)$$

a. 1.0 M HCl
b. A solution having pH 4
c. Pure water
d. 1.0 M NaOH

13 For the cell, $Zn \mid Zn^{2+}\,(1M) \parallel H^+\,(1M) \mid \underset{1\,bar}{Pt\,(H_2)}, E^\circ_{cell} = 0.76$ V and for the cell

$Pt(H_2) \mid H^+\,(1M) \parallel Ag^+\,(1M) \mid Ag,\ E^\circ_{cell} = 0.80$ V

Thus, E_{cell} for the

$Ag \mid Ag^+(0.1\,M) \parallel Zn^{2+}(0.1M) \mid Zn$ is ....and cell reaction is ....

a. 1.44 V, spontaneous
b. 0.4V, spontaneous
c. – 1.44 V, non-spontaneous
d. –1.53V non-spontaneous

14 Consider the following cell reaction,

$$2Fe\,(s) + O_2(g) + 4H^+\,(aq) \longrightarrow 2Fe^{2+}(aq) + 2H_2O(l),\ E^\circ = 1.67\text{ V}$$

At $[Fe^{2+}] = 1 \times 10^{-3}$M, $p_{O_2} = 0.1$ atm and pH = 3, the cell potential at 298 K is

a. 1.47 V
b. 1.77 V
c. 1.87 V
d. 1.57 V

15 For the following cell with gas electrodes at p_1 and p_2 as shown:

$$\underset{\text{at } p_1 \text{ bar}}{Pt\,(Cl_2)} \mid \underset{(1M)}{HCl} \mid \underset{\text{at } p_2 \text{ bar}}{Pt\,(Cl_2)}$$

Cell reaction is spontaneous, if

a. $p_1 = p_2$
b. $p_1 > p_2$
c. $p_1 < p_2$
d. $p_2 = 0$

16 For the half-cell, $Cl^- \mid Hg_2Cl_2, Hg\,(l)$, $E^\circ = 0.280$V at 298 K electrode potential has maximum value when KCl used is

a. 1 M
b. 0.1 M
c. 0.01 M
d. 10 M

17 Given at 298 K standard oxidation potential of quinhydrone electrode = – 0.699 V
Standard oxidation potential of calomel electrode = – 0.268 V
Thus, emf of the cell at 298 K is

Calomel electrode with 0.1 M KCl	Buffer pH = 3.5	Quinhydrone electrode

a. 0.82 V
b. 0.16 V
c. – 0.17 V
d. – 0.82 V

18 E°_{red} (standard reduction electrode potentials) of different half-cell are given

$$E^\circ_{Cu^{2+}/Cu} = 0.34\ V$$
$$E^\circ_{Zn^{2+}/Zn} = -0.76\ V$$
$$E^\circ_{Ag^{+}/Ag} = 0.80\ V$$
$$E^\circ_{Mg^{2+}/Mg} = -2.37\ V$$

In which cell, is ΔG° most negative?

a. $Zn|Zn^{2+}$ (1M)$||Mg^{2+}$ (1M)$|$Mg
b. $Zn|Zn^{2+}$ (1M)$||Ag^{+}$ (1M)$|$Ag
c. $Cu|Cu^{2+}$ (1M)$||Ag^{+}$ (1M)$|$Ag
d. $Ag|Ag^{+}$ (1M)$||Mg^{2+}$ (1M)$|$Mg

19 Given the following half-cell reactions and corresponding reduction potentials:

	E°
I. $A + e^- \longrightarrow A^-$	– 0.24 V
II. $B + e^- \longrightarrow B^{2-}$	+ 1.25 V
III. $C^- + 2e^- \longrightarrow C^{3-}$,	– 1.25 V
IV. $D + 2e^- \longrightarrow D^{2-}$	0.68 V
V. $E + 4e^- \longrightarrow E^{4-}$,	– 0.38 V

Which combination of two half-cell would result in a cell with largest ($E^\circ_{cell} > 0$)?

a. II and III
b. I and II
c. I and III
d. II and IV

20 For

$$Ag \longrightarrow Ag^+ + e^-, \quad E^\circ = -0.800\ V$$
$$V^{2+} + VO^{2+} + 2H^+ \longrightarrow 2V^{3+} + H_2O, \quad E^\circ = 0.616\ V$$
$$V^3 + Ag^+ + H_2O \longrightarrow VO^{2+} + 2H^+ + Ag, \quad E^\circ = 0.439\ V$$

Then E° for the reaction $V^{3+} + e^- \longrightarrow V^{2+}$ is

a. + 0.255 V
b. – 0.255 V
c. + 1.055 V
d. – 1.055 V

21 Following half-cell, Pt $(H_2)|H_2O$ behaves as SHE at a pressure of

a. 1 bar
b. 10^{-14} bar
c. 10^7 bar
d. 10^{14} bar

22 Electrode potential of the following half-cell is dependent on

$$Hg, HgO\ |\ OH^-(aq)$$

a. temperature
b. pH
c. Both (a) and (b)
d. None of these

One or More than One Options Correct Type

Working Space

Direction *(Q. Nos. 23-26) This section contains 4 multiple choice questions. Each question has four choices (a), (b), (c) and (d), out of which ONE or MORE THAN ONE are correct.*

23 Select the correct statement(s).

a. If salt-bridge is removed, potential falls to zero

b. Quinhydrone electrode is reversible to H^+ ion

c. Liquid-junction potential developed across the boundary of the two solutions of different concentration is minimised if concentration cells are used

d. Calomel electrode contains calcium chloride solution in contact with Pt electrode

24 In the following electrochemical cell.

$$Zn \text{ cell } Zn^{2+}||H^+\ |Pt(H_2)$$

$E_{cell} = E^\circ_{cell}$ if

a. $[Zn^{2+}] = [H^+] = 1M$ and $pH_2 = 1$ atm

b. $[Zn^{2+}] = 0.01M$, $[H^+] = 0.1$ M and $pH_2 = 1$ atm

c. $[Zn^{2+}] = 1M$, $[H^+] = 0.1M$ and $pH_2 = 0.01$ atm

d. $[Zn^{2+}] = [H^+] = 0.1M$ and $pH_2 = 1$ atm

25 For Daniell cell $E^\circ_{cell} = 1.10$ V. If state of equilibrium is attained, then

a. $\log_{10} K_c = 1.1$

b. $\log_{10} K_c = \dfrac{1.1}{0.0591}$

c. $\log_{10} K_c = \dfrac{2.2}{0.0591}$

d. $\dfrac{2.303\,RT}{2F} \log K_c = 11$

26 Select the half-cell for the half-cell reaction(s) to be spontaneous

a. $\underset{1\text{ bar}}{Pt(H_2)}\,|\,pH = 5$

b. $\underset{0.5\text{ bar}}{Pt(H_2)}\,|\,0.01M\ HCl$

c. $\underset{1\text{ bar}}{Pt(H_2)}\,|\,0.01M\ CH_3COOH\ pK_a = 4.74$

d. $\underset{4.0\text{ bar}}{Pt(H_2)}\,|\,1.0\,M\ HCl$

Comprehension Type

Direction *(Q. Nos. 27-30) This section contains 4 paragraphs, each describing theory, experiments, data, etc. Four questions related to the paragraphs have been given. Each question has only one correct answer among the four given options (a), (b), (c) and (d).*

Passage I

1.05 g of lead ore containing impurity of Ag was dissolved in HNO_3 and the volume was made 350 mL. A silver electrode was dipped in the solution and E_{cell} of $Pt(H_2)\,|H^+(1M)||\,Ag^+|Ag$ was 0.500 V at 298 K. $E^\circ_{Ag^+/Ag} = 0.80$ V

27 Pure $[Ag^+]$ in the ore is

a. 8.4×10^{-6} M

b. 2.9×10^{-6} M

c. 7.0×10^{-11} M

d. 4.2×10^{-6}M

28 Percentage of silver in the sample is

a. 0.03%

b. 0.06%

c. 3.4%

d. 1.20%

Working Space

Passage II

For the following, $Pt(H_2) \underset{1\,bar}{} \mid HA\,(aq) \underset{0.02\,M}{} \mid\mid H^+\,(0.01M) \mid Pt\,(H_2) \underset{1\,bar}{}$, $E_{cell} = 0.19$ V.

29 pH of the solution in the half-cell containing 0.02 HA is (HA is a weak monobasic acid)

a. 5.21
b. 3.21
c. 1.00
d. 1.70

30 pK_a of the weak monobasic acid is

a. 6.72
b. 4.50
c. 7.00
d. 8.72

One Integer Value Correct Type

Direction *(Q. Nos. 31-33) This section contains 3 questions. When worked out will result in an integer from 0 to 9 (both inclusive).*

31 For the following cell with metal X electrodes, $X \mid X^{n+}$ (0.5 M) $\mid\mid X^{n+}$ (0.05)$\mid X$ $E_{cell} = -0.028$ V at 290 K, if there is no liquid junction potential, valency of X is...... .

32 Osmotic pressure of a 0.001 M weak monobasic acid (HA) at 300 K is 2.5×10^{-2} atm.

Thus, emf of the following in decivolt is $Pt(H_2) \underset{1\,bar}{} \mid 0.001M\,HA$ ....

33 pH of the solution in the anode compartment of the following cell at 298 K is x when $E_{cell} - E^\circ_{cell} = 0.0591$ V. $Pt\,(H_2) \mid pH = x \mid\mid Ni^{2+}$ (1M)$\mid$ Ni.

Derive the value of x.

DPP-4 Thermodynamics of Electrochemistry

Subjective Questions

Direction (Q. No. 1) *This question is subjective in nature, need to be solved completely on notebook.*

Working Space

1 For the given data at 298 K

$$E^\circ_{Cr^{3+}/Cr} = -0.74\,V, \quad E^\circ_{Cd^{2+}/Cd} = -0.40\,V$$

Derive the value of ΔG° and K_{eq} (equilibrium constant) for the reaction,

$$2Cr(s) + 3Cd^{2+}(aq) \rightleftharpoons 2Cr^{3+}(aq) + 3Cd(s)$$

Only One Option Correct Type

Direction (Q. Nos. 2-16) *This section contains 15 multiple choice questions. Each question has four choices (a), (b), (c) and (d), out of which ONLY ONE is correct.*

2 Temperature coefficient of EMF of a cell in terms of entropy change is

a. $\Delta S / nF$ b. $-\Delta S / nF$
c. $n / \Delta S$ d. $nF / \Delta S$

3 The standard reduction potential at 298 K of the reaction,

$$2H_2O + 2e^- \rightleftharpoons H_2 + 2OH^- \text{ is } -0.8277\,V$$

Thus, thermodynamic equilibrium constant for the reaction,

$$2H_2O \rightleftharpoons H_3O^+ + OH^- \text{ at 298K is}$$

a. 1.0×10^{-14} b. 1.0×10^{14}
c. 1.0×10^{-7} d. 1.0×10^{7}

4 Given,

$$Fe^{3+} + e^- \longrightarrow Fe^{2+}, \quad E^\circ = 0.77\,V$$

$$\frac{1}{2}I_2 + e^- \longrightarrow I^-, \quad E^\circ = 0.54\,V$$

Thus, equilibrium constant for the reaction in terms of $\log k$ is

$$2Fe^{3+} + 3I^- \rightleftharpoons 2Fe^{2+} + I_3^-$$

a. 3.89 b. −7.78
c. 7.78 d. −3.89

5 Given, $E^\circ_{Cu^{2+}/Cu^+} = 0.15\,V$, $E^\circ_{In^{3+}/In^+} = -0.42\,V$, $E^\circ_{In^{2+}\,In^+} = -0.40\,V$

Thus, $(\log K_{eq})$ for the reaction, $Cu^{2+} + In^{2+} \rightleftharpoons Cu^+ + In^{3+}$ is

a. 9.2 b. 7.8
c. 11.2 d. 10.0

6 An excess of liquid mercury is added to an acidified solution of 1.0×10^{-3} M Fe^{3+}. Thus, $E^\circ_{(Hg_2^{2+}/Hg)}$ is if 5% of Fe^{3+} remains at equilibrium at 298 K

$$2Hg + 2Fe^{3+}(aq) \longrightarrow Hg_2^{2+}(s) + 2Fe^{2+}(aq), \; E^\circ_{(Fe^{3+}/Fe^{2+})} = 0.77\,V$$

a. −0.793 V b. +0.793 V c. −0.397 V d. +0.397 V

7 EMF of the following cell is 0.2905 V

$$Zn/Zn^{2+}\ (a = 0.1M)\ ||\ Fe^{2+}\ (a = 0.01M)|\ Fe$$

The equilibrium constant for the cell reaction is **[IIT JEE 2004]**

a. $10^{0.32/0.059}$ b. $10^{0.32/0.0295}$
c. $10^{0.26/0.0295}$ d. $10^{0.32/0.295}$

8 The half-cell reactions for rusting of iron are

$$2H^+ + 2e^- + \frac{1}{2}O_2 \longrightarrow H_2O(l), \quad E^\circ = +1.23\,V$$

$$Fe^{2+} + 2e^- \longrightarrow Fe(s), \quad E^\circ = -0.44\,V$$

ΔG° (in kJ) for the reaction is **[IIT JEE 2005]**

a. −76 b. −322
c. −122 d. −176

9 The Gibbs free energy for the decomposition of Al_2O_3 at 800 K is as follows :

$$2Al_2O_3 \longrightarrow 4Al + 3O_2, \quad \Delta_r G = 2898\,kJ\,mol^{-1}$$

The potential difference needed for electrolytic reduction of Al_2O_3 is at least

a. 4.5 V b. 3.0 V
c. − 2.5 V d. 5.0 V

10 Given, $E^\circ_{Fe^{2+}/Fe} = -0.439\,V$, $E^\circ_{Fe^{3+}/Fe} = -0.036\,V$

The value of standard electrode potential for the half-reaction is

$$Fe^{3+}(aq) + e^- \longrightarrow Fe^{2+}(aq)$$ **[AIEEE 2009]**

a. $-0.072\,V$ b. 0.385 V
c. 0.770 V d. $-0.270\,V$

11 For the reaction, $2H_2(g) + O_2(g) \longrightarrow 2H_2O(l)$, $E^\circ_{cell} = 1.23\,V$ at 298 K and $\Delta H^\circ_f\,(H_2O) = -285.8\,kJ\,mol^{-1}$. Thus, ΔS°(standard entropy change) is

a. $-0.325\,kJK^{-1}$ b. $+3.5\,kJK^{-1}$
c. $0.63\,kJK^{-1}$ d. $-0.63\,kJK^{-1}$

12 For the reaction,

$$MnO_4^- + 8H^+ + 5e^- \longrightarrow Mn^{2+} + 4H_2O, \quad E^\circ = 1.51\,V$$

$$MnO_2 + 4H^+ + 2e^- \longrightarrow Mn^{2+} + 2H_2O, \quad E^\circ = 1.23\,V$$

Thus, for the reaction

$$MnO_4^- + 4H^+ + 3e^- \longrightarrow MnO_2 + 2H_2O,\ E^\circ \text{ is}$$

a. 1.70 V b. 5.09 V
c. 0.28 V d. 0.84 V

13 Consider the following equations for a cell reaction.

$$A + B \rightleftharpoons C + D, \quad E^\circ = x\text{ volt}, K_{eq} = K_1$$

$$2A + 2B \rightleftharpoons 2C + 2D, \quad E^\circ = y\text{ volt}, K_{eq} = K_2$$

Then,

a. $x = y, K_1 = K_2$ b. $x = 2y, K_1 = 2K_2$
c. $x = y, K_1^2 = K_2$ d. $x^2 = y, K_1^2 = K_2$

14 Which of the following statements about the spontaneous reaction occurring in a galvanic cell is always true?

a. $E^\circ_{cell} > 0, \Delta G^\circ < 0, Q < K_{eq}$ b. $E^\circ_{cell} > 0, \Delta G^\circ < 0, Q > K_{eq}$
c. $E^\circ_{cell} > 0, \Delta G^\circ > 0, Q > K_{eq}$ d. $E_{cell} > 0, \Delta G < 0, Q < K_{eq}$

Working Space

15 For a (Ag - Zn) button cell, the net reaction is

$$Zn(s) + Ag_2O(s) \longrightarrow ZnO(s) + 2Ag(s)$$

$$\Delta G^\circ_f (Ag_2O) = -11.21 \text{ kJ mol}^{-1}, \quad \Delta G^\circ_f (ZnO) = -318.3 \text{ kJ mol}^{-1}$$

Hence, E°_{cell} of the button cell is

a. 3.182 V b. 1.71 V c. −1.591V d. 1.591 V

16 For the reaction, $4Al + 3O_2 + 6H_2O + 4OH^- \longrightarrow 4Al(OH)_4^-, E^\circ_{cell} = 2.73 V$

$$\Delta G^\circ_f (OH^-) = -57 \text{ kJmol}^{-1}, \quad \Delta G^\circ_f (H_2O) = -237.2 \text{ kJmol}^{-1}$$

Thus, $\Delta G^\circ_f [Al(OH)_4^-]$ is (in $kJmol^{-1}$)

a. -5.21×10^3 b. -1.303×10^3 c. -4.81×10^3 d. -1.438×10^3

Matching List Type

Direction *(Q. Nos. 17 and 18) Choices for the correct combination of elements from Column I and Column II are given as options (a), (b), (c) and (d) out of which one is correct.*

17 The standard reduction potential data at 298 K is given below :

$$E^\circ_{(Fe^{3+}/Fe^{2+})} = +0.77 V; \quad E^\circ_{(Fe^{2+}/Fe)} = -0.44 V;$$

$$E^\circ_{(Cu^{2+}/Cu)} = +0.34 V; \quad E^\circ_{(Cu^+/Cu)} = +0.52 V;$$

$$E^\circ(O_2(g) + 4H^+ + 4e^- \longrightarrow 2H_2O) = +1.23 V$$

$$E^\circ(O_2(g) + 2H_2O(+ 4e^-) \longrightarrow 4OH^-) = +0.40 V$$

$$E^\circ_{(Cr^{3+}/Cr)} = -0.74 V, \quad E^\circ_{(Cr^{2+}/Cr)} = +0.91 V$$

Match E° of a redox pair in Column I with the values given in Column II and select the correct answer using the code given below.

	Column I		Column II
i.	$E^\circ_{(Fe^{3+}/Fe)}$	p.	− 0.18 V
ii.	$E^\circ(4H_2O \rightleftharpoons 4H^+ + 4OH^-)$	q.	− 0.40 V
iii.	$E^\circ(Cu^{2+} + Cu \longrightarrow 2Cu^+)$	r.	− 0.04 V
iv.	$E^\circ_{(Cr^{3+}/Cr^{2+})}$	s.	− 0.83 V

[JEE Advanced 2013]

Codes

	i	ii	iii	iv		i	ii	iii	iv
a.	s	p	q	r	b.	q	r	s	p
c.	p	q	r	s	d.	r	s	p	q

18 The standard potential of the following cell is 0.23 V at 288 K and 0.21V at 308 K.

$$E^\circ_{Ag^+/Ag} = 0.80 \text{ at } 298 \text{ K}$$

$$Pt(H_2) | HCl(aq) || AgCl(s) | Ag(s)$$

Match the parameters in Column I with their values in Column II and select the answer from the codes given below the list.

	Column I		Column II
i.	Temperature coefficient of EMF of the cell (in mV) K^{-1}	p.	− 9.81
ii.	Entropy change (in JK^{-1}) in the temperature range 288 K and 308 K.	q.	− 96.5
iii.	Enthalpy change (in $kJ\ mol^{-1}$) in the temperature range 288 K and 308 K.	r.	− 50.0
iv.	Solubility product of AgCl in water and 298 K in terms of (log)	s.	− 1.0

Codes

	i	ii	iii	iv		i	ii	iii	iv
a.	p	q	s	r	b.	q	s	p	r
c.	s	q	r	p	d.	s	p	r	q

Working Space

Comprehension Type

Direction (Q. Nos. 19-22) *This section contains 2 paragraphs, each describing theory, experiments, data, etc. Four questions related to the paragraphs have been given. Each question has only one correct answer among the four given options (a), (b), (c) and (d).*

Passage I

The concentration of potassium ions inside a biological cell is at least twenty times higher than the outside. The resulting potential difference across the cell is important in several process such as transmission of nerve impulses and maintaining the ion balance. A simple model for such a concentration cell involving a metal M is

$$M(s) \mid M^+ (aq, 0.05\text{M}), \mid\mid M^+(aq)\, 1\,\text{M}) \mid M(s) \mid E_{cell} \mid = 77\text{mV}]$$ **[IIT JEE 2010]**

19 For the above cell,

a. $E_{cell} < 0, \Delta G < 0$ b. $E_{cell} > 0, \Delta G < 0$
c. $E_{cell} < 0, \Delta G > 0$ d. $E_{cell} > 0, \Delta G < 0$

20 If 0.05 M solution of M^+ is replaced by 0.0025 M solution of M^+, then $|E_{cell}|$ would be

a. 35 mV b. 70 mV
c. 154 mV d. 700 mV

Passage II

Given, ΔG°_f (AgCl) = -109 kJmol^{-1}, ΔG°_f (Cl^-) = -129kJmol^{-1}, ΔG°_f (Ag^+) = 77kJmol^{-1},

21 Thus, E°_{cell} of the cell reaction is

$$Ag^+(aq) + Cl^-(aq) \longrightarrow AgCl(s) \text{ is}$$

a. $+0.59$ V b. -0.59 V
c. $+0.29$ V d. -0.29 V

22 K_{sp} of AgCl is thus,

a. 10^{10} b. 10^{-10}
c. -5 d. -10

One Integer Value Correct Type

Direction (Q. Nos. 23-25) *This section contains 3 questions. When worked out will result in an integer from 0 to 9 (both inclusive).*

23 A platinum electrode is immersed in a solution containing 0.1 M Fe^{2+} and 0.1 M Fe^{3+}. It is coupled with SHE. Concentration of Fe^{3+} is increased to 1.0 M without change in [Fe^{2+}], then change in EMF (in centivolt) is ......

24 Using $Cr_2O_7^{2-}$ aqueous, solution E°_{red} = 1.33 V and $\Delta G^\circ = -770.07$ kJmol^{-1}. What is the valency of the ion formed after reduction?

25 $E^\circ_{Cu^{2+}/Cu} = 0.3385$ V, 0.34 V, $E^\circ_{Fe^{3+}/Fe^{2+}} = 0.77$ V.

Equilibrium constant of the cell reaction,

$$Cu + 2Fe^{3+} \rightleftharpoons 2Fe^{2+} + Cu^{2+}$$ is $y \times 10^{14}$. What is the value of y?

DPP-5

Applications of EMF-Measurements

Subjective Questions

Working Space

Direction (Q. Nos. 1 and 2) *These questions are subjective in nature, need to be solved completely on notebook.*

1 The EMF of a cell corresponding to the reaction.

$Zn(s) + 2H^+(aq) \longrightarrow Zn^{2+}(0.1M) + H_2(g, 1\,atm)$ is 0.28 V at 298 K.

i. Write the half-cell reaction.
ii. Draw the electrochemical cell.
iii. Calculate pH of the solution at the hydrogen electrode.

$$E^\circ_{Zn^{2+}/Zn} = -0.76\,V$$
$$E^\circ_{H^+/H_2} = 0.00\,V$$

2 For the cell at 298 K, $\underset{\ominus}{Ag} \mid Ag^+ \parallel I^- \mid \underset{\oplus}{AgI, Ag}$

$$E^\circ_{Ag^+/Ag} = 0.80\,V$$

and $$E^\circ_{I^-|AgI, Ag} = -0.15\,V$$

Derive the value of ($\log K_{sp}$).

Only One Option Correct Type

Direction (Q. Nos. 3-13) *This section contains 11 multiple choice questions. Each question has four choices (a), (b), (c) and (d), out of which ONLY ONE is correct.*

3 For, $Pt(H_2)/H_2O$, electrode potential at 298 K and 1 bar is

a. $-0.2364\,V$ b. $0.4137\,V$
c. $-0.4137\,V$ d. $0.00\,V$

4 For the cell reaction at 298 K, $Cu^{2+}(aq) + 2e^- \longrightarrow Cu(s)$

Variation of $\log[Cu^{2+}]$ with $E_{Cu^{2+}/Cu}$ is a straight line of intercept 0.34 V on $E_{Cu^{2+}/Cu}$ axis. Then electrode potential of the half-cell. Cu/Cu^{2+} (0.1 M) will be

a. $\left(0.34 + \frac{0.0591}{2}\right)V$ b. $\left(-0.34 - \frac{0.0591}{2}\right)V$

c. 0.34 V d. $\left(-0.34 + \frac{0.0591}{2}\right)V$

5 Given, $Cu^{2+} + 2e^- \longrightarrow Cu$, $E^\circ = +0.34\,V$,

K_{sp} of $Cu(OH)_2 = 1.0 \times 10^{-19}$

What is E°_{red} of Cu^{2+}/Cu couple at pH = 12 ?

a. 0.78 V b. 0.44 V
c. $-0.10\,V$ d. $+0.10\,V$

Working Space

6 EMF of the following cell is 0.67 V at 298K.

$$Pt(\underset{1\,bar}{H_2}) \mid H^+(pH = X) \mid\mid 1MKCl \mid Hg_2Cl_2(s), Hg$$

Thus, pH of the solution $E^\circ_{Cl^-/Hg_2Cl_2,Hg} = 0.28$ V

a. 6.6
b. 4.2
c. 7.0
d. 1.0

7 Given, $A^{3+} + e^- \rightleftharpoons A^{2+}$, $E^\circ = 1.42$ V

$B^{4+} + 2e^- \rightleftharpoons B^{2+}$, $E^\circ = 0.40$ V

In the potentiometric titration of B^{2+} with A^{3+} the potential at the equivalence point is

a. 0.61 V
b. 0.74 V
c. 0.91 V
d. 1.08 V

8 For the cell of 298 K,

$$Zn(s) + Cu^{2+}(aq) \rightleftharpoons Cu(s) + Zn^{2+}(aq)$$

Variation of E_{cell} with $\log Q$ (where Q is reaction quotient) is of the type

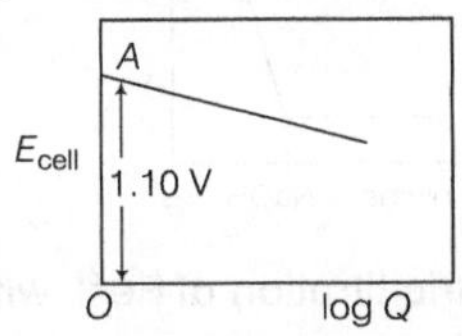

At what value of ratio of molar concentration of ions, E_{cell} would be 1.1591?

a. $\frac{[Cu^{2+}]}{[Zn^{2+}]} = 0.01$
b. $\frac{[Zn^{2+}]}{[Cu^{2+}]} = 0.01$
c. $\frac{[Zn^{2+}]}{[Cu^{2+}]} = 0.10$
d. $\frac{[Zn^{2+}]}{[Cu^{2+}]} = 1.00$

9 Given, $Ag^+ + e^- \longrightarrow Ag$, $E^\circ_{Ag^+/Ag} = 0.80$ V

$$[Ag(NH_3)_2^+] \rightleftharpoons Ag^+ + 2NH_3, K = 6 \times 10^{-8}$$

Thus, E° of the following half-cell reaction at 298 K.

$$Ag(NH_3)_2^+ + e^- \longrightarrow Ag + 2NH_3 \text{ is}$$

a. 0.80 V
b. −0.37 V
c. 0.30 V
d. 0.37 V

10 Ag^+/Ag electrode is immersed in 1.00 M KCl at 298 K. K_{sp} (AgCl) = 1.0×10^{-10}

$$E^\circ_{Ag^+/Ag} = 0.80 \text{ V}$$

Thus, EMF of the cell set up is

a. 0.209 V
b. −0.209 V
c. 0.505 V
d. −0.505 V

11 For the following cell, $Pt(H_2) \mid HCl(aq) \mid\mid AgCl \mid Ag$

$E_{cell} = 0.2650$ V at 298 K and $E_{cell} = 0.2595$ V at 308 K

Thus, heat of reaction at 298 K is

a. 0.00
b. −82.8 kJ
c. +82.8 kJ
d. −41.4 kJ

12 Quinhydrone electrode is an indicator electrode in contact with platinum metal and can be used in acid-base potentiometric titration.

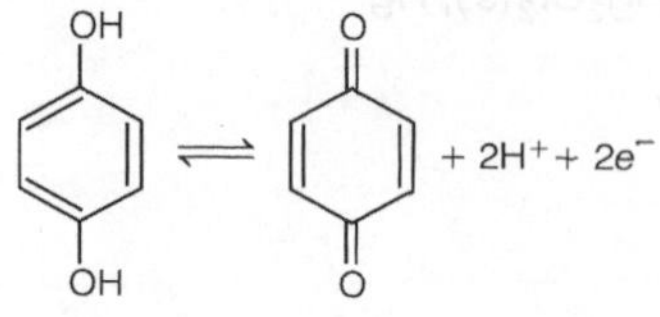

(A) (B)

At the equivalence point, $[A] = [B]$ and thus, EMF of the cell is dependent on pH. When NaOH is added to HCl in potentiometric titration, potentiogram is

a. E_{cell} 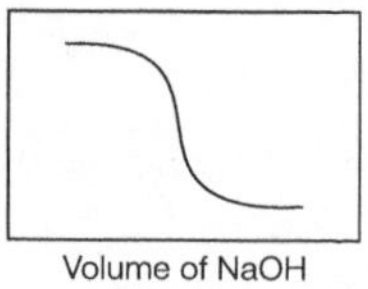

b. E_{cell} 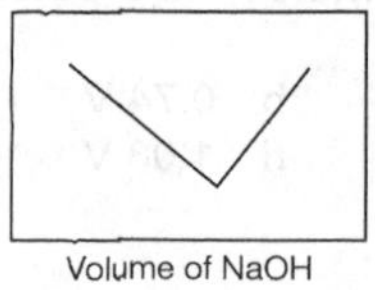

c. E_{cell} 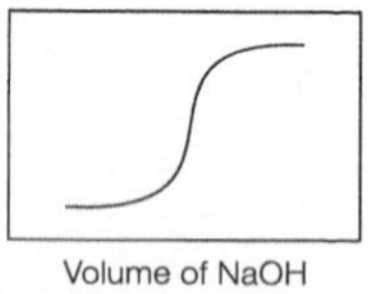

d. E_{cell} 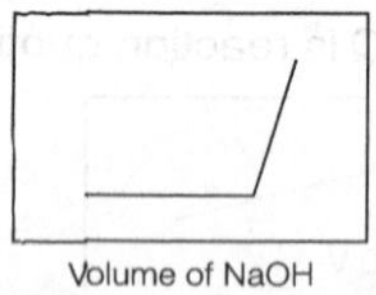

13 Potential at the equivalence point in the potentiometric titration of Fe^{2+} with MnO_4^- in acidic medium is

$$Fe^{3+} + e^- \longrightarrow Fe^{2+}, \quad E_1^\circ$$

$$MnO_4^- + 8H^+ + 5e^- \longrightarrow Mn^{2+} + 4H_2O, \quad E_2^\circ$$

a. $E_{eq} = \dfrac{E^\circ_1 + 5E_2^\circ}{6} - 0.08\,\text{pH}$ b. $E_{eq} = \dfrac{5E_1^\circ + E_2^\circ}{6} + 0.47\,\text{pH}$

c. $E_{eq} = \dfrac{E^\circ_1 + E_2^\circ}{6} - 0.08\,\text{pH}$ d. $E_{eq} = \dfrac{E_1^\circ + E_2^\circ}{6} - 0.47\,\text{pH}$

Comprehension Type

Direction (Q. Nos. 14-19) *This section contains 3 paragraphs, each describing theory, experiments, data, etc. Six questions related to the paragraph have been given. Each question has only one correct answer among the four given options (a), (b), (c) and (d).*

Passage I

The electrochemical cell shown below is a concentration cell :

M / M^{2+} (saturated solution of a sparingly soluble salt (MX_2) $\|$ $M^{2+}(0.001\text{ mol dm}^{-3})|\,M$.

The EMF of the cell depends on the difference in concentration of M^{2+} ions at the two electrodes. The EMF of the cell at 298 K is 0.059 V.

14 K_{sp} mol^3 dm^{-9} (the solubility product) of MX_2 at 298 K is ($2.303\,RT/F = 0.059\,V$ at 298 K)

a. 1×10^{-15} b. 4×10^{-15}

c. 1×10^{-12} d. 4×10^{-12}

15 The value of ΔG (kJ mol^{-1}) is ($1F = 96500$ C mol^{-1}) **[IIT JEE 2012]**

a. -5.7 b. 5.7

c. 11.4 d. -11.4

Working Space

Passage II

The standard half-cell reduction potential of $Fe^{3+}(aq)|Fe$ is -0.036 V and that of $OH^-|Fe(OH_3)(s)|Fe$ is -0.786 V.

16 For the determination of solubility product of $Fe(OH)_3$ (K_{sp}) the appropriate cell representation and its EMF respectively are

a. $Fe|Fe(OH)_3(s)||OH^-(aq)Fe^{3+}(aq)|Fe$ -0.750 V

b. $Fe|Fe^{3+}(aq), OH^-(aq)||Fe(OH)_3(s),|Fe$ -0.750 V

c. $Fe|Fe(OH_3)(s)||OH^-(aq), Fe^{3+}(aq)|Fe$ $+0.750$ V

d. $Fe|Fe^{3+}(aq), OH^-(aq)|Fe(OH_3)(s)|Fe$ -0.822 V

17 The value of $\log_e K_{sp}$ for $Fe(OH)_3$ at 298 K is

a. -38.1
b. $+87.7$
c. -96.0
d. -87.7

Passage III

Tollen's reagent is used for the detection of aldehydes. When a solution of $AgNO_3$ is added to glucose with NH_4OH, then gluconic acid is formed.

$$Ag^+ + e^- \longrightarrow Ag, \qquad E°_{red} = 0.80\text{ V}$$

$$\underset{\text{Glucose}}{C_6H_{12}O_6} + H_2O \longrightarrow \underset{\text{Gluconic acid}}{C_6H_{12}O_6} + 2H^+ + 2e^-, \qquad E°_{ox} = -0.05\text{ V}$$

$$Ag(NH_3)_2^+ + e^- \longrightarrow Ag(s) + 2NH_3, \qquad E°_{red} = 0.33\text{ V}$$

$$\left(\text{Use} = \frac{2.303}{F}RT = 0.0592\text{ V and } \frac{F}{RT} = 38.92 \text{ at } 298\text{ K}\right)$$

[IIT JEE 2006]

18 For the reaction,

$2Ag^+ + C_6H_{12}O_6 + H_2O \longrightarrow 2Ag(s) + C_6H_{12}O_7 + 2H^+$ $\log_e K$ is

a. 66.13
b. 58.45
c. 28.30
d. 46.29

19 In the presence of NH_3, pH is raised to 11.Thus,

a. E_{ox} will increase by a factor of 0.65 from $E°_{ox}$

b. E_{ox} will decrease by a factor of 0.65 from $E°_{ox}$

c. E_{red} will increase by a factor 0.65 from $E°_{red}$

d. E_{red} will decrease by a factor of 0.65 from $E°_{red}$

One Integer Value Correct Type

Direction (Q. Nos. 20 and 21) *This section contains 2 questions. When worked out will result in an integer from 0 to 9 (both inclusive).*

20 For the following half-cell, $\underset{1\text{bar}}{Pt(H_2)}|H^+$ (pH = x)

EMF at 298 K is 0.2364 V. What is pH?

21 For Cu $(OH)_2|Cu$ hall-cell at pH= 12, electrode potential is 0.0455 at 298 K.

For $Cu^{2+}(aq) + 2e^- \longrightarrow Cu$, $E° = 0.34$ V. K_{sp} value of $Cu(OH)_2$ is $x \times 10^{-4}$. What is the value of x?

DPP-6 Battery and Corrosion

Only One Option Correct Type

Direction (*Q. Nos.* 1-17) *This section contains 17 multiple choice questions. Each question has four choices (a), (b), (c) and (d), out of which ONLY ONE is correct.*

Working Space

1 In corrosion of iron

a. an electrochemical cell (galvanic cell) is formed in which Fe acts as anode and where O_2 is reduced, is the cathode
b. electrons flow from anode to cathode through the metal while ions flow through the water droplets
c. dissolved O_2 oxidises Fe^{2+} to Fe^{3+} before it is deposited as rust ($Fe_2O_3 \cdot H_2O$)
d. All of the above are observed

2 Select the correct statement(s).

a. In lead-storage battery, galvanic cells are linked in series
b. Cathode and anode compartments are not separated in a battery as oxidant and reductant both are solids
c. Recharging of a storage battery is a non-spontaneous ($\Delta G° > 0$)
d. All of the above

3 "Maintenance - free" batteries now in use in place of common batteries, have

a. electrodes made of lead-lead oxide
b. electrodes made of calcium-containing lead alloy
c. non-aqueous solvents as medium
d. platinum electrodes

4 In the following batteries, alkaline electrolytes are used

I. Mercury
II. Nickel-cadmium
III. Modified Leclanche cell

Cell potential is found to be independent of $[OH^-]$ in

a. I, II, III   b. I, II
c. II, III   d. Only III

5 Following are certain facts about batteries.

I. They are classified as primary, secondary and flow batteries.
II. Leclanche cell is a primary battery.
III. Lead-acid (storage) battery is a secondary cell.

Select the correct facts.

a. I, II   b. I, III
c. II, III   d. I, II, III

6 In an atmosphere with industrial smog, copper corrodes to a basic sulphate (and also to a basic carbonate). This basic sulphate is

a. $Cu_2(OH)_2SO_4$ b. $Cu(OH)_2(SO_4)_2$
c. $Cu(OH)SO_4$ d. $Cu(OH)(SO_4)_2$

7 In which of the following, reactants are not contained with the cell but are continuously supplied from the external source?

a. Lithium battery
b. Fuel cell
c. Dry cell
d. Lead-storage battery

8 When a buried pipeline is protected from corrosion by connecting to Mg block, it is known as

a. sacrified cathodic protection
b. sacrified anodic protection
c. impressed voltage protection
d. None of the above

9 The rate of corrosion is more in the environment which is

a. dry
b. humid
c. salty
d. normal

10 During galvanic corrosion, the more noble metal acts as

a. anode
b. cathode
c. anode and cathode
d. corroding metal

11 Some or more factors/species can affect corrosion.

I. Temperature
II. pH
III. Impurity of metal
IV. Conc. HNO_3

Select the correct factors/species.

a. All
b. All except I
c. All except II, IV
d. All except I, IV

12 When a lead storage battery is discharged,

a. SO_2 gas is evolved
b. lead is formed
c. lead sulphate is consumed
d. sulphuric acid is consumed

13 Select the correct statement(s) about corrosion.

a. It is the oxidative deterioration of a metal
b. It is an electrochemical phenomenon
c. Impurities present in iron enhance rusting by setting a number of miniature cells
d. All of the above

14 Rusting of iron or corrosion of active metal is initiated by

a. cathodic reaction when oxygen gas which is dissolved in H_2O, is reduced to OH^-
b. acid rain which oxidises metal to metal ion
c. Both (a) and (b)
d. None of the above

15 Rusting of iron takes place in

a. 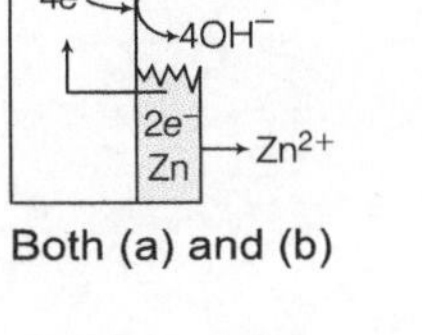

b. 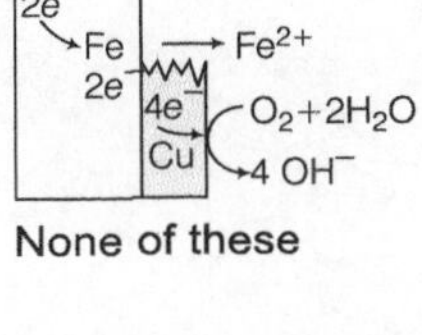

c. Both (a) and (b)
d. None of these

Working Space

16 The non-chromium redox battery makes use of the reaction

$$Cr^{2+}(aq) + Fe^{3+}(aq) \longrightarrow Cr^{3+}(aq) + Fe^{2+}(aq)$$

Thus, cell diagram is

a. $Cr^{2+}|Cr^{3+}(C_1)||Fe^{3+}(C_2)|Fe^{2+}$

b. $Cr^{3+}|Cr^{2+}(C_1)||Fe^{2+}(C_2)|Fe^{3+}$

c. $Fe|Fe^{2+}, Fe^{3+}||Cr^{3+}, Cr^{2+}|Cr$

d. $Cr|Cr^{2+}, Cr^{3+}||Fe^{3+}, Fe^{2+}|Fe$

17 Some statement(s) about dry-cell batteries are

I. Anode is a zinc metal and cathode is inert graphite in a paste of MnO_2 and C-black.

II. Electrolyte is a moist paste of NH_4Cl and $ZnCl_2$.

III. Oxidation product is Zn^{2+} and reduction product is Mn_2O_3.

Select the correct statement(s).

a. I, II and III
b. I and II
c. I and III
d. II and III

One or More than One Options Correct Type

***Direction** (Q. Nos. 18-20) This section contains 3 multiple choice questions. Each question has four choices (a), (b), (c) and (d), out of which ONE or MORE THAN ONE are correct.*

18 In the protection of iron from corrosion, galvanisation is one of the methods. In such case

a. as long as active metal remains, the iron is protected

b. it is called cathodic protection

c. active metal is called sacrified anode

d. active metal is called sacrified cathode

19 The rechargeable cell used in calculators and battery-operated tools are based on nickel and cadmium electrodes. It may involve following cell reactions to make cell reaction spontaneous

a. $Cd(s) + 2OH^-(aq) \longrightarrow Cd(OH)_2(s) + 2e^-$, $E° = 0.40\,V$

b. $Cd(OH)_2(s) + 2e^- \longrightarrow Cd(s) + 2OH^-(aq)$, $E° = -0.40\,V$

c. $Ni(OH)_2(s) + 2OH^-(aq) \longrightarrow NiO_2(s) + 2H_2O(l) + 2e^-$, $E° = -0.49\,V$

d. $NiO_2(s) + 2H_2O(l) + 2e^- \longrightarrow Ni(OH)_2(s) + 2OH^-(aq)$, $E° = +0.49\,V$

20 For Edison storage battery,

$$Fe(s)|FeO(s)|KOH(aq)|Ni_2O_3(s)|Ni(s)$$

The half-cell reactions are

$$Ni_2O_3(s) + H_2O(l) + 2e^- \rightleftharpoons 2NiO(s) + 2OH^-(aq); \quad E° = 0.40\,V$$

$$FeO(s) + H_2O(l) + 2e^- \rightleftharpoons Fe(s) + 2OH^-(aq); \quad E° = -0.87\,V$$

Select the correct statement(s) about the cell.

a. Battery emf is dependent on $[OH^-]$

b. Battery emf is independent of $[OH^-]$

c. About 245 kJ of energy is obtained per mole of Ni_2O_3

d. No work is obtained

Working Space

Comprehension Type

Direction (Q. Nos. 21-24) *This section contains 2 paragraphs, each describing theory, experiments, data, etc. Four questions related to the paragraphs have been given. Each question has only one correct answer among the four given options (a), (b), (c) and (d).*

Passage I

Silver tarnish is mainly Ag_2S, which has the reduction potential

$$Ag_2S(s) + 2e^- \longrightarrow 2Ag(s) + S^{2-}(aq);\ E^\circ = -0.691\text{V}$$

A tarnish silver spoon is placed in contact with a commercial available metallic product in a glass baking dish. Hot $NaHCO_3$ solution is added and tarnish part from silver spoon is removed.

21 What metal or metals do you think are in the product?

a. Cu
b. Cu and Zn
c. Au
d. Al

22 Metal used converts

a. Ag_2S into Ag_2SO_4
b. Ag_2S into Ag_2CO_3 and S
c. Ag_2S into Ag and S^{2-}
d. Ag_2S into $AgHCO_3$

Passage II

The electrodes in a lead-storage battery are made of Pb and PbO_2. During discharge following reaction occurs :

$$Pb + PbO_2 + 2H_2SO_4 \longrightarrow 2PbSO_4 + 2H_2O\ (Pb = 207);\ E^\circ = 2.00\text{V}$$

23 $[SO_4^{2-}]$ due to 4.8 M H_2SO_4 in lead-storage battery is

a. 1.0×10^{-12}M
b. 4.8 M
c. 3.3×10^{-12}M
d. 9.6 M

24 Under zero load, average voltage is 2.00 V, then ΔG° is

a. 386 kJ
b. −386kJ
c. −193kJ
d. − 572kJ

One Integer Value Correct Type

Direction (Q. No. 25) *This section contains 1 question. When worked out will result in an integer from 0 to 9 (both inclusive).*

25 An expensive but lighter alternate to the lead storage battery is the silver-zinc battery.

$$Ag_2O(s) + Zn(s) + H_2O(l) \longrightarrow Zn(OH)_2(s) + 2Ag(s); \qquad E^\circ = 1.5544\text{V}$$

Numerical value of ΔG° in scientific notation is $x \times 10^y$ J. What is the value of $x + y$?

DPP-7 Electrolysis and Faraday's Laws

Subjective Questions

Working Space

Direction *(Q. Nos. 1 and 2) These questions are subjective in nature, need to be solved completely on notebook.*

1 A solution $I^-(aq)$ was electrolysed for a period of 6.244 h using a current of 0.043 A to produce 1.27 g I_2.

$$2I^- \longrightarrow I_2 + 2e^-$$

What is the experimental value of electronic charge e?

2 i. What is the charge on one electron?

ii. A lamp draws a current of 1.0 A. How much charge does a lamp use in 10 minutes?

iii. How many kJ of energy is expended when a current of 2.0 A passes through 200 s under a potential difference of 230 V?

Only One Option Correct Type

Direction *(Q. Nos. 3-18) This section contains 16 multiple choice questions. Each question has four choices (a), (b), (c) and (d), out of which ONE or MORE THAN ONE is correct.*

3 During the electrolysis of aqueous $Zn(NO_3)_2$ solution

a. Zn plates out at the cathode
b. Zn plates out at the anode
c. H_2 gas is evolved at the anode
d. O_2 gas is evolved at the anode

4 Schematic diagram of an electrolytic cell is

a. 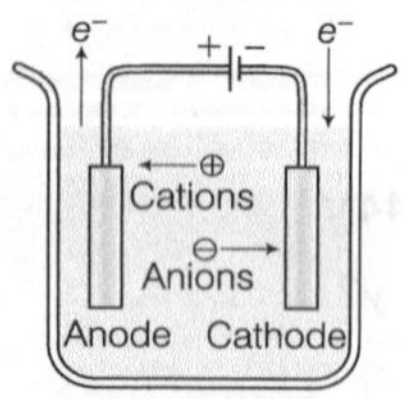

b. 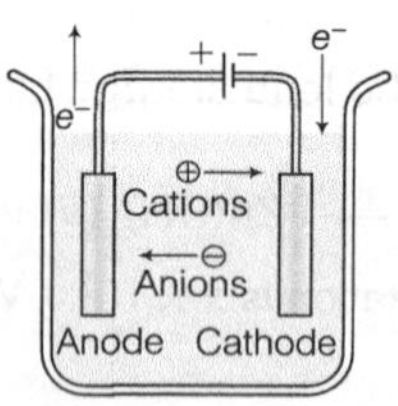

c. 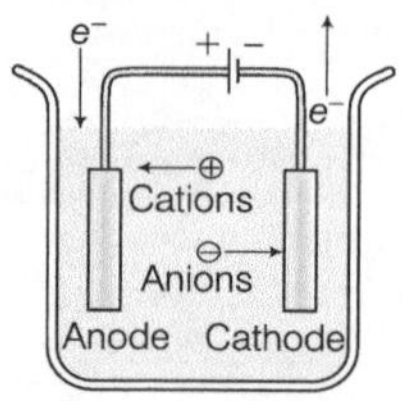

d. None of these is the correct presentation

Working Space

5 The metal that cannot be obtained by electrolysis of an aqueous solution of its salt is **[JEE Main 2014]**

a. Ag
b. Ca
c. Cu
d. Cr

6 When during electrolysis of a solution of $AgNO_3$, 9650 C of charge pass through the electroplating bath, the mass of silver deposited on the cathode will be

a. 1.08 g
b. 10.8 g
c. 21.6 g
d. 1.8 g

7 Refining of impure metal is done by electrolysis using impure metal as anode. Select the correct statement about this refining.

a. E°_{cell} of the net reaction is zero
b. metal to be refined is dissolved in the solution and is deposited at the cathode
c. Both (a) and (b)
d. None of the above

8 W g of Ag is deposited at the cathode of one electrolytic cell due to passage of 1A of current for 1h. Time required for passage of current to deposit W g of Mg by the same value of current is

a. 9 h
b. 4.5 h
c. 1 h
d. 3.0 h

9 In the refining of silver by electrolytic method, current of 5A is passed for 2 h using 100 g of impure anode of silver (of 95% purity). Weight of silver anode after electrolysis is

a. 40.29 g
b. 59.71 g
c. 57.59 g
d. 42.41 g

10 In the electrorefining of metals, impure metal is

a. cathode and oxidation taken place
b. anode and oxidation takes place
c. anode and reduction takes place
d. cathode and reduction takes place

11 In the electrosynthesis, potassium manganate (VII) is converted to manganese (IV) dioxide. By passage of 1F of electrolysis, one mole of potassium manganate (VII) will form manganese dioxide.

a. 1.00 mol
b. 0.50 mol
c. 0.66 mol
d. 0.33 mol

12 When Al_2O_3 is electrolysed, cation and anions are discharged. For a given quantity of electricity, ratio of number of moles of Al and O_2 gas is

a. 1 : 1
b. 2 : 1
c. 2 : 3
d. 4 : 3

13 1 Faraday of electricity is passed through the solution containing 1 mole each of $AgNO_3$, $CuSO_4$, $AlCl_3$ and $SiCl_4$. Elements are discharged at the cathode. Number of moles of Ag, Cu, Al and Si formed will be in the ratio of

a. 1 : 1 : 1 : 1
b. 1 : 2 : 3 : 4
c. 6 : 4 : 3 : 1
d. 12 : 6 : 4 : 3

14 During electrolysis of acidified water, O_2 gas is formed at the anode. To produce O_2 gas at the anode at the rate of 0.224 cc per second at STP, current passed is

a. 0.224 A
b. 2.24 A
c. 9.64 A
d. 3.86 A

Working Space

15 A solution of copper (II) suphate (VI) is electrolysed between copper electrodes by a current of 10.0 A for exactly 9650 s. Which remains unchanged?

a. Molar concentration of solution
b. Cathodic plate
c. Anodic plate
d. All of these

16 A 300 mL solution of NaCl was electrolysed for 60.0 min. If the pH of the final solution was 12.24, average current used is

a. 1.40 A
b. 1.27 A
c. 0.70 A
d. 0.14 A

17 100 mL of a buffer of 1 M $NH_3(aq)$ and 1M NH_4^+ (aq) are placed in two volatic cells separately. A current of 1.5 A is passed through both cells for 20 min. If electrolysis of water only takes place

$$2H_2O + O_2 + 4e^- \longrightarrow 4OH^- \text{ (RHS)}$$

$$2H_2O \longrightarrow 4H^+ + O_2 + 4e^- \text{ (LHS)}$$

then pH of the

a. LHS half-cell will increase
b. RHS half-cell will increase
c. both half-cells will increase
d. both half-cells will decrease

18 What product are formed during the electrolysis of a concentrated aqueous solution of sodium chloride using an electrolytic cell in which electrodes are separated by a porous pot?

I. $Cl_2(g)$
II. $NaOH(aq)$
III. $H_2(g)$
IV. $NaClO(aq)$
V. $NaClO_3(aq)$

Select the correct choice.

a. I and II
b. I, II and V
c. I, III and V
d. I, II and III

One or More than One Options Correct Type

Direction (Q. Nos. 19-23) *This section contains 5 multiple choice questions. Each question has four choices (a), (b), (c) and (d), out of which ONE or MORE THAN ONE are correct.*

19 Select the correct point(s) of distinction between a volatic cell and electrolysis cell.

	Volatic cell	Electrolysis cell
a.	Oxidation takes place at anode.	Oxidation takes place at anode.
b.	Oxidation takes place at cathode.	Oxidation takes place at anode.
c.	Reduction takes place at cathode.	Reduction takes place at cathode.
d.	Anode is (– ve) terminal and cathode is (+ ve) terminal.	Anode is (+ ve) terminal and cathode is (– ve) terminal.

20 In the electrolysis of aqueous sodium chloride solution, two types of reactions can take place at anode :

I. $2Cl^-(aq) \longrightarrow Cl_2(g) + 2e^-$

II. $2H_2O(l)g \longrightarrow O_2(g) + 4H^+(aq) + 4e^-$

Select the correct statement(s) about these.

a. Electrode potentials of I and II are quite similar
b. I is favoured in higher (NaCl) and II is favoured in higher pH
c. Because of overvoltage, Cl_2 and not O_2 is formed
d. Electrode potentials of I and II are different

Working Space

21 Select the correct statement(s) about electrolysis of aqueous $CuSO_4$ solution.

a. Anodic reaction is $2H_2O(l) \longrightarrow 4H^+(aq) + O_2(g) + 4e^-$

b. Cathodic reactions $2Cu^{2+}(aq) + 4e^- \longrightarrow 2Cu(s)$

c. Due to the deposition of Cu at the cathode, this is the basis of electroplating of copper

d. Blue colour of the solution fades

22 Select the correct observation about electrolysis.

a. Electric current is used to drive a non-spontaneous reaction

b. E_{cell} is positive and ΔG is negative

c. Cations and anions are discharged at the cathode and anode respectively

d. Overvoltage is responsible that a particular reaction takes place

23 Which pair of electrolysis could be distinguished by the products of electrolysis using inert electrodes?

a. 1 M $CuSO_4$, 1 M $CuCl_2$ b. 1M KCl, 1M KI

c. 1M $AgNO_3$, 1M $CuSO_4$ d. 1M $CuBr_2$, 1M $NiBr_2$

Matching List Type

Direction *(Q. No. 24) Choices for the correct combination of elements from Column I and Column II are given as options (a), (b), (c) and (d), out of which one is correct.*

24 Four electrolytic cells containing solution of $ZnSO_4$, $AgNO_3$, $CuSO_4$ and dil. H_2SO_4 are connected in series. A steady current of 1.5 A was passed through them until 77 mL of O_2 gas at STP is formed in anode compartment of H_2SO_4 solution. Match the different elements/gas formed in Column I with their values in Column II and select the answer from the codes given below.

(Cu = 63.5, Zn = 65, Ag = 108, S = 32, H = 1, O = 16)

	Column I		Column II
i.	Zn	p.	0.45 g
ii.	Ag	q.	1.485 g
iii.	Cu	r.	0.014 g
iv.	H_2	s.	0.44 g

Codes

	i	ii	iii	iv		i	ii	iii	iv
a.	p	q	s	r	b.	q	r	p	s
c.	s	q	p	r	d.	p	r	q	s

Comprehension Type

Direction *(Q. Nos. 25-28) This section contains 2 paragraphs, each describes theory, experiments, data, etc. Four questions related to the paragraphs have been given. Each question has only one correct answer among the four given options (a), (b), (c) and (d).*

Passage I

A constant current of 30 A is passed through an aqueous solution of NaCl for a time of 1.00 h.

25 Thus, NaOH formed due to electrolysis is

a. 40.2 g b. 44.8 g

c. 31.1 g d. 22.4 g

26 Cl_2 formed under STP condition is

a. 12.54 L b. 11.22 L c. 22.40 L d. 5.60 L

Working Space

Passage II

In hydrogen economy fuel-cell, anodic and cathodic reactions are

Anodic $H_2 + 2OH^- \longrightarrow 2H_2O + 2e^-$

Cathodic $O_2 + 2H_2O + 2e^- \longrightarrow 4OH^-$

67.2 L H_2 at STP react in 15 min. Entire current is used for electro deposition of copper from copper (II) sulphate solution.

27 Average current produced in fuel cell is

a. 610.00 A
b. 620.00 A
c. 617.8 A
d. 643.3 A

28 Copper deposited would be

a. 1.0 mol
b. 2.0 mol
c. 3.0 mol
d. 4.0 mol

One Integer Value Correct Type

Direction *(Q. Nos. 29-34) This section contains 6 questions. When worked out will result in an integer from 0 to 9 (both inclusive).*

29 M^{n+} solution was electrolysed for 75 min by a current of 0.30 A. Mass of the metal M deposited at the cathode was 9.958 g. M^{n+} contains 68 protons and 69 neutrons. What is the value of n?

30 An electrochemical cell was based on the following reaction:

$$Mn(OH)_2(s) + H_2O_2(aq) \longrightarrow MnO_2(s) + 2H_2O(e)$$

During the operation of this for 1 min, 0.135 g of MnO_2 was produced. What is the average electric current (in ampere) produced by the cell?

31 In a fuel cell, following reactions takes place and electricity is produced.

Anodic $H_2 + 2OH^- \longrightarrow 2H_2O + 2e^-$

Cathodic $O_2 + 2H_2O + 4e^- \longrightarrow 4OH^-$

If 100.8 L of H_2 at STP reacts in 96500 s, what is the average current produced (in ampere) ?

32 How long (in minutes) a current of 1.56 A has to be passed through a solution of $AgNO_3$ to coat a metal surface of 80 cm^2 with a 0.010 mm thick layer? Density of silver is 10.5 g cm^{-3}.

33 A fully charged battery contains 500 mL of 5.00 M H_2SO_4. What is the concentration of H_2SO_4 in the battery after 6.0 A of current is drawn from the battery for 13.40 h?

34 A hydrogen-oxygen fuel cell operates on the simple reaction:

$$H_2(g) + \frac{1}{2}O_2(g) \longrightarrow H_2O(l)$$

If the cell is designed to produce 1.46A of current and if the H_2 is contained in a 1.0L tank at 2.00 atm pressure at 298 K, how long can the fuel cell operate (in hour) before the H_2 runs out. (Assume unlimited supply of oxygen)

Revisal Problems for JEE Main

SET - 1

Only One Option Correct Type

Working Space

Direction *(Q. Nos. 1-35) This section contains 35 multiple choice questions. Each question has four choices (a), (b), (c) and (d), out of which ONLY ONE is correct.*

1 Some of the batteries are rechargeable :

I. Dry cell, II. Lead-storage,
III. Nickel-Cadmium, IV. Lithium,
V. Fuel cell

Select the batteries which can be recharged

a. I, II and III b. II, III and IV
c. III, IV and V d. II, III and V

2 Consider the following facts about the electrolysis of $CuSO_4$ solution:

I. Copper will deposit at cathode.
II. Copper will deposit at anode.
III. Oxygen will be released at anode.
IV. Copper will dissolve at anode.

Thus, with platinum electrodes observed facts are

a. I and III b. II and III
c. I, II and III d. II and IV

3 In a galvanic cell, the salt-bridge **[JEE Advanced 2014]**

a. does not participate chemically in the cell reaction
b. stops the diffusion of ions from one electrode to another
c. is necessary for the occurrence of the cell reaction
d. ensures mixing of the two electrolytic solutions

4 For aqueous $NH_3(NH_4OH)$, at 298 K

Molar conductance at infinite dilution $= 2.38 \times 10^{-2}$ S m^2 mol^{-1}.

K_b (ionisation constant) $= 1.6 \times 10^{-5}$

What is the molar conductance at 0.01 M NH_4OH?

a. 9.52×10^{-3} S m^2 mol^{-1} b. 3.81×10^{-5} S m^2 mol^{-1}
c. 9.52×10^{-4} S m^2 mol^{-1} d. 3.81×10^{-3} S m^2 mol^{-1}

5 Electrode potential for Mg electrode varies according to the equation at 298 K

$$E^\circ_{Mg^{2+}/Mg} = E^\circ_{Mg^{2+}/Mg} - \frac{0.0591}{2} \log \frac{1}{[Mg^{2+}]}$$

The graph of $E_{Mg^{2+}/Mg}$ *vs* log $[Mg^{2+}]$

a.

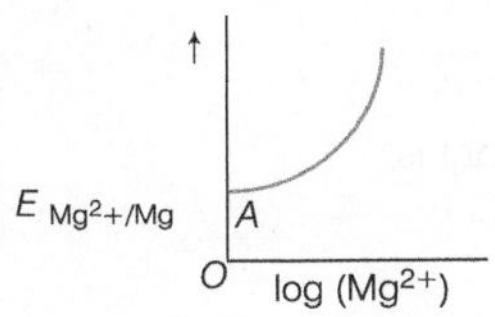

b.

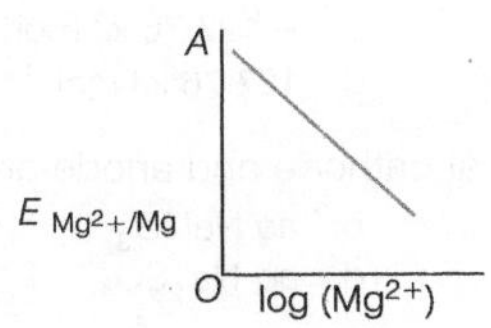

c.

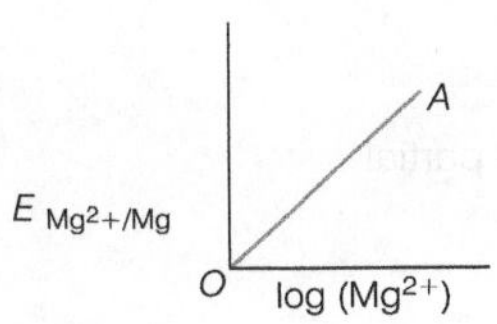

d.

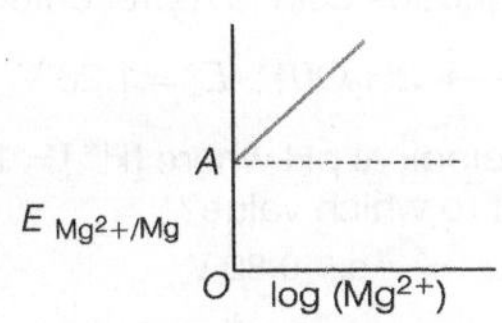

6 ($E_{cell} - E^\circ_{cell}$) has maximum value for spontaneous reaction in

a. $Pt(H_2)_{1\,bar} \mid pH = 1 \parallel pH = 2 \mid Pt(H_2)_{1\,bar}$

b. $Pt(Cl_2)_{1\,bar} \mid Cl^- = 0.1\ M \parallel Cl^- = 0.01\ M \mid Pt(Cl_2)_{2\,bar}$

c. $Pt \left| \begin{matrix} Fe^{2+}(0.1\ M) \\ Fe^{3+}(0.2\ M) \end{matrix} \right\| \begin{matrix} Sn^{2+}(0.1\ M) \\ Sn^{4+}(0.2\ M) \end{matrix} \right| Pt$

d. Ag, AgCl | $Cl^- = 0.01$ M $\parallel Cl^{-1} = 1$ M | Ag, AgCl

7 For strong electrolytes, the values of molar conductivity at infinite dilution are given below :

Electrolyte	Λ°_m (S m^2 mol^{-1})
$BaCl_2$	280.0×10^{-4}
NaCl	126.5×10^{-4}
NaOH	248.0×10^{-4}

Thus, Λ° of $Ba(OH)_2$ is (in S m^2 mol^{-1})

a. 401.5×10^{-4} b. 523.0×10^{-4}
c. 275.0×10^{-4} d. 654.5×10^{-4}

8 The reduction potential of hydrogen half-cell will be negative if **[AIEEE 2011]**

	pH/atm	$[H^+]$/M
a.	1.0	2.0
b.	1.0	1.0
c.	2.0	1.0
d.	2.0	2.0

9 At 298 K, given

$$Cu^{2+} + 2e^- \longrightarrow Cu; \qquad E^\circ = 0.34\ V$$
$$[Cu(NH_3)_4^{2+}] + 2e^- \longrightarrow Cu + 4NH_3; \qquad E^\circ = -0.12\ V$$

Thus, stability constant of the complex $[Cu(NH_3)_4]^{2+}$ is

a. 2.92×10^{-15} b. 2.92×10^{15}
c. 3.42×10^{-16} d. 3.71×10^{15}

10 For the half-cell

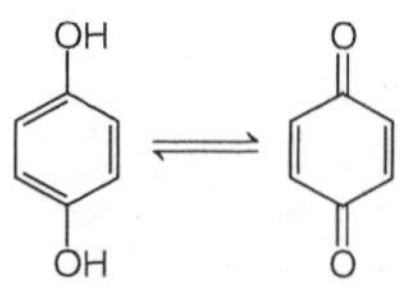

$+ 2H^+ + 2e^-$, $E^\circ = 1.30$ V

At pH = 2, electrode potential is

a. 1.36 V b. 1.30 V
c. 1.42 V d. 1.20 V

11 The hydrazine fuel cell is based on the reaction,

$$N_2H_4(aq) + O_2(g) \longrightarrow N_2(g) + 2H_2O(l), \quad E^\circ_{cell} = 1.56\ V$$

Given, ΔG°_f $(H_2O) = -237.2$ kJ mol^{-1}.

Thus, ΔG°_f (N_2H_4) is

a. -364.96 kJ mol^{-1} b. -127.76 kJ mol^{-1}
c. $+364.96$ kJ mol^{-1} d. 127.76 kJ mol^{-1}

12 Electrolysis products in gaseous state at cathode and anode are same when reactant is

a. *aq* NaCl b. *aq* $NaNO_3$
c. *aq* NaH d. *aq* Na_2SO_4

13 The reduction of oxygen gas in 1 M aqueous acid is represented by

$$O_2(g) + 4H^+(aq) + 4e^- \longrightarrow 2H_2O(l); \quad E^\circ = 1.23\ V$$

The potential for this reaction at physiological pH where $[H^+] = 1 \times 10^{-7}$ M, and the partial pressure of O_2 is 1 atm, will be closest to which value?

a. 0.18 V b. 0.41 V c. 0.82 V d. 1.64 V

Working Space

14 K_{sp} of $Al(OH)_3 = 1.0 \times 10^{-33}$ and $E^\circ_{Al^{3+}/Al} = -1.66$ V

Reduction potential of the above couple at pH = 14 is

a. − 2.31 V
b. + 2.31 V
c. + 1.01 V
d. − 1.01 V

15 Consider the half-cell reactions:

$$H_2O_2 \longrightarrow O_2^- + 2H^+ + 2e^-; \quad E^\circ = -0.69 \text{ V}$$
$$H_2O_2 + 2H^+ + 2e^- \longrightarrow 2H_2O; \quad E^\circ = 1.77 \text{ V}$$
$$Fe^{2+} \longrightarrow Fe^{3+} + e^-; \quad E^\circ = 0.77 \text{ V}$$
$$Fe \longrightarrow Fe^{2+} + 2e^-; \quad E^\circ = 0.44 \text{ V}$$

If Fe^{2+} and H_2O_2 are mixed, which reaction is more likely?

a. Oxidation of Fe^{2+} to Fe^{3+}
b. Reduction of Fe^{2+} to Fe
c. Both (a) and (b)
d. None of the above

16 If for the half-cell reactions, E° values are known

$$Cu^{2+} + e^- \longrightarrow Cu^+; \quad E^\circ = 0.15 \text{ V}$$
$$Cu^{2+} + 2e^- \longrightarrow Cu; \quad E^\circ = 0.34 \text{ V}$$

Predict which exists in aqueous solution?

a. Copper (I) sulphate
b. Copper (II) sulphate
c. Both (a) and (b)
d. None of these

17 For the half-cell $Cu^+/Cu, E^\circ = +0.52$ V; $Cu^{2+}/Cu^+, E^\circ = +0.16$ V

Hence, E° for the disproportionation reaction $2Cu^+ \longrightarrow Cu^{2+} + Cu$ would be

a. 0.36 V
b. 0.78 V
c. 0.84 V
d. − 0.36 V

18 Consider the following half-cell reactions :

I. $A + e^- \longrightarrow A^-; \quad E^\circ = 0.96$ V

II. $B^- + e^- \longrightarrow B^{2-}; \quad E^\circ = -0.12$ V

III. $C^+ + e^- \longrightarrow C; \quad E^\circ = +0.18$ V

IV. $D^{2+} + 2e^- \longrightarrow D; \quad E^\circ = -1.12$ V

What combination of two half-cells would result in a cell with the largest potential?

a. I and II
b. I and III
c. I and IV
d. II and IV

19 In a cell that utilises the reaction

$$Mg(s) + 2H^+(aq) \longrightarrow Mg^{2+}(aq) + H_2(g)$$

addition of dil. H_2SO_4 to cathode compartment will

a. lower the E and shift equilibrium to the left
b. lower the E and shift the equilibrium to the right
c. increase the E and shift the equilibrium to the right
d. increase the E and shift the equilibrium to the left

20 The standard reduction potentials for Zn^{2+}/Zn, Ni^{2+}/Ni and Fe^{2+}/Fe are − 0.76, − 0.23 V and − 0.44 V respectively. The reaction

$$X + Y^{2+} \longrightarrow X^{2+} + Y$$

will be spontaneous when **[AIEEE 2012]**

a. X = Ni, Y = Fe
b. X = Ni, Y = Zn
c. X = Fe, Y = Zn
d. X = Zn, Y = Ni

21 The cell $Zn | Zn^{2+}(1M) || Cu^{2+}(1M) | Cu$, $E^\circ_{cell} = 1.10$ V was allowed to be discharged at 298 K. Thus, at a time when, $E_{cell} = 0.8634$ V, $\log [Zn^{2+}] / [Cu^{2+}]$ is

a. 4.00
b. 10.00
c. 8.00
d. 7.00

Working Space

22 Given the data at 298 K,

$$Ag + I^- \longrightarrow AgI + e^-; \quad E^\circ = 0.150\ V$$
$$Ag \longrightarrow Ag^+ + e^-; \quad E^\circ = -0.800\ V$$

Thus, solubility product of AgI(s) is (given, $\frac{2.303RT}{F} = 0.0592$ at 298 K)

a. 1×10^4 b. 1×10^{-12}
c. 1×10^{-10} d. 1×10^{-16}

23 The conductivity of a solution of 0.001 mol dm^{-3} Na_2SO_4 is 2.6×10^{-2} S m^{-1}. Given that the molar conductivity of Na^+ is 5.0×10^{-3} S m^2 mol^{-1}, thus molar conductivity of sulphate ion is (in S m^2 mol^{-1})

a. 2.4×10^{-3} b. 7.6×10^{-3}
c. 21.0×10^{-3} d. 16×10^{-3}

24 At 25° C, limiting equivalent conductances (in Ω^{-1} cm^2 $equiv^{-1}$) of $BaCl_2$, H_2SO_4 and HCl respectively are 100, 250 and 150. Specific conductance (conductivity)of saturated aqueous $BaSO_4$ solution is 1×10^{-6} Ω^{-1} cm^{-1}. Thus, solubility product of $BaSO_4$ solution is

a. 1×10^{-10} b. 4×10^{-10}
c. 0.25×10^{-10} d. 2.5×10^{-10}

25 Cell constant (*C*) dependent of resistance (*R*) between two electrodes is shown by which of the following graph in terms of log *R* and log *C*?

a.
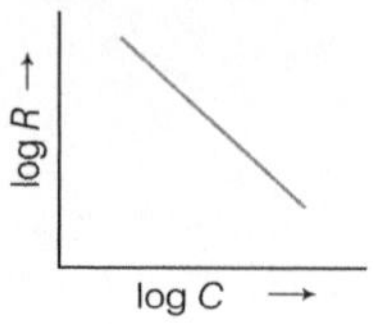

b.
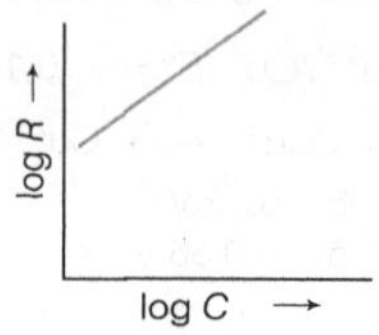

c.
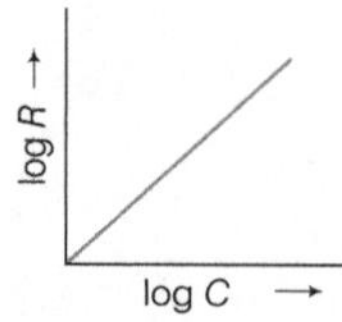

d.
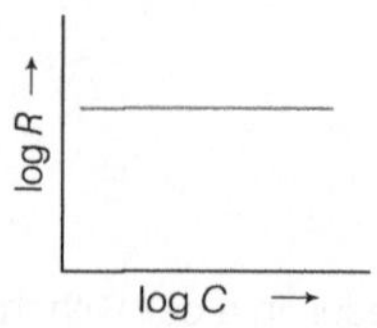

26 The limiting molar conductivities Λ° for NaCl, KBr and KCl are 126, 152 and 150 S cm^2 mol^{-1}, respectively. The Λ° for NaBr is (in S cm^2 mol^{-1})

a. 128 b. 176
c. 278 d. 302

27 Following solutions having equal ionisation have been provided

I. 0.01 M HCl,
II. 0.01 MH_2SO_4,
III. 0.01 M H_3PO_4

Maximum value of molar conductance (Λ_m) and equivalent conductance (Λ_{equiv}) are of

	Λ_m	Λ_{equiv}
a.	III,	III
b.	I,	II
c.	I,	III
d.	Equal of each	I

28 The resistivity of aluminium is 2.824×10^{-8} Ω m. Thus, conductance across a piece of aluminium wire, that is 2.0 mm in diameter and 1.00 m long is (assume current = 1.25 A)

a. 111.0 S b. 1.11 S
c. 0.111 S d. 1111 S

Working Space

29 Aluminium oxide may be electrolysed at 1270 K to form Al metal (atomic mass = 27 u, 1F = 96500 C)

$$Al^{3+} + 3e^- \longrightarrow Al$$

To prepare 5.12 kg of Al metal by this method, it would require

a. 5.49×10^2 C of electricity
b. 5.49×10^4 C of electricity
c. 1.83×10^7 C of electricity
d. 5.49×10^7 C of electricity

30 For an electrosynthesis of a substance, sign of ΔG and E_{cell} should be

	ΔG	E_{cell}
a.	+	−
b.	−	+
c.	−	−
d.	+	+

31 96500 coulombs of electricity deposits 108 g of silver. If electronic charge is 1.6×10^{-19} C, the Avogadro's number is

a. 6.030×10^{-23} mol^{-1}
b. 6.030×10^{23} mol^{-1}
c. 6.25×10^{18} mol^{-1}
d. 6.75×10^{20} mol^{-1}

32 9.65 C of electric current is passed through fused anhydrous $MgCl_2$. The magnesium metal thus obtained is completely converted into a Grignard's reagent (R—Mg—X). The number of moles of Grignard's reagent obtained is

a. 5×10^{-4}
b. 1×10^{-4}
c. 5×10^{-3}
d. 5×10^{-5}

33 Cost of electricity for the production of X L H_2 gas at STP at cathode is ₹ x, then cost of electricity for the production of X L O_2 gas at STP at the anode is (assume 1 mole of electrons as one unit electricity).

a. 2X
b. 4X
c. 16X
d. 32X

34 Given below are the half-cell reactions

$$Mn^{2+} + 2e^- \longrightarrow Mn;\quad E^\circ = -1.18\ V$$

$$Mn^{3+} + e^- \longrightarrow Mn^{2+};\quad E^\circ = 1.51\ V$$

[JEE Main 2014]

Thus, E° for the reaction

$$3Mn^{2+} \longrightarrow Mn + 2Mn^{3+}$$

will be

a. − 2.69 V, the reaction will not occur
b. − 2.69 V, the reaction will occur
c. − 3.33 V, the reaction will not occur
d. − 0.33 V, the reaction will occur

35 The formation of rust on the surface of iron occurs through the reaction(s)

a. $Fe(s) \longrightarrow Fe^{2+}(aq) + 2e^-$ (at anode)
b. $O_2(g) + 4H^+(aq) + 4e^- \longrightarrow 2H_2O\ (l)$ (at cathode)
c. $Fe^{2+}(aq) + O_2(g) + 4H_2O(l) \longrightarrow 2Fe_2O_3(s) + 8H^+$
d. All of the above

Statement Type

Working Space

Direction (Q. Nos. 36-40) *This section is based on Statement I and Statement II. Select the correct answer from the codes given below.*

a. Both Statement I and Statement II are correct and Statement II is the correct explanation of Statement I
b. Both Statement I and Statement II are correct but Statement II is not the correct explanation of Statement I
c. Statement I is correct but Statement II is incorrect
d. Statement II is correct but Statement I is incorrect

36 **Statement I** For the following cell

$$Zn\,|\,Zn^{2+}(0.01M)\,||\,Zn^{2+}(0.1M)\,|\,Zn$$

$$E_{cell} > 0 \text{ and spontaneous}$$

Statement II Spontaneous change in a concentration cell always occurs in the direction in which the concentration solution becomes more dilute and the dilute solution becomes more concentrated.

37 **Statement I** Iron can be protected from rusting by connecting it to the more active metals, as zinc.

Statement II It is a type of galvanisation.

38 **Statement I** In an electrochemical cell, two half-cells are connected with a salt-bridge that allows cations and anions to move between two half-cells.

Statement II Electrolyte used in salt-bridge can be NaCl or KCl with silver electrodes.

39 **Statement I** $CuSO_4$ solution can be stored in a vessel made of zinc.

Statement II $E^\circ_{Cu^{2+}/Cu} = 0.34\,V$, $E^\circ_{Zn^{2+}/Zn} = -0.76\,V$

40 **Statement I** If

$$Fe^{3+} + 3e^- \longrightarrow Fe^{2+}; \quad E^\circ = x_1\,V$$

$$Fe^{2+} + 2e^- \longrightarrow Fe; \quad E^\circ = x_2\,V$$

then,

$$Fe^{3+} + 3e^- \longrightarrow Fe^{2+}; \quad E^\circ = x_1 - x_2$$

Statement II $\Delta G_3 = |\Delta G_1| + |\Delta G_2|$

Revisal Problems
for JEE Advanced

Only One Option Correct Type

Working Space

Direction (Q. Nos. 1-12) *This section contains 12 multiple choice questions. Each question has four choices (a), (b), (c) and (d), out of which ONLY ONE is correct.*

1 For the following potentiometric titration,

$$2MnO_4^- + 6H^+ + 5H_2C_2O_4 \rightleftharpoons 2Mn^{2+} + 8H_2O + 10\,CO_2$$

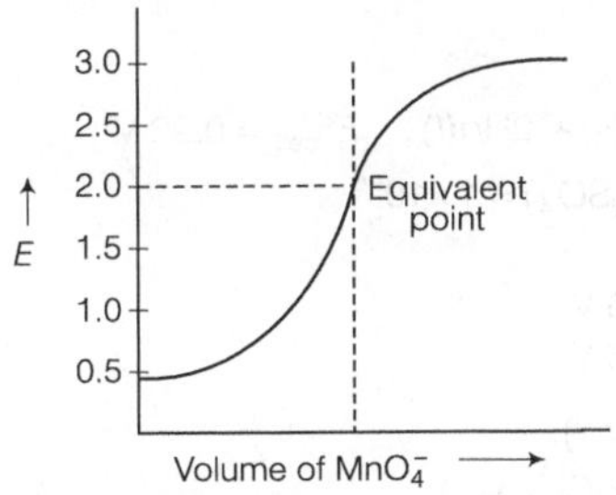

EMF at the equivalent point is given by the graph.
What is the value of ($\log K_{eq}$)?

a. 1/0.0591 b. 10/0.0591
c. 5/0.0591 d. 20/0.0591

2 Given at 298 K,

$$NiO_2(s) + 4H^+ + 2e^- \longrightarrow Ni^{2+} + 2H_2O; \quad E^\circ = 1.678\text{ V}$$
$$NiO_2(s) + 2H_2O + 2e^- \longrightarrow Ni(OH)_2 + 2OH^-; \quad E^\circ = -0.490\text{ V}$$

Thus, ΔG° (Gibbs free energy) for the reaction,

$$Ni^{2+} + 2\,H_2O \longrightarrow Ni(OH)_2 + 2H^+ \text{ is}$$

a. + 836.8 kJ b. − 418.4 kJ
c. + 418.4 kJ d. + 209.2 kJ

3 One molal solution of $AgNO_3$ containing 1 kg of water is electrolysed in two electrolytic cells containing Ag and Pt electrodes. 2 A current is passed for 2 h in each of them. After electrolysis $AgNO_3$ is

	Electrolytic cell with Ag electrodes	Electrolytic cell with Pt electrodes
a.	1 molal	0.85 molal
b.	0.85 molal	1.00 molal
c.	1.15 molal	0.85 molal
d.	0.85 molal	0.85 molal

4 Following two electrolytic cells are taken and current is passed as indicated at 298 K.

I. $Ag\,|\,\underset{1\text{ L}}{Ag^+(1\text{ M})}$, 0.5 F

II. $\underset{1\text{ L}}{Ag^+(2\text{ M})}\,|\,Ag$, 1.0 F

After current is passed, they were connected using a salt-bridge. Change in emf of the cell is

a. 0.0591 log 2
b. 0.0591 log 5
c. −0.0591 log 2
d. No change

Working Space

5 When temperature of the following cell Cu | Cu^{2+}(1 M) || Ag^+(1 M) | Ag is increased by 10°, E°_{cell} becomes 0.48 V.
Given, $E^\circ_{Cu^{2+}/Cu} = 0.34$ V, $E^\circ_{Ag^+/Ag} = 0.80$ V. Thus, entropy change is

a. 193 JK^{-1}
b. 386 JK^{-1}
c. 579 JK^{-1}
d. – 193 JK^{-1}

6 Select the correct statement based on following half-cell reactions.

$$Fe^{2+} \longrightarrow Fe^{3+} + e^-, \quad E^\circ = -0.77\text{ V}$$
$$O_2 + 4H^+ + 4e^- \longrightarrow 2H_2O, \quad E^\circ = 1.23\text{ V}$$
$$Fe \longrightarrow Fe^{2+} + 2e^-, \quad E^\circ = 0.44\text{ V}$$

a. Fe^{2+} solutions are not stable in air
b. Fe nails can prevent oxidation of Fe^{2+} to Fe^{3+} by air
c. Corrosion of iron takes place by air
d. All of the above

7 For the following cell reaction,

$$Pb(s) + Hg_2SO_4(s) \rightleftharpoons PbSO_4(s) + 2Hg(l); \quad E^\circ_{cell} = 0.92\text{ V}$$
$$K_{sp}(PbSO_4) = 2\times10^{-8}, \quad K_{sp}(Hg_2SO_4) = 1\times10^{-6}$$

Thus, E_{cell} is

a. 0.92 V
b. 0.89 V
c. 1.04 V
d. 0.95 V

8 Given, limiting equivalent conductance (S cm^2 $equiv^{-1}$).

$$BaCl_2 = x_1, \quad H_2SO_4 = x_2, \quad HCl = x_3$$

Specific conductance of saturated $BaSO_4 = y$ S cm^{-1}
Thus, K_{sp} (solubility product) of $BaSO_4$ is

a. $10^3 y/2(x_1 + x_2 - 2x_3)$
b. $\dfrac{10^6 y^2}{(x + x_2 - 2x_3)^2}$
c. $\dfrac{10^6 y^2}{4(x_1 + x_2 - 2x_3)^2}$
d. $\dfrac{x_1 + x_2 - 2x_3}{10^6 y^2}$

9 Given, $\lambda^\circ_{Ag^+} = 6\times10^{-3}$ S m^2 mol^{-1}
$\lambda^\circ_{Br^-} = 8\times10^{-3}$ S m^2 mol^{-1}
$\lambda^\circ_{NO_3^-} = 7\times10^{-3}$ S m^2 mol^{-1}
K_{sp} of AgBr $= 12\times10^{-14}$.

1×10^{-7} M $AgNO_3$ solution is added to 1 L saturated solution of AgBr.
Then specific conductance of AgBr solution is

a. 5.5×10^{-6} S m^{-1}
b. 5.5×10^{-8} S m^{-1}
c. 5.5×10^{-7} S m^{-1}
d. 5.5×10^{-9} S m^{-1}

10 In the electrolysis of aqueous sodium chloride solution, following reactions may take place at cathode (reduction):

I. $2Na^+ + 2e^- \longrightarrow 2Na(s)$; $\quad E^\circ = -2.71$ V
II. $2H_2O(l) + 2e^- \longrightarrow H_2(g) + 2OH^-(aq)$, $\quad E^\circ = -0.83$ V

Select the correct option(s).

a. Reduction of $H_2O(l)$ takes place more rapidly than that of water
b. If liquid mercury is used as cathode, Na^+ ion is reduced in preference to H_2O, Na being soluble in Hg by forming amalgam
c. Both (a) and (b)
d. None of the above

Working Space

11 For the following half-cell reaction,

$$OX_1 + pe^- \rightleftharpoons RED_1, E°_1$$
$$OX_2 + qe^- \rightleftharpoons RED_2, E°_2$$

Potential at the equivalence point is

a. $\dfrac{E°_1 + E°_2}{p+q}$

b. $\dfrac{qE°_1 + pE°_2}{p+q}$

c. $\dfrac{pE°_1 + qE°_2}{p+q}$

d. $pE°_1 + qE°_2$

12 For a thermocell, $\underset{T_1}{Cl_2(g)} \mid AgCl(s) \mid \underset{T_2}{Cl_2(g)}$

operating between 800 K and 1200 K,

$$\frac{dE}{dT} = -7.0 \times 10^{-4}\ VK^{-1}$$

Thus, emf in the operating temperature 850 K and 1050 K is

a. 0.14 V b. − 0.14 V c. 0.29 V d. 1.4 V

Statement Type

Direction (Q. Nos. 13 and 14) ***This section is based on Statement I and Statement II. Select the correct answer from the codes given below.***

a. Both Statement I and Statement II are correct and Statement II is the correct explanation of Statement I
b. Both Statement I and Statement II are correct but Statement II is not the correct explanation of Statement I
c. Statement I is correct but Statement II is incorrect
d. Statement II is correct but Statement I is incorrect

13 **Statement I** In the conductometric titration of CH_3COOH with NaOH, conductance first increases slowly and after complete reaction of CH_3COOH, it increases rapidly as shown

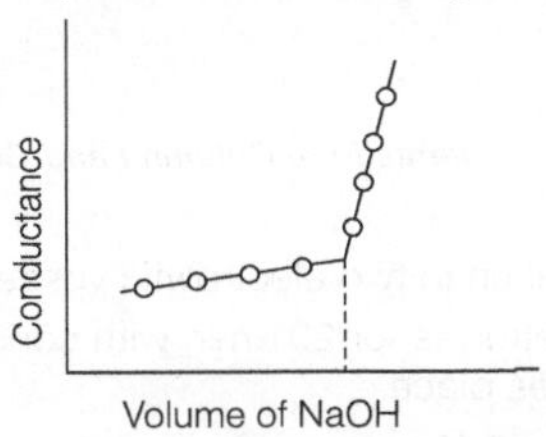

Statement II Initial increase in conductance is due to hydrolysis of salt formed.

14 **Statement I** Electrolysis of NaCl solution gives chlorine at anode instead of O_2.

Statement II Formation of oxygen at anode requires over voltage.

One or More than One Options Correct Type

Direction (Q. Nos. 15-19) ***This section contains 5 multiple choice questions. Each question has four choices (a), (b), (c) and (d), out of which ONE or MORE THAN ONE are correct.***

15 Standard electrode potential data are useful for understanding the suitability of an oxidant in a redox titration. Some half-cell reactions and their standard potentials are given below :

$$MnO_4^-(aq) + 8H^+(aq) + 5e^- \longrightarrow Mn^{2+}(aq) + 4H_2O(l); \quad E° = 1.51\ V \quad ...(i)$$
$$Cr_2O_7^{2-}(aq) + 14H^+(aq) + 6e^- \longrightarrow 2Cr^{3+}(aq) + 7H_2O(l); \quad E° = 1.38\ V \quad ...(ii)$$
$$Fe^{3+}(aq) + e^- \longrightarrow Fe^{2+}(aq); \quad E° = 0.77\ V \quad ...(iii)$$
$$Cl_2(g) + 2e^- \longrightarrow 2Cl^-(aq); \quad E° = 1.40\ V \quad ...(iv)$$

Identify correct statement regarding the quantitative estimation of aqueous $Fe(NO_3)_2$.

a. MnO_4^- can be used in aqueous HCl
b. $Cr_2O_7^{2-}$ can be used in aqueous HCl
c. MnO_4^- can be used in aqueous H_2SO_4
d. $Cr_2O_7^{2-}$ can be used in aqueous H_2SO_4

Working Space

16 For a given KCl solution, molar conductance $\Lambda_m = 0.0147 \text{ S m}^2 \text{ mol}^{-1}$.

L (Conductance) $= 10 L$ (H_2O), (L in S) of solution.

For pure of solution water κ (Specific conductance $= 5.7 \times 10^{-6} \text{ S m}^{-1}$. Thus, molar conductance of KCl solution is

a. 3.5×10^{-3} mol m^{-3}
b. 3.5×10^{-6} mol L^{-1}
c. 3.5×10^{-3} mol dm^{-3}
d. 3.5×10^{-6} mol cm^{-3}

17 Electrolytic conductance of an electrolyte is decreased due to

a. dilution
b. increase in temperature
c. interionic attraction in higher concentration range
d. interionic attraction in lower concentration range

18 For the reduction of NO_3^- in an aqueous solution $E°$ is 0.96 V. Values of $E°$ for some metals are given below :

$$V^{2+}(aq) + 2e^- \longrightarrow V, \quad E° = -1.19 \text{ V}$$
$$Fe^{3+}(aq) + 3e^- \longrightarrow Fe, \quad E° = -0.04 \text{ V}$$
$$Au^{3+}(aq) + 3e^- \longrightarrow Au, \quad E° = +1.40 \text{ V}$$
$$Hg^{2+}(aq) + 2e^- \longrightarrow Hg, \quad E° = +0.80 \text{ V}$$

The pair(s) of metals that is (are) oxidised by NO_3^- ion in aqueous solution is (are)

a. V and Hg
b. Hg and Fe
c. Fe and Au
d. Fe and V

19 For the reaction, $2F^{3+} + Fe \longrightarrow 3Fe^{2+}, E° = 1.21$ V. Thus,

a. the reaction occurs in the absence of a second half-cell or current
b. in a complete cell, Fe is anode but Pt is the cathode
c. Fe cannot be used in cathode, because it would directly react with Fe^{3+}, thus short-circuiting the cell
d. cell set up is $Fe \mid Fe^{2+} \parallel Fe^{3+} \mid Fe$

Matching List Type

Direction *(Q. Nos. 20-22) Choices for the correct combination of elements from Column I and Column II are given as options (a), (b), (c) and (d) out of which one is correct.*

20 1 L buffer of 1.0 M NaH_2PO_4 and 1.0 M Na_2HPO_4 are taken in two electrolytic vessel separately. Electrolysis was carried using platinum electrodes for 201 min with constant current of 1.25 A. Assume that electrolysis of water takes place.

$$H_3PO_4 \rightleftharpoons H_2PO_4^- + H^+, \ pK_a = 2.15$$

Match the parameters given in Column I with their values in Column II and select the answer from the codes given below the list.

	Column I		Column II
i.	pH of the electrolytic cell before current is passed	p.	2.29
ii.	pH in the anodic side after current is passed	q.	0.156
iii.	pH in the cathodic side after current is passed	r.	2.01
iv.	H^+ ion formed (in terms of molarity)	s.	2.15

Codes

	i	ii	iii	iv
a.	p	q	r	s
b.	s	r	p	q
c.	q	p	r	s
d.	r	p	q	s

Working Space

21 For H_2O at 298 K, $\lambda^\circ(H^+) = 3.5 \times 10^{-2}$ S m² mol⁻¹, $\lambda^\circ(OH^-) = 2.0 \times 10^{-2}$ S m² mol⁻¹

Specific conductance $(\kappa) = 5.7 \times 10^{-6}$ S m⁻¹,

Density of water = 1.0 g cm⁻³

Match the parameters in Column I (in terms of −log (value) i.e. p_{value}) with their values in Column II.

	Column I		Column II
i.	$p_x = -\log x$ (x = degree of dissociation)	p.	10.0
ii.	p_{Λ_c} (Λ_c = molar conductance)	q.	8.73
iii.	pH	r.	14.0
iv.	pK_w (K_w ionic product)	s.	7.00

Codes

	i	ii	iii	iv
a.	p	q	r	s
b.	s	q	r	p
c.	q	p	s	r
d.	q	s	p	r

22 $\Lambda^\circ_m(CH_3COOH) = 390.5$ S cm² mol⁻¹, $K_a(CH_3COOH) = 1.8 \times 10^{-5}$, $[CH_3COOH] = 0.040$ M

Cell constant of the cell used = 0.206 cm⁻¹

Match the parameter given in Column I with the values in Column II and select answer from the codes given at the end.

	Column I		Column II
i.	Λ_m (molar conductance at given concentration in S m² mol⁻¹)	p.	8.28×10^{-4}
ii.	κ (conductivity in S m⁻¹)	q.	1.61×10^{-3}
iii.	Ω (resistance in kilo ohm)	r.	3.312×10^{-4}
iv.	S (conductance) ohm⁻¹	s.	0.622

Codes

	i	ii	iii	iv
a.	p	r	s	q
b.	p	q	s	r
c.	q	s	p	r
d.	p	q	r	s

Comprehension Type

***Direction** (Q. Nos. 23-26) This section contains 2 paragraphs, each describing theory, experiments, data, etc. Four questions related to the paragraphs have been given. Each question has only one correct answer among the four given options (a), (b), (c) and (d).*

Passage I

During the discharge of a lead storage battery, the density of sulphuric acid fell from 1.294 g/mL (39% H_2SO_4 by weight) to 1.139 g/mL (20% H_2SO_4 by weight).

The battery holds 3.5 L of the acid and the volume remains practically constant during discharge.

23 Total H_2SO_4 used is

a. 9.89 mol
b. 5.15 mol
c. 2.32 mol
d. 2.82 mol

24 Total number of ampere-hour for which battery must have been used is

a. 205.1
b. 265.1
c. 138.1
d. 67.0

Working Space

Passage II

Variation of molar conductance Λ_m with molar concentration C is given following Debye-Huckel Onsager equation.

$$\Lambda_m = \Lambda^\circ_m - (A + B\,\Lambda^\circ_m)\sqrt{C}$$

With A and B as Debye-Huckel constants and Λ°_m is the molar conductance of infinitely diluted (≈ zero concentration) solution.

25 Select the correct option.

a. $\Lambda^\circ_m > \Lambda_m$
b. $\Lambda_m = \Lambda^\circ_m$ when $C \longrightarrow 0$
c. Both A and B are dependent on temperature T and dielectric constant of the medium
d. All of the above

26 Following are some facts about electrolytic conduction :

I. Λ_m decreases with $\sqrt{C}$ due to increase in interionic attraction between ions.
II. For weak electrolytes, there is rapid increase in the degree of ionisation with dilution, hence, interionic attraction decreases and Λ_m increases rapidly in lower concentration range.
III. Λ°_m can be obtained by extrapolation of graph of Λ_m and $\sqrt{C}$ to zero concentration for strong electrolytes.

Correct facts are

a. I, II and III
b. I and III
c. I and II
d. II and III

One Integer Value Correct Type

***Direction** (Q. Nos. 27-30) This section contains 4 questions. When worked out will result in an integer from 0 to 9 (both inclusive).*

27 A cell, at 298 K is set up $Ag \mid Ag^+$ (1 M) || Cu^{2+}(1 M) | Cu

A current of 9.65 A is passed for 1h. What is change in EMF (in centivolt)?

28 An acidic solution of Cu^{2+} salt containing 0.4 g of Cu^{2+} is electrolysed until all the copper is deposited. Calculate the volume of gas liberated at the other electrode (Express result in centilitre at STP).

29 At 298 K, the EMF of the cell, $Pt(H_2)$ 1 bar | H^+ (pH = x) || $\underset{1\,M}{Cl^-}$ | Hg_2Cl_2, Hg is 0.7530 V.

$$E^\circ_{Hg,\, Hg_2Cl_2/Cl^-} = -0.2802\ V$$

If liquid junction potential is zero, the pH of the solution x is

30 Silver is electro-deposited on a metallic vessel surface of area 800 cm^2 by passing a current of 0.20 A for 3 h. Thickness of the silver deposited was $x \times 10^{-4}$ cm. What is the value of x? (Density of Ag = 10.20 g cm^{-3})

JEE Main & AIEEE Archive

(Compilation of Last 13 Years' Questions)

1 Two Faraday of electricity is passed through a solution of $CuSO_4$. The mass of copper deposited at the cathode is (atomic mass of Cu = 63.5 amu) **(JEE Main 2015)**

a. 127 g b. 0 g
c. 63.5g d. 2 g

2 Resistance of 0.2 M solution of an electrolyte is 50 Ω. The specific conductance of the solution is $1.4\,Sm^{-1}$. The resistance of 0.5 M solution of the same electrolyte is 280 Ω. The molar conductivity of 0.5 M solution of the electrolyte in $S\,mol^{-1}$ is **(JEE Main 2014)**

a. 5×10^{-4} b. 5×10^{-3}
c. 5×10^{3} d. 5×10^{2}

3 The equivalent conductance of NaCl at concentration C and at infinite dilution are λ_c and λ_∞, respectively. The correct relationship between λ_c and λ_∞ is given as **(JEE Main 2014)**

a. $\lambda_c = \lambda_\infty + (B)C$
b. $\lambda_c = \lambda_\infty - (B)C$
c. $\lambda_c = \lambda_\infty - (B)\sqrt{C}$
d. $\lambda_c = \lambda_\infty + (B)\sqrt{C}$

4 Given below are the half-cell reactions

$Mn^{2+} + 2e^- \longrightarrow Mn,\ E° = -1.18\,V$

$2(Mn^{3+} + e^- \longrightarrow Mn^{2+}),\ E° = +1.51\,V$

The $E°$ for $3Mn^{2+} \longrightarrow Mn + 2Mn^{3+}$ will be

a. –2.69V, the reaction will not occur **(JEE Main 2014)**
b. –2.69 V the reaction will occur
c. –0.33V, the reaction will not occur
d. –0.33 V, the reaction will occur

5 Given,

$$E^\circ_{Cr^{3+}/Cr} = -0.74\,V$$

$$E^\circ_{MnO_4^-/Mn^{2+}} = 1.51\,V$$

$$E^\circ_{Cr_2O_7^{2-}/Cr^{3+}} = 1.33\,V$$

$$E^\circ_{Cl/Cl^-} = 1.36\,V$$

Based on the data given above, strongest oxidising agent will be **(AIEEE 2013)**

a. Cl b. Cr^{3+}
c. Mn^{2+} d. MnO_4^-

6 The standard reduction potentials for Zn^{2+} / Zn, Ni^{2+} /Ni and Fe^{2+} / Fe are – 0.76, – 0.23 and – 0.44 V, respectively. The reaction $X + Y^{2+} \longrightarrow X^{2+} + Y$ will be spontaneous when **(AIEEE 2012)**

a. X = Ni, Y = Fe
b. X = Ni, Y = Zn
c. X = Fe, Y = Zn
d. X = Zn, Y = Ni

7 The reduction potential of hydrogen half-cell will be negative if **(AIEEE 2011)**

a. $p(H_2)$ = 1 atm and $[H^+]$ = 2.0 M
b. $p(H_2)$ = 1 atm and $[H^+]$ = 1.0 M
c. $p(H_2)$ = 2 atm and $[H^+]$ = 1.0 M
d. $p(H_2)$ = 2 atm and $[H^+]$ = 2.0 M

8 Resistance of 0.2 M solution of an electrolyte is 50 Ω. The specific conductance of the solution is $1.3\,S\,m^{-1}$. If resistance of the 0.4 M solution of the same electrolyte is 260 Ω, its molar conductivity is **(AIEEE 2011)**

a. $6250\,S\,m^2\,mol^{-1}$
b. $6.25\times10^{-4}\,S\,m^2\,mol^{-1}$
c. $625\times10^{-4}\,S\,m^2\,mol^{-1}$
d. $62.5\,S\,m^2\,mol^{-1}$

9 The Gibbs energy for the decomposition of Al_2O_3 at 500°C is as follows **(AIEEE 2010)**

$$\frac{2}{3}Al_2O_3 \longrightarrow \frac{4}{3}Al + O_2$$

$$\Delta_rG = +966\,kJ\,mol^{-1}$$

The potential difference needed for electrolytic reduction of Al_2O_3 at 500°C is at least

a. 4.5 V b. 3.0 V
c. 2.5 V d. 5.0 V

10 Given,

$$E^\circ_{Fe^{3+}/Fe} = -0.036\,V,$$

$$E^\circ_{Fe^{2+}/Fe} = -0.439\,V$$

The value of standard electrode potential for the charge,

$$Fe^{3+}(aq) + e^- \longrightarrow Fe^{2+}(aq)$$

will be **(AIEEE 2009)**

a. – 0.072 V
b. 0.385 V
c. 0.770 V
d. – 0.270 V

11 Given,

$$E^\circ_{Cr^{3+}/Cr} = -0.72\,V,$$

$$E^\circ_{Fe^{2+}/Fe} = -0.42\,V.$$

The potential for the cell

$Cr|Cr^{3+}\ (0.1M)\,||\,Fe^{2+}\ (0.01M)\,|\,Fe$ is **(AIEEE 2008)**

a. 0.26 V
b. 0.399 V
c. – 0.339 V
d. – 0.26 V

12 The equivalent conductances of two strong electrolytes at infinite dilution in H_2O (where ions move freely through a solution) at 25°C are given below

$$\Lambda^\circ_{CH_3COONa} = 91.0 \text{ S cm}^2 / \text{equiv}$$
$$\Lambda^\circ_{HCl} = 426.2 \text{ S cm}^2 / \text{equiv}$$

What additional information/quantity one needs to calculate Λ° of an aqueous solution of acetic acid? **(AIEEE 2007)**

a. Λ° of NaCl
b. Λ° of CH_3COOK
c. The limiting equivalent conductance of H^+ $(\lambda^\circ_{H^+})$
d. Λ° of chloroacetic acid ($ClCH_2COOH$)

13 The cell, $Zn | Zn^{2+}(1M) || Cu^{2+}(1M) | Cu$ ($E^\circ_{cell} = 1.10$ V), was allowed to be completely discharged at 298 K. The relative concentration of Zn^{2+} to Cu^{2+} $\left(\frac{[Zn^{2+}]}{[Cu^{2+}]}\right)$ is **(AIEEE 2007)**

a. antilog (24.08)
b. 37.3
c. $10^{37.3}$
d. 9.65×10^4

14 Resistance of a conductivity cell filled with a solution of an electrolyte of concentration 0.1 M is 100 Ω. The conductivity of this solution is 1.29 S m^{-1}. Resistance of the same cell when filled with 0.2 M of the same solution is 520 Ω. The molar conductivity of 0.2 M solution of the electrolyte will be **(AIEEE 2006)**

a. 124×10^{-4} S m^2 mol^{-1}
b. 1240×10^{-4} S m^2 mol^{-1}
c. 1.24×10^{-4} S m^2 mol^{-1}
d. 12.4×10^{-4} S m^2 mol^{-1}

15 The molar conductivities Λ°_{NaOAc} and Λ°_{HCl} at infinite dilution in water at 25°C are 91.0 and 426.2S cm^2/mol, respectively. To calculate Λ°_{HOAc}, the additional value required is **(AIEEE 2006)**

a. $\Lambda^\circ_{H_2O}$ b. Λ°_{KCl}
c. Λ°_{NaOH} d. Λ°_{NaCl}

16 Given the data at 25°C,

$$Ag + I^- \longrightarrow AgI + e^-; \quad E^\circ = 0.152 \text{ V}$$
$$Ag \longrightarrow Ag^+ + e^-; \quad E^\circ = -0.800 \text{ V}$$

What is the value of $\log K_{sp}$ for AgI?

$\left(2.303\dfrac{RT}{F} = 0.059 \text{ V}\right)$ **(AIEEE 2006)**

a. − 8.12
b. + 8.612
c. −37.83
d. −16.13

17

Electrolyte	KCl	KNO_3	HCl	NaOAc	NaCl
Λ^∞ (S cm^2 mol^{-1})	149.9	145.0	426.2	91.0	126.5

Calculate Λ^∞_{HOAc} using appropriate molar conductances of the electrolytes listed above at infinite dilution in H_2O at 25°C. **(AIEEE 2005)**

a. 217.5 b. 390.7
c. 552.7 d. 517.2

18 Aluminium oxide may be electrolysed at 1000°C to furnish aluminium metal (atomic mass = 27 u; 1 Faraday = 96500 C). The cathode reaction is

$$Al^{3+} + 3e^- \longrightarrow Al^0$$

To prepare 5.12 kg of aluminium metal by this method would require **(AIEEE 2005)**

a. 5.49×10^1 C of electricity
b. 5.49×10^4 C of electricity
c. 1.83×10^7 C of electricity
d. 5.49×10^7 C of electricity

19 The limiting molar conductivities Λ° for NaCl, KBr and KCl are 126, 152 and 150 S cm^2 mol^{-1}, respectively. The Λ° for NaBr is **(AIEEE 2004)**

a. 128 S cm^2 mol^{-1} b. 176 S cm^2 mol^{-1}
c. 278 S cm^2 mol^{-1} d. 302 S cm^2 mol^{-1}

20 The $E^\circ_{M^{3+}/M^{2+}}$ values for Cr, Mn, Fe and Co are − 0.41, + 1.57, + 0.77 and + 1.97 V, respectively. For which one of these metals, the change in oxidation state from + 2 to + 3 is easiest? **(AIEEE 2004)**

a. Cr b. Mn
c. Fe d. Co

21 Consider the following E° values :

$$E^\circ_{Fe^{3+}/Fe^{2+}} = +0.77 \text{ V}$$
$$E^\circ_{Sn^{2+}/Sn} = -0.14 \text{ V}$$

Under standard conditions, the potential for the reaction,

$Sn(s) + 2Fe^{3+}(aq) \rightarrow 2Fe^{2+}(aq) + Sn^{2+}(aq)$

is **(AIEEE 2004)**

a. 1.68 V b. 1.40 V
c. 0.91 V d. 0.63 V

22 In a cell that utilises the reaction,

$Zn(s) + 2H^+(aq) \rightarrow Zn^{2+}(aq), + H_2(g)$

addition of H_2SO_4 to cathode compartment will **(AIEEE 2004)**

a. lower the E and shift equilibrium to the left
b. lower the E and shift the equilibrium to the right
c. increase the E and shift the equilibrium to the right
d. increase the E and shift the equilibrium to the left

23 The standard emf of a cell, involving one electron change is found to be 0.591 V at 25° C. The equilibrium constant of the reaction is ($F = 96500$ C mol^{-1}) **(AIEEE 2004)**

a. 1.0×10^1 b. 1.0×10^5
c. 1.0×10^{10} d. 1.0×10^{30}

24 When during electrolysis of a solution of $AgNO_3$, 9650 C of charge pass through the electroplating bath, the mass of silver deposited on the cathode will be **(AIEEE 2003)**

a. 1.08 g b. 10.8 g
c. 21.6 g d. 108 g

25 Standard reduction electrode potentials of three metals A, B and C are + 0.5 V, – 3.0 V and – 1.2 V, respectively. The reducing power of these metals are **(AIEEE 2003)**

a. $B > C > A$
b. $A > B > C$
c. $C > B > A$
d. $A > C > B$

26 For a cell reaction involving a two electron change, the standard emf of the cell is found to be 0.295 V at 25°C. The equilibrium constant of the reaction at 25°C will be **(AIEEE 2003)**

a. 1×10^{-10} b. 29.5×10^{-2}
c. 10 d. 1×10^{10}

27 Conductivity (Seimen's S) is directly proportional to area of the vessel and the concentration of the solution in it and is inversely proportional to the length of the vessel, then constant of proportionality is expressed in **(AIEEE 2002)**

a. S m mol^{-1} b. S^2 m^2 mol^{-2}
c. S m^2 mol^{-1} d. S^2 m^2 mol

28 For a cell given, $\underset{-}{Ag} \mid Ag^+ \mid\mid Cu^{2+} \mid \underset{+}{Cu}$

$$Ag^+ + e^- \longrightarrow Ag;\ E° = x$$
$$Cu^{2+} + 2e^- \longrightarrow Cu;\ E° = y$$

$E°$ cell is **(AIEEE 2002)**

a. $x + 2y$ b. $2x + y$
c. $y - x$ d. $y - 2x$

29 For the following cell with hydrogen electrodes at two different pressures p_1 and p_2

$$\underset{p_1}{Pt(H_2)} \mid \underset{1M}{H^+(aq)} \mid \underset{p_2}{Pt(H_2)}$$

emf is given by **(AIEEE 2002)**

a. $\frac{RT}{F} \log_e \frac{p_1}{p_2}$
b. $\frac{RT}{2F} \log_e \frac{p_1}{p_2}$
c. $\frac{RT}{F} \log_e \frac{p_2}{p_1}$
d. $\frac{RT}{2F} \log_e \frac{p_2}{p_1}$

JEE Advanced & IIT JEE Archive

(Compilation of Last 10 Years' Questions)

1 In a galvanic cell, the salt-bridge

(2014 Adv., One or More than One Options Correct Type)

a. does not participate chemically in the cell reaction

b. stops the diffusion of ions from one electrode to another

c. is necessary for the occurrence of the cell reaction

d. ensures mixing of the two electrolytic solutions

2 An aqueous solution of *X* is added slowly to an aqueous solution of *Y* as shown in Column I. The variation in conductivity of these reactions is given in Column II. Match Column I with Column II and select the correct answer using the codes given below. **(2013 Adv., Matching Type)**

	Column I		Column II
i.	$(C_2H_5)_3N + CH_3COOH$ X Y	p.	Conductivity decreases and then increases.
ii.	KI(0.1 M) + $AgNO_3$(0.01 M) X Y	q.	Conductivity decreases and then does not change much.
iii.	CH_3COOH + KOH X Y	r.	Conductivity increases and then does not change much.
iv.	NaOH + HI X Y	s.	Conductivity does not change much and then increases.

Codes

	i	ii	iii	iv
a.	r	s	q	p
b.	s	r	q	p
c.	q	r	s	p
d.	p	s	r	q

3 The standard reduction potential data at 25°C is given below.

$$E^\circ_{(Fe^{3+}/Fe^{2+})} = +0.77\text{ V}$$
$$E^\circ_{(Fe^{2}/Fe)} = -0.44\text{ V}$$
$$E^\circ_{(Cu^{2+}/Cu)} = +0.34\text{ V}$$
$$E^\circ_{(Cu^{+}/Cu)} = +0.52\text{ V}$$
$$E^\circ(O_2(g) + 4H^+ + 4e^-) \longrightarrow 2H_2O) = +1.23\text{ V}$$
$$E^\circ(O_2(g) + 2H_2O + 4e^-) \longrightarrow 4OH) = +0.40\text{ V}$$
$$E^\circ_{(Cr^{3+}/Cr)} = -0.74\text{ V}$$
$$E^\circ_{(Cr^{2+}/Cr)} = +0.91\text{ V}$$

Match E° of the rebox pair in Column I with the values given in Column II and select the correct answer using the codes given below.

(2013 Adv., Matching Type)

	Column I		Column II
i.	$E^\circ_{(Fe^{3+}/Fe)}$	p.	− 0.18 V
ii.	$E^\circ(4H_2O \rightleftharpoons 4H^+ + 4OH^-)$	q.	− 0.4 V
iii.	$E^\circ(Cu^{2+} + Cu \longrightarrow 2Cu^+)$	r.	− 0.04 V
iv.	$E^\circ_{(Cr^{3+}, Cr^{2+})}$	s.	− 0.83 V

Codes

	i	ii	iii	iv
a.	s	p	q	r
b.	q	r	s	p
c.	p	q	r	s
d.	r	s	p	q

Passage for Q. Nos. (4-5)

The electrochemical cell shown below is a concentration cell. $M|M^{2+}$ (saturated solution of a sparingly soluble salt, MX_2) $||M^{2+}$ (0.001 mol dm^{-3}) $|M$.

The emf of the cell depends on the difference in concentration of M^{2+} ions at the two electrodes. The emf of the cell at 298K is 0.059 V.

(2012, Comprehension Type)

4 The solubility product (K_{sp} ; mol^3dm^{-9}) of MX_2 at 298 K based on the information available the given concentration cell is (take $2.303 \times R \times 298 / F = 0.059$ V)

a. 1×10^{-15} b. 4×10^{-15}

c. 1×10^{-12} d. 4×10^{-12}

5 The value of ΔG (kJ mol^{-1}) for the given cell is (take 1 F = 96500 C mol^{-1})

a. − 5.7 b. 5.7

c. 11.4 d. −11.4

6 Consider the following cell reaction,

$$2Fe(s) + O_2(g) + 4H^+(aq) \longrightarrow 2Fe^{2+}(aq) + 2H_2O(l), \quad E^\circ = 1.67\text{ V}$$

At $[Fe^{2+}] = 10^{-3}$ M, $p(O_2) = 0.1$ atm and pH = 3, the cell potential at 25° C is

(2011, Only One Option Correct Type)

a. 1.47 V b. 1.77 V

c. 1.87 V d. 1.57 V

7 $AgNO_3$ (aqueous) was added to an aqueous KCl solution gradually and the conductivity of the solution was measured. The plot of conductance (Λ) *versus* the volume of $AgNO_3$ is

(2011, Only One Option Correct Type)

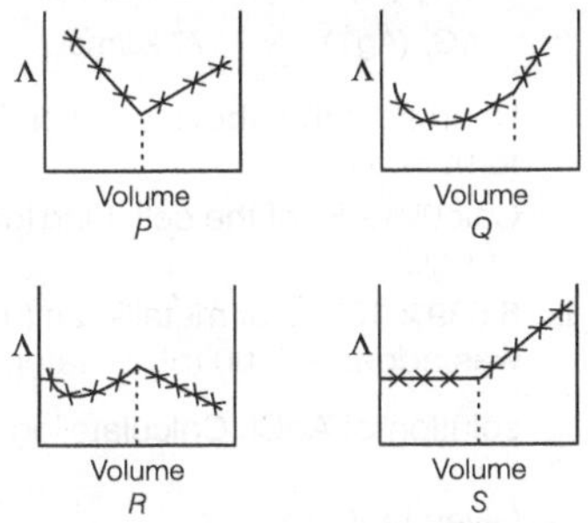

a. *P* b. *Q*
c. *R* d. *S*

Passage for Q. Nos. (8-9)

The concentration of potassium ions inside a biological cell is at least twenty times higher than the outside. The resulting potential difference across the cell is important in several processes such as transmission of nerve impulses and maintaining the ion balance. A simple model for such a concentration cell involving a metal *M* is

$M(s) | M^+ (aq; 0.05 \text{ molar}) || M^+ (aq; 1 \text{molar}) | M(s)$.

For the above electrolytic cell the magnitude of the cell potential $|E_{cell}| = 70$ mV.

(2010, Comprehension Type)

8 For the above cell

a. $E_{cell} < 0; \Delta G > 0$
b. $E_{cell} > 0; \Delta G < 0$
c. $E_{cell} < 0; \Delta G^\circ > 0$
d. $E_{cell} > 0; \Delta G^\circ < 0$

9 If the 0.05 molar solution of M^+ is replaced by a 0.0025 molar M^+ solution, then the magnitude of the cell potential would be

a. 35 mV b. 70 mV
c. 140 mV d. 700 mV

10 For the reduction of NO_3^- ion in an aqueous solution E° is + 0.96 V. Values of E° for some metal ions are given below

$V^{2+} (aq) + 2e^- \longrightarrow V$; $E^\circ = -1.19$ V

$Fe^{3+} (aq) + 3e^- \longrightarrow Fe$; $E^\circ = -0.04$V

$Au^{3+} (aq) + 3e^- \longrightarrow Au$; $E^\circ = +1.40$ V

$Hg^{2+} (aq) + 2e^- \longrightarrow Hg$; $E^\circ = +0.86$V

The pair(s) of metals that is(are) oxidised by NO_3^- in aqueous solution is(are)

(2009, One or More than One Options Correct Type)

a. V and Hg b. Hg and Fe
c. Fe and Au d. Fe and V

11 Electrolysis of dilute aqueous NaCl solution was carried out by passing 10 mA current. The time required to liberate 0.01 mole of H_2 gas at the cathode is (1 F = 96500 C mol^{-1})

(2008,Only One Option Correct Type)

a. 9.65×10^4 s
b. 19.3×10^4 s
c. 28.95×10^4 s
d. 38.6×10^4 s

Passage for Q. Nos. (12-13)

Redox reaction play a vital role in chemistry and biology. The values of standard redox potential (E°) of two half-cell reactions decide which way the reaction is expected to proceed. A simple example is a Daniell cell in which zinc goes into solution and copper gets deposited. Given below are a set of half-cell reactions (acidic medium) along with their E° (*V* with respect to normal hydrogen electrode) values.

Using this data obtain the correct explanations to Questions 15-17.

$I_2 + 2e^- \rightarrow 2I^-$ $E^\circ = 0.54$

$Cl_2 + 2e^- \rightarrow 2Cl^-$ $E^\circ = 1.36$

$Mn^{3+} + e^- \rightarrow Mn^{2+}$ $E^\circ = 1.50$

$Fe^{3+} + e^- \rightarrow Fe^{2+}$ $E^\circ = 0.77$

$O_2 + 4H^+ + 4e^- \rightarrow 2H_2O$ $E^\circ = 1.23$

(2007, Comprehension Type)

12 Among the following, identify the correct statement.

a. Chloride ion is oxidised by O_2
b. Fe^{2+} is oxidised by iodine
c. Iodide ion is oxidised by chlorine
d. Mn^{2+} is oxidised by chlorine

13 While Fe^{3+} is stable, Mn^{3+} is not stable in acid solution because

a. O_2 oxidises Mn^{2+} to Mn^{3+}
b. O_2 oxidises both Mn^{2+} to Mn^{3+} and Fe^{2+} to Fe^{3+}
c. Fe^{3+} oxidises H_2O to O_2
d. Mn^{3+} oxidises H_2O to O_2

Passage for Q. Nos. (14-16)

Tollen's reagent is used for the detection of aldehydes. When a solution of $AgNO_3$ is added to glucose with NH_4OH, then gluconic acid is formed.

$Ag^+ + e^- \longrightarrow Ag$; $E^\circ_{red} = 0.80$ V

$C_6H_{12}O_6 + H_2O \longrightarrow \underset{\text{Gluconic acid}}{C_6H_{12}O_7} + 2H^+ + 2e^-$; $E^\circ_{oxi} = -0.05$ V

$[Ag(NH_3)_2^+] + e^- \longrightarrow Ag(s) + 2NH_3$; $E^\circ_{red} = 0.337$ V

[Use $2.303 \times \frac{RT}{F} = 0.0592$ and $\frac{F}{RT} = 38.92$ at 298 K]

(2006, Comprehension Type)

14 $2Ag^+ + C_6H_{12}O_6 + H_2O \longrightarrow 2Ag(s) + C_6H_{12}O_7 + 2H^+$

Find $\ln K$ of this reaction

a. 66.13 b. 58.38
c. 28.30 d. 46.29

15 When ammonia is added to the solution, pH is raised to 11. Which half-cell reaction is affected by pH and by how much?

a. E_{oxi} will increase by a factor of 0.65 from E°_{oxi}
b. E_{oxi} will decrease by a factor of 0.65 from E°_{oxi}
c. E_{red} will increase by a factor of 0.65 from E°_{red}
d. E_{red} will decrease by a factor of 0.65 from E°_{red}

16 Ammonia is always added in this reaction. Which of the following must be incorrect?

a. NH_3 combines with Ag^+ to form a complex
b. $[Ag(NH_3)_2^+]$ is a stronger oxidising reagent than Ag^+
c. In the absence of NH_3 silver salt of gluconic acid is formed
d. NH_3 has affected the standard reduction potential of glucose/gluconic acid electrode

17 We have taken a saturated solution of AgBr, K_{sp} is 12×10^{-14}. If 10^{-7} M of $AgNO_3$ are added to 1 L of this solution, find conductivity (specific conductance) of this solution in terms of 10^{-7} Sm^{-1} units.

Given,

$\lambda^\circ_{(Ag^+)} = 6 \times 10^{-3} Sm^2 mol^{-1}$,

$\lambda^\circ_{(Br^-)} = 8 \times 10^{-3} Sm^2 mol^{-1}$,

$\lambda^\circ_{(NO_3^-)} = 7 \times 10^{-3} Sm^2 mol^{-1}$. **(2006, Subjective Type)**

18 i. Calculate ΔG°_r of the following reaction

$Ag^+(aq) + Cl^-(aq) \longrightarrow AgCl(s)$

Given,

ΔG°_f (AgCl)	−109 kJ/mol
ΔG°_f $(Cl)^-$	−129 kJ/mol
ΔG°_f (Ag^+)	77 kJ/mol

Represent the above reaction in the form of a cell.

Calculate E° of the cell. Find $\log_{10} K_{sp}$ of AgCl.

ii. 6.539×10^{-2} g of metallic Zn (u = 65.39) was added to 100 mL of saturated solution of AgCl. Calculate $\log_{10} \frac{[Zn^{2+}]}{[Ag^+]^2}$.

Given that

$Ag^+ + e^- \longrightarrow Ag$ $E^\circ = 0.80$ V

$Zn^{2+} + 2e^- \longrightarrow Zn$ $E^\circ = -76$ V

Also, find how many moles of Ag will be formed? **(2005, Subjective Type)**

19 The half-cell reactions for rusting of iron are:

$2H^+ + 2e^- + \frac{1}{2}O_2 \longrightarrow H_2O(l)$, $E^\circ = +1.23$ V

$Fe^{2+} + 2e^- \longrightarrow Fe(s)$, $E^\circ = -0.44$ V

ΔG° (in kJ) for the reaction is **(2005, Only One Option Correct Type)**

a. −76
b. −322
c. −122
d. −176

JEE Archive

4. Chemical Kinetics

DPP
DAILY PRACTICE PROBLEMS

DPP-1 Rate Law, Molecularity and Order of Reaction

Subjective Questions

Working Space

Direction *(Q. Nos. 1-3) These questions are subjective in nature, need to be solved completely on notebook.*

1 The rate of the reaction,

$$A + 2B \longrightarrow 3C + D$$

was reported as 1.0 Ms^{-1}. State the rates of consumption and formation of the reactants and the products.

2 The rate of the gaseous reaction,

$$A(g) + B(g) \longrightarrow \text{Product}$$

Is given by the expression: $\frac{dx}{dt} = k[A][B]$.

If the volume of the reaction vessel is suddenly reduced to one-fourth of the initial volume, find the ratio of the final rate to the initial rate.

3 The experimental rate law for the reaction,

$$2NO_2^- + 4H^+ + 2I^- \longrightarrow I_2 + 2NO + 2H_2O,$$

is rate $= k[NO_2^-][H^+]^2[I^-]$. How would the rate of the reaction be altered, if

i. $[H^+]$ and $[I^-]$ were kept constant but $[NO_2^-]$ was doubled?

ii. $[I^-]$ and $[NO_2^-]$ were kept constant but $[H^+]$ was doubled?

iii. $[I^-]$ and $[NO_2^-]$ were kept constant and pH was increased by one unit?

iv. All the concentrations were doubled?

Only One Option Correct Type

Direction *(Q. Nos. 4-18) This section contains 15 multiple choice questions. Each question has four choices (a), (b), (c) and (d), out of which ONLY ONE is correct.*

4 In the following reaction, which has maximum rate w.r.t. rate of disappearance of NH_3?

$4NH_3 + 5O_2 \longrightarrow 4NO + 6H_2O$

a. O_2 b. NO

c. H_2O d. Equal

5 The rate of formation of $NO(g)$ in the reaction,

$$2NOBr(g) \longrightarrow 2NO(g) + Br_2(g)$$

was reported as $1.6 \times 10^{-4} \text{ Ms}^{-1}$. Thus, rate of the reaction is

a. $1.6 \times 10^{-4} \text{ Ms}^{-1}$ b. $8.0 \times 10^{-5} \text{ Ms}^{-1}$

c. $3.2 \times 10^{-4} \text{ Ms}^{-1}$ d. $6.4 \times 10^{-4} \text{ Ms}^{-1}$

Working Space

6 Rate of formation of SO_3 in the following reaction:

$$2SO_2 + O_2 \longrightarrow 2SO_3$$

is 100 kg min^{-1}. Hence, rate of disappearance of SO_2 will be

a. 100 kg min^{-1}
b. 80 kg min^{-1}
c. 64 kg min^{-1}
d. 32 kg min^{-1}

7 For a reversible reaction, net rate is

$$\left(\frac{dx}{dt}\right) = k_1 [A]^2 [B]^{-1} - k_2[C]$$

hence given reaction is

a. $2A + \frac{1}{B} \underset{k_2}{\overset{k_1}{\rightleftharpoons}} C$
b. $2A - B \underset{k_2}{\overset{k_1}{\rightleftharpoons}} C$
c. $2A \underset{k_2}{\overset{k_1}{\rightleftharpoons}} C + B - 1$
d. None of these

8 Following reaction can take place in both directions

$$A \underset{k_2}{\overset{k_1}{\rightleftharpoons}} B$$

For the forward reaction, rate varies with concentration of A as and for the backward reaction

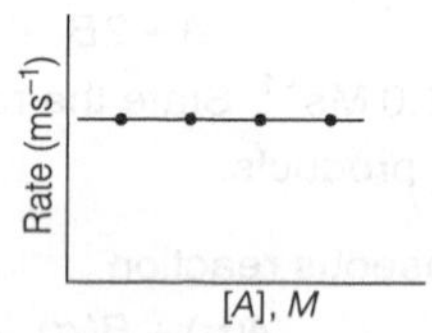

[B]	Rate
0.01 M	2×10^{-2} Ms^{-1}
0.02 M	4×10^{-2} Ms^{-1}

Hence, net reaction rate is

a. $= k_1[A] - k_2[B]$
b. $= (k_1 - k_2[B])$
c. $= (k_1[A] - k_2)$
d. $= (k_1 - k_2)$

9 For a reaction, $xA \longrightarrow yB$, rate of disappearance of 'A' is related to the rate of appearance of 'B' by the equation $-\frac{d[A]}{dt} = \frac{1}{4}\frac{d[B]}{dt}$

Thus, x and y respectively are

a. 1, 2
b. 2, 4
c. $\frac{1}{2}$, 2
d. 4, 1

10 For the reaction,

$$H_2(g) + A_2(g) \underset{k_b}{\overset{k_f}{\rightleftharpoons}} 2H(g) + A_2(g)$$

$k_f = 2.2 \times 10^4$ L mol^{-1}s^{-1} and $K_c = 1.00 \times 10^{-4}$ at 3000 K. Thus, k_b is

a. 2.2×10^8 L^{2}mol^{-2}s^{-1}
b. 0.45×10^{-8} L^{-2}mol^{2}s^{-1}
c. 2.2 L mol^{-1}s^{-1}
d. 0.45 L^{-1}mol s^{-1}

11 For a reaction,

$$xA + yB \longrightarrow zC$$

$$-\frac{d[A]}{dt} = -\frac{d[B]}{dt} = \frac{1.5d[C]}{dt}$$

then x, y and z are

a. 1, 1, 1
b. 3, 2, 3
c. 3, 3, 2
d. 2, 2, 3

12 For the reaction,

$$2N_2O_5(g) \longrightarrow 4NO_2(g) + O_2(g)$$

$$-\frac{\Delta[N_2O_5]}{\Delta t} = k_1[N_2O_5],\ +\frac{[NO_2]}{\Delta t} = k_2[N_2O_5] \text{ and } +\frac{\Delta[O_2]}{\Delta t} = k_3[N_2O_5]$$

Thus,

a. $k_1 = k_2 = k_3$
b. $2k_1 = k_2 = 4k_3$
c. $2k_1 = 4k_2 = k_3$
d. None of these

13 The decomposition of acetaldehyde is given by the following reaction:

$$CH_3CHO(g) \longrightarrow CH_4(g) + CO(g)$$

Rate of the reaction with respect to the pressure of the reactant is

a. $-\frac{dp}{dt}$
b. $-\frac{1}{V}\frac{dp}{dt}$
c. $-\frac{1}{RT}\frac{dp}{dt}$
d. $-\frac{dp}{dt}$

14 For the reaction, $2A + 3B \longrightarrow$ Product, A is present in excess. When concentration of B is changed from 0.01 M to 0.04 M, rate is doubled. Thus, rate law is

a. $\left(\frac{dx}{dt}\right) = k[A]^2[B]^3$
b. $\left(\frac{dx}{dt}\right) = k[A][B]$
c. $\left(\frac{dx}{dt}\right) = k[A]^0[B]$
d. $\left(\frac{dx}{dt}\right) = k[A]^0[B]^{1/2}$

15 For the reaction, $A \longrightarrow B,\ -\frac{d[A]}{dt} = \frac{2d[B]}{dt}$, then rate law is

a. $-\frac{d[A]}{dt} = k[A]^2$
b. $\frac{d[B]}{dt} = k[A]$
c. $\frac{d[A]}{dt} = k[A]^2$
d. $-\frac{d[B]}{dt} = k[B]^2$

16 For gaseous reaction, the rate can be expressed as

a. $\frac{1}{RT}\left(\frac{dp}{dt}\right) = \frac{1}{V}\left(\frac{dn}{dt}\right) = \left(\frac{dC}{dt}\right)$
b. $\left(\frac{dC}{dt}\right) = \left(\frac{dn}{dt}\right) = \left(\frac{dp}{dt}\right)$
c. $\left(\frac{dC}{dt}\right) = \left(\frac{dn}{dt}\right) = \frac{V}{RT}\left(\frac{dp}{dT}\right)$
d. None of these

17 The initial rates of reaction for the equation, $2A + B \longrightarrow$ Products.
Products were determined under various initial concentrations of reactants.

$[A]_0$	$[B]_0$	$-d[B]/dt$
0.10 M	0.10 M	0.25 M s^{-1}
0.20 M	0.10 M	0.50 M s^{-1}
0.10 M	0.20 M	0.25 M s^{-1}

Thus, rate law is equal to

a. $k[A][B]$
b. $k[A][B]^2$
c. $k[A]$
d. $k[B]$

18 For the reaction, $A + 2B \longrightarrow$ Product, the reaction rate was halved on doubling the concentration of A. Thus, order w.r.t. A is

a. 1
b. $\frac{1}{2}$
c. $-\frac{1}{2}$
d. -1

Comprehension Type

Working Space

Direction (Q. Nos. 19 and 20) *This section contains a paragraph, which describing theory, experiments, data, etc. Two questions related to the paragraph have been given. Each question has only one correct answer among the four given options (a), (b), (c) and (d).*

Passage

The decomposition of NO_2 at 400 K proceeds at a of rate of 5.4×10^{-5} mol $L^{-1} s^{-1}$ when $[NO_2] = 0.01$ mol L^{-1}.

$$2NO_2(g) \longrightarrow 2NO(g) + O_2(g)$$

19 What is the rate law when observed rate is 1.35×10^{-5} mol L^{-1} s^{-1} at $[NO_2] = 0.005$ mol L^{-1}?

a. $k[NO_2]$
b. $k[NO_2]^2$
c. $k[NO_2]^3$
d. $k[NO_2]^0$

20 Rate constant of the reaction will be

a. $0.54\ L\ mol^{-1} s^{-1}$
b. $5.4 \times 10^{-3}\ L\ mol^{-1} s^{-1}$
c. $5.4\ s^{-1}$
d. $0.54\ s^{-1}$

One Integer Value Correct Type

Direction (Q. Nos. 21-23) *This section contains 3 questions. When worked out will result in one integer from 0 to 9 (both inclusive).*

21 In the following reaction;

$$xA \longrightarrow yB$$

$$\log\left[-\frac{d[A]}{dt}\right] = \log\left[\frac{d[B]}{dt}\right] + 0.3$$

Where negative sign indicates rate of disappearance of the reactant. What is the value of x/y?

22 Consider the following reaction,

$$N_2(g) + 3H_2(g) \longrightarrow 2NH_3(g)$$

$$-\frac{d[H_2]}{dt} = 6\text{g min}^{-1}$$

Derive the value of $\frac{d[NH_3]}{dt}$ in mol min^{-1}.

23 In a certain polluted atmosphere containing O_3 at a steady concentration of 2.0×10^{-6} M, the hourly production of O_3 by all sources was estimated as 7.2×10^{-15} M. If the only mechanism for destruction of O_3 in the second-order reaction is

$$2O_3 \longrightarrow 3O_2$$

then rate constant for destruction reaction, defined by the rate law for $-\Delta[O_3]/\Delta t$ is $x \times 10^{-7}\ M^{-1} s^{-1}$. What is the value of x?

DPP-2 Zeroth-Order and First-Order Reactions

Only One Option Correct Type

Working Space

Direction *(Q. Nos. 1-21) This section contains 21 multiple choice questions. Each question has four choices (a), (b), (c) and (d), out of which ONLY ONE is correct.*

1 For zeroth order reaction,

$$A \longrightarrow B$$

$[A]_0 = 0.01\,\text{M}$, $[A]_t = 0.008\,\text{M}$ after 10 min.
Thus, half-life is

a. 5 min
b. 12 min
c. 16 min
d. 25 min

2 A certain zeroth order reaction has $k = 0.025\ \text{M s}^{-1}$ for the disappearance of A. What will be the concentration of A after 15 s, if the initial concentration is 0.50 M?

a. 0.50 M
b. 0.375 M
c. 0.125 M
d. 0.060 M

3 For zeroth order reaction, variation of x with time is shown as

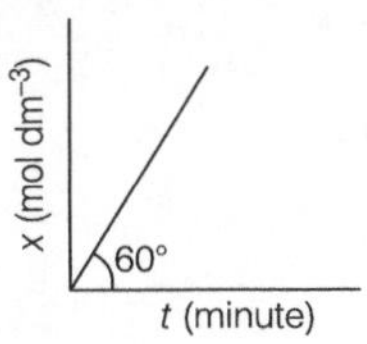

At initial concentration of the reactant as 17.32 mol dm^{-1}, half-life period is

a. 5.0min
b. 10.0min
c. 8.66min
d. 15.0min

4 For the first order reaction, rate constant k is given by

$$kt = \log C_0 - \log C_t$$

where, C_0 is the initial concentration and C_t is the concentration at time t. Graph between $\log C_t$ and time t is

a. $\log C_t$ 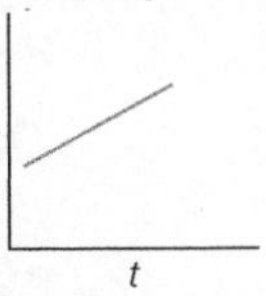

b. $\log C_t$ 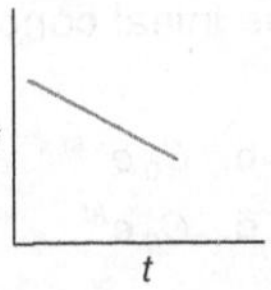

c. $\log C_t$ 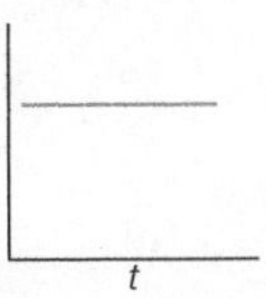

d. $\log C_t$ 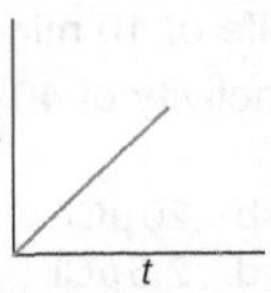

Working Space

5 In a first order reaction, the concentration of the reactant decreases from 0.8 M to 0.4 M is 15 min. The time taken for the concentration to change from 0.1 M to 0.025 M is

a. 30 min
b. 15 min
c. 7.5 min
d. 60 min

6 For the first order reaction, concentration of the reactant after two average lives is reduced to

a. 25%
b. 75%
c. $\frac{100}{e}\%$
d. $\frac{100}{e^2}\%$

7 The half-life period of a first order chemical reaction is 6.93 min. The time required for the completion of 99% of the chemical reaction will be ($\log 2 = 0.3010$) **[AIEEE 2009]**

a. 230.3 min
b. 23.03 min
c. 46.06 min
d. 460.6 min

8 For the first order reaction,

$$A \longrightarrow \text{Product}$$

The concentration of A changes from 0.1 M to 0.025 M in 40 min. The rate of the reaction when the concentration of A is 0.01 M is **[AIEEE 2012]**

a. $1.73 \times 10^{-6}\ \text{M min}^{-1}$
b. $3.47 \times 10^{-4}\ \text{M min}^{-1}$
c. $3.47 \times 10^{-5}\ \text{M min}^{-1}$
d. $1.73 \times 10^{-6}\ \text{M min}^{-1}$

9 For the reaction,

$$A \longrightarrow \text{Products}, \quad -\frac{d[A]}{dt} = k$$

Concentration of A at different time intervals are given :

Time/min	[A]/M
0	20
5	18
10	16
15	14
20	–

At 20 min, rate will be

a. $12\ \text{M min}^{-1}$
b. $10\ \text{M min}^{-1}$
c. $8\ \text{M min}^{-1}$
d. $0.4\ \text{M min}^{-1}$

10 For the first order reaction with, C_0 as the initial concentration and C at time $t, (C_0 - C) =$

a. $C_0[1 - e^{-kt}]$
b. $C_0 e^{-kt}$
c. $C_0[1 + e^{kt}]$
d. $C_0 e^{kt}$

11 The radioisotope N-13 which has a half-life of 10 min is used to image organs in the body. If the injected sample has an activity of 40 μCi (microcurie), what is the activity after 30 min?

a. 10 μCi
b. 20 μCi
c. 5 μCi
d. 2.5 μCi

Working Space

12 The following data were obtained during the first order decomposition of

$$2A(g) \longrightarrow B(g) + C(s)$$

at constant volume and at a particular temperature :

S. No.	Time	Total pressure (in pascal)
1.	After 10 min	300
2.	After complete decomposition	200

Thus, rate constant is

a. $0.0693\ \text{min}^{-1}$ b. $69.3\ \text{min}^{-1}$
c. $6.93\ \text{min}^{-1}$ d. $6.93 \times 10^{-4}\ \text{min}^{-1}$

13 A G.M. Counter is used to study the process of first order of radioactive disintegration. In the absence of radioactive substance A, it counts 3 dps (disintegration per second). When A is placed in G.M Counter it records 23 dps at the start and 13 dps after 10 min. It records x dps after next 10 min. Half-life period of A is y min. Thus, x and y are

a. 8 dps, 10 min b. 5 dps, 10 min
c. 5 dps, 20 min d. 5 dps, 5 min

14 Inversion of sucrose ($C_{12}H_{22}O_{11}$) is a first order reaction and is studied by measuring angle of rotation at different interval of time

$$\underset{\substack{\text{Sucrose}\\ d-}}{C_{12}H_{22}O_{11}} + H_2O \xrightarrow{H^+} \underset{\substack{\text{Glucose}\\ d-}}{C_6H_{12}O_6} + \underset{\substack{\text{Fructose}\\ l-}}{C_6H_{12}O_6}$$

r_0 = angle of rotation at the start, r_t = angle of rotation at time t
r_∞ = angle of rotation at the complete reaction
There is 50% inversion, when

a. $r_0 = 2r_t - r_\infty$ b. $r_0 = r_t - r_\infty$
c. $r_0 = r_t - 2r_\infty$ d. $r_0 = r_t + r_\infty$

15 Following radioactive disintegration follows first order kinetics :

$$^{M}_{Z}A \longrightarrow {}^{M-8}_{Z-4}B + \alpha\text{-particles}$$

Starting with 1 mol of A in a 1L flask at 300 K, pressure set up after two half-lifes is approximately

a. 25 atm b. 12 atm c. 37 atm d. 40 atm

16 For the first order reaction T_{av} (average life), T_{50} (half-life) and T_{75} (time 75% reaction) are in the ratio of

a. 1 : 2 : 3 b. 1.44 : 2 : 3
c. 1.44 : 1 : 2 d. 1 : 2 : 1.44

17 For the first order reaction, $\dfrac{-d[A]}{d} = k[A]$

	A	$\longrightarrow$ Product
at $t = 0$	a	0
at time t	$(a-x)$	x

Following equations are given :

I. $x = a(1 - e^{-kt})$ II. $(a-x) = a\left(\dfrac{1}{2}\right)^y$, where $y = t/t_{50}$

III. $k = \dfrac{2.303}{t}\log\left(\dfrac{a}{a-x}\right)$

Select the correct equations.

a. I and II b. II and III c. I and III d. I, II and III

Working Space

18 Rate constant of a first order reaction ($A \longrightarrow B$) is 0.0693 min^{-1}. If initial concentration is 1.0 M, rate after 30 min is

a. 0.0693 M min^{-1}
b. 8.66×10^{-3} M min^{-1}
c. 0.0233 M min^{-1}
d. 0.0466 M min^{-1}

19 Following reaction is pseudo-unimolecular w.r.t. $C_6H_5N_2Cl(A)$

$$\underset{(A)}{C_6H_5N_2Cl} + H_2O \xrightarrow{\Delta} C_6H_5OH + N_2 + HCl$$

50 mL of 1M benzene diazonium chloride (A) is taken. After 1 h, 1.226 L of N_2 gas at 1 atm and 300 K is obtained. Thus, half-life of the reaction is ($\log 250 = 2.40$)

a. 7.53 min
b. 0.092 min
c. 10.86 min
d. None of these

20 Minimum half-life of an isotope needed so that not more than 0.1% of the nuclei undergo decay during a 3.0 h laboratory experiment is

a. 72 days
b. 80 days
c. 80.7 days
d. 86.5 days

21 Naturally occurring potassium consists of 0.01% at 40 K which has a half-life of 1.28×10^9 yr. Activity of 1.00 g sample of KCl is

a. 8×10^{17} yr^{-1}
b. 4×10^{8} yr^{-1}
c. 4.37×10^{8} yr^{-1}
d. 2.286×10^{-9} yr^{-1}

Comprehension Type

Direction *(Q. Nos. 22-27) This section contains 3 paragraphs, each describing theory, experiments, data, etc. Six questions related to the paragraphs have been given. Each question has only one correct answer among the four given options (a), (b), (c) and (d).*

Passage I

Isotope of oxygen with mass number less than 16 undergoes β^+-emission. Assume an equimolar mixture of ^{14}O and ^{15}O. Also t_{50} (^{14}O) = 71s, $t_{50}(^{15}O)$ = 124 s.

22 Ratio of rate constants of ^{14}O to ^{15}O is

a. 1.75
b. 0.573
c. 1.14
d. 0.873

23 Ratio of ^{14}O to ^{15}O left after 1.00 h is

a. 3.31×10^6
b. 3.02×10^{-7}
c. 3.02×10^7
d. 4.12×10^6

Passage II

Consider acid-hydrolysis of ester which is first order reaction.

$$CH_3COOC_2H_5 + H_2O \xrightarrow{H^+} CH_3COOH + C_2H_5OH$$

Kinetics is studied by titration of the acid with NaOH at a given time. V_0 = Volume of NaOH at start, V_t = Volume of NaOH at time t, V_∞ = Volume of NaOH at the end of the complete hydrolysis.

24 Rate constant k is given by

a. $k = \frac{2.303}{t} \log \frac{V_\infty}{(V_\infty - V_t)}$
b. $k = \frac{2.303}{t} \log \frac{V_\infty - V_0}{V_\infty - V_t}$
c. $k = \frac{2.303}{t} \log \frac{V_0}{V_\infty - V_t}$
d. $k = \frac{2.303}{t} \log \frac{V_0 - V_\infty}{V_t}$

Working Space

25 If ester is 50% hydrolysed, then

a. $V_\infty = V_t$ b. $V_\infty = V_t - V_0$
c. $V_\infty = 2V_t - V_0$ d. $V_\infty = 2V_t + V_0$

Passage III

I, II and III are three vessels of equal volumes containing molecules of A as shown and in each A decomposes into products.

$$-\frac{d[A]}{dt} = k[A]$$

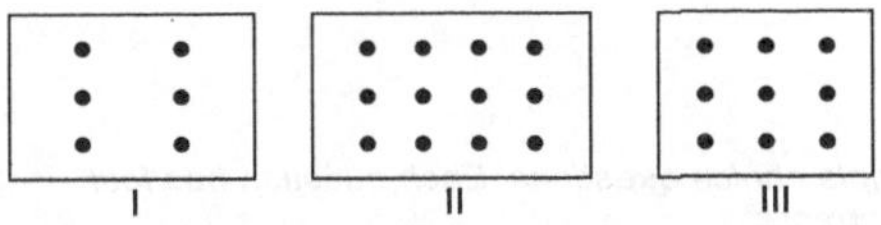

26 Initial rate of decomposition of A will be in order in these three vessels as

a. 1 : 2 : 1.5 b. 1 : 1 : 1
c. 1 : 3.5 : 2 d. 2 : 4 : 1

27 Time of 50% decomposition is in order

a. 1 : 1 : 1 b. 1 : 2 : 1.5
c. 9 : 4.5 : 1 d. 2 : 1 : 3

One Integer Value Correct Type

Direction *(Q. Nos. 28-30) This section contains 3 questions. When worked out will result in one integer from 0 to 9 (both inclusive).*

28 A radioactive element gets spilled over the floor of a room. Its half-life period is 30 days. If the initial activity is ten times the permissible value, it will be safer to enter the room after 10^n days. What is the value of n?

29 For the first order reaction, variation of $\log_{10}(a-x)$ *vs* time (t) is shown below.

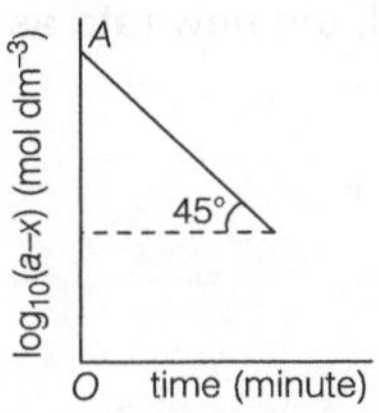

Its half-life period is $\log x$. What is the value of x?

30 If $OA = 1.3010$ (in Q. 29), then reactant left after 0.6021 min is ...... unit.

DPP-3 Determination of Order of Reaction

Only One Option Correct Type

Working Space

Direction (Q. Nos. 1-13) This section contains 13 multiple choice questions. Each question has four choices (a), (b), (c) and (d), out of which ONLY ONE is correct.

1 For a given reaction,

$$A \longrightarrow \text{Product},$$

$$\text{rate} = 1\times 10^{-4}\ \text{Ms}^{-1} \quad \text{at} \quad [A] = 0.01\,\text{M}$$

and $$\text{rate} = 1.41\times 10^{-4}\ \text{Ms}^{-1} \quad \text{at} \quad [A] = 0.02\,\text{M}$$

Hence, rate law is

a. $-\frac{d[A]}{dt} = k[A]^2$

b. $-\frac{d[A]}{dt} = k[A]$

c. $-\frac{d[A]}{dt} = \frac{k}{4}[A]$

d. $-\frac{d[A]}{dt} = k[A]^{1/2}$

2 The rate law for a reaction between the substances A and B is given by

$$\text{rate} = k[A]^n\,[B]^m$$

If concentration of A is doubled and that of B is halved, the new rate as compared to the earlier rate would be

a. $\frac{1}{2^{m+n}}$ b. $(m+n)$

c. $(n-m)$ d. $2^{(n-m)}$

3 The rate equation for the reaction,

$$2A + B \longrightarrow C$$

is found to be, $\text{rate} = k[A]\,[B]$

The correct statement in relation to this reaction is that the

a. unit of k must be s^{-1}

b. $t_{1/2}$ is constant

c. rate of formation of C is half of the rate of disappearance of A

d. value of k is independent of the initial concentration of A and B

4 For the reaction,

$$2A + B \longrightarrow \text{Product}$$

The half-life period was independent of concentration of B. On doubling the concentration A, rate increases two times. Thus, unit of rate constant for this reaction is

a. $\text{L mol}^{-1}\text{s}^{-1}$ b. $\text{mol L}^{-1}\text{s}^{-1}$

c. unitless d. s^{-1}

Working Space

5 Which of the following statements appears to the first order reaction?

$$2N_2O_5(CCl_4) \longrightarrow 4NO_2(CCl_4) + O_2(g)$$

I. $-\dfrac{d[N_2O_5]}{dt} = k[N_2O_5]$

II. $t_{50} = \dfrac{\log_e 2}{k}$

III. $[N_2O_5] = -kt + [N_2O_5]_0$

IV. $\dfrac{-d[N_2O_5]}{dt} = k[N_2O_5]^2$

V. $\log_e [N_2O_5] = -kt + \log_e [N_2O_5]_0$

VI. $t_{50} = [N_2O_5]_0 / 2k$

a. I, II and V
b. I,II and VI
c. II, V and VI
d. I, II and III

6 For a reaction, time of 75% reaction is thrice of time of 50% reaction. Thus, order of the reaction is

a. 0
b. 1
c. 2
d. 3

7 For nth order reaction,

$$\left(\frac{dx}{dt}\right) = \text{rate} = k[A]^n$$

Graphs between log (rate) and $\log[A_0]$ are of the type ($[A_0]$ is the initial concentration)

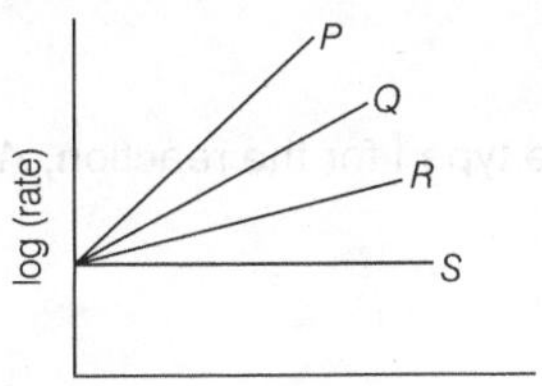

Lines P, Q, R and S are for the order

	P	Q	R	S
a.	0	1	2	3
b.	3	2	1	0
c.	1	2	3	0
d.	0	3	2	1

8 $nA \longrightarrow$ Product

$$\frac{dx}{dt} = k[A]^n$$

For the reaction, rate constant and rate of the reaction are equal, then on doubling the concentration of A, rate becomes,

a. doubled
b. four times
c. halved
d. constant

9 For the simple reaction,

$$A \longrightarrow B$$

When $[A]$ was changed from 0.502 mol dm^{-3} to 1.004 mol dm^{-3}, half-life dropped from 52 s to 26 s at 300 K. Thus, order of the reaction is

a. 0
b. 1
c. 2
d. 3

10 For the following reaction,

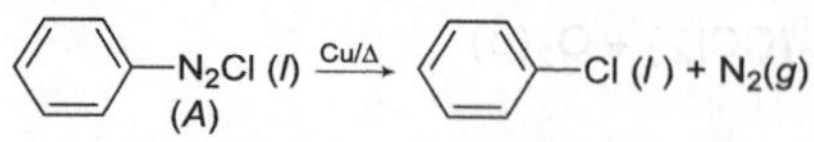

Variation of T_{50} with $[A]$ is shown

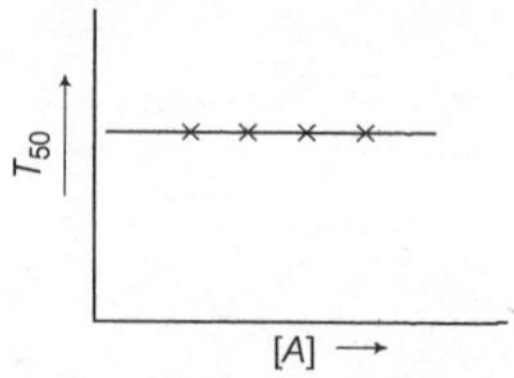

After 10 min volume of $N_2(g)$ is 10 L and after complete reaction, volume of $N_2(g)$ is 50 L. Thus, T_{50} is

a. 10 min
b. 20 min
c. 31 min
d. 41 min

11 Kinetics of the following reaction,

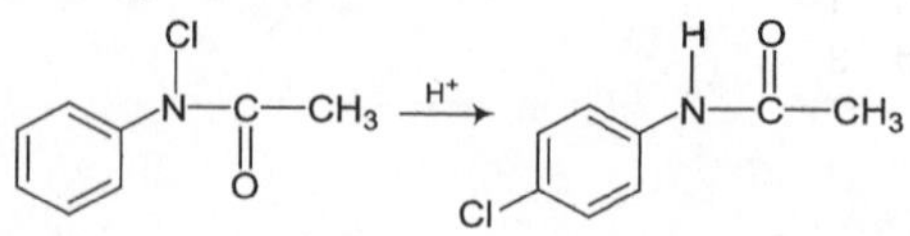

can be studied by

a. measurement of pH
b. titration with hypo after adding KI
c. titration with $AgNO_3$ solution
d. All of the above

12 Graph between $[B]$ and time t of the reaction of the type I for the reaction, $A \longrightarrow B$

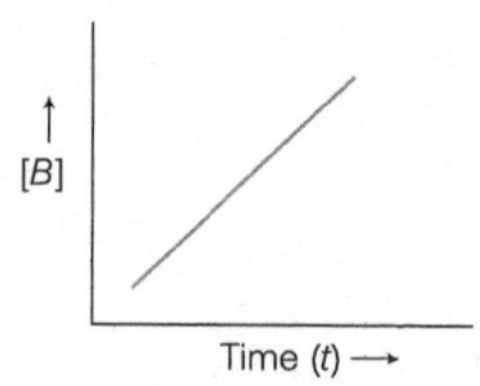

Hence, graph between $\frac{-d[A]}{dt}$ and time t will be of the type

a. 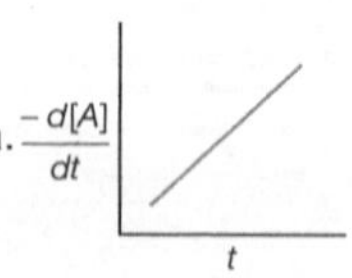

b. 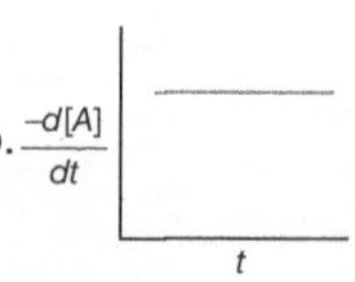

c. 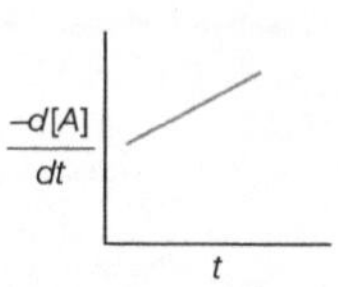

d. 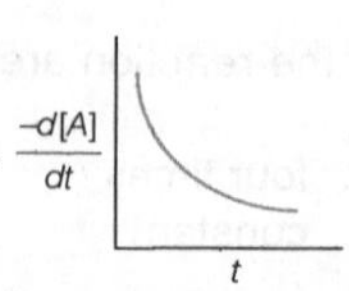

13 If in the fermentation of sugar in an enzymatic solution that is 0.12 M, the concentration of sugar is reduced to 0.06 M in 10 h and to 0.03 M in 20 h. Thus, order of the reaction is

a. 3
b. 2
c. 1
d. 0

Working Space

Matching List Type

Direction *(Q. No. 14) Choices for the correct combination of elements from Column I and Column II are given as options (a), (b), (c) and (d). Out of which one is correct.*

14 Match the graphical study in Column I with the order of the reactions in Column II and select the answer from the codes given :

	Column I		Column II
i.	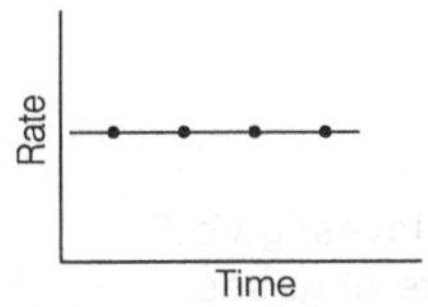	p.	3
ii.	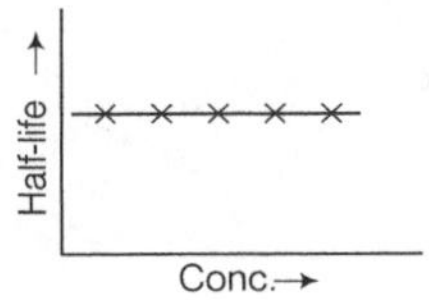	q.	2
iii.	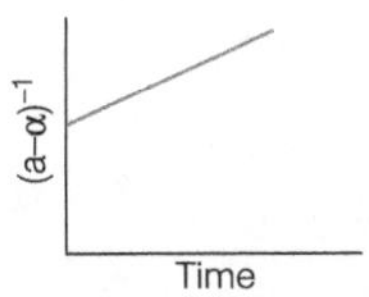	r.	1
		s.	0

Codes

	i	ii	iii		i	ii	iii
a.	p	q	r	b.	r	q	p
c.	s	r	p	d.	q	r	s

Comprehension Type

Direction *(Q. Nos. 15-20) This section contains 3 paragraphs, each describing theory, experiments, data, etc. Six questions related to the paragraphs have been given. Each question has only one correct answer among the four given options (a), (b), (c) and (d).*

Passage I

Peroxy acetyl nitrite (PAN), $CH_3\overset{O}{\overset{\|}{C}}OONO_2$ is an air pollutant, as it decomposes into radicals :

$$\underset{PAN}{CH_3\overset{O}{\overset{\|}{C}}OONO_2} \longrightarrow CH_3\overset{O}{\overset{\|}{C}}O\dot{O} + NO_2$$

A sample of polluted air is analysed for its PAN content which is reported as molecules per litre of air at 298 K.

Time (min)	**Molecules** $\times 10^{14}$**/L** (air)
0.0	5.0
10.0	4.0
20.0	3.2
30.0	2.5
40.0	2.1
50.0	1.7
60.0	1.25

15 Determine the order of the PAN decomposition

a. 0 b. 1
c. 2 d. 3

16 Rate constant of the reaction is

a. 0.0231 min^{-1}
b. 0.0067 L $mol^{-1} min^{-1}$
c. 0.009 min^{-1}
d. 0.0231 L $mol^{-1} min^{-1}$

Passage II

The reaction between nitric oxide (NO) and hydrogen (H_2) has been investigated by measuring the initial rate of decrease of pressure in known mixture of gases. The following results were obtained at 1000 K.

S.No.	Initial p_{NO}	pressure (atm) p_{H_2}	Initial rate of decrease of pressure/atm min^{-1}
1.	0.25	0.2	0.0012
2.	0.50	0.1	0.0024
3.	0.50	0.2	0.0048

17 Overall order of the reaction is

a. 1 b. 2
c. 3 d. 0

18 Rate constant of the overall reaction is

a. 0.0048 $atm^{-2} min^{-1}$
b. 0.0012 $atm^{-1} min^{-1}$
c. 0.096 $atm^{-2} min^{-1}$
d. 0.096 $atm^{-1} min^{-1}$

Passage III

A reaction between substances *A* and *B* is represented as:

$$A + B \longrightarrow C$$

Observations on the rate of this reaction are obtained as :

S.No.	Initial concentration $[A]_0$, M	$[B]_0$, M	Duration of experiment $\Delta t/h$	Final concentration $[A]_f$, M
1.	0.1	1.0	0.50	0.0975
2.	0.1	2.0	0.50	0.0900
3.	0.05	1.0	2.00	0.0450

19 Order of the reaction w.r.t. *A* and *B* respectively are

a. 1, 1 b. 1, 2
c. 2, 1 d. 2, 2

20 Rate constant of the overall reaction is

a. $5.0 \times 10^{-2} M^{-1} h^{-1}$
b. $5.0 \times 10^{-2} M^{-2} h^{-1}$
c. $5.0 \times 10^{-2} h^{-1}$
d. $5.0 \times 10^{-2} M h^{-1}$

One Integer Value Correct Type

Working Space

Direction *(Q. Nos. 21-24) This section contains 4 questions. When worked out will result in one integer from 0 to 9 (both inclusive).*

21 Graph between $\log T_{50}$ (T_{50} is half-life period) and $\log\left(\frac{1}{a}\right)$ (a = initial concentration) for nth order reaction is of the type.
What is the value of n?

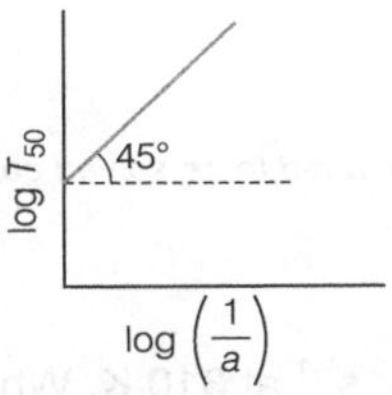

22 The acid catalysed hydrolysis of an organic compound A at 30° C has a time for half change of 100 mm when carried out in a buffer solution at pH 5 and 10 mm when carried out at pH 4. Half-life in both cases are independent of initial concentration of A. If rate law is $\frac{-d[A]}{dt} = k[A]^a [H^+]^b$, then what is the value of $(a + b)$?

23 For a reaction, $A \rightarrow B$, half-life time for a reaction at $[A] = 0\cdot1$M is 200 s and at $[A] = 0.4$M it is 50 s. Thus, order of the reaction is ......

24 For the reaction,

$$A + B \longrightarrow \text{Product}$$

Following four experiments were carried out:

Exp. No.	[A]	[B]	$\frac{-d[A]}{dt}$
1.	1M	0.01 M	1×10^{-4} Ms^{-1}
2.	2M	0.01 M	2×10^{-4} Ms^{-1}
3.	2M	0.02 M	8×10^{-4} Ms^{-1}
4.	0.2M	0.002 M	8×10^{-y} Ms^{-1}

Derive the value of y.

DPP-4 Energy of Activation

Subjective Questions

Working Space

Direction *(Q. Nos. 1and 2) These questions are subjective in nature, need to be solved completely on notebook.*

1 For a reaction, $A \longrightarrow B$

rate constant is $2.0 \times 10^{-5}\ s^{-1}$ at 300 K and $4.0 \times 10^{-5}\ s^{-1}$ at 310 K. What is the rate constant at 320 K? Also, calculate the energy of activation.

2 Reaction between NH_3 and NO_2 follows second order kinetics. Given

Temperature	Rate constants
600 K	$3.80 \times 10^2\ dm^3\ mol^{-1}\ s^{-1}$
700 K	$1.52 \times 10^3\ dm^3\ mol^{-1}\ s^{-1}$

Calculate

i. activation energy

ii. Arrhenius pre-exponential factor

Only One Option Correct Type

Direction *(Q. Nos. 3-14) This section contains 12 multiple choice questions. Each question has four choices (a), (b), (c) and (d), out of which ONLY ONE is correct.*

3 The rate of a reaction doubles when its temperature changes from 300 K to 310 K. Activation energy of such a reaction will be (Given, $R = 8.314\ JK^{-1}\ mol^{-1}$, $\log 2 = 0.3010$). **[JEE Main 2013]**

a. $53.6\ kJ\ mol^{-1}$ b. $48.6\ kJ\ mol^{-1}$

c. $58.5\ kJ\ mol^{-1}$ d. $60.5\ kJ\ mol^{-1}$

4 In the following reaction,

$$N_2O \xrightarrow{Ar} N_2 + O$$

$k = (5.0 \times 10^{11}\ L\ mol^{-1}\ s^{-1})e^{-29000R/T}$, E_a (energy of activation) is

a. $5.0 \times 10^{11}\ kcal\ mol^{-1}$

b. $29000\ kcal\ mol^{-1}$

c. $58.0\ kcal\ mol^{-1}$

d. $-29000\ cal\ mol^{-1}$

5 Rate of a reaction can be expressed by Arrhenius equation as $k = Ae^{-Ea/RT}$. In this equation, E_a represents

a. the energy above which all the colliding molecules will react

b. the energy below which colliding molecules will not react

c. the total energy of the reacting molecules at a temperature T

d. the fraction of molecules with energy greater than the activation energy of the reaction

Working Space

6 In the upper atmosphere, formation of H_2O_2 takes place as

$$H_2O + O \longrightarrow 2OH \longrightarrow H_2O_2$$

$\Delta H^\circ = 72.0\ \text{kJ mol}^{-1}$; $E_a = 77.0\ \text{kJ mol}^{-1}$

E_a for the bimolecular recombination of two OH radicals to form H_2O and O is

a. 149 kJ mol^{-1}
b. −149 kJ mol^{-1}
c. 5 kJ mol^{-1}
d. −5 kJ mol^{-1}

7

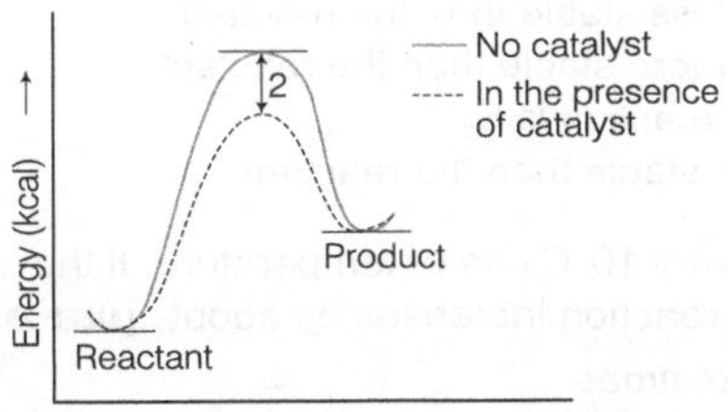

How much rate is affected in the presence of catalyst at 300 K?

a. 20 times
b. 14 times
c. 28 times
d. 2 times

8 For a reaction,

$$A \longrightarrow \text{Product}$$

$t_{50} = 100$ min at temperature T_1
and $t_{50} = 80$ min at temperature T_2.

Thus,
a. $T_1 > T_2$ and reaction is endothermic
b. $T_2 > T_1$ and reaction exothermic
c. $T_1 > T_2$ and reaction is exothermic
d. $T_2 > T_1$ and reaction can be exothermic or endothermic

9 Rate constant (k) of a reaction has least value at

a. high T and high E_a
b. high T and low E_a
c. low T and low E_a
d. low T and high E_a

10 Two reactions with different activation energies have the same rate at room temperature. Which statement correctly describes the rates of these two reactions at the same higher temperature?

a. The reaction with the greater activation energy will be faster
b. The reaction with the smaller activation energy will be faster
c. The two reactions will have the same rate at higher temperature also
d. Temperature range is also required

11 The activation energy of the reaction at a given temperature is found to be $(2.303\ RT)$ J mol^{-1}. The ratio of rate constant to the Arrhenius factor is

a. 0.01
b. 0.1
c. 0.02
d. 0.001

Working Space

12 From the given figure, select the correct statement about activation energy.

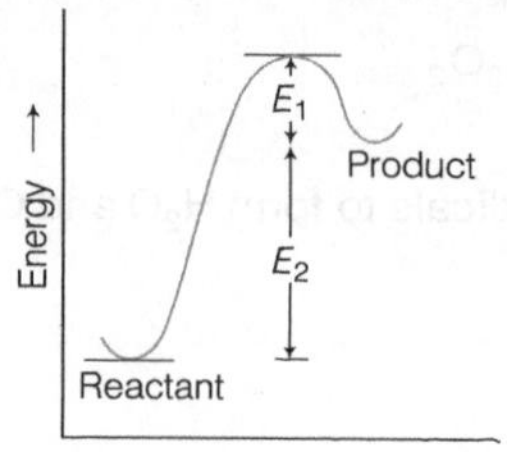

Activation energy of

a. forward reaction is $(E_1 + E_2)$ and product is less stable than the reactant
b. forward reaction is $(E_1 + E_2)$ and product is more stable than the reactant
c. backward reaction is E_2 and that of forward reaction is E_1
d. backward reaction is E_1 and product is more stable than the reactant

13 The rate of a chemical reaction doubles for every 10°C rise of temperature. If the temperature is raised by 50°C, the rate of the reaction increases by about **[AIEEE 2011]**

a. 10 times
b. 24 times
c. 32 times
d. 60 times

14 A reactant A forms two products B and C as given

$$A \xrightarrow{k_1} B, \text{ Activation energy } E_{a_1}$$

$$A \xrightarrow{k_2} C, \text{ Activation energy } E_{a_2}$$

If $E_{a_2} = 2E_{a_1}$, then k_1 and k_2 are related as **[AIEEE 2011]**

a. $k_1 = 2k_2 e^{Ea_2/RT}$
b. $k_1 = k_2\, e^{Ea_1/RT}$
c. $k_2 = k_1\, e^{Ea_2/RT}$
d. $k_1 = A'k_2 e^{Ea_1/RT}$

One or More than One Options Correct Type

Direction (Q. Nos. 15 and 16) This section contains 2 multiple choice questions. Each question has four choices (a), (b), (c) and (d), out of which ONE or MORE THAN ONE are correct.

15 Consider the endothermic reaction, $X \longrightarrow Y$

Heat of reaction = ΔH with E_f = activation energy of the forward reaction.
E_b = activation energy of the backward reaction.
Then,

a. $E_b > E_f$
b. $E_b = \Delta H + E_f$
c. $\Delta H + E_b = E_f$
d. $E_f > E_b$

16 Select the correct statement(s) about collision theory of reaction rates.

a. It explains that the effect of temperature on rate of reaction
b. It assumes that the reactants must be in correct orientation to react
c. It predicts that rate depends on the frequency at which reactants collide
d. It explains that the collisions having energy higher than the threshold value will give successful reactions

Working Space

Matching List Type

Direction *(Q. No. 17) Choices for the correct combination of elements from Column I and Column II are given as options (a), (b), (c) and (d), out of which one is correct.*

17 $A \longrightarrow$ Product, $-d[A]/dt = k[A]^n$. Rate constant (k) of a reaction is given by

$$\log_{10} k(\text{min}^{-1}) = 5 - \left(\frac{2 \times 10^3 K}{T}\right)$$

Match the given property in Column I with its related value in Column II and select answer from codes given:

	Column I		Column II
i.	Arrhenius constant (min^{-1})	p.	1.0
ii.	Order of the reaction	q.	2.0
iii.	At 500 K, rate of reaction at $[A]_0 = 0.2\text{M}$	r.	10^5
iv.	E_a (in kcal mol^{-1})	s.	9.2

Codes

	i	ii	iii	iv		i	ii	iii	iv
a.	p	q	r	s	b.	q	p	r	s
c.	r	p	q	s	d.	p	r	s	q

Comprehension Type

Direction *(Q. Nos. 18-21) This section contains 2 paragraphs, which describing theory, experiments, data, etc. Four questions related to the paragraphs have been given. Each question has only one correct answer among the four given options (a), (b), (c) and (d).*

Passage I

For the displacement reaction,

$$[Co(NH_3)_5Cl]^{2+}(aq) + H_2O(l) \longrightarrow [Co(NH_3)_5H_2O]^{3+} + Cl^-(aq)$$

The rate constant is given by

$$\log_{10} k(\text{min}^{-1}) = \frac{-11067K}{T} + 31.330$$

18 Rate constant at 25°C is

a. $5.10 \times 10^{-5}\ \text{s}^{-1}$ b. $2.60 \times 10^{-8}\ \text{s}^{-1}$

c. $3.06 \times 10^{-3}\ \text{s}^{-1}$ d. $2.60 \times 10^{-5}\ \text{s}^{-1}$

19 Energy of activation for this reaction is

a. $211.90\ \text{kJ mol}^{-1}$ b. $92.02\ \text{kJ mol}^{-1}$

c. $-211.90\ \text{kJ mol}^{-1}$ d. $-92.02\ \text{kJ mol}^{-1}$

Passage II

For the opposing reaction,

$$A + B \underset{k_2}{\overset{k_1}{\rightleftharpoons}} C + D$$

E_a (forward reaction) = $100\ \text{kJ mol}^{-1}$

Pre-exponential factor $A = 1.0 \times 10^{10}\ \text{M}^{-1}\text{s}^{-1}$

At equilibrium $[A] = 1.0\ \text{M}$

At 700° K $[B] = 2.0\text{M}$

$[C] = 5.0\text{M}$

$[D] = 4.0\text{A}$

Working Space

20 The value of k_1 and k_2 at 700K respectively are

a. $20M^{-1}s^{-1}$ and $2.0M^{-1}s^{-1}$
d. $345M^{-1}s^{-1}$ and $34.5M^{-1}s^{-1}$
c. $34.5M^{-1}s^{-1}$ and $3.45M^{-1}s^{-1}$
d. $200M^{-1}s^{-1}$ and $20M^{-1}s^{-1}$

21 The rate constant (k_1) for the forward reaction at 1000 K is

a. $5.97 \times 10^4\ M^{-1}\ min^{-1}$
b. $5.97 \times 10^2\ M^{-1}s^{-1}$
c. $1.00 \times 10^3\ M^{-1}s^{-1}$
d. $5.97 \times 10^4\ M^{-1}s^{-1}$

One Integer Value Correct Type

Direction *(Q. Nos. 22-24) This section contains 3 questions. When worked out will result in one integer from 0 to 9 (both inclusive).*

22 At room temperature (27°C) milk turns sour in about 64 h. In a refrigerator at −3°C, milk can be stored x times as long before it sours. The activation energy of the reaction that causes the souring of milk is 11.231 kcal mol^{-1}.
What is the value of x?

23 The Arrhenius equations for the rate constant of decomposition of methyl nitrite and ethyl nitrite are

$$k_1\,(s^{-1}) = 10^{13}\ \exp\left(-\frac{152.3\ kJ\ mol^{-1}}{RT}\right)$$

$$k_2\,(s^{-1}) = 10^{14}\ \exp\left(-\frac{157.7\ kJ\ mol^{-1}}{RT}\right)$$

Find the temperature (in °C), at which rate contants are equal.

24 For the following sequence of reactions

	Rate constant	E_a
$A \longrightarrow B$	k_1	10 kJ mol^{-1}
$B \longrightarrow C$	k_2	6 kJ mol^{-1}
$C \longrightarrow D$	k_2	7 kJ mol^{-1}

Observed rate constant is $k_3\left(\frac{k_1}{k_2}\right)^{\frac{1}{2}}$.

Thus, observed E_a is .... kJ mol^{-1}.

DPP-5 Collision Theory and Reaction Mechanism

Only One Option Correct Type

Working Space

Direction *(Q. Nos. 1-14) This section contains 14 multiple choice questions. Each question has four choices (a), (b), (c) and (d), out of which ONLY ONE is correct.*

1 A chemical reaction proceeds through the following steps :

***Step* I** $2A \rightleftharpoons X$ (Fast)

***Step* II** $X + B \rightleftharpoons Y$ (Slow)

***Step* III** $Y + B \rightleftharpoons$ Product (Fast)

The rate law for the overall reaction is

a. $k'[A]^2$ b. $k'[B]^2$ c. $k'[A][B]$ d. $k'[A]^2[B]$

2 The following reaction obey rate law

$$2NO_2(g) + O_3(g) \longrightarrow N_2O_5(g) + O_2(g)$$

$$-\frac{d[NO_2]}{dt} = k[NO_2][O_3]$$

It can follow mechanism

I. $NO_2 + O_3 \xrightarrow{\text{Slow}} NO_3 + O_2$, $NO_3 + NO_2 \xrightarrow{\text{Fast}} N_2O_5$

II. $O_3 \underset{}{\overset{\text{Fast}}{\rightleftharpoons}} O_2 + O$

$NO_2 + O \xrightarrow{\text{Slow}} NO_3$

$NO_3 + NO_2 \xrightarrow{\text{Fast}} N_2O_5$

Select the correct mechanism.

a. Only I b. Only II
c. Both I and II d. None of these

3 A chemical reaction is said to take place through the various stages with $\Delta G°$ values indicated by the graph:

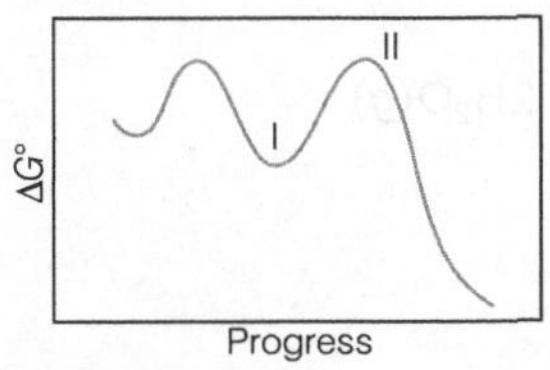

Stages I and II represent respectively

	I	II
a.	activated complex	activated complex
b.	reaction intermediate	activated complex
c.	activated complex	reaction intermediate
d.	reaction intermediate	reaction intermediate

Working Space

4 Following reaction takes place by mechanism:

$$NO_2 + CO \longrightarrow NO + CO_2$$

***Step* I** $NO_2 + NO_2 \xrightarrow{k_1, \text{Slow}} NO_3 + NO$

***Step* II** $NO_3 + CO \xrightarrow{k_2, \text{Fast}} NO_2 + CO_2$

Hence, $\left(\frac{dx}{dt}\right)$ for the given reaction is

a. $k_1[NO_2]^2 - k_2[NO_3][CO]$
b. $k_1[NO_2]^2 + k_2[NO_3][CO]$
c. $k_1[NO_2]^2$
d. $k_1[NO_2] + k_2[CO]$

5 For the reaction, $2NO_2(g) + F_2(g) \longrightarrow 2NO_2F(g)$
Steps are

I. $NO_2(g) + F_2(g) \underset{k_2}{\overset{k_1}{\rightleftharpoons}} NO_2F_2(g)$ (Fast)

II. $NO_2F_2(g) \xrightarrow{k_3} NO_2F(g) + F(g)$ (Slow)

III. $F(g) + NO_2(g) \xrightarrow{k_4} NO_2F(g)$ (Fast)

Select the correct statement(s).

a. Rate law is in agreement with net stoichiometry of the reaction
b. Rate law is in agreement with slow step
c. Rate law is in agreement with net stoichiometry of the reaction as well as with slow step
d. None of the above

6 For the reaction,

$$A + 2B \longrightarrow C + D$$

The rate law is $k[A][B]$.
Select the correct statement.

a. Following mechanism is consistent with the stoichiometry of the net reaction and with the rate law

$$A + B \longrightarrow I \text{ (Slow)}$$
$$I + B \longrightarrow C + D \text{ (Fast)}$$

b. Following mechanism is consistent with the stoichiometry of the net reaction but not with the rate law

$$2B \underset{k_2}{\overset{k_1}{\rightleftharpoons}} B_2 \text{ (Fast)}$$
$$A + B_2 \xrightarrow{k_3} C + D \text{ (Slow)}$$

c. Both (a) and (b)
d. None of the above

7 For the reaction, $2H_2(g) + 2NO(g) \longrightarrow N_2(g) + 2H_2O(g)$

Rate law is $\left(\frac{dx}{dt}\right) = k[NO]^2[H_2]$

Mechanism is given by

***Step* I** $2NO \rightleftharpoons N_2O_2$

***Step* II** $N_2O_2 + H_2 \longrightarrow N_2O + H_2O$

***Step* III** $N_2O + H_2 \longrightarrow N_2O + H_2O$

Rate law is true, if

a. step I is the slow step
b. step II is the slow step
c. step III is the slow step
d. step I and step II are the slow steps

Working Space

8 For the following S_N reaction, $(CH_3)_3CCl + H_2O \longrightarrow (CH_3)_3COH + HCl$

Rate-determining step is

a. $(CH_3)_3CCl \longrightarrow (CH_3)_3C^+ + Cl^-$

b. $(CH_3)_3CCl + H_2O \longrightarrow (CH_3)_3COH + HCl$

c. $(CH_3)_3C^+ + H_2O \longrightarrow (CH_3)_3COH + H^+$

d. $H^+ + Cl^- \longrightarrow HCl$

9 Reaction between nitric oxide and hydrogen takes place by the following mechanism:

***Step* I** $2NO(g) \underset{k_{-1}}{\overset{k_1}{\rightleftharpoons}} N_2O_2(g)$ (Fast)

***Step* II** $N_2O_2(g) + H_2(g) \xrightarrow[\text{Slow}]{k_2} N_2O(g) + H_2O(g)$

***Step* III** $N_2O(g) + H_2(g) \xrightarrow[\text{Fast}]{k_3} N_2(g) + H_2O(g)$

Rate of the overall reaction in terms of the concentration of the reactants is

a. $\frac{k_1 k_2}{k_{-1}}[NO]^2[H_2]$ b. $\frac{k_1 k_2}{k_{-1}}[NO][H_2]^2$ c. $k_1 k_2 [NO]^2[H_2]$ d. $\frac{k_2}{k_1}[NO]^2[H_2]$

10 Consider the following reaction,

$$Cl_2(aq) + H_2S(aq) \longrightarrow S(s) + 2H^+(aq) + 2Cl^-(aq)$$

The rate equation for this reaction is, rate $= k[Cl_2][H_2S]$

Which of these mechanisms is/are consistent with this rate equation?

I. $Cl_2 + H_2S \xrightarrow{\text{Slow}} H^+ + Cl^- + Cl^+ + HS^-$, $Cl^+ + HS^- \xrightarrow{\text{Fast}} H^+ + Cl^- + S$

II. $H_2S \overset{\text{Fast}}{\rightleftharpoons} H^+ + HS^-$ (equilibrium), $Cl_2 + HS^- \xrightarrow{\text{Slow}} 2Cl^- + H^+ + S$ **[AIEEE 2010]**

a. Only II b. Only I

c. Both (a) and (b) d. Neither (a) nor (b)

11 The following mechanism has been proposed for the reaction,

$$2NO(g) + Br_2(g) \longrightarrow 2NOBr(g)$$

It involves the following steps:

$$NO(g) + Br_2(g) \overset{\text{Fast}}{\rightleftharpoons} NOBr_2(g)$$

$$NOBr_2(g) + NO(g) \xrightarrow{\text{Slow}} 2NOBr(g)$$

The order of the reaction w.r.t. $NO(g)$ is

a. 1 b. 2 c. 3 d. zero

12 Consider the three following proposed mechanism for the overall equation

$$2A + B \longrightarrow 2C$$

Mechanism I $2A + B \xrightarrow{k} 2C$

Mechanism II $2A \rightleftharpoons A_2, K_{eq} = K$

$A_2 + B \xrightarrow{k} 2C$

Mechanism III $2A \underset{k_2}{\overset{k_1}{\rightleftharpoons}} A_2$

$A_2 + B \xrightarrow{k_3} 2C$

Which mechanism predicts a third order reaction?

a. Only I b. I and II

c. II and III d. I, II and III

Working Space

13 Consider the following proposed mechanism:

$$A_2 \rightleftharpoons 2A, \qquad K_C = 4$$
$$A + B \rightleftharpoons C, \qquad K_C' = 16$$
$$A_2 + C \longrightarrow D + A, \qquad k = 1 \times 10^{-2}$$

Thus, $\frac{d[D]}{dt}$ is

a. $1.6 \times 10^{-2} [A_2]^{3/2}[B]$

b. $6.4 \times 10^{-2} [A_2]^{3/2}[B]$

c. $0.32[A_2]^{3/2}[B]$

d. $1.6 \times 10^{-3} [A_2]^{3/2}[B]$

14 Based on the following steps:

***Step* I** $NO(g) + O_2(g) \underset{k_2}{\overset{k_1}{\rightleftharpoons}} NO_3(g)$ (Fast)

***Step* II** $NO_3(g) + NO(g) \xrightarrow{k_3} 2NO_2(g)$ (Slow)

Rate law is

a. $\frac{k_1 k_3}{k_2} [NO]^2 [O_2]$

b. $k_3 [NO_3] [NO]$

c. $\frac{k_1 k_2}{k_3} [NO] [O_2]$

d. $\frac{k_1}{k_2} [NO] [O_2]$

Comprehension Type

Direction *(Q. Nos. 15-20) This section contains 3 paragraphs, each describing theory, experiments, data, etc., Six questions related to the paragraphs have been given. Each question has ONLY ONE correct answer among the four given options (a), (b), (c) and (d).*

Passage I

The ozone in the earth's ozone layer decomposes according to the equation,

$$2O_3(g) \longrightarrow 3O_2(g)$$

It involves

***Step* I** $O_3(g) \overset{\text{Fast}}{\rightleftharpoons} O_2(g) + O(g), \qquad K_{eq} = 5.00$

***Step* II** $O_3(g) + O(g) \xrightarrow{\text{Slow}} 2O_2, \qquad k_1 = 2.00 \times 10^{-3}$

(Time is in seconds and concentration in mol L^{-1})

15 Rate of disappearance of O_3 is

a. $k[O_3]^2[O_2]$

b. $k[O_3]^2/[O_2]$

c. $k[O_3]$

d. $k[O_3][O_2]^2$

16 What is the rate at a time of 50% decomposition of O_3, when $[O_3]_0 = 1.0$ M?

a. 6.67×10^{-3} mol $L^{-1}s^{-1}$

b. 1.00×10^{-2} mol $L^{-1}s^{-1}$

c. 5.00×10^{-2} mol $L^{-1}s^{-1}$

d. 6.67×10^{-2} mol $L^{-1}s^{-1}$

Passage II

N_2O_5 decomposes according to the following equation:

$$N_2O_5 \xrightarrow{k} 2NO_2 + \frac{1}{2}O_2$$

The following mechanism has been given:

$$N_2O_5 \overset{K_{eq}}{\rightleftharpoons} NO_2 + NO_3$$
$$NO_2 + NO_3 \xrightarrow{k_1} NO_2 + O_2 + NO$$
$$NO + NO_3 \xrightarrow{k_2} 2NO_2$$

Working Space

17 Rate of formation of O_2 is given by

a. $\frac{d[O_2]}{dt} = k_1 K_{eq}[N_2O_5]$

b. $\frac{d[O_2]}{dt} = k_1 k_2 [NO_2][NO_3]$

c. $\frac{d[O_2]}{dt} = k_1 [N_2O_5]$

d. None of these

18 Half-life period is independent of the concentration of N_2O_5. Thus,

a. $\frac{-d[N_2O_5]}{dt} = k$

b. $\frac{-d[N_2O_5]}{dt} = k[N_2O_5]$

c. $\frac{-d[N_2O_5]}{dt} = K_{eq}[N_2O_5]$

d. $\frac{-d[N_2O_5]}{dt} = K_{eq}[NO_2][NO_3]$

Passage III

The reaction, $I^-(aq) + OCl^-(aq) \longrightarrow IO^-(aq) + Cl^-(aq)$

is believed to proceed through the following three step mechanism:

***Step* I** $OCl^- + H_2O \underset{k_{-1}}{\overset{k_1}{\rightleftharpoons}} HOCl + OH^-$ (Fast equilibrium)

***Step* II** $I^- + HOCl \overset{k_2}{\rightleftharpoons} HOI + Cl^-$ (Slow equilibrium)

***Step* III** $HOI + OH^- \overset{k_3}{\rightleftharpoons} H_2O + IO^-$ (Fast equilibrium)

19 Identify the false statement.

a. The molecularity of each step is 2
b. The limiting step is II
c. HOCl is only the reaction intermediate
d. The reaction is a complex reaction

20 What is the rate equation of the reaction?

a. $r = \left(\frac{k_1 k_2}{k_{-1}}\right)\frac{[I^-][OCl^-]}{[OH^-]}$

b. $r = k_2[I^-][HOCl]$

c. $r = \left(\frac{k_2 k_{-1}}{k_1}\right)\frac{[I^-][OCl^-]}{[OH^-]}$

d. None of these

One Integer Value Correct Type

Direction *(Q. Nos. 21 and 22) This section contains 2 questions. When worked out will result in one integer from 0 to 9 (both inclusive).*

21 Consider the following proposed mechanism:

$$2A \rightleftharpoons A_2, \quad (K_{eq} = k_1)$$

$$A + B \rightleftharpoons C \quad (K_{eq} = k_2)$$

$A_2 + C \longrightarrow D + 2A$, (rate constant $= k$) to describe the overall equations

$$A + C \longrightarrow D$$

What is the order predicted by the proposed mechanism?

22 For the overall reaction between A and B to yield C and D, two mechanisms are proposed.

I. $A + B \longrightarrow AB^* \longrightarrow C + D, k_1 = 2 \times 10^{-5} M^{-1}s^{-1}$

II. $A \longrightarrow A^* \longrightarrow E, k_2 = 1 \times 10^{-4} s^{-1}$

$E + B \longrightarrow C + D, \quad k_2' = 1 \times 10^{10} M^{-1}s^{-1}$

At what concentration of A and/or B will the inherent rates be equal?

SET - 1

Revisal Problems for JEE Main

Only One Option Correct Type

Working Space

Direction *(Q. Nos. 1-24) This section contains 24 multiple choice questions. Each question has four choices (a), (b), (c) and (d), out of which ONLY ONE is correct.*

1 At a given temperature $k_1 = k_2$ for the reaction,

$$A + B \underset{k_2}{\overset{k_1}{\rightleftharpoons}} C + D$$

For which case, a steady state is obtained?

	$[A]$	$[B]$	$[C]$	$[D]$
a.	0.1 M	0.2 M	0.3 M	0.4 M
b.	0.4 M	0.25 M	0.2 M	0.5 M
c.	0.2 M	0.2 M	0.3 M	0.2 M
d.	0.2 M	0.2 M	0.4 M	0.2 M

2 In the reaction, $P + Q \longrightarrow R + S$, the time taken for 75% reaction of P is twice the time taken for 50% reaction of P. The concentration of Q varies with the reaction time as shown in the figure. The overall order of the reaction is **[JEE Advanced 2013]**

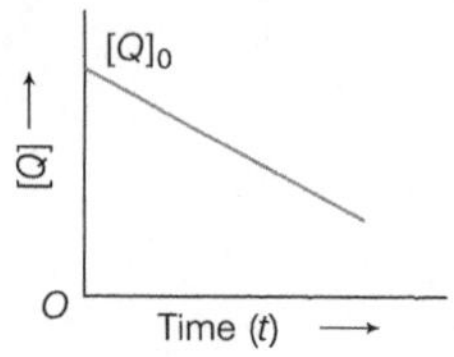

a. 2
b. 3
c. 0
d. 1

3 The time for half-life period of a certain reaction, $A \longrightarrow$ Products is 1 h. When the initial concentration of the reactant A is 2.0 mol L^{-1}, how much time does it take for its concentration to come from 0.50 to 0.25 mol L^{-1}, if it is a zero order reaction? **[AIEEE 2010]**

a. 4 h
b. 0.5 h
c. 0.25 h
d. 1 h

4 A radioactive substance decays 20% in 10 min. If at the start, there are 5×10^{20} atoms present, after what time will the number of atoms be reduced to 5×10^{17} atoms?

a. 310 min
b. 155 min
c. 100 min
d. 80 min

5 Rate constant of a reaction with a virus is $3.3 \times 10^{-4}\ s^{-1}$. Time required for a virus to become 75% inactivated is

a. 35 min
b. 70 min
c. 105 min
d. 17.5 min

6 For the first order reaction, $2\,N_2O_5(g) \longrightarrow 4\,NO_2(g) + O_2(g)$,

total pressure is 350 mm after 30 min and 500 mm after complete reaction. Thus, rate constant is

a. $2.31 \times 10^{-2}\ min^{-1}$
b. $2.31 \times 10^{-3}\ min^{-1}$
c. $4.62 \times 10^{-2}\ min^{-1}$
d. $1.73 \times 10^{-2}\ min^{-1}$

7 $t_{1/4}$ can be taken as the time taken for the concentration of a reactant to drop to $\frac{3}{4}$ of the initial value. If the rate constant for a first order reaction is k, then $t_{1/4}$ can be written as

a. $0.75/k$
b. $0.69/k$
c. $0.29/k$
d. $0.010/k$

Working Space

8 For the following first order gaseous phase reaction,

$$\underset{\text{Acetone}}{CH_3COCH_3(g)} \longrightarrow C_2H_6(g) + CO(g), \quad t_{50} = 810\text{ s}$$

time taken for acetone vapour pressure to drop to 0.30 atm from 2 atm is

a. 37 min b. 40 min c. 18.5 min d. 16 min

9 The rate constant for a first order reaction is 60 s^{-1}. How much time will it take to reduce the initial concentration of the reactant to its (1/16)th value?

a. 0.1732 s b. 0.0861 s
c. 0.3464 s d. 0.0462 s

10 The half-life of a reaction was doubled when initial concentration was doubled. Thus, order of the reaction is

a. zero b. one
c. two d. three

11 Rate law of the reaction $A \longrightarrow$ Product is
rate = $k[A]$
Graphically, it is represented as

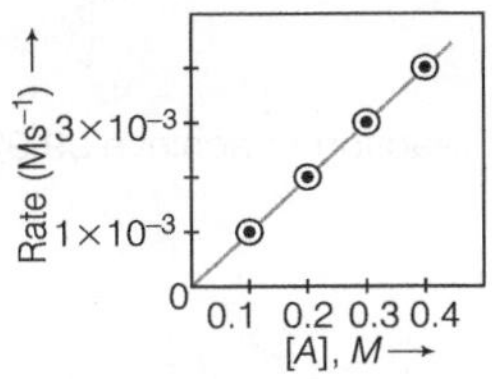

Hence, rate constant is

a. $3\times10^{-4}\, s^{-1}$ b. $1\times10^{-2}\, s^{-1}$
c. $3\times10^{-2}\, s^{-1}$ d. $1\times10^{-4}\, s^{-1}$

12 For the non-stoichiometric reaction, $2A + B \longrightarrow C + D$
the following kinetic data were obtained in three separate experiments, all at 298 K.

	Initial concentration of *A*	Initial concentration of *B*	Initial rate of formation of *C* (in mol $L^{-1}\, s^{-1}$)
I.	0.1 M	0.1 M	1.2×10^{-3}
II.	0.1 M	0.2 M	1.2×10^{-3}
III.	0.2 M	0.1 M	2.4×10^{-3}

The rate law for the formation of *C* is **[JEE Main 2014]**

a. $\frac{dC}{dt} = k[A][B]$ b. $\frac{dC}{dt} = k[A]^2[B]$
c. $\frac{dC}{dt} = k[A][B]^2$ d. $\frac{dC}{dt} = k[A]$

13 A hydrogenation reaction is carried out at 500 K.

$$CH_2 = CH_2 + H_2 \xrightarrow[\text{no catalyst}]{500\text{ K}} CH_3 - CH_3$$

Activation energy = E_a kJ mol^{-1}

$$CH_2 = CH_2 + H_2 \xrightarrow{\text{Pd, 400 K}} CH_3 - CH_3$$

Activation energy = $(E_a - 20)$ kJ mol^{-1}

If rate remains constant, then E_a is

a. 100 kJ mol^{-1} b. 120 kJ mol^{-1}
c. 80 kJ mol^{-1} d. 20 kJ mol^{-1}

14 Rate of a particular reaction increases by a factor 2 when the temperature is increased from 27°C to 37°C. Hence, activation energy of the reaction is

a. 12.9 kcal mol^{-1} b. 0.14 kcal mol^{-1}
c. 1.10 kcal mol^{-1} d. None of these

15 Interconversion of boat form to chair form of cyclohexane is first order (in both sides).

Working Space

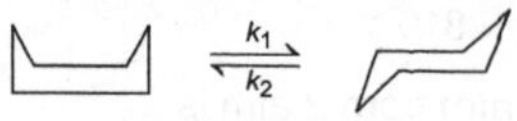

Equilibrium constant, $K_C = 10^4$

Also, energy of activation of conversion of chair to boat form is 42 kJ mol^{-1} in the following equation $k\,(s^{-1}) = 10^{12} e^{-E_a/RT}$

Observed reaction rate constant at 298 K is

a. 4.34×10^{4} s^{-1} b. 4.34×10^{8} s^{-1}

c. 4.34×10^{-4} s^{-1} d. 4.34×10^{-8} s^{-1}

16 In the reaction, $A \longrightarrow$ Product

$$\frac{-d[A]}{dt} = k\,[A]$$

if we start with $[A]_0 = 10\,\text{M}$, then after one natural life time, concentration of A decreased to

a. 5 M b. 2.5 M

c. $\frac{10}{e}$ d. $\frac{10}{e^2}$

17 Number of natural life times (T_{av}) required for a first order reaction to achieve 99.9% level of completion is

a. 2.3 b. 6.9

c. 9.2 d. 0.105

18 For the reaction, $A(g) \longrightarrow B(g) + C(g)$

rate law is $\frac{-d[A]}{dt} = k\,[A]$

At the start, pressure is 100 mm and after 10 min, pressure is 120 mm. Thus, rate constant (in min^{-1}) is

a. $\frac{2.303}{10}\log\frac{6}{5}$ b. $\frac{2.303}{10}\log 5$

c. $\frac{2.303}{10}\log\frac{5}{4}$ d. $\frac{2.303}{10}\log\frac{5}{6}$

19 $X \xrightarrow{HA} Y, \quad \frac{-d[X]}{dt} = 2 \times 10^{-3}\,\text{min}^{-1}[X],$

$X \xrightarrow{HB} Y, \quad \frac{-d[X]}{dt} = 1 \times 10^{-3}\,\text{min}^{-1}[X]$

HA and HB are two strong acids. Relative strength of strong acids is

a. 0.5 b. 0.002

c. 0.001 d. 2.0

20 $A \longrightarrow$ Product, $\frac{-d[A]}{dt} = k[A]$

Initial concentration of $A = 2\text{M}$ and after 10 min, reaction is 10% completed. Thus, half-life period is

a. 50 min b. 66.0 min

c. 69.3 min d. 6.93 min

21 The decomposition of N_2O to N_2 and O_2 follows rate law in gaseous phase as

$$N_2O \longrightarrow N_2 + \frac{1}{2}O_2$$

$$\frac{-d\,[N_2O]}{dt} = k\,[N_2O]$$

At 1000 K, half-life of the reaction is 3.58×10^3 min. Starting with initial pressure of N_2O as 2.20 atm, total pressure after one half-life is

a. 2.10 atm b. 3.15 atm

c. 2.625 atm d. None of these

Working Space

22 The rate of a first order reaction is 0.04 mol L^{-1} s^{-1} at 10 min and 0.03 mol L^{-1} s^{-1} at 20 min. Thus, half-life of the reaction is

a. 24.1 min
b. 34.8 min
c. 50.1 min
d. 48.2 min

23 For the non-equilibrium process,

$$A + B \longrightarrow \text{Products}$$

the rate is first order w.r.t. A and second order w.r.t. B.

When $[A] = 1.0$ M, $[B] = 1.0$ M, rate $= 1.0 \times 10^{-2}$ M s^{-1}

When 50% of each of reactant has been converted into products, rate is

a. 1.25×10^{-3} M s^{-1}
b. 1.0×10^{-2} M s^{-1}
c. 2.50×10^{-3} M s^{-1}
d. 2.0×10^{-2} M s^{-1}

24 For a reaction of the type $X \underset{k_2}{\overset{k_1}{\rightleftharpoons}} y$,

$[X]_0$ is the concentration of X at time $t = 0$ and $[X]$ is the concentration of X at time $t = t$

Thus, correct rate expression is

a. $\frac{-d[X]}{dt} = k_1[X]_0 - (k_1 + k_2)[X]$

b. $\frac{-d[X]}{dt} = (k_1 + k_2)[X] - k_2[X]_0$

c. $\frac{-d[X]}{dt} = (k_1 + k_2)[X]_0 - k_2[X]$

d. $\frac{-d[X]}{dt} = (k_1 - k_2)[X] - k_2[X]_0$

Statement Type

Direction *(Q. Nos. 25-27)* ***This section is based on Statement I and Statement II. Select the correct answer from the codes given below.***

a. Both Statement I and Statement II are correct and Statement II is the correct explanation of Statement I
b. Both Statement I and Statement II are correct but Statement II is not the correct explanation of Statement I
c. Statement I is correct but Statement II is incorrect
d. Statement II is correct but Statement I is incorrect

25 **Statement I** The rate of chemical reaction generally increases rapidly for a small temperature increase.

Statement II Increase in temperature increases fraction of molecules with energies in excess of activation energy.

26 **Statement I** For a certain reaction, large fraction of molecules has energy more than the threshold energy, yet the rate of reaction is very slow.

Statement II Collisions are only effective if they have proper orientation. In the absence of proper orientation rate becomes, slow.

27 **Statement I** Certain reactions are thermodynamically feasible but their rates are very slow.

Statement II Energy of activation may be high for such reactions.

Revisal Problems for JEE Advanced

Only One Option Correct Type

Direction (Q. Nos. 1-12) *This section contains 12 multiple choice questions. Each question has four choices (a), (b), (c) and (d), out of which ONLY ONE is correct.*

Working Space

1 For the first order gaseous phase reaction, $A \longrightarrow B + C$

initial pressure is 200 mm and after 20 min, total pressure is 250 mm. Thus, half-life period is

a. 48 min
b. 24 min
c. 12 min
d. 96 min

2 For the elementary reaction ($M \longrightarrow N$), the rate of disappearance of M increases by a factor of 8 upon doubling the concentration of M. The order of the reaction with respect to M is **[JEE Advanced 2014]**

a. 4
b. 3
c. 2
d. 1

3 For the reaction,

$$P + Q + R \longrightarrow S$$

experimental data for the measured initial rates is given below.

Expt.	(P)	(Q)	R	Initial rate (in Ms^{-1})
1.	0.2M	0.5M	0.4M	8.0×10^{-5}
2.	0.4M	0.5M	0.4M	3.2×10^{-4}
3.	0.4M	2.0M	0.4M	1.28×10^{-3}
4.	0.1M	0.25M	1.6M	4.0×10^{-8}

The order of the reaction with respect of P, Q and R respectively is

a. 2, 2, 1
b. 2, 1, 2
c. 2, 1, 1
d. 1, 1, 2

4 Reaction coordinate diagram for the reaction

$$COCl_2(g) \longrightarrow CO(g) + Cl_2(g), \qquad \Delta H = 103.1 \text{ kJ}$$

with $\log k\,(\text{min}^{-1}) = \dfrac{-11420\text{K}}{T} + 15.154$ is

a. 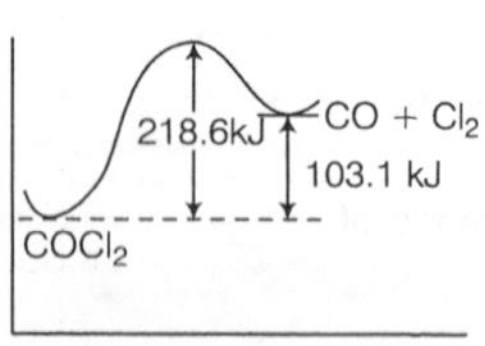

b. 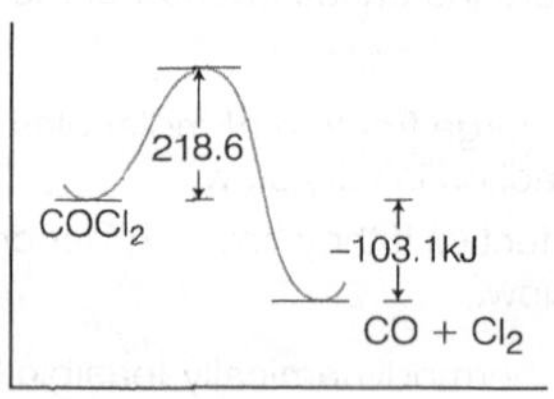

c. Both (a) and (b)
d. None of these

5 For a reaction, $k = 1.2\times10^{3}\ \text{mol}^{-1}\ \text{L s}^{-1}$ and $E_a = 2.0\times10^{2}\ \text{kJ mol}^{-1}$

When $T \longrightarrow \infty$, Arrhenius pre-exponential factor is

a. $2.0\times10^{2}\ \text{kJ mol}^{-1}$
b. $1.2\times 10^{3}\ \text{mol}^{-1}\ \text{L s}^{-1}$
c. $1.2\times 10^{3}\ \text{mol L}^{-1}\ \text{s}^{-1}$
d. $2.4\times 10^{5}\ \text{kJ mol}^{-1}\ \text{s}^{-1}$

6 Half-life period of the first order reaction is 100 days. After 144.3 days, concentration of the reactant is reduced to ...... of the original value.

a. $\dfrac{1}{e}$
b. $\dfrac{1}{e^2}$
c. 10%
d. 20%

Working Space

7 For the following chemical equation, a first order reaction, $A(g) \longrightarrow nB(g)$ total pressure of the system as a function of time is

a. $p_0[n + (1 - n)e^{-kt}]$
b. $np_0[1 - e^{-kt}]$
c. Both (a) and (b)
d. None of these

8 What function of [X], plotted against time will give a straight line for a second order reaction?

a. $[X]$
b. $[X]^2$
c. $\log[X]$
d. $\frac{1}{[X]}$

9 The rate constant at 298 K for the reaction of NH_4^+ and OH^+ to form NH_4OH is 4.0×10^{10} M^{-1} s^{-1}. Thus, rate constant for proton transfer to NH_3 is
(Given, $NH_4OH \rightleftharpoons NH_4^+ + OH^-, K_b = 1.8 \times 10^{-5}$)

a. 5.55×10^5 s^{-1}
b. 1.8×10^5 M^{-1} s^{-1}
c. 7.20×10^5 s^{-1}
d. 1.8×10^{-5} s^{-1}

10 For a reaction $A \longrightarrow$ Product

$$\frac{-d[A]}{dt} = k[A]^2$$

$[A]_0$ = Initial concentration, $[A]$ = Concentration of the reactant A after time t.
Then

a. $\frac{1}{[A]} = \frac{1}{[A]_0} + kt$
b. $[A] = \frac{[A]_0}{1 + k + [A]_0}$
c. $t_{50} = \frac{1}{k[A]_0}$
d. All of these

11 For a second order reaction,

$$A + B \longrightarrow \text{Product}$$

$$k = \frac{2.303}{t(a-b)} \log \frac{b(a-x)}{a(b-x)}$$

where $a \neq b$ and a and b are the initial concentrations of A and B. Reaction would be pseudo-unimolecular when

a. $a >> b$
b. $t \rightarrow \infty$
c. $t \rightarrow 0$
d. $x \rightarrow a$

12 A certain reaction, $A + B \longrightarrow$ Product
is first order w.r.t. each reactant with $k = 5.0 \times 10^{-3}$ M^{-1} s^{-1}.
Given, $[A]_0 = 0.1$ M, $[B]_0 = 6.0$ M
then concentration of A remaining after 100 s is

a. 0.005 M
b. 0.050 M
c. 0.020 M
d. 0.0025 M

One or More than One Options Correct Type

Direction (Q. Nos. 13-16) ***This section contains 4 multiple choice questions. Each question has four choices (a), (b), (c) and (d), out of which ONE or MORE THAN ONE are correct.***

13 Rate constant k varies with temperature as given by the equation

$$\log_{10} k\,(\text{min}^{-1}) = 5 - \frac{2000\,\text{K}}{T}$$

Select the correct statement(s) about this equation.

a. It follows first order kinetics
b. Pre-exponential factor is 5
c. Activation energy is 9.212 kcal mol^{-1}
d. Variation of $\log_{10} k$ with $\frac{1}{T}$ is linear

Working Space

14 Which is/are the graphical representation for the zeroth order reaction?

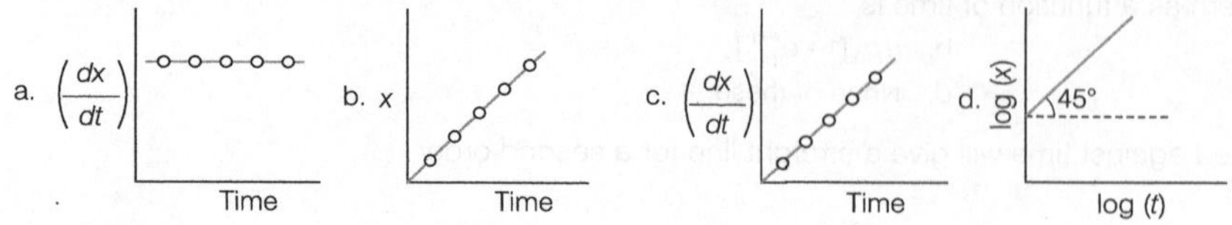

15 Select the correct statement(s) about half-life period.

a. It is proportional to initial concentration for zeroth order reaction
b. Average life = 1.44 half-life for first order reaction
c. Time of 75% reaction is thrice of half-life period in second order reaction
d. When rate constant is 0.0693 min^{-1}. 99.9% reaction takes place in 100 min

16 For the reaction,

$$3BrO^- \longrightarrow BrO_3^- + 2Br^-$$

In alkaline aqueous solution, the value of the second order (in BrO^-), rate constant at 80°C in the rate law for $-\frac{\Delta[BrO^-]}{\Delta t}$ was found to be 0.056 L mol^{-1} s^{-1}.

Select the correct alternate reaction.

a. for the rate law $\frac{\Delta[BrO_3^-]}{\Delta t}$ rate constant is 0.019 L mol^{-1} s^{-1}
b. for the rate law $\frac{\Delta[Br^-]}{\Delta t}$, rate constant is 0.037 L mol^{-1} s^{-1}
c. it is a redox reaction
d. it is a disproportionation reaction

Matching List Type

***Direction** (Q. No. 17) Choices for the correct combination of elements from Column I and Column II are given as options (a), (b), (c) and (d), out of which one is correct.*

17 Rate law of the *n*th order reaction is given by

$$\left(\frac{dx}{dt}\right) = k\,[A]^x$$

Match the different properties in Column I with their related values of half-life in Column II and select the correct answer from the codes given.

	Column I		Column II
i.	Graph between $\log\left(\frac{dx}{dt}\right)$ and $\log[A]$ is a straight line of slope 45°.	p.	Half-life increase as $[A]$ increases.
ii.	Graph between $\log\left(\frac{dx}{dt}\right)$ and $\log[A]$ is a straight line parallel to $\log[A]$ axis.	q.	Half-life period is independent of concentration.
iii.	If $[A] = [H^+]$ rate becomes 100 times on decreasing pH by 1 unit.	r.	Half-life is 100 min at 1M and 25 min at 2M.
iv.	On doubling rate becomes eight times.	s.	Half-life $\propto \frac{1}{[A]}$

Codes

	i	ii	iii	iv
a.	p	q	r	s
b.	q	p	r	s
c.	s	r	q	p
d.	q	p	s	r

Comprehension Type

Working Space

***Direction** (Q. Nos. 18-27) This section contains 5 paragraphs, each describing theory, experiments, data, etc. Ten questions related to the paragraphs have been given. Each question has only one correct answer among the four given options (a), (b), (c) and (d).*

Passage I

A 22.4 L flask contains 0.76 mm of O_3 at 298 K. It reacts with oxygen atoms as $O + O_3 \longrightarrow 2O_2, k = 1.5 \times 10^7$ L mol^{-1} s^{-1} at the rate of 0.15 mol L^{-1} s^{-1}.

18 What is the concentration of oxygen atoms needed to maintain this rate?

a. 2.45×10^{-4} mol L^{-1} b. 4.087×10^{-5} mol L^{-1}
c. 0.30 mol^{-1} d. None of these

19 What is the rate of formation oxygen (O_2) under these conditions?

a. 2.45×10^{-4} mol L^{-1} b. 4.087×10^{-5} mol L^{-1}
c. 0.30 mol L^{-1} d. None of these

Passage II

The complex ion of Fe^{2+} with the chelating agent diphyridyl (abbreviated as dipy) has been studied kinetically in both the forward and the reverse directions.

Given $Fe^{2+} + 3\ \text{dipy} \longrightarrow [Fe(dipy)_3]^{2+}$

$$\text{rate} = 2.44 \times 10^{13}\ L^3\ mol^{-3}\ s^{-1}\ [Fe^{2+}]\ [dipy]^3$$

and $[Fe(dipy)_3]^{2+} \longrightarrow Fe^{2+} + 3\ \text{dipy}$

$$\text{rate} = 1.22 \times 10^{-4}\ s^{-1}\ [Fe\ (dipy)_3]^{2+}$$

20 Half-life of the reverse reaction is

a. 1.578 h b. 2.28 h
c. 0.693 h d. 1.44 h

21 Stability constant in terms of *pk* is

a. 17.30 b. −17.30
c. 0.0578 d. − 0.0578

Passage III

$$A \underset{t_{50} = t_1}{\longrightarrow} \text{Product} \quad -\frac{d[A]}{dt} = k_1 [A]$$

$$B \underset{t_{50} = t_2}{\longrightarrow} \text{Product} \quad -\frac{d[B]}{dt} = k_2 [B]^2$$

Also, $t_1 = t_2$ and $[A]_0 = [B]_0 = a$

22 The ratio of the rates of these two reactions at the start of the reaction is

a. 2.303 log2 b. $t_1 k_1$
c. 0.693 d. All of these

23 The ratio of the rates of these two reaction after a lapse of one half-life is

a. $2t_1 k_1$ b. 1.386
c. Both a and b d. None of these

Passage IV

The rate constant for the decomposition of ethylene oxide into CH_4 and CO may be described by

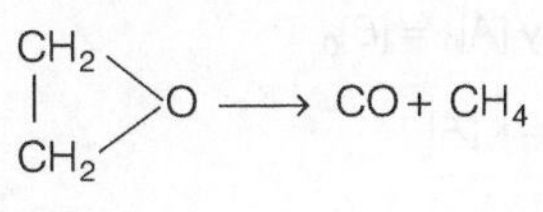

$$\log_{10} k\ (s^{-1}) = 14.34 - \frac{1.25 \times 10^4\ K}{T}$$

24 Energy of activation for this reaction is

a. 239.3 kJ mol^{-1} b. 239.3 kcal mol^{-1}
c. − 239.3 kJ mol^{-1} d. − 239.3 kcal mol^{-1}

25 What is the rate constant at 670 K?

a. 4.8×10^{-5} Ms^{-1} b. 4.8×10^{-5} s^{-1}

c. 4.8×10^{-2} Ms^{-1} d. 4.8×10^{-2} s^{-1}

Working Space

Passage V

The following first order reaction is conducted in $CCl_4(l)$ at 45°.

$$N_2O_5(g) \longrightarrow N_2O_4(g) + \frac{1}{2}O_2(g), \qquad k = 6.2 \times 10^{-4}\ s^{-1}$$

An 80.0 g sample of $N_2O_5(g)$ is dissolved in $CCl_4(l)$ and allowed to decompose at 450°C.

26 Time taken for the quantity of N_2O_5 to be reduced to 2.5 g is

a. 1.553 h b. 0.674 h

c. 3.577 h d. None of these

27 Volume of O_2 produced at STP at this point is

a. 0.7176 L b. 8.04 L

c. 16.08 L d. 4.02 L

One Integer Value Correct Type

Direction (Q. Nos. 28-37) *This section contains 10 questions. When worked out will result in one integer from 0 to 9 (both inclusive).*

28 The concentration of R in the reaction,

$$R \longrightarrow P$$

was measured as a function of time and the following data is obtained :

time (min)	$[R]/M$
0	1.0
0.05	0.75
0.12	0.40
0.18	0.10

What is the order of reaction? **[IIT JEE 2010]**

29 For the reaction system,

$$2NO(g) + O_2(g) \longrightarrow 2NO_2(g)$$

volume is suddenly reduced to half its value by increasing the pressure on it. If the reaction is of first order w.r.t. O_2 and the second order w.r.t. NO, the rate of reaction will increase ...... times of its initial value.

30 For the reaction, $A(g) + 2B(g) \longrightarrow C(g) + D(g)$

$$\frac{dx}{dt} = k[A][B]^2$$

Initial pressure of A and B are respectively 0.60 atm and 0.80 atm. At a time when pressure of C is 0.20 atm, ratio of initial rate and final rate is y. What is the value of y?

31 The rate constant for the reaction $CO_2 + OH^- \longrightarrow HCO_3^-$ in aqueous solution is 4.0×10^3 L mol^{-1} s^{-1} when

$$[CO_2] = 10^{-6}\ mol\ L^{-1} \text{ and } [OH^-] = 0.1\ mol\ L^{-1}.$$

Number of moles of HCO_3^- formed per second is $x \times 10^{-4}$ mol L^{-1}. What is the value of x?

32 A flask contains a mixture of compounds A and B initially $[A]_0 = [B]_0$

$$A \xrightarrow{t_{50}\,=\,54\text{ min}} \text{Product} \quad -\frac{d\,[A]}{dt} = k\,[A]$$

$$b \xrightarrow{t_{50}\,=\,18\text{ min}} \text{Product} \quad -\frac{d\,[B]}{dt} = k\,[B]$$

After 54 minutes, $[A] = x\,[B]$. What is the value of x?

Working Space

33 For the bimolecular reaction $O(g) + OH(g) \longrightarrow O_2(g) + H(g)$

The rate constant is $k = 1.3 \times 10^{10}$ L mol^{-1} s^{-1}.

Compare the initial value of $\frac{d}{dt}[O]$ for $[O] = [OH] = 1.0 \times 10^{-4}$ mol L^{-1} to that $[O] = 1.0 \times 10^{-4}$ mol L^{-1} and $[OH] = 2.0 \times 10^{-4}$ mol L^{-1}

34 A chemical reaction is known to be zeroth order with $k = 5 \times 10^{-8}$ mol s^{-1}. The reaction takes t_1 seconds to decrease concentration from 4×10^{-4} mol dm^{-3} to 2×10^{-4} mol dm^{-3} and t_2 seconds to decrease concentration from 2×10^{-4} mol dm^{-3} to 1×10^{-4} mol dm^{-3}. What is the ratio of t_1 to t_2?

35 T_{50} (Radium-226) = 1620 years, One year = 3.154×10^7 s

Calculate the weight of Ra-226 which has activity equal to 5 curie.

36 For the zero order reaction, sets I and II are given.

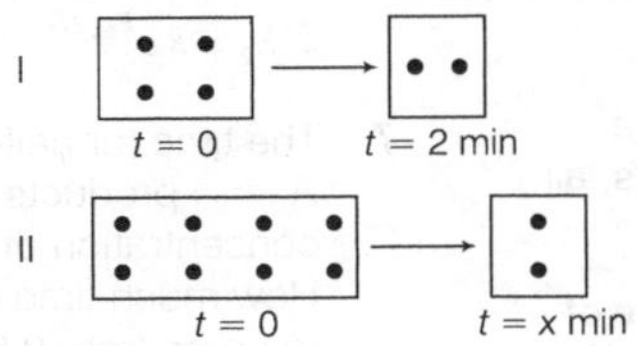

What is the value of x?

37 An organic compound undergoes first order decomposition. The time taken for its decomposition to $\frac{1}{8}$ and $\frac{1}{10}$ of its initial concentration are $t_{1/8}$ and $t_{1/10}$ respectively. What is the value of $\left(\frac{10\,t_8}{t_{10}}\right)$? **[IIT JEE 2012]**

JEE Main & AIEEE Archive

(Compilation of Last 13 Years' Questions)

1 Higher order (>3) reactions are rare due to

a. loss of active species on collision
b. low probability of simultaneous collision of all the reacting species
c. increase in entropy and activation energy as more molecules are involved
d. shifting of equilibrium towards reactants due to elastic collisions **(JEE Main 2015)**

2 For the non-stoichiometric reaction, $2A + B \longrightarrow C + D$, the following kinetic data were obtained in three separate experiments, all at 298 K.

Initial concentration (A)	Initial concentration (B)	Initial rate of formation of C (mol L^{-1} s^{-1})
0.1 M	0.1 M	1.2×10^{-3}
0.1 M	0.2 M	1.2×10^{-3}
0.2 M	0.1 M	2.4×10^{-3}

The rate law for the formation of C is **(JEE Main 2014)**

a. $\frac{dc}{dt} = k\,[A]\,[B]$
b. $\frac{dc}{dt} = k\,[A]^2\,[B]$
c. $\frac{dc}{dt} = k\,[A]\,[B]^2$
d. $\frac{dc}{dt} = k\,[A]$

3 The rate of a reaction double when its temperature changes from 300 K to 310 K. Activation energy of such a reaction will be (given, $R = 8.314$ JK^{-1} mol^{-1} and $\log 2 = 0.301$) **(JEE Main 2013)**

a. 53.6 kJ mol^{-1} b. 48.6 kJ mol^{-1}
c. 58.5 kJ mol^{-1} d. 60.5 kJ mol^{-1}

4 For a first order reaction, $(A) \longrightarrow$ products the concentration of A changes from 0.1 M to 0.025 M in 40 min. The rate of reaction when the concentration of A is 0.01 M is **(AIEEE 2012)**

a. 1.73×10^{-5} M / min
b. 3.47×10^{-4} M / min
c. 3.47×10^{-5} M / min
d. 1.73×10^{-4} M / min

5 The rate of a chemical reaction doubles for every 10°C rise of temperature. If the temperature is raised by 50°C, the rate of the reaction increases by about **(AIEEE 2011)**

a. 10 times b. 24 times
c. 32 times d. 64 times

6 A reactant (A) forms two products

$A \xrightarrow{k_1} B$, Activation energy E_{a_1}

$A \xrightarrow{k_2} C$, Activation energy E_{a_2}

If $E_{a_2} = 2\,E_{a_1}$, then k_1 and k_2 are related as **(AIEEE 2011)**

a. $k_1 = 2k_2 e^{E_{a_2}/RT}$ b. $k_1 = k_2 e^{E_{a_1}/RT}$
c. $k_2 = k_1 e^{E_{a_2}/RT}$ d. $k_1 = Ak_2 e^{E_{a_1}/RT}$

7 The time for half-life period of a certain reaction, $A \longrightarrow$ products is 1 h, when the initial concentration of the reactant 'A' is 2.0 mol L^{-1}. How much time does it take for its concentration to come from 0.50 to 0.25 mol L^{-1}, if it is a zero order reaction? **(AIEEE 2010)**

a. 4 h b. 0.5 h
c. 0.25 h d. 1 h

8 Consider the reaction,

$Cl_2(aq) + H_2S(aq) \longrightarrow S(s) + 2H^+(aq) + 2Cl^-(aq)$

The rate equation for this reaction is,

$$\text{rate} = k[Cl_2]\,[H_2S]$$

Which of these mechanisms is/are consistent with this rate equation? **(AIEEE 2010)**

I. $Cl_2 + H_2S \longrightarrow H^+ + Cl^- + Cl^+ + HS^-$ (Slow)
$Cl^+ + HS^- \longrightarrow H^+ + Cl^- + S$ (Fast)

II. $H_2S \rightleftharpoons H^+ + HS^-$ (Fast equilibrium)
$Cl_2 + HS^- \longrightarrow 2Cl^- + H^+ + S$ (Slow)

a. Only II b. Both I and II
c. Neither I nor II d. Only I

9 The half-life period of a first order chemical reaction is 6.93 min. The time required for the completion of 99% of the chemical reaction will be (take, $\log 2 = 0.301$) **(AIEEE 2009)**

a. 230.3 min b. 23.03 min
c. 46.06 min d. 460.6 min

10 For a reaction $\frac{1}{2}A \longrightarrow 2B$, rate of disappearance of A is related to the rate of appearance of B by the expression **(AIEEE 2008)**

a. $-\frac{d[A]}{dt} = \frac{1}{2} \cdot \frac{d[B]}{dt}$
b. $-\frac{d[A]}{dt} = \frac{1}{4} \cdot \frac{d[B]}{dt}$
c. $-\frac{d[A]}{dt} = \frac{d[B]}{dt}$
d. $-\frac{d[A]}{dt} = 4 \cdot \frac{d[B]}{dt}$

11 Consider the reaction, $2A + B \rightarrow$ product. When concentration of B alone was doubled, the half-life did not change. When the concentration of A alone was doubled, the rate increased by two times. The unit of rate constant for this reaction is **(AIEEE 2007)**

a. L mol^{-1} s^{-1}
b. no unit
c. mol L^{-1} s^{-1}
d. s^{-1}

12 The energies of activation for forward and reverse reactions for $A_2 + B_2 \rightleftharpoons 2AB$ are 180 kJ mol^{-1} and 200 kJ mol^{-1}, respectively. The presence of a catalyst lowers the activation energy of both (forward and reverse) reactions by 100 kJ mol^{-1}. The enthalpy change of the reaction ($A_2 + B_2 \rightarrow 2AB$) in the presence of catalyst will be (in kJ mol^{-1}) **(AIEEE 2007)**

a. 300 b. 120
c. 280 d. 20

13 A radioactive element gets spilled over the floor of a room. Its half-life period is 30 days. If the initial activity is ten times the permissible value, after how many days will it be safe to enter the room? **(AIEEE 2007)**

a. 1000 days b. 300 days
c. 10 days d. 100 days

14 The following mechanism has been proposed for the reaction of NO with Br_2 to form NOBr

$$NO(g) + Br_2(g) \rightleftharpoons NOBr_2(g)$$
$$NOBr_2(g) + NO(g) \longrightarrow 2NOBr(g)$$

If the second step is the rate determining step, the order of the reaction with respect to NO (g) is **(AIEEE 2006)**

a. 1 b. 0
c. 3 d. 2

15 Rate of a reaction can be expressed by Arrhenius equation as $k = Ae^{-E/RT}$
In this equation, E represents **(AIEEE 2006)**

a. the energy above which all the colliding molecules will react
b. the energy below which colliding molecules will not react
c. the total energy of the reacting molecules at a temperature, T
d. the fraction of molecules with energy greater than the activation energy of the reaction

16 A reaction was found to be second order with respect to the concentration of carbon monoxide. If the concentration of carbon monoxide is doubled, with everything else kept the same, the rate of reaction will **(AIEEE 2006)**

a. remain unchanged
b. triple
c. increase by a factor of 4
d. double

17 A reaction involving two different reactants can never be **(AIEEE 2005)**

a. bimolecular reaction
b. second order reaction
c. first order reaction
d. unimolecular reaction

18 $t_{1/4}$ can be taken as the time taken for the concentration of a reactant to drop to $\frac{3}{4}$ of its initial value. If the rate constant for a first order reaction is k, the $t_{1/4}$ can be written as **(AIEEE 2005)**

a. 0.75 / k
b. 0.69 / k
c. 0.29 / k
d. 0.10 / k

19 Consider an endothermic reaction $X \rightarrow Y$ with the activation energies E_b and E_f for the backward and forward reactions respectively. In general **(AIEEE 2005)**

a. there is no definite relation between E_b and E_f
b. $E_b = E_f$
c. $E_b > E_f$
d. $E_b < E_f$

20 The rate equation for the reaction $2A + B \rightarrow C$ is found to be rate = k [A] [B]
The correct statement in relation to this reaction is that the **(AIEEE 2004)**

a. unit of k must be s^{-1}
b. $t_{1/2}$ is a constant
c. rate of formation of C is twice the rate of disappearance of A
d. unit of k is L mol^{-1} s^{-1}

21 In a first order reaction, the concentration of the reactant, decreases from 0.8 M to 0.4 M in 15 min. The time taken for the concentration to change from 0.1 M to 0.025 M is **(AIEEE 2004)**

a. 30 min b. 15 min
c. 7.5 min d. 60 min

22 In respect of the equation $k = Ae^{-E_a/RT}$ in chemical kinetics, which one of the following statements is correct? **(AIEEE 2003)**

a. k is equilibrium constant
b. A is adsorption factor
c. E_a is energy of activation
d. R is Rydberg constant

23 The rate law for a reaction between the substances A and B is given by rate = $k\,[A]^n[B]^m$. On doubling the concentration of A and halving the concentration of B, the ratio of the new rate to the earlier rate of the reaction will be as **(AIEEE 2003)**

a. $\frac{1}{2^{m+n}}$ b. $(m + n)$
c. $(n - m)$ d. $2^{(n-m)}$

JEE Archive

24 For the reaction system,

$$2NO(g) + O_2(g) \longrightarrow 2NO_2(g)$$

volume is suddenly reduced to half its value by increasing the pressure on it. If the reaction is of first order with respect to O_2 and second order with respect to NO, the rate of reaction will **(AIEEE 2003)**

a. diminish to one-fourth of its initial value
b. diminish to one-eighth of its initial value
c. increase to eight times of its initial value
d. increase to four times of its initial value

25 For the reaction, $H_2 + I_2 \longrightarrow 2HI$, the differential rate law is **(AIEEE 2002)**

a. $-\frac{d[H_2]}{dt} = -\frac{d[I_2]}{dt} = 2\frac{d[HI]}{dt}$

b. $-2\frac{d[H_2]}{dt} = -2\frac{d[I_2]}{dt} = \frac{d[HI]}{dt}$

c. $-\frac{d[H_2]}{dt} = -\frac{d[I_2]}{dt} = \frac{d[HI]}{dt}$

d. $-\frac{d[H_2]}{2dt} = -\frac{d[I_2]}{2dt} = \frac{d[HI]}{dt}$

26 For a reaction $A + 2B \longrightarrow C$, rate is given by $+\frac{d[C]}{dt} = k[A][B]$, hence the order of the reaction is **(AIEEE 2002)**

a. 3 b. 2
c. 1 d. 0

27 Rate constant, k of the first order reaction when initial concentration (C_0) and concentration (C_t) at time t, is given by equation $kt = \log C_0 - \log C_t$ Graph is a straight line, if we plot **(AIEEE 2002)**

a. t vs $\log C_0$ b. t vs $\log C_t$
c. t^{-1} vs $\log C_t$ d. $\log C_0$ vs $\log C_t$

28 Consider the following two reactions,

$$A \longrightarrow \text{Product}, \quad -\frac{d[A]}{dt} = k_1[A]^0$$

$$B \longrightarrow \text{Product}, \quad -\frac{d[B]}{dt} = k_2[B]$$

k_1 and k_2 are expressed in terms of molarity (mol L^{-1}) and time (s^{-1}) as **(AIEEE 2002)**

a. s^{-1}, $M\,s^{-1}\,L^{-1}$ b. $M\,s^{-1}$, $M\,s^{-1}$
c. s^{-1}, $M^{-1}\,s^{-1}$ d. $M\,s^{-1}$, s^{-1}

JEE Advanced & IITJEE Archive

(Compilation of Last 10 Years' Questions)

1 For the elementary reaction $M \to N$, the rate of disappearance of M increases by a factor of 8 upon doubling the concentration of M. The order of the reaction with respect to M is

(2014 Adv., Only One Option Correct Type)

a. 4   b. 3
c. 2   d. 1

2 In the reaction, $P + Q \longrightarrow R + S$
the time taken for 75% reaction of P is twice the time taken for 50% reaction of P. The concentration of Q varies with reaction time as shown in the figure. The overall order of the reaction is **(2013 Adv., Only One Option Correct Type)**

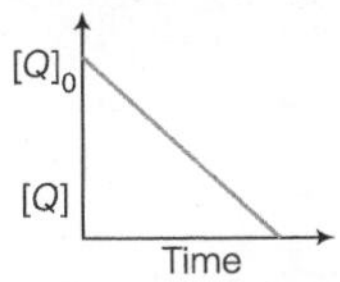

a. 2   b. 3
c. 0   d. 1

3 An organic compound undergoes first order decomposition. The time taken for its decomposition to 1/8 and 1/10 of its initial concentration are $t_{1/8}$ and $t_{1/10}$ respectively. What is the value of $\frac{[t_{1/8}]}{[t_{1/10}]} \times 10$? ($\log_{10} 2 = 0.3$)

(2012, Integer Type)

4 For the first order reaction,
$2N_2O_5(g) \longrightarrow 4NO_2(g) + O_2(g)$

(2011, One or More than One Options Correct Type)

a. the concentration of the reactant decreases exponentially with time
b. the half-life of the reaction decreases with increasing temperature
c. the half-life of the reaction depends on the initial concentration of the reactant
d. the reaction proceeds of 99.6% completion in eight half-life duration

5 Plots showing the variation of the rate constant (k) with temperature (T) are given below. The plot that follows Arrhenius equation is

(2010, Only One Option Correct Type)

a. 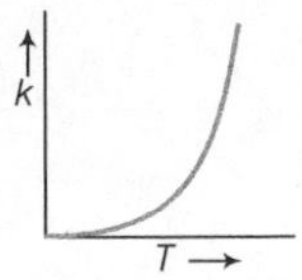

b. 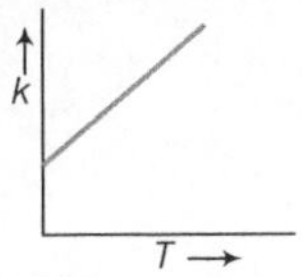

c. 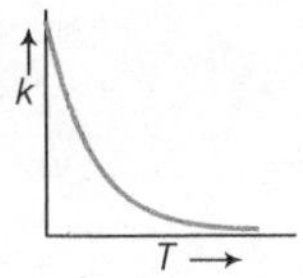

d. 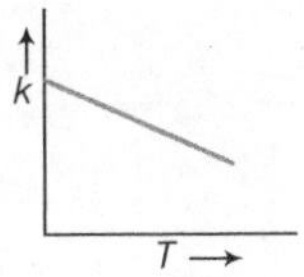

6 The concentration of R in the reaction $R \longrightarrow P$ was measured as a function of time and the following data is obtained :

$[R]$ (molar)	1.0	0.75	0.40	0.10
t (min)	0.0	0.05	0.12	0.18

The order of the reaction is **(2010,Integer Type)**

7 For a first order reaction $A \to P$, the temperature (T) dependent rate constant (k) was found to follow the equation

$$\log k = \frac{2000}{T} + 6.0$$

The pre-exponential factor A and the activation energy E_a, respectively, are

(2009, Only One Option Correct Type)

a. $1.0 \times 10^6\ s^{-1}$ and 9.2 kJ mol^{-1}
b. 6.0 s^{-1} and 16.6 kJ mol^{-1}
c. $1.0 \times 10^6\ s^{-1}$ and 16.6 kJ mol^{-1}
d. $1.0 \times 10^6\ s^{-1}$ and 38.3 kJ mol^{-1}

8 Under the same reaction conditions, initial concentration of 1.386 mol dm^{-3} of a substance becomes half in 40 s and 20 s through first order and zero order kinetics respectively. Ratio $\left(\frac{k_1}{k_0}\right)$ of the rate constants for first order (k_1) and zero order (k_0) of the reaction is

(2008, Only One Option Correct Type)

a. 0.5 $mol^{-1}\ dm^3$ b. 1.0 mol dm^{-3}
c. 1.5 mol dm^{-3} d. 2.0 $mol^{-1}\ dm^3$

9 Consider a reaction $aG + bH \to$ products. When concentration of both the reactants G and H is doubled, the rate increases by eight times. However, when concentration of G is doubled keeping the concentration of H fixed, the rate is doubled. The overall order of the reaction is

(2007, Only One Option Correct Type)

a. 0   b. 1
c. 2   d. 3

10 Which one of the following statements is incorrect about order of reaction?

(2005, Only One Option Correct Type)

a. Order of reaction is determined experimentally
b. Order of reaction is equal to sum of the power of concentration terms in differential rate law
c. It is not affected with stoichiometric coefficient of the reactants
d. Order cannot be fractional

Answers *with* Explanations

Answers

1. Solid State

DPP-1 Classification and Characteristics

1. *c* 2. *d* 3. *a* 4. *a* 5. *d* 6. *d* 7. *b* 8. *a* 9. *d* 10. *a*
11. *d* 12. *c* 13. *a,b,c* 14. *a,b,c* 15. *b,c,d* 16. *a,b,c,d* 17. *a,b,d* 18. *b,c*
19. *a,b,c,d* 20. *a* 21. *b* 22. *a* 23. *b* 24. *c* 25. *d* 26. *a* 27. 3
28. 4 29. 5 30. 6

DPP-2 Unit Cells, Lattices and Bragg's Equation

1. *d* 2. *c* 3. *c* 4. *c* 5. *d* 6. *c* 7. *a* 8. *b* 9. *c* 10. *a*
11. *a,c,d* 12. *a,d* 13. *a,b,c,d* 14. *b* 15. *c* 16. 6 17. 7 18. 1 19. 1
20. 3

DPP-3 Packing in Crystals

For (1), (2), (3) Refer detailed solution 4. *d* 5. *d* 6. *c* 7. *c* 8. *b* 9. *a*
10. *c* 11. *a* 12. *b* 13. *a* 14. *a* 15. *b* 16. *a* 17. *a,b,c* 18. *a,b*
19. *a, d* 20. *a* 21. *c* 22. *a* 23. *b* 24. 7 25. 4 26. 4

DPP-4 Imperfection in Crystals

1. *a* 2. *b* 3. *a* 4. *a* 5. *d* 6. *d* 7. *a* 8. *b* 9. *c* 10. *a*
11. *b* 12. *a* 13. *a* 14. *a,b* 15. *a,c* 16. *a,b,c* 17. *a,b,c,d* 18. *a,b,c,d*
19. *a,b* 20. *b* 21. *c* 22. *b* 23. *a* 24. *c* 25. 4

DPP-5 Magnetic and Electric Properties of Solids and Semiconductors

1. *d* 2. *d* 3. *b* 4. *c* 5. *d* 6. *a* 7. *c* 8. *b* 9. *a* 10. *c*
11. *b* 12. *b* 13. *a,b* 14. *a,b,c,d* 15. *a,b,d* 16. *a,b* 17. *a,c,d* 18. *c* 19. *c*
20. *b* 21. 5 22. 5 23. 7 24. 9

Revisal Problems (JEE Main)

1. *d* 2. *a* 3. *a* 4. *a* 5. *c* 6. *b* 7. *d* 8. *b* 9. *a* 10. *c*
11. *a* 12. *b* 13. *c* 14. *a* 15. *a* 16. *c* 17. *a* 18. *a* 19. *b* 20. *a*

Revisal Problems (JEE Advanced)

1. *d* 2. *a* 3. *a* 4. *d* 5. *c* 6. *a* 7. *c* 8. *d* 9. *a* 10. *a*
11. *b* 12. *c* 13. *c* 14. *c* 15. *b* 16. *a* 17. *b,c* 18. *a* 19. *c* 20. *a*
21. *d* 22. *c* 23. *a* 24. *a* 25. *a* 26. 6 27. 8 28. 8 29. 4 30. 7
31. 6 32. 4 33. 7

JEE Main & AIEEE Archive

1. *b* 2. *c* 3. *b* 4. *b* 5. *d* 6. *d* 7. *c* 8. *d* 9. *b* 10. *a*
11. *d* 12. *b* 13. *b* 14. *a* 15. *c*

JEE Advanced & IIT JEE Archive

1. *a* 2. *b* 3. 8 4. 7 5. *d* 6. *a* 7. 217 pm 8. 117 pm
9. *b*

2. Solutions

DPP-1 Different Methods of Expressing Concentration of Solutions

1. 5.96×10^{-3} 2. *a* 3. *b* 4. *b* 5. *b* 6. *c* 7. *b* 8. *a* 9. *c*
10. *a* 11. *a* 12. *b* 13. *b* 14. *a* 15. *a* 16. *c* 17. *b* 18. *d* 19. *c*
20. *a* 21. *c* 22. *c* 23. *b* 24. *d* 25. *b* 26. *a,b,c* 27. *a,b* 28. *b* 29. *c*
30. 8

DPP-2 Henry's Law

1. 0.123 2. *b* 3. *a* 4. *b* 5. *a* 6. *b* 7. *a* 8. *d* 9. *a* 10. *a*
11. *a* 12. *b* 13. *a,b,c* 14. *a,b,c* 15. *b* 16. *b* 17. *b* 18. *a* 19. *d* 20. *a*
21. *b* 22. 8 23. 2 24. 9

DPP-3 Colligative Properties-I (Lowering of Vapour Pressure)

1. 0.094 2. 39.22 3. 96.8 4. *d* 5. *c* 6. *a* 7. *d* 8. *c* 9. *b* 10. *d*
11. *b* 12. *a* 13. *c* 14. *a* 15. *a* 16. *b,c,d* 17. *a,c,d* 18. *b* 19. *a* 20. *a*
21. *b* 22. *c* 23. *c* 24. 4 25. 1 26. 3

DPP-4 Colligative Properties-II (Depression in Freezing Point)

1. *d* 2. *a* 3. *a* 4. *b* 5. *d* 6. *a* 7. *a* 8. *a* 9. *c* 10. *a*
11. *a* 12. *b* 13. *a,d* 14. *a,b,c,d* 15. *d* 16. *d* 17. *c* 18. 6 19. 4 20. 6

DPP-5 Colligative Properties-III (Elevation in Boiling Point)

1. *c* 2. *b* 3. *a* 4. *a* 5. *c* 6. *c* 7. *a* 8. *b* 9. *b* 10. *a*
11. *b* 12. *d* 13. *d* 14. *a,c,d* 15. *a,b,c,d* 16. *a* 17. *b* 18. *c* 19. 8 20. 4

DPP-6 Colligative Properties-IV (Osmotic Pressure)

1. *c* 2. *c* 3. *d* 4. *d* 5. *d* 6. *d* 7. *a* 8. *c* 9. *c* 10. *c*
11. *d* 12. *b* 13. *a* 14. *d* 15. *b* 16. *a* 17. *a* 18. *a* 19. *d* 20. *a*
21. *b* 22. 8 23. 6 24. 8

DPP-7 van't Hoff Factor

1. 36.945 2. (i) 1.01 (ii) 0.01 (iii) 3 (iv) 1.0×10^{-5} 3. *a* 4. *a* 5. *a* 6. *b* 7. *a*
8. *d* 9. *d* 10. *b,c,d* 11. *a,c* 12. *b* 13. *a* 14. *d* 15. *c* 16. *c* 17. *a*
18. *b* 19. *b* 20. *c* 21. 1 22. 8 23. 2 24. 2 25. 3

Revisal Problems (JEE Main)

1. *c* 2. *a* 3. *d* 4. *d* 5. *a* 6. *d* 7. *b* 8. *b* 9. *a* 10. *a*
11. *a* 12. *c* 13. *b* 14. *c* 15. *c* 16. *a* 17. *a* 18. *a* 19. *a* 20. *d*
21. *a* 22. *c* 23. *c* 24. *b*

Revisal Problems (JEE Advanced)

1. *d* 2. *b* 3. *b* 4. *b* 5. *d* 6. *a* 7. *a* 8. *b* 9. *a* 10. *c*
11. *d* 12. *d* 13. *a* 14. *d* 15. *b* 16. *c* 17. *d* 18. *a,c* 19. *b,d*
20. *a,b,c,d* 21. *a,c* 22. *b* 23. *a* 24. *b* 25. *d* 26. *d* 27. *a* 28. 4
29. 2 30. 2 31. 3 32. *d*

JEE Main & AIEEE Archive

1. *c* 2. *a* 3. *b* 4. *c* 5. *a* 6. *a* 7. *c* 8. *b* 9. *b* 10. *a*
11. *c* 12. *b* 13. *d* 14. *d* 15. *c* 16. *a* 17. *d* 18. *c* 19. *d* 20. *b*
21. *b* 22. *d* 23. *b* 24. *b* 25. *c* 26. *b* 27. *a* 28. *d* 29. *a* 30. *a*
31. *a*

JEE Advanced & IIT JEE Archive

1. *a* 2. *a* 3. *a* 4. *d* 5. *a* 6. *b* 7. *a* 8. 100 % 9. *c*

3. Electrochemistry

DPP-1 Conduction

1. $33.33\ m^{-1}, 0.025 S^{-1}, 0.833\ Sm^{-1}$ 2. $0.076\ Sm^{-1}$ 3. $240\ Scm^2\ mol^{-1}, 240\ Scm^2 equiv^{-1}$
4. $300\ Scm^2\ mol^{-1}, 150\ Scm^2 equiv^{-1}$ 5. 1.856×10^{-5} 6. *b* 7. *b* 8. *b* 9. *a*
10. *b* 11. *b* 12. *d* 13. *a* 14. *d* 15. *c* 16. *c* 17. *c* 18. *d* 19. *a*
20. *b* 21. *c* 22. *c* 23. *d* 24. *a* 25. *a,c* 26. *a,c,d* 27. *a,c,d* 28. *a* 29. *d*
30. *b* 31. *a* 32. *a* 33. *d* 34. 5 35. 7

DPP-2 Electrochemical Series

1. $1.65 \times 10^{-9} M$ 2. *a* 3. *d* 4. *c* 5. *b* 6. *d* 7. *a* 8. *c* 9. *b*
10. *a* 11. *d* 12. *d* 13. *a* 14. *c* 15. *a* 16. *a* 17. *b* 18. *d* 19. *a*
20. *b,d* 21. *a,c* 22. *b,c,d* 23. *a* 24. *a* 25. *b* 26. *b* 27. *a* 28. *b* 29. *d*
30. 4

DPP-3 Reference Half-Cells and Nernst Equations

1. 0.34 V, 1.07 V, 1.13 V, $E_{cell} > 0$ 2. −0.781, 0.65V, −1.26 V 3. 0.299V, 0.299 V
4. 0.89 V, −0.05 V 5. *c* 6. *b* 7. *c* 8. *c* 9. *b* 10. *b* 11. *a* 12. *d*
13. *d* 14. *d* 15. *c* 16. *c* 17. *b* 18. *b* 19. *a* 20. *b* 21. *b* 22. *c*
23. *a,b,c* 24. *a,b,c* 25. *c,d* 26. *a,b,c,d* 27. *a* 28. *a* 29. *a* 30. *d* 31. 2 32. 3
33. 1

DPP-4 Thermodynamics of Electrochemistry

1. 3.17×10^{34} 2. *a* 3. *a* 4. *c* 5. *d* 6. *a* 7. *b* 8. *b* 9. *c*
10. *c* 11. *a* 12. *a* 13. *c* 14. *d* 15. *d* 16. *c* 17. *d* 18. *c* 19. *b*
20. *c* 21. *a* 22. *b* 23. 6 24. 3 25. 4

DPP-5 Applications of EMF-Measurements

1. 8.6 2. − 16.07 3. *b* 4. *d* 5. *c* 6. *a* 7. *b* 8. *b* 9. *d* 10. *a*
11. *b* 12. *c* 13. *a* 14. *b* 15. *d* 16. *b* 17. *d* 18. *b* 19. *a* 20. 4
21. 1

DPP-6 Battery and Corrosion

1. *d* 2. *d* 3. *b* 4. *a* 5. *d* 6. *a* 7. *b* 8. *a* 9. *b* 10. *b*
11. *a* 12. *d* 13. *d* 14. *c* 15. *b* 16. *d* 17. *a* 18. *a,b,c* 19. *a,d* 20. *b,c*
21. *d* 22. *c* 23. *c* 24. *b* 25. 8

DPP-7 Electrolysis and Faraday's Laws

1. $1.606 \times 10^{-19} C$ 2. $1.603 \times 10^{-19} C$, 6000 C, 92.0 kJ 3. *d* 4. *b* 5. *b* 6. *b* 7. *c*
8. *a* 9. *c* 10. *b* 11. *d* 12. *d* 13. *d* 14. *d* 15. *a* 16. *a* 17. *b*
18. *d* 19. *a,c,d* 20. *a,b,c* 21. *a,b,c,d* 22. *a,c,d* 23. *b,c* 24. *a* 25. *b* 26. *a*
27. *d* 28. *c* 29. 2 30. 5 31. 9 32. 8 33. 2 34. 3

Revisal Problems (JEE Main)

1. *b* 2. *a* 3. *a* 4. *c* 5. *d* 6. *b* 7. *b* 8. *c* 9. *d* 10. *c*
11. *d* 12. *c* 13. *c* 14. *a* 15. *a* 16. *b* 17. *a* 18. *c* 19. *c* 20. *d*
21. *c* 22. *d* 23. *d* 24. *a* 25. *b* 26. *a* 27. *d* 28. *a* 29. *c* 30. *a*
31. *b* 32. *d* 33. *a* 34. *a* 35. *d* 36. *a* 37. *c* 38. *c* 39. *d* 40. *d*

Revisal Problems (JEE Advanced)

1. *d* 2. *c* 3. *a* 4. *d* 5. *b* 6. *d* 7. *d* 8. *c* 9. *a* 10. *c*
11. *c* 12. *a* 13. *a* 14. *a* 15. *b,c,d* 16. *a,b,c* 17. *c* 18. *a,b,d* 19. *a,b,c* 20. *b*
21. *c* 22. *a* 23. *a* 24. *b* 25. *d* 26. *a* 27. 1 28. 7 29. 8 30. 3

JEE Main & AIEEE Archive

1. c	2. a	3. c	4. a	5. d	6. d	7. c	8. b	9. c	10. c
11. a	12. a	13. c	14. d	15. d	16. d	17. b	18. d	19. a	20. a
21. c	22. c	23. c	24. b	25. a	26. d	27. c	28. c	29. b	

JEE Advanced & IIT JEE Archive

1. a	2. a	3. d	4. b	5. d	6. d	7. d	8. b	9. c	
10. a,b,d	11. b	12. c	13. d	14. b	15. c	16. d	17. 55	18. a	19. b

4. Chemical Kinetics

DPP-1 Rate Law, Molecularity and Order of Reaction

1. 1.0	2. 16	3. Doubled, four, $\frac{1}{100}$, 16	4. c	5. b	6. b	7. b	8. b		
9. c	10. a	11. c	12. b	13. c	14. d	15. a	16. a	17. c	18. d
19. b	20. a	21. 2	22. 2	23. 5					

DPP-2 Zeroth-Order and First-Order Reaction

1. d	2. c	3. a	4. b	5. a	6. d	7. c	8. b	9. d	10. a
11. c	12. a	13. a	14. a	15. c	16. c	17. d	18. b	19. a	20. d
21. c	22. a	23. b	24. b	25. c	26. a	27. a	28. 2	29. 2	30. 5

DPP-3 Determination of Order of Reaction

1. d	2. d	3. c	4. a	5. a	6. d	7. b	8. d	9. c	10. c
11. b	12. b	13. c	14. c	15. b	16. a	17. c	18. c	19. b	20. b
21. 2	22. 3	23. 2	24. 7						

DPP-4 Energy of Activation

1. 53.60	2. 6.225×10^6	3. a	4. c	5. b	6. c	7. c	8. d	9. d	
10. b	11. b	12. a	13. c	14. b	15. c,d	16. a,b,c,d	17. c	18. b	
19. a	20. b	21. d	22. 8	23. 9	24. 9				

DPP-5 Collision Theory and Reaction Mechanism

1. d	2. c	3. b	4. c	5. b	6. b	7. b	8. a	9. a	10. b
11. b	12. d	13. c	14. a	15. b	16. a	17. a	18. b	19. c	20. a
21. 4	22. 5								

Revisal Problems (JEE Main)

1. b	2. d	3. c	4. a	5. b	6. a	7. c	8. a	9. d	10. a
11. b	12. d	13. a	14. a	15. b	16. c	17. b	18. c	19. d	20. b
21. c	22. a	23. a	24. b	25. a	26. a	27. a			

Revisal Problems (JEE Advanced)

1. a	2. b	3. c	4. a	5. b	6. a	7. a	8. d	9. c	10. d
11. a	12. a	13. a,c,d	14. a,b,d	15. a,b,c,d	16. a,b,c,d	17. d	18. a		
19. c	20. a	21. b	22. d	23. c	24. a	25. b	26. a	27. b	28. 0
29. 8	30. 6	31. 4	32. 4	33. 2	34. 2	35. 5	36. 6	37. 9	

JEE Main & AIEEE Archive

1. b	2. d	3. a	4. b	5. c	6. b	7. c	8. d	9. c	10. b
11. a	12. d	13. d	14. d	15. b	16. c	17. d	18. c	19. d	20. d
21. a	22. c	23. d	24. c	25. b	26. b	27. b	28. d		

JEE Advanced & IIT JEE Archive

1. b	2. d	3. 9	4. a,b,d	5. a	6. 0	7. d	8. a	9. d	10. d

1. Solid State

DPP-1 Classification and Characteristics

1. (c)

(a) Weak cohesive forces keep the molecules far apart thus, liquid or gaseous state.

(b) High thermal energy tends to keep the molecules apart by making them move faster thus, liquid or gaseous state.

(c) Low temperature increases force of attraction and decreases thermal energy (αT) thus, solid state.

(d) High temperature increases the kinetic energy and set the molecules in motion thus, gaseous state.

2. (d)

	Crystalline solids	Amorphous solids
Shape	Definite	Irregular
Cleavage property	Split into two pieces, newly generated surfaces are plain and smooth.	Split into two pieces, newly generated surfaces are irregular.
Heat of fusion	Definite	Irregular

3. (a) In amorphous solids, arrangement of molecules is irregular in all the directions and thus, physical properties are same in any direction. This is called isotropic nature. In crystalline solids, this characteristic is not observed.

4. (a) I. Liquid $\xrightarrow{\text{Freezing point}}$ solid formed is crystalline in nature.

II. Ice is also a crystalline solid.

5. (d) Amorphous solids are not crystalline and do not have definite melting points. SiO_4^{2-} tetrahedra are randomly joined in the form of polymeric chains or three-dimensional sheets.

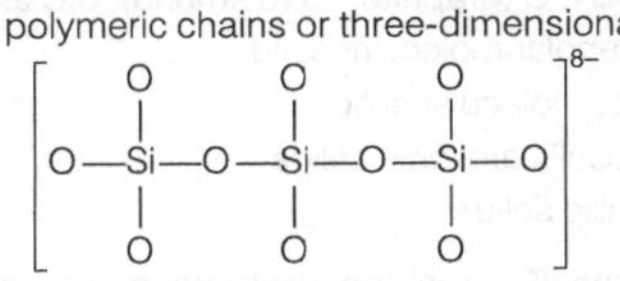

6. (d) In olive oil, forces holding the molecules together are dipole-dipole interaction and dispersion forces.

On cooling, it solidifies and changes to a molecular crystal.

7. (b) Dispersion forces hold the O_2 molecules together. On lowering the temperature, it changes to molecular crystalline form.

8. (a) sp^3-hybrid orbitals is formed by one s and three p-orbitals. These are of group 14 (carbon family) with four valence electrons.

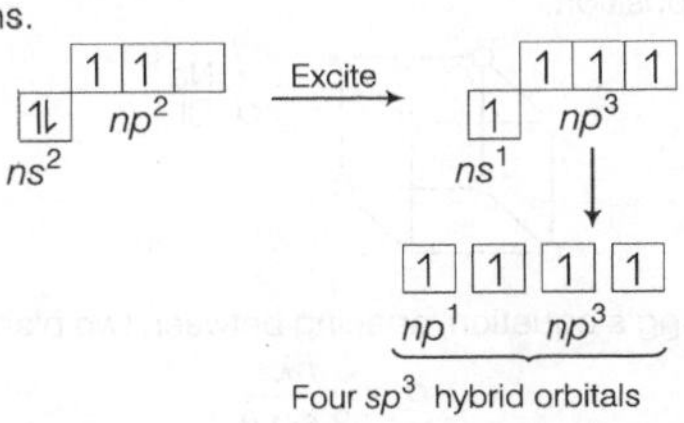

9. (d) Monoatomic gases like He, Ar and diatomic molecules like Cl_2, H_2, I_2 formed by non-polar covalent bonds are non-polar molecular solids and shows following charactertics :

(i) Atoms or molecules are held by weak dispersion forces true.

(ii) Being non-polar, they are non-conductors true.

(iii) Being covalent bonded by weak forces, they have low melting point and are liquid or gases at room temperature and pressure true.

10. (a) Such solid is said to be amorphous in nature. They have arrangement of atoms or molecules in irregular manner and there is no long range order. Thus, value of any physical property would be same along all directions. They have irregular shape hence, the solid is said to have isotropy.

11. (d) Amorphous solids have short range order of arrangement of molecules or atoms in the irregular manner. Thus, physical properties are same in all the directions hence, isotropic.

12. (c) In case of crystalline solids, constituent particles are arranged in a regular manner over a long distance.

13. (a,b,c)

(a) Xe being inert in nature have smallest dipole-dipole forces.

(b) H—F molecule is polar due to high electronegativity of F and H—F molecules are H-bonded.

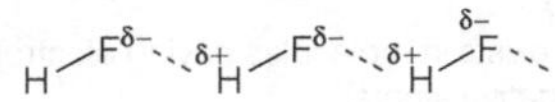

(c) Dispersion forces are attractive in nature and are important at short distances. They depend on the polarisability of the molecule.

14. (a,b,c) Amorphous solids have irregular arrangement of molecules. There is no long range order and structure is similar to that of liquids. They are called supercooled liquids or pseudo solids, e.g. glass, rubber or plastics and quartz glass.

(d) Sodium chloride is a perfect crystalline solid.

15. (b,c,d)

(a) Liquid and gases are called fluids because of their ability to flow.

(b), (c) In crystalline solids, constituent particles are arranged in a regular manner and can oscillate about their mean positions.

(d) As constituent particles are arranged in a perfect order thus, randomness decreases, i.e. $\Delta S < 0$.

Note *Randomness (disorder) increases entropy.*

16. (a,b,c,d)

(a) Molecules in solid state are arranged in a regular manner, have definite mass, shape and volume.

(b), (c) Molecules are clined to one another by strong intermolecular forces and thus, intermolecular distances are short.

(d) Being in a definite packing, molecules are incompressible and rigid.

17. (a,b,d) Polar molecular solids include SO_2, HCl, NH_3, etc.

(i) Bonding between two or more atoms is of covalent nature.

(ii) Molecules are held by dipole-dipole interactions ($\overset{\delta+}{H}—\overset{\delta-}{Cl}$).

(iii) Due to dipole-dipole interaction, attractive force is strong hence, higher melting point as compared to non-polar molecular solids.

18. (b,c) Quartz glass is an amorphous solid. Molecules/atoms are arranged in irregular manner over a short range distance. Thus, heat of fusion is not definite. It is called supercooled liquid and is isotropic in nature, i.e. physical properties are identical in all the directions.

19. (a,b,c,d) C-atoms in graphite as well as in diamond have covalent bonding. They are strong and directional in nature, thus, high melting point. Graphite shows sp^2-hybridisation and one remaining electron of each atom is present between different layers and is free to move about. These free electrons make graphite a good conductor of electricity. Different layers can slide one over the other. Thus, graphite is soft solid and a good solid lubricant. Diamond shows sp^3-hybridisation, no free electron, thus hard, insulator and high melting point.

20. (a) Thermal energy sets the molecules in motion and thus move them faster-a gaseous state. If temperature is low, thermal energy decreases being proportional to temperature. Molecules come closer and closer and are arranged in a regular manner, oscillate about their mean positions.

Thus, both the Statement I and Statement II are correct and Statement II is the correct explanation of Statement I.

21. (b) In amorphous solids, there is no long range order and arrangement is irregular along all the directions. Thus, physical properties are same along all directions hence, isotropic.

Thus, both Statement I and Statement II are correct but Statement II is not the correct explanation of Statement I.

22. (a) B_2O_3 is amorphous in nature with irregular arrangement of B—O network. Thus, it behaves like a liquid even though it looks like a solid.

Thus, both Statement I and Statement II are correct. Statement II is the correct explanation of Statement I.

23. (b) van der Waals' forces are weak in nature and exist in non-polar molecules as CO_2, H_2.

Thus, **(i) → (q)**

Ionic bonding exists in molecules having electropositive atoms and electronegative atoms.

Electropositive : Mg^{2+}, Na^{+} Electronegative : O^{2-}, Cl^{-}

Thus, **(ii) → (s)**

Metallic bonding is in metals which is a network of positive ions immersed in a sea of electrons.

Thus, **(iii) → (p)**

Covalent bonding is in quartz (SiO_2) and diamond.

Thus, **(iv) → (r)**

24. (c) Hydrogen bonded molecular solids are volatile as weak forces of attraction (as H-bond) makes the solid of lower boiling point (as in ice, water).

Thus, **(i) → (r)**

In solid state, ions are not free to move about, hence, they are insulators.

Thus, **(ii) → (p)**

Metallic solids are malleable (can be formed into sheets) and ductile (can be formed into wires).

Thus, **(iii) → (q)**

Covalent solids as SiO_2 and diamond are giant molecules.

Thus, **(iv) → (s)**

25. (d) $r_{Rb^+} > r_{Na^+}$ and $r_{Br^-} > r_{F^-}$

Thus, internuclear distance in $r_{NaF} < r_{RbBr}$ decreasing electrostatic forces of attraction between molecules in RbBr (force $\propto \frac{1}{r^2}$) thus, melting point of RbBr < NaF.

26. (a) (i) Force of attraction between cation and anion is

$$F = \frac{k\, z_1 z_2 e^2}{Dr^2}$$

where, k = constant

D = dielectric constant of solvent z_1 and z_2 are charges

r = internuclear distance

$z_1 = z_2 = \pm 1$ for both D (medium H_2O) same

$$F \propto \frac{1}{r^2}$$

Smaller the force of attraction, greater the solubility in water.

$$r\,(\text{RbBr}) > r\,(\text{NaF})$$

$$F\,(\text{RbBr}) < F\,(\text{NaF})$$

Thus, RbBr is more soluble in water than NaF.

(ii) Both are ionic compounds, thus (b) is incorrect.

(iii) Being ionic soluble in water, thus (c) is incorrect.

(iv) Both lack unpaired electrons in d-orbitals thus, colourless.

27. (3) Amorphous solids have irregular arrangement of molecules and thus are isotropic. Amorphous solids are quartz glass, rubber, plastic.

Like liquids, they have the tendency to flow.

28. (4) In crystalline solids, molecules are arranged in a regular manner and thus, physical properties differ point to point. They are called anisotropic, e.g. graphite, diamond, quartz crystalline, and sodium chloride.

29. (5) SiO_2, AlN, C (graphite), C (diamond), SiC are covalent solids.

CO_2 is non-polar molecular solid.

SO_2 is polar molecular solid.

MgO, ZnS, CaF_2 are ionic solids.

Mg is metallic solid.

30. (6) Fullerene (C_{60}) is the allotrope of carbon. It consists of spherical C_{60} molecules with the shape of a soccer ball having 12 pentagonal and 20 hexagonal faces with each atom sp^2 hybridised and bonded to three other atoms.

DPP-2 Unit Cells, Lattices and Bragg's Equation

1. (d) 14 types of three-dimensional Bravais lattices (primitive and centred unit cells).

2. (c) Lattice sites always contain atoms/molecules/ions. If atom is replaced by electron, then imperfection arises in the crystal and it would not be pure crystal.

3. (c)

Salt	Structure	Number of formula units in unit cell
MgO	fcc [rock salt (NaCl) type]	4
ZnS	fcc	4
Pt	fcc	4

4. (c)

Atom	Location	Number of atoms
A	Corner	$\frac{8}{8} = 1$
B	Face-centred	$\frac{6}{2} = 3$ (Two faces share one atom)

Thus, compound is AB_3.

5. (d) Every unit cell of NaCl (fcc) has four Na^+ and four Cl^- ions at alternate position.

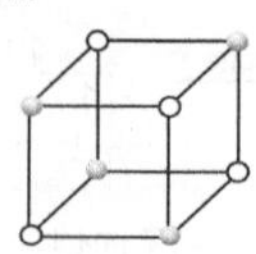

6. (c) By Bragg's equation, spacing between two planes is given by

$$d = \frac{n\lambda}{2 \sin \theta}$$

For first-order diffraction, $n = 1$

$$\therefore \quad d = \frac{\lambda}{2 \sin \theta}$$

If $d = \lambda$, then $2 \sin \theta = 1$

$$\sin \theta = \frac{1}{2}$$

$$\therefore \quad \theta = 30°$$

If $d = \frac{\lambda}{2}$, then $\sin \theta = 1 \Rightarrow \theta = 90°$

7. (a)

Atom	Location	Number of atom(s)
W	Corners	$\frac{8}{8}=1$
O	Edges (faces)	$\frac{6}{2}=3$ (One atom is common for two faces)

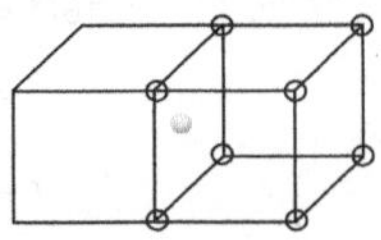

Thus, anion is WO_3^-

Charge on anion $= 5 - 6 = -1$

Thus, cation $=1$ (Na^+).

Thus, lattice is $NaWO_3$ (cubic).

8. (b) Number of atoms in the unit cell

$$z = \frac{n_c}{8} + \frac{n_f}{2} + \frac{n_i}{1}$$

where, n_c = number of atoms at the corners

n_f = number of atoms at the faces

n_i = number of atoms inside

	n_c	n_f	n_i	z
Primitive cubic	8	0	0	1
Body-centred cubic (bcc)	8	0	1	2
Face-centred cubic (fcc)	8	6	0	4

9. (c) By using Bragg's equation,

$$n\lambda = 2d \sin \theta$$

where, $n=2$ (second order diffraction)

$$\lambda = d \sin \theta$$

or $$\log \lambda = \log d + \log \sin \theta$$

d is constant (distance between two planes) as θ changes, λ also changes.

When $\theta = 90°$, $\sin 90° = 1$, hence $\log \sin 90° = 0$

$$OX = 0.3010 \text{Å}$$

$\therefore$ $$\log \lambda = 0.3010 \text{ Å}$$

$\therefore$ $$\lambda = 2.0 \text{ Å}$$

10. (a)

(i) Triclinic, $a \neq b \neq c$ and $\alpha \neq \beta \neq \gamma = 90°$

Thus, **(i) → (r)**

(ii) Hexagonal, $a = b \neq c$, $\alpha = \beta = 90°$, $\gamma = 120°$

Thus, **(ii) → (s)**

(iii) Cubic, $a = b = c$, $\alpha = \beta = \gamma = 90°$

Thus, **(iii) → (p)**

(iv) Tetragonal, $a = b \neq c$, $\alpha = \beta = \gamma = 90°$

Thus, **(iv) → (q)**

11. (a,c,d)

(a) Orthorhombic : $a \neq b \neq c$ (b) Tetragonal : $a = b \neq c$

(c) Monoclinic : $a \neq b \neq c$ (d) Triclinic : $a \neq b \neq c$

12. (a,d)

(a) Cubic : $a = b = c$ (b) Hexagonal : $a = b \neq c$

(c) Orthorhombic : $a \neq b \neq c$ (d) Rhombohedral : $a = b = c$

13. (a,b,c,d)

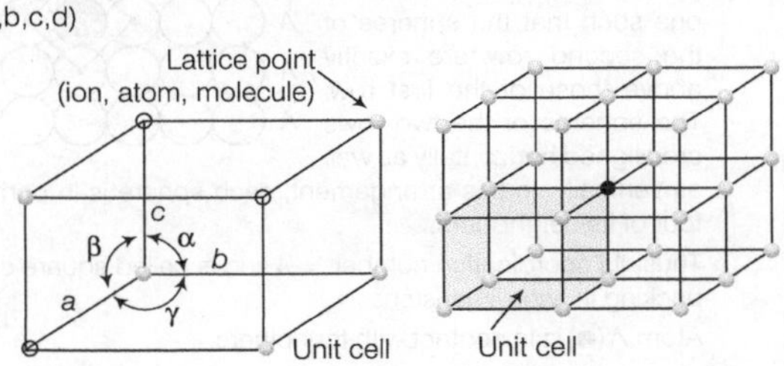

Explanations for Q. Nos. (14-15)

Calcium appears at eight corners, hence number of calcium atoms $= \frac{8}{8} = 1$

Oxygen appears on six faces and two faces share one oxygen atom hence, number of oxygen atoms $= \frac{6}{2} = 3$

Titanium appears at the centre, thus, number of Ti-atom $= \frac{1}{1} = 1$

Thus, A is $CaTiO_3$.

Let the oxidation number of Ti $= x$

$$CaTiO_3$$

$$(+2) + x + 3(-2) = 0$$

$$2 + x - 6 = 0$$

$\Rightarrow$ $$x = +4$$

14. (b) **15.** (c)

16. (6) Edge length : a, b, c and angles between the edges : α, β, γ.

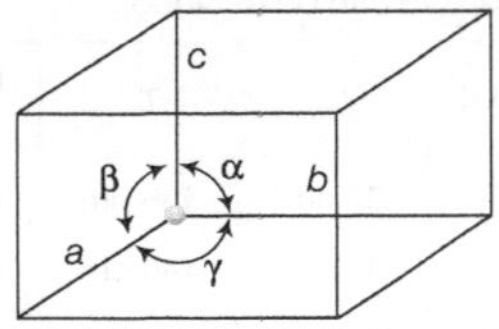

17. (7) Primitive unit cells are

(i) Cubic

(ii) Tetragonal

(iii) Orthorhombic

(iv) Hexagonal

(v) Rhombohedral

(vi) Monoclinic

(vii) Triclinic

18. (1)

Cubic	$a=b=c$	$\alpha=\beta=\gamma=90°$
Tetragonal	$a=b\neq c$	$\alpha=\beta=\gamma=90°$
Orthorhombic	$a\neq b\neq c$	$\alpha=\beta=\gamma=90°$
Hexagonal	$a=b\neq c$	$\alpha=\beta=90°, \gamma=120°$
Trigonal (rhombohedral)	$a=b=c$	$\alpha=\beta=\gamma\neq 90°$
Monoclinic	$a\neq b\neq c$	$\alpha=\gamma=90°, \beta\neq 120°$
Triclinic	$a\neq b\neq c$	$\alpha\neq\beta\neq\gamma=90°$

19. (1) By Bragg's equation,

$$2d \sin\theta = n\lambda$$

$\therefore$ $$\lambda = \frac{2d \sin\theta}{n}$$

$$d = 1.155 \text{ Å}$$

$$\sin\theta = \sin 60° = \sqrt{3}/2$$

$n = 2$ (second order diffraction)

$$\lambda = \frac{2 \times 1.155 \times \sqrt{3}/2}{2} \text{ Å}$$

$$= 1.155 \times \sqrt{3}/2$$

$$= 1.155 \times \frac{1.732}{2}$$

$$= 1.0 \text{ Å}$$

20. (3) $240 \text{ g } X = \frac{240}{40} = 6 \text{ mol} = 6 N_0$ atoms

Number of atoms in one unit cell of bcc structure $=2$

Thus, number of unit cells $= \frac{6N_0}{2} = 3 N_0$

Thus, $y = 3$

DPP-3 Packing in Crystals

1. Plan By the limiting radius ratio of cation to anion, one can decide the coordination number and thus, structure also.

r_+ / r_-	Coordination number	Structure
< 0.155	2	Linear
0.155 - 0.225	3	Planar triangle
0.225 - 0.414	4	Tetrahedral
0.414 - 0.732	6	Octahedral
0.732 - 0.999	8	Body-centred cubic
i. MgO 0.65/1.40 = 0.464	6	Octahedral
ii. MgS 0.65/1.84 = 0.353	4	Tetrahedral
iii. CsCl 1.69/1.81 = 0.933	8	Body-centred cubic

2. Plan From the value of radius ratio (r_+ / r_-), coordination number and type of structure can be decided.

	Crystal	r_+	r_-	Radius ratio	Type
i.	NaF	95 pm	136 pm	0.6985	NaCl
ii.	CoO	32 pm	140 pm	0.2286	ZnS
iii.	CsBr	169 pm	195 pm	0.8667	CsCl
iv.	BeO	31 pm	140 pm	0.2214	ZnS
v.	BaF_2	135 pm	136 pm	0.9926	CaF_2

3. Plan d = Internuclear distance of $(A—B) = r_{A^+} + r_{B^-}$

Given, $d\,(RbI) = 367\text{ pm} = r_{Rb^+} + r_{I^-}$

$r_{I^-} = 219\text{ pm}$

$\therefore\quad r_{Rb^+} = 367 - 219 = 148\text{ pm}$

$d\,(RbBr) = 344\text{ pm}$

$\therefore\quad r_{Br^-} = 344 - 148 = 196\text{ pm}$

Given, $d\,(KBr) = 329\text{ pm}$

$\therefore\quad r_{K^+} = 329 - 196 = 133\text{ pm}$

Also, $d\,(KCl) = 314\text{ pm}$

$\therefore\quad r_{Cl^-} = 314 - 133 = 181\text{ pm}$

Given, $d\,(NaCl) = 279\text{ pm}$

$\therefore\quad r_{Na^+} = 279 - 181 = 98\text{ pm}$

Since, $d\,(NaF) = 231\text{ pm}$

$\therefore\quad r_{F^-} = 231 - 98 = 133\text{ pm}$

4. (d) CsCl has bcc structure.

Closest distance between Cs^+ and Cl^- is the sum of radii of Cs^+ and $Cl^- = 169 + 181 = 350$ pm

Cubic diagonal $= \sqrt{3}a$

Closest distance is one-half of the cubic diagonal

$\therefore\quad \frac{\sqrt{3}a}{2} = 350$

$\therefore\quad a = \frac{700}{\sqrt{3}}\text{ pm} = 404\text{ pm}$

5. (d) AB_2 crystallises as cubic close packed (ccp).

Thus, A^{2+} is surrounded by eight B^- and B^- is surrounded by four A^{2+}.

Thus, coordination number of $A^{2+} = 8$

coordination number of $B^- = 4$

Number of atoms in tetrahedral in small cube of unit cell of atom $B = 4$

Thus, fraction occupied $= \frac{4}{4} = 1$ (100%)

6. (c) $AB = a,\ AC = a$

$BC^2 = AB^2 + AC^2 = 2a^2$

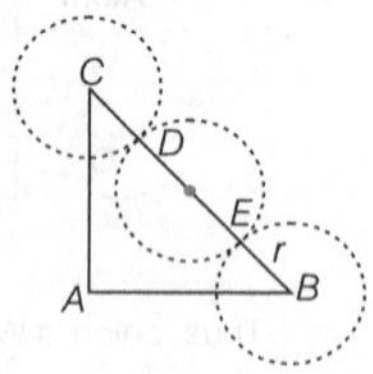

Radius $= r$,

then $BC = CD + DE + EB$

$= r + 2r + r = 4r$

$\therefore\quad 16r^2 = 2a^2$

$r^2 = \frac{a^2}{8}$

$\therefore\quad r = \frac{a}{2\sqrt{2}} = \frac{407}{2\sqrt{2}} = 143.9\text{ pm}$

Note *Detailed calculations to relate radius with edge length have been given. Examinee can use direct relation.*

Simple cubic $r = \frac{a}{2}$, *hcp/ccp* $r = \frac{a}{2\sqrt{2}}$, *bcc* $r = \frac{\sqrt{3}}{4}a$

7. (c) Tetrahedral holes = 8, Octahedral holes = 4

Cobalt atoms $= \frac{8}{8} = 1$

Oxide ions $= \frac{4}{2} = 2$

Thus, formula is CoO_2.

8. (b) Unit cell of fcc is divided into eight small cubes. Each cube has four atoms at alternate positions. When joined to each other, they make a regular tetrahedron. Thus, number of tetrahedral void in small cube = 1 and in eight cubes, voids = 8.

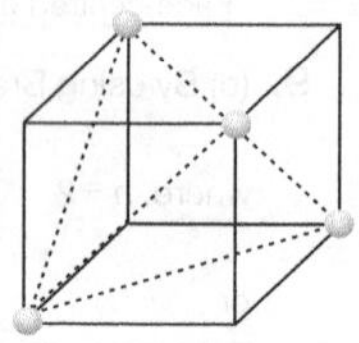

One tetrahedral void in a small cube

Note *If number of atoms in a cube = N, then, number of octahedral voids = N and number of tetrahedral voids = 2N*

9. (a) Edge length = a, Volume of one face = a^3

Volume of six faces of a cube = $6a^3 = V$

Radius = r

Structure	r related to a	Volume v (of sphere)	z	Packing efficiency $= \frac{zv}{V} \times 100$
Simple cubic	$r = \frac{a}{2}$	$\frac{4}{3}\pi\left(\frac{a}{2}\right)^3$	1	52%
Body-centred cubic (bcc)	$r = \frac{\sqrt{3}a}{4}$	$\frac{4}{3}\pi\left(\frac{\sqrt{3}a}{4}\right)^3$	2	68%
Face-centred cubic (fcc)	$r = \frac{a}{2\sqrt{2}}$	$\frac{4}{3}\pi\left(\frac{a}{2\sqrt{2}}\right)^3$	4	74%

fcc can be hexagonal closest packed (hcp) or cubic closest packed (ccp)

z = number of atoms in unit cell.

10. (c) Close packing in two-dimensions can be

Type I ***AAA*****......type**

The second row may be placed in contact with the first one such that the spheres of the second row are exactly above those of the first row. The spheres of the two rows are aligned horizontally as well as vertically. In this arrangement, each sphere is in contact with four of its neighbours.

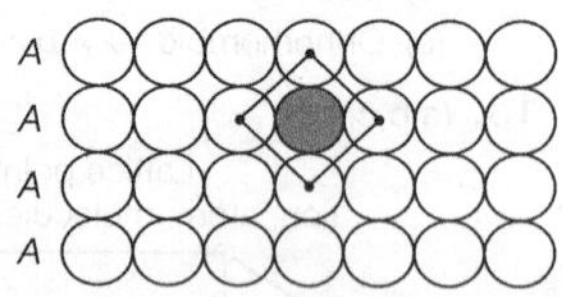

Thus, its coordination number = 4 and is called square close packing in two-dimensions.

Atom A (●) is in contact with four others.

Type II *ABAB* ... type

The second row may be placed above the first in a staggered manner such that its spheres fit in the depressions of the first row.

First row arrangement : *A* type

Second row arrangement : *B* type

Third row arrangement : *A* type and so on

There is less free space and packing is more efficient than the square close packing. Each sphere is in contact with six of its neighbours and coordination number = 6 in two-dimensions. It is called two dimensional hexagonal close-packing. Voids formed in this case are triangular in shape. If the apex of the triangles downward in the first row, it is upward in the second row.

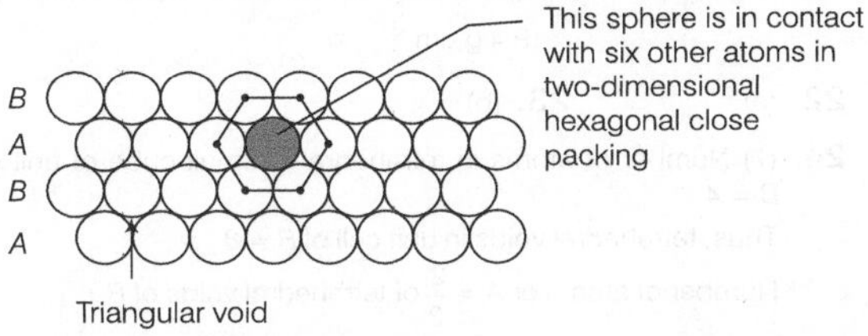

11. (a) In a bcc structure, the atoms touch each other along the body diagonal.

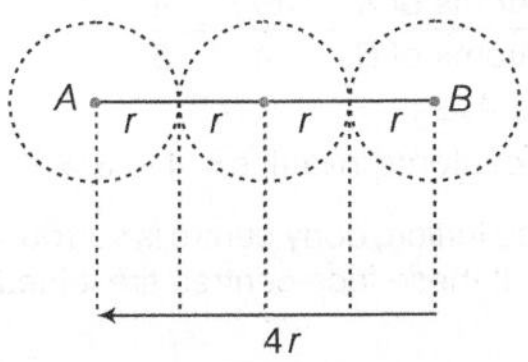

Radius of the atom $= r$

$$AB^2 = BC^2 + AC^2$$

$$BC^2 = BD^2 + CD^2 = a^2 + a^2 = 2a^2$$

$\therefore \quad AB^2 = 2a^2 + a^2 = 3a^2$

$\therefore \quad AB = \sqrt{3}a$

But $\quad AB = 4r$

$\therefore \quad 4r = \sqrt{3}a \Rightarrow r = \sqrt{3}a/4$

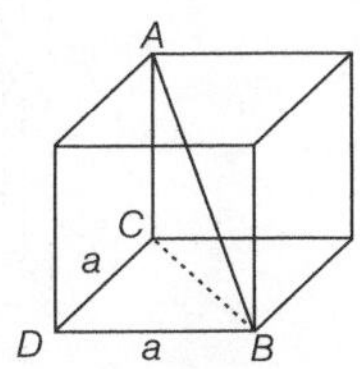

$$r = \sqrt{3} \times \frac{352}{4} \text{ pm}$$

$$= \sqrt{3} \times \frac{352}{4} \times 10^{-12} \text{m}$$

$$= \sqrt{3} \times \frac{352}{4} \times 10^{-10} \text{cm} = 152.42 \times 10^{-10} \text{cm}$$

Volume of one atom $= \frac{4}{3}\pi r^3 = \frac{4}{3}\pi\,(152.42 \times 10^{-10})^3 \text{ cm}^3$

$$= 1.48 \times 10^{-23} \text{cm}^3$$

Volume of N_o atoms in one mole of the metal

$$= 6.02 \times 10^{23} \times 1.48 \times 10^{-23} \text{cm}^3 = 8.93 \text{ cm}^3$$

12. (b) Structure is cubic close packing.

Number of atoms of $M = \frac{1}{4} \times 4$ (from edge centre)
+ 1 (from body centre)
$= 2$

Number of atoms of $X = \frac{1}{8} \times 8$ (from corners of the cube)
$+ \frac{1}{2} \times 6$ (from face centre)
$= 4$

Thus, $\quad M : X = 2 : 4 = 1 : 2$

Hence, the empirical formula of the compound is MX_2.

13. (a) Ratio of r_+ / r_- will decide the type of unit cell.

For *AX* $\quad \frac{r_{A^+}}{r_{X^-}} = \frac{1}{1} = 1.0$ hence, *AX* has bcc structure.

$\therefore \quad 2(r_{A^+} + r_{X^-}) = a\sqrt{3}$

$$a = \frac{2}{\sqrt{3}}(r_{A^+} + r_{X^-})$$

$$= 2 \times \frac{2}{\sqrt{3}} = \frac{4}{\sqrt{3}} \text{ pm}$$

$\therefore \quad$ Volume $(AX) = a^3 = \left(\frac{4}{\sqrt{3}}\right)^3 = \frac{64}{3\sqrt{3}}$

For *AY* $\quad \frac{r_{A^+}}{r_{Y^-}} = \frac{1}{2} = 0.5$

Thus, octahedral or square planar structure.

$$r = \frac{a}{2}$$

$\therefore \quad a = 2r = 2(r_{A^+} + r_{Y^-}) = 2(1 + 2) = 6 \text{ pm}$

Volume $(AY) = a^3 = 216 \text{ pm}^3$

$\therefore \quad$ Ratio $= \frac{\text{Volume } (AX)}{\text{Volume } (AY)} = \frac{64}{3\sqrt{3} \times 216} = 0.057$

14. (a) Along the edge, $\quad AB = AC = 2(r_{Cl^-} + r_{Na^+}) = a$

$\therefore \quad r_{Na^+} = \left(\frac{a}{2} - r_{Cl^-}\right)$

Also, $\quad AC^2 + AB^2 = BC^2$

$\Rightarrow \quad a^2 + a^2 = (4r_{Cl^-})^2$

$\therefore \quad a = 2\sqrt{2}\, r_{Cl^-}$

$\therefore \quad r_{Na^+} = \left(\frac{2\sqrt{2}\, r_{Cl^-}}{2} - r_{Cl^-}\right) = r_{Cl^-}(\sqrt{2} - 1) = r_{Cl^-}(1.414 - 1)$

$\therefore \quad \frac{r_{Na^+}}{r_{Cl^-}} = 0.414$

15. (b)

(i) **Body-centred cubic** Every cation (Cs^+) is coordinated to eight anions (Br^-) and every anion (Br^-) is coordinated to eight cations (Cs^+).
Thus, **(i) → (s)**

(ii) **Face-centred cubic** Every cation (Na^+) is coordinated to six anions (Cl^-) and every anion (Cl^-) is coordinated to six cations (Na^+).
Thus, **(ii) → (q)**

(iii) **Face-centred cubic** (tetrahedral void) Every cation (Zn^{2+}) is coordinated to four anions (S^{2-}) and *vice-versa*.
Thus, **(iii) → (p)**

(iv) **Face-centred cubic** (antifluorite) Every Ca^{2+} is coordinated to eight fluoride and every F^- is coordinated to four Ca^{2+} (to balance charge).

$$4Ca^{2+} + 8F^-$$

Thus, **(iv) → (r)**

16. (a) Difference of internuclear distance between NaI and LiI
$= 0.317 - 0.310$
$= 0.007$ nm

between NaCl and LiCl $= 0.279 - 0.249$
$= 0.030$ nm

0.03 nm > 0.007 nm

Thus, iodide-iodide contact of LiI and NaI makes closer approach impossible.

Thus, both Statements I and II are correct and Statement II is the correct explanation of Statement I.

17. (a,b,c)

(a) r = radius, a = edge length

then for hcp, $r = \frac{a}{2\sqrt{2}}$

Volume of one spherical atom $= \frac{4}{3}\pi r^3 = \frac{4}{3}\pi\left(\frac{a}{2\sqrt{2}}\right)^3$

Number of atoms in one unit cell = 4

$\therefore$ Volume of four atoms $= \frac{16}{3}\pi\left(\frac{a}{2\sqrt{2}}\right)^3$

Volume of cube $= a^3$

$\therefore$ Packing efficiency $= \dfrac{\frac{16}{3}\pi\left(\frac{a}{2\sqrt{2}}\right)^3}{a^3} = 0.74$

Thus, correct.

(b) In hcp coordination number = 12, six at the diagonal and six at the edge length thus, correct.

(c) A tetrahedral void is formed when a sphere of a second layer is above the void of the first layer or *vice-versa* thus, correct.

18. (a, b)

(a) A tetrahedral void is formed when a sphere of a second layer is placed above triangular void in the first layer thus, correct.

(b) All the triangular voids are not covered by this arrangement thus, correct.

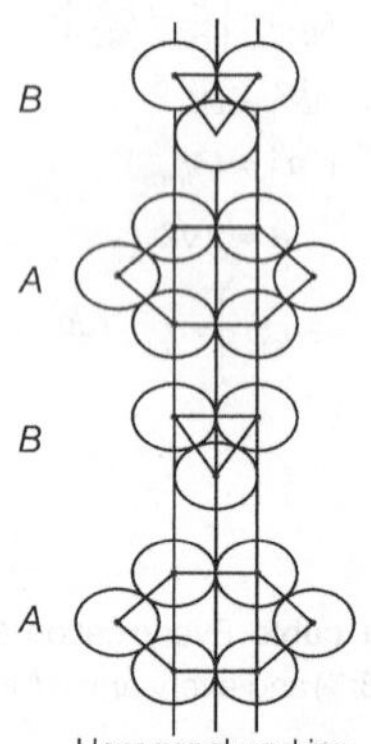

Hexagonal packing

19. (a, d) Octahedral voids are formed in

(i) hexagonal close packing

(ii) face-centred cubic

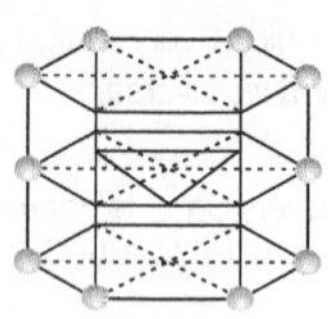

hcp

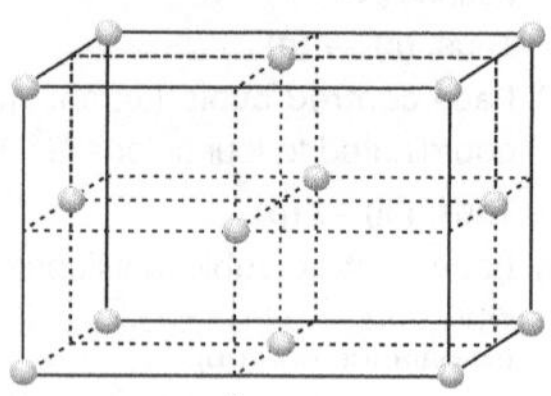

fcc

Explanations for Q. Nos. (20-21)

Nearest neighbours are along the diagonal at the half distance apart.

Diagonal $AB = \sqrt{3}a$

$\therefore$ Distance of the nearest neighbour

$= \frac{\sqrt{3}a}{2}$

$= \sqrt{3} \times \frac{520}{2}$

$= 450$ pm

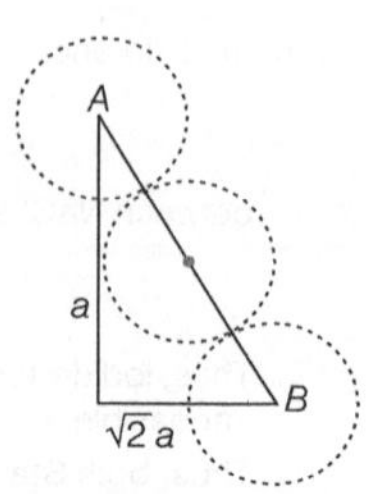

Next nearest neighbours are along the cell edge = 520 pm

K-atom at the centre is surrounded by eight K-atoms as nearest neighbours.

In bcc, coordination number is six as next nearest neighbours.

20. (a) **21.** (c)

Explanations for Q. Nos. (22-23)

Closest distance between two gold atoms is half of the diagonal.

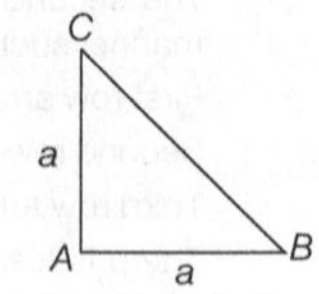

$$BC = \sqrt{a^2 + a^2} = a\sqrt{2}$$

$\therefore$ Closest distance $= \frac{a\sqrt{2}}{2} = \frac{407}{2}\sqrt{2}$ pm

$= 287.8$ pm

Note *There are total of 12 nearest neighbours (six along diagonal and six along the edges).*

$$\text{Density}, d = \frac{zM}{N_0 a^3} = \frac{4(\text{for fcc}) \times 197 \text{ g mol}^{-1}}{6.02 \times 10^{23}\text{mol}^{-1} \times (407 \times 10^{-10})^3\text{cm}^3}$$

$= 19.4 \text{ g cm}^{-3}$

22. (a) **23.** (b)

24. (7) Number of atoms in tetrahedral in small cube of unit cell of $B = 4$

Thus, tetrahedral voids in unit cell of $B = 8$.

Number of atoms of $A = \frac{2}{3}$ of tetrahedral voids of B

$= \frac{2}{3} \times 8 = \frac{16}{3}$

Thus, ratio : $\frac{\text{Atoms of } A}{\text{Atoms of } B} = \frac{16/3}{4} = \frac{4}{3}$

Thus, lattice is A_4B_3.

Total number of atoms in lattice = 4 + 3 = 7

25. (4) In fcc or hcp lattice, body centre is surrounded by six atoms on face centres. If these face centres are joined, one octahedron is formed.

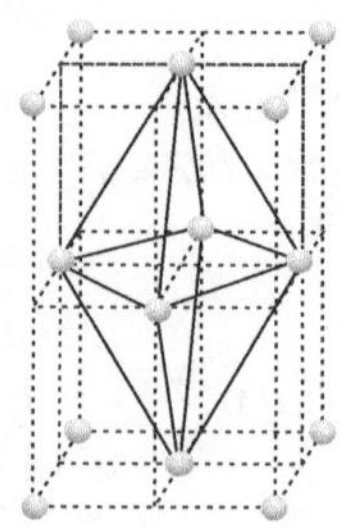

(i)

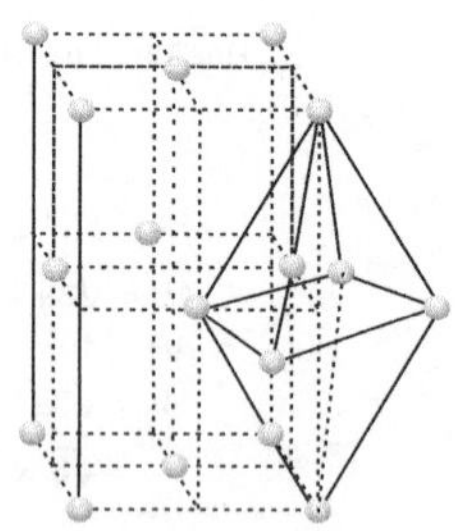

(ii)

Thus, number of octahedral void in the body centre of the cube = 1

There is also one octahedral void at the centre of each of the 12 edges. It is surrounded by six atoms, three belonging to the same unit cell and three belonging to the two adjacent unit cells.

Each edge of the cube is shared between four adjacent unit cells, thus, number of octahedral voids $= \frac{12}{4} = 3$.

Thus, total octahedral voids = 1 + 3 = 4

26. (4) Density $(d) = \frac{zM}{N_0 a^3}$

where, M = Atomic mass = 48 g mol^{-1}

N_0 = Avogadro's number $= 6.02 \times 10^{23}$ mol^{-1}

z = Number of atoms in the unit cell

a^3 = Volume

$z = \frac{d N_0 a^3}{M}$

$d = 4.54$ g cm^{-3}

$a = 412.6 \text{ pm} = 412.6 \times 10^{-12} \text{ m} = 412.6 \times 10^{-10}$ cm

$\therefore \quad z = \frac{4.54 \times 6.02 \times 10^{23} \times (412.6 \times 10^{-10})^3}{48} = 4$

Note *$z = 4$ indicates that* Ti *has fcc structure.*

DPP-4 Imperfection in Crystals

1. (a) A Schottky defect consists of a pair of holes in the crystal lattice. One positive ion and one negative ion are absent.

Note *This sort of defect occurs mainly in highly ionic compounds where cation and anion are of similar size, hence the coordination number is high* (8, 6) *as in* NaCl, CsCl, KCl, KBr.

2. (b) It is clear from the figure that there is an excess of cation and electrical neutrality is made by the interstitial electron. This is thus metal excess defect and lattice may be represented as $A_{1+\delta}$.

3. (a) If there is no deficiency of cation or anion but cation migrates to interstitial site, it is called dislocation. It is observed in the crystals with smaller cation and larger anion, so that cation can dislocate easily. It is called Frenkel defect.

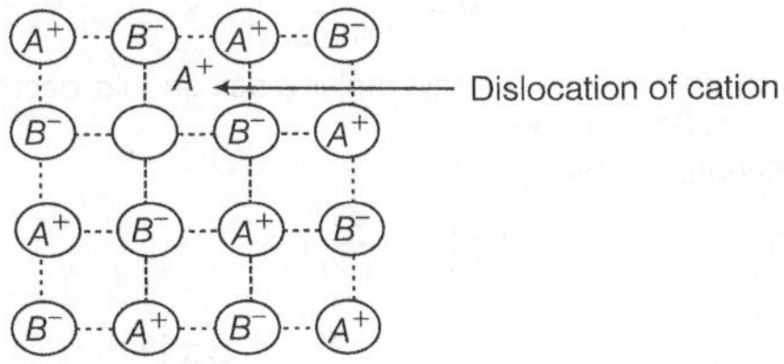

4. (a) Frenkel defect is observed in the crystals in which cation is much smaller than the anion so that cation can migrate to interstitial position.

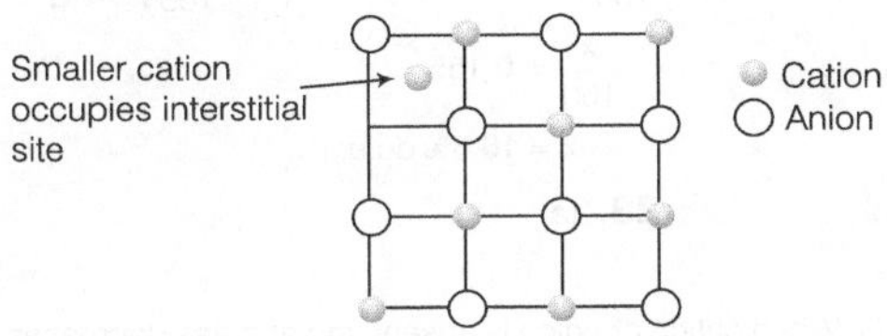

5. (d)

(a) Non-stoichiometric compounds are also called Berthollide compounds thus, correct.

(b) Since, composition of the constituent elements is not fixed, thus law of constant composition is not followed as is $Fe_{0.98}O$ thus, correct.

(c) If cation (say A^+) is missing then, charge is balanced by the cation A^{2+}. If anion (say B^-) is missing, then charge is balanced by inserting electron (e^-) thus, correct.

6. (d)

Atom	Location	Number of atom(s)	Simple ratio
A	Corners	$\frac{8}{8}=1$	2
B	Faces	$(6-1)\frac{1}{2}=5/2$ (Since, one B is removed to make defect)	5

Thus, compound is A_2B_5.

7. (a) For fcc structure,

$$\left(\frac{r_+}{r_-}\right) = 0.414 \text{ for octahedral void}$$

$$= 0.225 \text{ for tetrahedral void}$$

Thus, maximum packing can be for octahedral void

$$\therefore \quad r_+ = 0.414\, r_-$$

Also, $$r_- = \frac{a}{2\sqrt{2}}$$

$$\therefore \quad r_+ = 0.414 \frac{a}{2\sqrt{2}} = 0.414 \times \frac{400}{2\sqrt{2}}$$

$$= 58.55 \text{ pm}$$

8. (b) $M_{0.98}O$

It represents stoichiometric defect.

If number of moles of $M^{2+} = x$

then $M^{3+} = (0.98 - x)$

Thus, by charge balance of $M_{0.98}O$

$(x \times 2) + (0.98 - x) \times 3 - (1 \times 2) = 0$

$-x = 2 - 0.98 \times 3$

$x = 0.94$

$\therefore \quad M^{3+} = 0.98 - 0.94 = 0.04$

$\therefore \quad$ Percentage of $M^{3+} = \frac{0.04}{0.98} \times 100$

$= 4.08\%$

9. (c) Due to the addition of $SrCl_2$, each Sr^{2+} ion replaces two Na^+ ions, but occupies one Na^+ lattice point. Thus, this exchange of Na^+ ion by Sr^{2+} ion makes one cationic vacancy.

$SrCl_2$ doped $= 10^{-3}$ mol per 100 mol

$= 10^{-5}$ mol per 1 mol

$\therefore$ Cation vacancies $= 10^{-5}$ mol per 1 mol

$= 10^{-5} \times N_0 \text{ mol}^{-1}$

$= 10^{-5} \times 6.02 \times 10^{23}$

Total $= 6.02 \times 10^{18}$ cationic vacancies mol^{-1}

10. (a) ZnO has Zn^{2+} ion. It has no unpaired electrons in d-orbital ($[Ar]3d^{10}$), hence it is colourless (white). Thus, Statement I is correct.

$$ZnO \xrightarrow{\text{Heat}} Zn^{2+} + \frac{1}{2}O_2 + 2e^-$$

Due to heating, there is excess of Zn^{2+} ions in the crystal and its formula changes to $Zn_{1+x}O$ (non-stoichiometric). The excess Zn^{2+} ions move to interstitial sites and the electrons to neighbouring interstitial sites and impart yellow colour due to defect.

Thus, Statement II is also correct and is the correct explanation of Statement I.

11. (b) When crystal of NaCl is heated in the atmosphere of sodium vapour, Na-atoms are deposited on the surface of crystal and Cl^- ions diffuse on the surface and NaCl is formed.

$$Na(g) \longrightarrow Na^+(g) + e^-$$

Electrons released diffuse into the crystal and occupy interstitial sites. These anionic sites occupied by unpaired electrons are called F-centres (colour imparting) and appears yellow. Thus, both statements I and II are correct. The colour results by excitation of these electrons when they absorb energy from the visible light falling on the crystals. Thus, Statement II is not the correct explanation of Statement I.

12. (a) Schottky defect is shown by ionic compounds having cations and anions of comparable size so that if pair of cation and anion is missing, mechanical distortion and charge displacement is minimised. Thus, both Statements I and II are correct and Statement II is the correct explanation of Statement I.

13. (a) Larger the anion, smaller the cation, larger the polarisability. Frenkel defect is shown by ionic solids with smaller cation so that, it can migrate to interstitial site. Thus, repulsive effect is minimised. Both Statements I and II are correct and Statement II is the correct explanation of Statement I.

14. (a, b) AgBr is a covalent compound with smaller Ag^+ ($\approx$ 115 pm) and larger Br^- ($\approx$ 196 pm) thus, Ag^+ ion can be migrated to interstitial site showing Frenkel defect. In AgBr, there is also Schottky defect per 10^{16} ions.

15. (a,c) Schottky defect is shown by ionic solids having cation/anion of equal size (as NaCl). Equal number of cation and anion are absent, thus density is decreased.

One mole of ions (6.02×10^{23} ions) has about 6×10^7 defects thus, one defect occurs in 1×10^{16} ions.

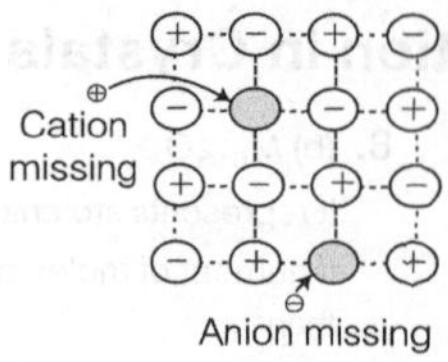

16. (a,b,c)

(a) It is a metal deficiency defect. If A^+ is missing, charge is balanced by A^{2+} (divalent cation).

(b) It represents Schottky defect. A pair of cation (A^+) and anion (B^-) is missing.

(c) It represents metal excess defect. It is like Frenkel defect. Due to the smaller size of cation (A^+), it is at interstitial position. Charge is balanced by interstitial electrons.

(d) It represents a case in which large anion (B^-) is interstitial position. Charge is balanced by replacing A^+ by A^{2+}. No crystal of such type is possible as larger B^- ion cannot be set into interstitial site.

17. (a,b,c,d)

(a) Since, number of cations > number of anions, hence it represents metal-excess defect. It is similar to Schottky defect in which equal number of cations and anions are missing thus, correct.

(b) Charge is balanced by electrons, these are called *F*-centres (farbe, colour producing). Compounds with *F*-centres are coloured. Greater the number of *F*-centres, greater the intensity thus, correct.

(c) Due to the unpaired electrons these are paramagnetic thus, correct.

(d) When NaCl, KCl, etc., are strongly heated in flame, they become deficient in anions and represented as $AB_{1-\delta}$ where, δ is small fraction.

Non-stoichiometric form of NaCl is yellow and that of KCl is blue-lilac thus, correct.

18. (a,b,c,d)

(a) correct

(b) As cation (A^+) occupies interstitial position, thus cation is smaller than anion, thus correct.

(c) Due to the smaller cation and larger anion, coordination number is usually low (4, 6), thus correct.

(d) By Fajans' rule, smaller cations polarise larger anions thus, compounds are covalent, thus correct.

19. (a, b)

(a) When cation, anion or both are missing, molar mass is decreased. Since, density $= \frac{\text{Mass}}{\text{Volume}}$, hence density decreases, thus correct.

(b) When there is insertion of cation (A^+) in the interstitial site, it causes metal-excess defect. Thus, molar mass increases, hence density increases, thus correct.

(c) Presence of impurity increases molar mass (as doping of metal with P or Si), hence density changes, thus incorrect.

(d) In Frenkel defect, cation is dislocated without affecting molar mass thus, no change in density, thus incorrect.

Explanations for Q. Nos. (20-21)

AB is rock salt like structure, thus number of *AB* molecules per unit cell = 4, $a = 2y^{1/3}$ nm

$$\text{Density, } d(\text{theoretical}) = \frac{Mz}{N_0 a^3} = \frac{6.02y \times 4}{6.02 \times 10^{23}(2y^{1/3})^3 \times 10^{-27}\ \text{m}^3}$$

$$= 5 \times 10^3\ \text{g m}^{-3} = 5\ \text{kg m}^{-3}$$

Observed value of d (20 kg m^{-3}) >> 5 kg m^{-3}

Thus, there is excess of metal ions which increases mass hence, the density.

20. (b) **21.** (c)

Explanations for Q. Nos. (22-23)

Rock salt (NaCl) is fcc type structure, $z = 4$

$M = 64$ g mol^{-1}, $a = 418$ pm

$$\text{Theoretical density } (d) = \frac{Mz}{N_0 a^3} = \frac{64 \times 4}{6.02 \times 10^{23} \times (418 \times 10^{-10}\ \text{cm})^3}$$

$$= 5.82\ \text{g cm}^{-3}$$

Experimental density = 4.92 g cm^{-3}

Thus, experimental density << theoretical density.

Thus, there is vacancy in the crystal.

$$d = \frac{\text{Mass}}{\text{Volume}}$$

Due to decrease in density, molar mass should decrease. This indicates Schottky defect.

Let Schottky defect be x%.

Then, $$z = \left(4 - \frac{4x}{100}\right)$$

$$d(\text{experimental}) = 4.92 = \frac{64\left(1 - \frac{x}{100}\right)4}{6.02 \times 10^{23} \times (418 \times 10^{-10})^3}$$

$$\therefore \quad \left(1 - \frac{x}{100}\right)4 = 3.38 \Rightarrow \left(1 - \frac{x}{100}\right) = \frac{3.38}{4} = 0.845$$

$$\therefore \quad \frac{x}{100} = 0.155$$

$$x = 15.5\%\ \text{defect}$$

22. (b) **23.** (a)

24. (c)

(i) When cation or anion is absent, molar mass decreases hence, density decreases.

Thus, **(i)** $\rightarrow$ **(q)**

(ii) When cation is packed in the interstitial site, molar mass increases. Hence, density increases.

Thus, **(ii)** $\rightarrow$ **(s)**

(iii) When smaller cation migrates into the interstitial site, Frenkel defect is observed. There is no change in molar mass and thus, no change in density.

Thus, **(iii)** $\rightarrow$ **(p)**

(iv) As in NaCl, when cation and anion are missing (Schottky defect), molar mass and thus, density decreases.

Thus, **(iv)** $\rightarrow$ **(r)**

25. (4) Let Schottky defect be x%

For fcc crystal, $z = 4$

$$\therefore \text{Number of unit cells} = \left(4 - \frac{4x}{100}\right)$$

$$z = 4\left(1 - \frac{x}{100}\right)$$

$$a^3 = (0.556 \times 10^{-9})^3\text{m}^3 = (0.556 \times 10^{-7})^3\text{cm}^3$$

$$\text{Density, } d = \frac{Mz}{N_0 a^3}$$

$$1.4848 = \frac{40 \times 4\left(1 - \frac{x}{100}\right)}{6.02 \times 10^{23} \times (0.556 \times 10^{-7})^3}$$

$$\left(1 - \frac{x}{100}\right) = 0.96$$

$$\therefore \quad \frac{x}{100} = 0.04 \Rightarrow x = 4$$

Thus, 4% Schottky defect.

DPP-5 Magnetic and Electric Properties of Solids and Semiconductors

1. (d)

I. Mg (metal) conducts electricity in solid state as well as in molten state due to available free valence electrons.

II. TiO behaves like metals and is a good conductor of electricity.

III. I_2 is a bad conductor.

IV. H_2O (ice) is a bad conductor.

2. (d)

(a) When smaller cation migrates to interstitial site, it is called dislocation defect. It is also called Frenkel defect.

(b) When equal number of cation and anion are missing or one cation (A^+) is missing and charge is balanced by A^{2+}, then this is called Schottky defect.

(c) as in (a)

(d) Electronic defects (*n*-type, *p*-type) is due to doping of impurity.

3. (b)

(a) When Si is doped with electron-deficient impurity -(*p*-type)

(b) When Si is doped with electron-rich impurity-(*n*-type)

(c) In case of semiconductors, the gap between the valence band and conduction band is small. Some electrons may jump to conduction band and thus, conduction is observed. Semiconductors of this type are called intrinsic semiconductors.

(d) Insulators are non-conducting with high energy gap.

4. (c) When polar crystal is subjected to a mechanical stress, electricity is produced. It is called piezoelectricity.

Piezoelectric crystals with permanent dipoles (domains aligned in same direction) are said to have ferroelectricity as in (a) and with zero dipoles (domains aligned equally in opposite direction) are said to have anti-ferroelectricity as in (b).

5. (d) Metals are best conductors of electricity and conductivities vary between 10^4 to 10^7 $ohm^{-1}\ m^{-1}$.

Semiconductors have conductivity values between 10^{-6} to 10^4 $ohm^{-1}\ m^{-1}$. These are *n*-type and *p*-type.

Insulators are non-conducting and have least conductivity value ≈ 0.

6. (a) $26.8\ mg = \dfrac{26.8}{23} \times 10^{-3}$ mol

$= \dfrac{0.0268}{23} N_0$ sodium atoms

Electrons in 3 *s* orbital in one sodium atom (conduction band) = 1.

Thus, total conduction bands (energy levels) $= \dfrac{0.0268}{23} N_0$

$= 7.01 \times 10^{20}$

7. (c) Ti_2O_3, conducting below 410 K and insulating at about 150 K.

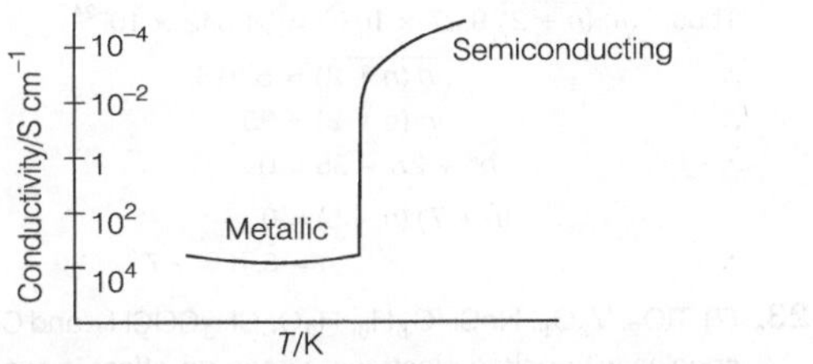

8. (b) Graphite has four valence electrons.

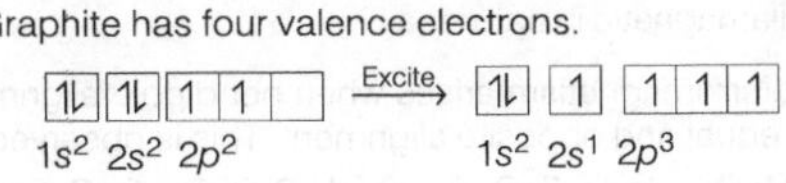

Three electrons are involved in covalent bonding with neighbouring atoms giving it a hexagonal type packing (layers) such that one layer can slide over the other. The fourth electron makes the graphite conducting. Hexagonal sheets are with sp^2- hybridised carbon.

Thus, both Statements I and II are correct but Statement II is not the correct explanation of Statement I.

9. (a) In solid state, metal ions of ferromagnetic substances are grouped together into small regions called domains. Each domain acts as a tiny magnet. In an unmagnetised ferromagnetic substance, domains are randomly oriented and their magnetic moments are cancelled. In a magnetic field, all the domains are oriented in the direction of the magnetic field producing a strong magnetic effect and magnetism persists permanently even after magnetic field is removed.

↑ ↑ ↑ ↑ ↑

Thus, both Statements I and II are correct and Statement II is the correct explanation of Statement I.

10. (c) Anti-ferromagnetic substances (as MnO) have following domain arrangements.

↑ ↓ ↑ ↓ ↑ ↓

These domains are oppositely oriented and cancel out each other's magnetic moment and making it zero (even in the absence of magnetic field).

Thus, Statement I is correct and Statement II is incorrect.

11. (b) Increase in temperature causes thermal agitation of metal ions and thus, electrical conductivity is decreased. Thus, Statement I is correct. In the presence of electric field, flow of electrons is impeded. Thus, Statement II is correct but is not the correct explanation of Statement I.

12. (b) Semiconductors have small energy gap between conduction band and valence band.

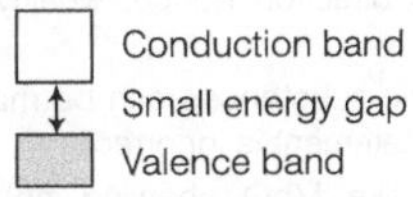

On increasing temperature, electron from valence band can jump to conduction band and thus, electrical conductivity is increased. Thus, both Statements I and II are correct but Statement II is not the correct explanation of Statement I.

13. (a,b) When Si or Ge is doped with a group 13 elements (B, Al, Ga) then, one valence electron of Si makes a hole or vacancy. When electric field is applied, electrons move towards positively charged plate through electronic holes. It would appear that electronic holes are positively charged and are moving towards negatively charged plate.

14. (a,b,c,d) Si (14) – [Ne] $3s^2\ 3p^2$ (group 14 element having four valence electrons). Thus, each atom of Si is covalently bonded to four neighbouring atoms.

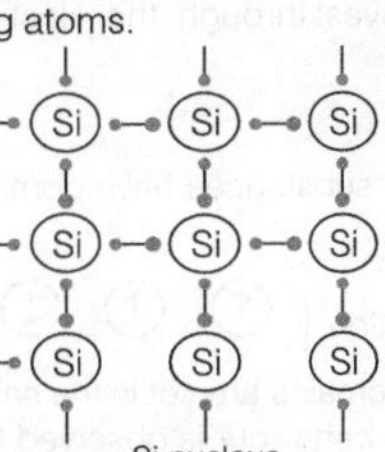

$\left.\begin{matrix}\text{As (33)}\\ \text{P (15)}\end{matrix}\right\} ns^2np^3$ (five valence electrons)

Group 15 elements Four electrons are involved in bonding with four electrons of Si and one remaining electron becomes delocalised and increases the conductivity of Si. Thus, increase in the conductivity is due to doped negatively charged electron. This is called *n*-type semiconductor.

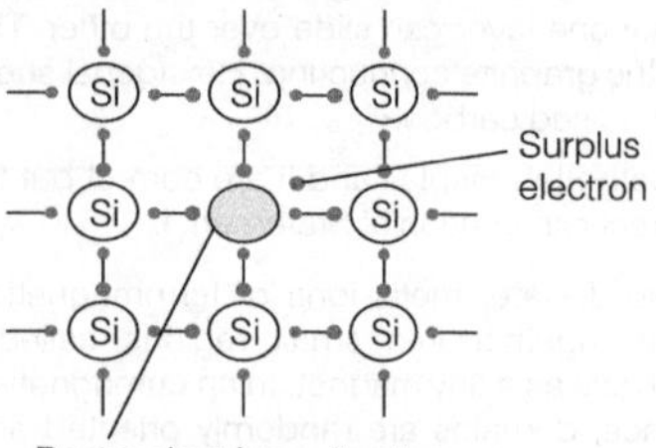

Group 13 elements

[B, Al, Ga] ns^2np^1

In this case, B, Al or Ga is short of one electron (as compared to Si). This is called **electron hole** or **electron vacancy**. Under the influence of electric field, electron hole moves towards the positively charged plate and is called *p*-type semiconductor.

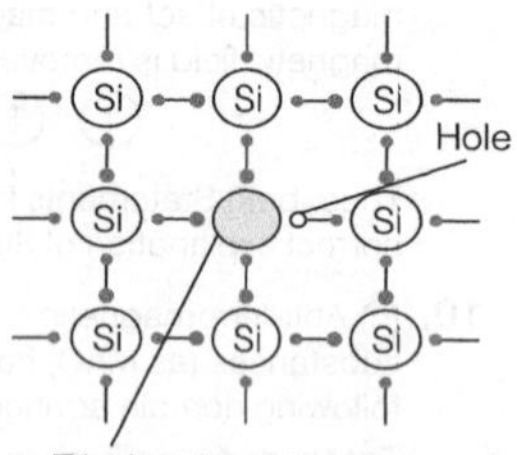

15. (a,b,d)

(a) | ⇅ | 1 | 1 | 1 | 1 |

Paramagnetic due to four unpaired electrons (*n*) and magnetic moment = $\sqrt{n(n+2)}$ BM

| ⇅ | ⇅ | ⇅ | | | All paired thus,

$n = 0$, Magnetic moment = 0

Thus, (a) is correct.

(b) Paramagnetic substances are magnetised in a magnetic field in the same direction hence, weakly attracted thus, (b) is correct.

(c) Ferromagnetic substances can be magnetised permanently, thus, given statement is incorrect.

(d) Substances like MnO showing anti-ferromagnetism have domain structure similar to ferromagnetic substances, but their domains are oppositely oriented and cancel out each other's magnetic moment.

↑ ↑ ↑ ↑ ↑ ↑ Ferromagnetic

↑ ↓ ↑ ↓ ↑ ↓ Anti-ferromagnetic

↑ ↑ ↓ ↓ ↑ ↑ Ferrimagnetic

16. (a,b) I and III have been doped with pentavalent impurity (P and As). There is one surplus electron and this results in the formation of *n*-type semiconductor. Surplus electron is responsible for conduction. II and IV have been doped with trivalent impurity (B and Al).

There is one electron short for bonding with Ge, hence a hole is formed. Hole moves through the electric field. This is *p*-type semiconductor.

17. (a,c,d)

(a) Ferrimagnetic substances have domains oriented in different direction.

Domains in different direction

On heating, domains are set in the same direction and paramagnetic behaviour is observed thus, correct.

Domains in same direction

(c) On heating, a small electric current is produced in polar compounds (it is called pyroelectricity) thus, correct.

(d) On heating, NaCl changes to CsCl structure with 8 : 8 coordination thus, correct.

18. (c)

(i) Si doped with As has one extra electron for conduction, thus *n*-type semiconductor.

(ii) Ge doped with B has short of one electron thus, a hole is formed, thus a *p*-type semiconductor.

(iii) Metals have sea of electrons responsible for electronic conduction.

(iv) NaCl (fused) $\longrightarrow Na^+ + Cl^-$

At anode, $Cl^- \longrightarrow Cl + e^-$

$$Cl \longrightarrow \frac{1}{2} Cl_2$$

At cathode, $Na^+ + e^- \longrightarrow Na$

Electron flow is due to electrolysis. Thus, electrolytic conduction.

Thus, (i) → (s), (ii) → (r), (iii) → (q), (iv) → (p).

Explanations for Q. Nos. (19-20)

P indicates that bonding molecular orbitals are empty.

Q indicates that bonding molecular orbitals are filled.

R indicates that bonding and antibonding molecular orbitals are filled.

Highest melting point is for the metal having filled bonding molecular orbitals (they have maximum bond order). Lowest melting point is for the metal having filled bonding and anti-bonding molecular orbitals thus, bond order is zero.

Increasing order of hardness of the metals is

$$R < P < Q$$

Hardness increases with increasing bond order.

19. (c) **20.** (b)

21. (5) Species with atleast one or more unpaired electron(s) are attracted by the magnetic field and is said to be paramagnetic.

Na^+, Zn^{2+}, Cu^+ and O_2^{2-}

No unpaired electron hence, diamagnetic.

O_2 – Two unpaired electrons in anti bonding molecular orbitals.

Fe^{2+} – Four unpaired electrons in 3*d*-orbital.

Cr^{3+} – Three unpaired electrons in 3*d*-orbital.

Cu^+ – One unpaired electron in 3*d*-orbital.

O_2^- – One unpaired electron in antibonding molecular orbitals.

22. (5) Magnetic moment = $\sqrt{n(n+2)} \neq \mu_B$

where, n = number of unpaired electrons

μ_B = Bohr magneton = 9.27×10^{-24} Am^2

Thus, $\sqrt{n(n+2)}\, 9.27 \times 10^{-24} = 54.842 \times 10^{-24}$

$\therefore \quad \sqrt{n(n+2)} = 5.916$

$\therefore \quad n(n+2) = 35$

$\therefore \quad n^2 + 2n - 35 = 0$

$(n+7)(n-5) = 0$

$\therefore \quad n = 5, n \neq -7$

23. (7) TiO_2, V_2O_5, NaCl, C_6H_6, H_2O, CH_3COCH_3 and CuCl are the species with paired electrons, hence no effect in magnetic field thus, diamagnetic in nature.

24. (9) Antiferromagnetism arises when net dipole alignment is zero due to equal and opposite alignment. This is observed in MnO, MnO_2, Mn_2O, FeO, Fe_2O_3, NiO, Cr_2O_3, CoO, Co_3O_4.

Revisal Problems (JEE Main)

1. (d) In ccp (cubic close packing),
number of atoms per unit cell = $N = 4$
Thus, tetrahedral voids $= 2N = 8$

2. (a) For CsCl, $\frac{r_+}{r_-} = 0.732$ to 1.000

$$r_+ = (0.732)r_- \text{ to } (1.000)r_-$$
$$= 0.732 \times 0.181 \text{ to } 1.000 \times 0.181$$
$$= 0.132 \text{ nm to } 0.181 \text{ nm}$$

3. (a) For octahedral holes,

$$\frac{r_{M^+}}{r_-} = 0.414$$

$\therefore$ $$r_{M^+} = 0.414 \times r_-$$
$$= 0.414 \times 500 = 207 \text{ pm}$$

4. (a) $\frac{r(Mn^{2+})}{r(S^{2-})} = \frac{0.080 \text{ nm}}{0.184 \text{ nm}} = 0.435$

For NaCl, $\frac{r_+}{r_-} = 0.414$ to 0.732

Thus, structure of the cubic unit cell is that of NaCl.

Note

Radius ratio	Coordination number	Structure
0.225-0.414	4	ZnS
0.414-0.732	6	NaCl
0.732-1.000	8	CsCl

5. (c)

$$d(\text{density}) = \frac{Mz}{N_0 a^3}$$

$\therefore$ $$z = \frac{dN_0 a^3}{M}$$

$$= \frac{3.18 \text{ g cm}^{-3} \times 6.02 \times 10^{23} \text{ mol}^{-1} \times (0.5451 \times 10^{-7} \text{ cm})^3}{78 \text{ g mol}^{-1}}$$

$$= 3.975 \approx 4$$

Unit cell contains 4 CaF_2 molecules.

Thus, $Ca^{2+} = 4$

$F^- = 8$

6. (b)

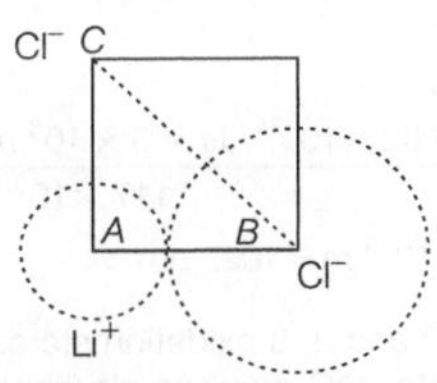

Li^+ ion is very small thus,

$$r_{Li^+} + r_{Cl^-} = a = 514 \text{ pm}$$

$\therefore$ $$a = \frac{514}{2} = 257 \text{ pm}$$

BC = distance between the centres of two Cl^- ions

$$= \sqrt{AB^2 + AC^2}$$
$$= \sqrt{a^2 + a^2} = \sqrt{2}a$$

$\therefore$ $$r_{Cl^-} = \frac{BC}{2} = \frac{\sqrt{2}a}{2} = \frac{a}{\sqrt{2}}$$

$$= \frac{257}{\sqrt{2}} = 182 \text{ pm}$$

7. (d) For nth order diffraction, $n\lambda = 2a \sin\theta$

$\therefore$ $$\sin\theta = \frac{n\lambda}{2a}$$
$$n = 2$$

$\therefore$ $$\sin\theta = \frac{\lambda}{a} = \frac{200 \text{ pm}}{400 \text{ pm}} = \frac{1}{2}$$

$\therefore$ $$\theta = \sin^{-1}\left(\frac{1}{2}\right) = 30°$$

8. (b) In closed packed structure lattice,
number of octahedral voids = 1
number of tetrahedral voids = 2

	Void	Ions	Simple ratio $(A_2B)^{x+}$
A^{3+}	1	1/2	2
B^{2+}	2	2/8	1

Charge on $(A_2B)^{x+} = (2 \times 3 + 2 \times 1) = +8$

Thus, number of O-atoms $= \frac{8}{2} = 4$ (to balance charge)

Lattice is A_2BO_4.

9. (a)

Atom	Location	Number of atoms
X	Corners	$\frac{8}{8} = 1$
Y	Centre	$\frac{1}{1} = 1$

Thus, XY

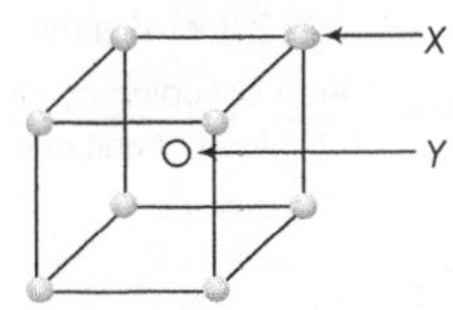

10. (c) Let the number of atoms in cubic close packing be x thus, number of tetrahedral voids $= 2x$
Number of atoms of $Q = x$ (due to cubic close packing)
P atoms occupy one-third of tetrahedral voids, thus number of P atoms $= \frac{2}{3}x$

Thus, $$\frac{P \text{ atoms}}{Q \text{ atoms}} = \frac{2/3\, x}{x} = \frac{2}{3}$$

Thus, compound is P_2Q_3.

11. (a) NaCl has fcc structure.
Thus, number of NaCl molecules in a unit cell = 4

$$\text{Density} = \frac{\text{Mass of 4(NaCl) molecules}}{\text{Volume of cube}}$$

Molar mass of NaCl = 58.5 g mol^{-1} = 58.5×10^{-3} kg mol^{-1}

$$\text{Mass of one molecule} = \frac{58.5 \times 10^{-3}}{6.02 \times 10^{23}} \text{ kg}$$

$\therefore$ $$2.17 \times 10^3 \text{ kg m}^{-3} = \frac{4 \times 58.5 \times 10^{-3}}{6.02 \times 10^{23} \times a^3}$$

$\therefore$ $$a^3 = \frac{4 \times 58.5 \times 10^{-3}}{6.02 \times 10^{23} \times 2.17 \times 10^3}$$
$$= 1.791 \times 10^{-28} \text{ m}^3$$
$$a = 5.64 \times 10^{-10} \text{ m} = 564 \times 10^{-12} \text{ m}$$
$$= 564 \text{ pm}$$

12. (b) Frenkel defects do not change the value of the theoretical density because the atom is only occupying an interstitial position instead of lattice position. But in case of Schottky defect,

$$z = 4\left(1 - \frac{0.1}{100}\right) = 3.996$$

$$d\,(\text{density}) = \frac{M\,z}{N_0 a^3}$$

$\frac{M}{N_0 a^3}$ is constant, thus $d \propto z$

$$z(\text{Frenkel}) = 4$$

$$z(\text{Schottky}) = 3.996$$

Thus, $d_f > d_s$

13. (c) For similar structures,

$$\frac{l(\text{B—N})}{l(\text{C—C})} = \frac{a(\text{B—N})}{a(\text{C—C})}$$

$$l(\text{B—N}) = \frac{a(\text{B—N})}{a(\text{C—C})} \times l(\text{C—C})$$

$$\therefore \quad l(\text{B—N}) = \frac{0.251}{0.2455} \times 0.1415$$

$$= 0.145 \text{ nm}$$

14. (a) Density $(d) = \frac{Mz}{N_0 a^3}$

$$\therefore \quad M = \frac{dN_0 a^3}{z}$$

$$= 3.048 \times 10^4 \text{ g mol}^{-1}$$

Since, protein fraction is 0.75, hence

$$M(\text{protein}) = 3.048 \times 10^4 \times 0.75$$

$$= 2.286 \times 10^4$$

$$\simeq 2.3 \times 10^4 \text{ g mol}^{-1}$$

15. (a) Packing fraction $= \frac{\text{Area occupied by circles}}{\text{Area of unit cell}}$

Let edge length $= l$

Thus, radius $(r) = \frac{l}{2}$

From figure, it is clear that parallelogram has one complete circle. Thus,

$$\text{Area of circle} = \pi r^2 = \pi\left(\frac{l}{2}\right)^2$$

Area of parallelogram = 2(area of ΔABC)

$$\text{Area of } \Delta ABC = \frac{1}{2} BC \times AD$$

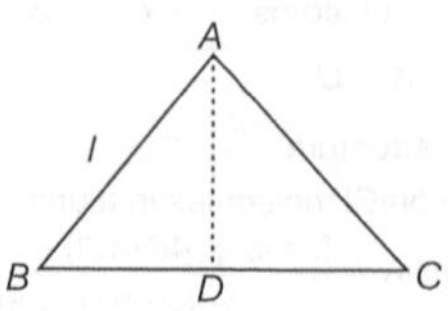

$$AD^2 = AB^2 - BD^2$$

$$= l^2 - \frac{l^2}{4} = \frac{3l^2}{4}$$

$$\therefore \quad AD = \sqrt{3}\left(\frac{l}{2}\right)$$

$$\therefore \quad \text{Area of } \Delta ABC = \frac{1}{2}\left(l \times \sqrt{3}\frac{l}{2}\right) = \sqrt{3}\frac{l^2}{4}$$

$$\therefore \quad \text{Area of parallelogram} = 2 \times \text{Area of } \Delta ABC$$

$$= 2\sqrt{3}\frac{l^2}{4} = \sqrt{3}\frac{l^2}{2}$$

$$\therefore \quad \text{Packing fraction} = \frac{\pi\, l^2/4}{\sqrt{3}\, l^2/2}$$

$$= \frac{\pi}{2\sqrt{3}} = 0.907$$

16. (c) In bcc structure, Cl^- ions lie at the corners of simple cubic and Cs^+ at the body centre. Hence, along the body diagonal, Cs^+ and Cl^- touch each other.

$$r_{Cs^+} + r_{Cl^-} = 2r$$

But $\quad r = \frac{\sqrt{3}}{4}a$

$$\therefore \quad r_{Cs^+} + r_{Cl^-} = 2\frac{\sqrt{3}}{4}a = \frac{\sqrt{3}}{2}a$$

17. (a) In MT, $(T^-—T^-) = 2.40$ Å

$\therefore \quad r_{T^-} = 1.20$ Å

If $\quad r_{M^+} + r_{T^-} = 1.70$ Å

then $\quad r_{M^+} = 0.50$ Å

In MQ, $\quad (Q^-—Q^-) = 1.63$ Å, $r_{Q^-} = 0.815$ Å

$r_{M^+} = 0.335$ Å

In RT, $\quad (T^-—T^-) = 2.66$ Å, $r_{T^-} = 1.33$ Å

$\therefore \quad r_{R^+} = 0.55$ Å (in RT)

and $\quad r_{M^+} = 1.70 - 1.33$ (in MT) $= 0.37$ Å

In RQ, $\quad (Q^-—Q^-) = 2.09$, $r_{Q^-} = 1.045$ Å

$r_{R^+} = 1.48 - 1.045 = 0.435$ Å (in RQ)

$r_{M^+} = 1.15 - 1.045 = 0.105$ Å (in MQ)

M^+ (in MT) $> M^+$ (in MQ)

T^- (in MT) $< T^-$ (in RT)

Thus, M^+ is not touching T^-.

Thus, (a) is correct.

18. (a) $E_g = 117 \text{ kJ mol}^{-1}$

$$= \frac{117 \times 10^3}{6.02 \times 10^{23}} \text{ per electron}$$

$$E_g = h\nu = \frac{hc}{\lambda}$$

$$\therefore \quad \lambda = \frac{hc}{E_g} = \frac{6.62 \times 10^{-34} \text{ Js} \times 3 \times 10^8 \text{ ms}^{-1} \times 6.02 \times 10^{23}}{117 \times 10^3 \text{ J}}$$

$$= 1022 \times 10^{-9} \text{ m} = 1022 \text{ nm}$$

19. (b) Ni shows + 2 and + 3 oxidation states while Na shows only +1 oxidation state. NiO can contain some Ni^{2+} ion vacancies with two remaining Ni^{2+} ions oxidised to Ni^{3+} to compensate for each Ni^{2+} vacancy. Thus, NiO contains more defects than NaCl. Thus, Statement I is correct. NaCl is an ionic compound. Thus, Statement II is also correct but not the correct explanation of Statement I.

20. (a) hcp and ccp for a given element have same coordination number hence, the same packing efficiency.

Thus, both Statements I and II are correct and Statement II is the correct explanation of Statement I.

Revisal Problems (JEE Advanced)

1. (d) By Bragg's equation,

$$2a \sin \theta = n\lambda$$

Given, $n = 2$, $\sin \theta = \sin 22.2 = 0.378$

$$\lambda = 154 \times 10^{-12} \text{ m}$$
$$= 154 \times 10^{-10} \text{ cm}$$
$$z = 4 \text{ (for fcc)}$$

$$\therefore \quad a = \frac{n\lambda}{2 \sin \theta} = \frac{2 \times 154 \times 10^{-10}}{2 \times 0.378}$$
$$= 4.074 \times 10^{-8} \text{ cm}$$
$$d = \frac{Mz}{N_0 a^3} = \frac{197 \times 4}{6.02 \times 10^{23} \times (4.074 \times 10^{-8})^3}$$
$$= 19.36 \text{ g cm}^{-3}$$

2. (a) Given structure represents octahedral void for which

$$\frac{r_+}{r_-} = 0.414$$
$$\therefore \quad r(A^+) = 0.414\, r(X^-)$$
$$= 0.414 \times 250 \text{ pm}$$
$$= 104 \text{ pm}$$

3. (a) V is in + 5 oxidation (highest) state in V_2O_5 and is likely to undergo reduction.

$$V^{5+} + ne^- \longrightarrow V^{(5+ne^-)}$$

Thus, it forms n-type semiconductor.
Co is in +2 oxidation (lowest) state in CoO and thus, can undergo oxidation.

$$Co^{2+} \longrightarrow Co^{3+} + e^-$$

Thus, holes are formed in the conduction band of the metal and the charge carriers are now positive, it is a case of p-type semiconductor.

4. (d) $r_n = \frac{n^2 a_0}{Z_{\text{eff}}}$, where, n = orbit, a_0 = Bohr's radius

Thus, $$r_n \propto \frac{1}{Z_{\text{eff}}}$$

($n = 2$ both Na^+ and F^- being isoelectronic)

$$\frac{r(Na^+)}{r(F^-)} = \frac{Z_{\text{eff}}(F^-)}{Z_{\text{eff}}(Na^+)} = \frac{9 - 4.15}{11 - 4.15}$$
$$= \frac{4.85}{6.85} = 0.708$$

Note *Due to shielding effect, $Z_{\text{eff}} = (Z - \text{shielding factor})$*

5. (c) Structure is cubic closest packed (fcc).

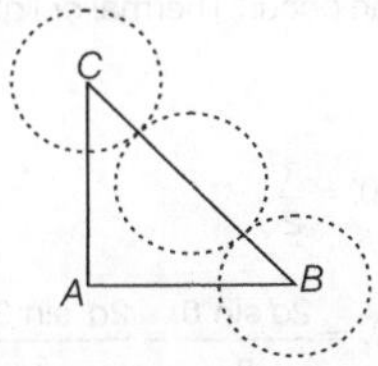

Thus, $$z = 4$$
$$BC = 4r$$
$$AB^2 + AC^2 = BC^2$$
$$a^2 + a^2 = (4r)^2$$
$$2a^2 = 16r^2$$
$$\therefore \quad a^2 = 8r^2$$
$$\therefore \quad a = 2\sqrt{2}r$$
$$= 2 \times \sqrt{2} \times 141.4 \text{ pm} = 400 \text{ pm}$$

6. (a) For fcc structure, radius (r) is related to the edge length by

$$r = \frac{a}{2\sqrt{2}}$$
$$\text{Density } (d) = \frac{Mz}{N_0 a^3}$$

For fcc, number of atoms per unit cell = 4

$$\therefore \quad a^3 = \frac{Mz}{N_0 d} = \frac{197 \times 4}{6.02 \times 10^{23} \times 19.3} \text{ cm}^3$$
$$= 6.782 \times 10^{-23} \text{ cm}^3$$
$$a = 4.08 \times 10^{-8} \text{ cm} = 4.08 \times 10^{-10} \text{ m}$$
$$= 408 \times 10^{-12} \text{ m} = 408 \text{ pm}$$
$$\therefore \quad r = \frac{a}{2\sqrt{2}} = \frac{408}{2\sqrt{2}} = 144.24 \text{ pm} \approx 144 \text{ pm}$$

7. (c) In bcc structure, radius (r) is related to edge length by

$$r = \frac{\sqrt{3}}{4} a = \frac{\sqrt{3}}{4} \times 429 \text{ pm} = 185.76 \approx 186 \text{ pm}$$

8. (d) Contribution of circles on the corners of the square $= \frac{1}{4}$

Number of circles per square to be taken in packing

$$= 4 \text{ (corner)} \times \frac{1}{4} + 1 \text{ (centre)} = 2$$

Area occupied by two circles $= 2\pi r^2$

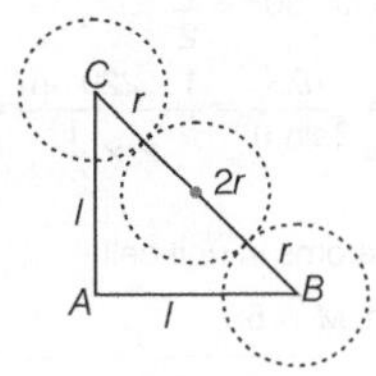

Area of square $= l^2$

But $l = AB$

Also, $$AB^2 + AC^2 = BC^2$$
$$\therefore \quad 2l^2 = 16r^2$$
$$\therefore \quad l^2 = 8r^2$$

Thus, packing efficiency

$$= \frac{\text{Area of two circles}}{\text{Area of square}} = \frac{2\pi r^2}{8r^2} = \frac{\pi}{4}$$
$$= 0.7854 = 78.54\%$$

9. (a) $BC = a$ = edge length

$AB = 2R = a$

where, R = radius

AE is a line through the triangular hole.

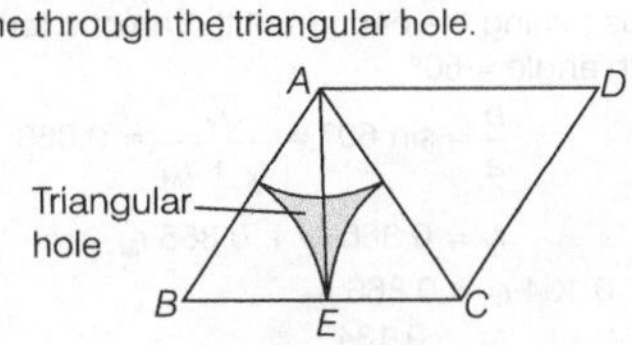

$$BE = \frac{a}{2}$$
$$AE^2 = AB^2 - BE^2$$
$$= a^2 - \frac{a^2}{4} = \frac{3a^2}{4}$$
$$\therefore \quad AE = \sqrt{3}\,\frac{a}{2}$$

The triangular hole is at the distance of $\frac{2}{3}(AE)$.

Thus, $r = \frac{2}{3}(AE) - R \Rightarrow \frac{2}{3}\sqrt{3}\frac{a}{2} - R \Rightarrow \frac{a}{\sqrt{3}} - R$

$$= \frac{2R}{\sqrt{3}} - R \Rightarrow R\left(\frac{2}{\sqrt{3}} - 1\right) \Rightarrow R\left(\frac{2}{1.732} - 1\right)$$

$$= R\,(1.1547 - 1) = 0.1547\,R$$

10. (a) By Bragg's equation,

$$2a\sin\theta = n\lambda$$

Since, $n = 1$ (first order diffraction)

$\therefore \quad \lambda = 2a\sin\theta$

$\therefore \quad = 2 \times 264 \times 10^{-12}\text{ m} \times \sin 7.75$

$= 2 \times 264 \times 10^{-12}\text{ m} \times 0.135$

$= 7.128 \times 10^{-11}\text{ m}$

By energy relation, $\Delta E = h\nu = \frac{hc}{\lambda}$

$$= \frac{6.62 \times 10^{-34}\text{ Js} \times 3 \times 10^{8}\text{ ms}^{-1}}{7.128 \times 10^{-11}\text{ m}}$$

$$= 27.86 \times 10^{-16}\text{ J}$$

11. (b) By Bragg's equation,

$$2a\sin\theta = n\lambda$$

$n = 1$ (first order diffraction)

$\lambda = 229$ pm

$\sin\theta = \sin 30° = \frac{1}{2}$

$\therefore \quad a = \frac{n\lambda}{2\sin\theta} = \frac{1 \times 229\text{ pm}}{2 \times \frac{1}{2}} = 229\text{ pm}$

12. (c) z = number of atoms in unit cell

a = edge length, $M = 56$

Type	z	a related to r	$d = \frac{Mz}{a^3 N_0}$
bcc	2	$a = \frac{4}{\sqrt{3}}r$	$d(\alpha) = \frac{M \times 2}{\left(\frac{4}{\sqrt{3}}r\right)^3 N_0}$
fcc	4	$a = 2\sqrt{2}r$	$d(\gamma) = \frac{M \times 4}{(2\sqrt{2}r)^3 N_0}$

$$\therefore \quad \frac{d(\alpha)}{d(\gamma)} = \frac{2\,(2\sqrt{2})^3}{4\left(\frac{4}{\sqrt{3}}\right)^3}$$

$$= \frac{1}{2}\left(\frac{\sqrt{6}}{2}\right)^3 = 0.92$$

13. (c) The lines joining the centres of the anions form an equilateral triangle with angle = 60°.

Thus, $\frac{b}{a} = \sin 60° = \frac{r_X}{r_X + r_M} = 0.866$

$\therefore \quad r_X = 0.866\,r_X + 0.866\,r_M$

$0.134\,r_X = 0.866\,r_M$

$\therefore \quad \frac{r_M}{r_X} = \frac{0.134}{0.866} = 0.155$

14. (c) Fraction is expressed in terms of

$$\frac{n}{N} = \exp\left(-\frac{E}{RT}\right)$$

$E = 112.01 \times 10^3\text{ J mol}^{-1} \times \frac{50}{100}$

$R = 8.3143\text{ J mol}^{-1}\text{ K}^{-1}$

$T = 298$ K

$$\therefore \quad \frac{n}{N} = \exp\left(-\frac{112.01 \times 10^3 \times 0.5}{8.3143 \times 298\text{ K}}\right)$$

$= \exp(-22.604)$

$= 1.53 \times 10^{-10}$

15. (b) $d_{hkl}^2 = \frac{a^2}{(h^2 + k^2 + l^2)}$

$\therefore \quad a^2 = d_{hkl}^2\,(h^2 + k^2 + l^2)$

$= (0.3080)^2\,(1^2 + 1^2 + 1^2)$

$= (0.3080)^2 \times 3$

$\therefore \quad a = 0.533$ nm

By Bragg's equation,

$$2a\sin\theta = n\lambda$$

$\therefore \quad \sin\theta = \frac{n\lambda}{2a} = \frac{1 \times 0.533\text{ nm}}{2 \times 0.533\text{ nm}} = \frac{1}{2} = 0.5$

16. (a) If B-atoms are in fcc lattice in boron phosphide (BP), then the P-atom must be in $\frac{1}{2}$ of the tetrahedral holes.

17. (b, c) By Bragg's equation,

$$2d\sin\theta = n\lambda$$

$\therefore \quad \sin\theta = \frac{n\lambda}{2d}$

$\sin\theta = \frac{n\lambda}{2d_{hkl}}$

(a) $\sin\theta = \frac{1 \times \lambda}{2 \times 0.321} = \frac{\lambda}{0.642}$

(b) $\sin\theta = \frac{2\lambda}{2 \times 0.161} = \frac{\lambda}{0.161}$

(c) $\sin\theta = \frac{1 \times \lambda}{0.161} = \frac{\lambda}{0.161}$

(d) $\sin\theta = \frac{2\lambda}{2 \times 0.321} = \frac{\lambda}{0.321}$

18. (a)

I. An *n*-type semiconductor operates by thermally promoting an electron from an impurity level to an empty metal band so that conduction can occur.

II. A *p*-type semiconductor operates by thermally promoting an electron from a completely filled band to a partially empty purity level so that conduction can occur in the originally filled band. The concept of hole is often used to describe conduction in this type of semiconductor.

III. If sufficient energy is available to promote an electron from the completely filled band to the completely empty band, conduction can occur. Thermal or light energy is the source of energy.

19. (c)

(i) $\theta = 30°$, $\sin 30° = \frac{1}{2}$

$n = 1$

$\therefore \quad \lambda = \frac{2d\sin\theta}{n} = \frac{2d\sin 30°}{1} = 2d \times \frac{1}{2} = d$

Thus, **(i) → (q)**

(ii) $\theta = 45°$, $\sin 45° = \frac{1}{\sqrt{2}}$

$n = 2$

$\therefore \quad \lambda = \frac{2d\sin\theta}{n} = \frac{2 \times \frac{1}{\sqrt{2}}d}{2}$

$\therefore \quad d = \sqrt{2}\lambda$

Thus, **(ii) → (r)**

(iii) $\theta = 60°$, $\sin 60° = \frac{\sqrt{3}}{2}$

$n = 3$

$$\lambda = \frac{2d \sin \theta}{n} = \frac{2d \times \sqrt{3}/2}{3} = \frac{d}{\sqrt{3}}$$

$\therefore$ $d = \sqrt{3}\lambda$

Thus, **(iii) → (s)**

(iv) $\theta = 90°$, $\sin 90° = 1$

$n = 4$

$\therefore$ $\lambda = \frac{2d \sin \theta}{n} = \frac{2d \times 1}{4} = \frac{d}{2}$

$\therefore$ $d = 2\lambda$

Thus, **(iv) → (p)**

Explanations for Q. Nos. (20-21)

20. (a) λ is related to electrical potential by de-Broglie's equation,

$$\lambda = \frac{h}{\sqrt{2m\,(KE)}}$$

$KE = 150 \text{ eV} = 150 \times 1.602 \times 10^{-19}$ J

$m_e = 9.11 \times 10^{-31}$ kg

$$\therefore \quad \lambda = \frac{6.6 \times 10^{-34}}{\sqrt{2 \times 9.11 \times 10^{-31} \times 1.602 \times 10^{-19} \times 150}}$$

$= 9.97 \times 10^{-11}$ m

$= 99.7 \times 10^{-12}$ m

$= 99.7$ pm

21. (d) $n\lambda = 2d \sin \theta$

$$\therefore \quad d = \frac{n\lambda}{2 \sin \theta} = \frac{1 \times 99.7 \times 10^{-12} \text{ m}}{2 \times 0.33}$$

$= 1.51 \times 10^{-10} \text{ m} = 151 \times 10^{-12}$ m

$= 0.151 \times 10^{-9}$ m

$= 1.51 \text{ Å} = 151 \text{ pm} = 0.151$ nm

Explanations for Q. Nos. (22-23)

Divalent cation Mg^{2+} ions are in $\frac{1}{8}$ of the eight possible tetrahedral holes and trivalent cation Al^{3+} ions are in $\frac{1}{2}$ of the four octahedral holes in $MgAl_2O_4$ ($MgO \cdot Al_2O_3$).

Similarly, in $FeCr_2O_4$ ($FeO \cdot Cr_2O_3$).

22. (c) **23.** (a)

Explanations for Q. Nos. (24-25)

Density = 4.0 g cm^{-3}

1 mole = 6.02×10^{23} molecules

Hence, mass of one molecule = $\frac{168.56}{6.02 \times 10^{23}}$ g $molecule^{-1}$

$$\text{Density} = \frac{\text{Mass}}{\text{Volume}}$$

$\therefore$ Volume of one pair of Cs^+Cl^- = $\frac{168.56}{6.02 \times 10^{23} \times 4.0}$

$= 7.0 \times 10^{-23}$ cm^3

Thus, number of Cs^+Cl^- molecules/cm^3 = $\frac{1}{7.02 \times 10^{-23}}$

$= 1.4286 \times 10^{22}$

N = number of Cs^+Cl^- molecules cm^{-1}

$= \sqrt[3]{1.4286 \times 10^{22}}$

$= 2.426 \times 10^7$ units

Thus, a (edge length) = $\frac{1}{N} = \frac{1}{2.426 \times 10^7}$

$a = 4.122 \times 10^{-8}$ cm

The Cs to Cl distance is the distance from the corner to the centre of the cube of edge 4.122×10^{-8} cm

$= \frac{\sqrt{3}}{2}(4.122 \times 10^{-8})$

$= 3.57 \times 10^{-8} = 3.57$ Å

24. (a) **25.** (a)

26. (6) Let molar mass be M.

$$\text{Moles} = \frac{200}{M}$$

$$\text{Atoms} = \frac{200\,N_0}{M}$$

Also, $\quad$ density (d) = $\frac{Mz}{N_0 a^3}$

This unit cell is simple cubic, thus number of atoms in unit cell, $z = 1$

$$13.82 = \frac{M}{N_0} \frac{1}{2.4 \times 10^{-23}}$$

$$\frac{N_0}{M} = \frac{1}{13.82 \times 2.4 \times 10^{-23}}$$

$\therefore$ Number of atoms $= 200 \frac{N_0}{M}$

$= \frac{200 \times 1}{13.82 \times 2.4 \times 10^{-23}}$

$= 6.03 \times 10^{23}$

$= y \times 10^{23}$

$\therefore$ $y = 6$

27. (8) Volume $= 12 \times 15 \times 20 \times 10^{-24}$

$= 3600 \times 10^{-24}$ cm^3

Density (d) = $\frac{Mz}{N_0 a^3}$

$\therefore$ $z = \frac{d\,N_0 a^3}{M}$

$= \frac{1.17 \times 6.02 \times 10^{23} \times 3600 \times 10^{-24}}{316.5}$

$= 8.01 \approx 8$

28. (8) As is clear from the figure,

Location (share)	Number of ions	Contribution
Body (1)	4	$4 \times 1 = 4$
Face $\left(\frac{1}{2}\right)$	6	$6 \times \frac{1}{2} = 3$
Edge $\left(\frac{1}{4}\right)$	0	$0 \times \frac{1}{4} = 0$
Corner $\left(\frac{1}{8}\right)$	8	$8 \times \frac{1}{8} = 1$
		Total = Eight

29. (4) I to IV are possible close-packing schemes. If the positions of the spheres in the first layer of a close-packed arrangement are designated *A*, those of adjacent layers can only be *B* or *C*, with the spheres of the adjacent layer in the indentations of the first layer. Thus, (V) and (VI) repeating the same positions (*BB*) and (*CC*) are not acceptable.

30. (7) There are seven primitive unit cells in a crystal.

System	Unit cell		Example (Common name)
	Lengths	Angles	
Cubic	$a=b=c$	$\alpha=\beta=\gamma=90°$	NaCl (rock salt)
Tetragonal	$a=b\neq c$	$\alpha=\beta=\gamma=90°$	TiO_2 (rutile)
Orthorhombic	$a\neq b\neq c$	$\alpha=\beta=\gamma=90°$	$MgSO_4\cdot 7H_2O$ (epsomite)
Monoclinic	$a\neq b\neq c$	$\alpha=\gamma=90°$, $\beta\neq 90°$	$CaSO_4\cdot 2H_2O$ (gypsum)
Triclinic	$a\neq b\neq c$	$\alpha\neq\beta\neq\gamma\neq 90°$	$K_2Cr_2O_7$ (potassium dichromate)
Hexagonal	$a=b\neq c$	$\alpha=\beta=90°$, $\gamma=120°$	SiO_2 (silica)
Rhombohedral	$a=b=c$	$\alpha=\beta=\gamma\neq 90°$	$CaCO_3$ (calcite)

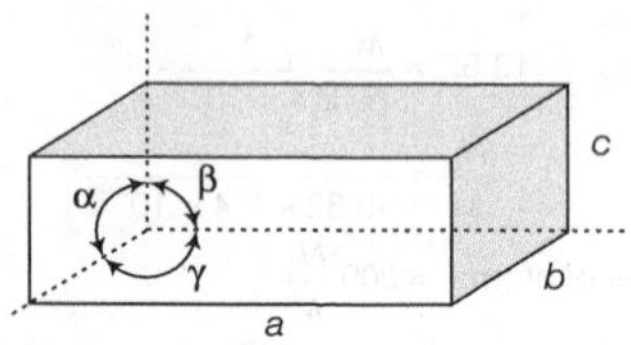

Unit cell can be
primitive or simple unit cell denoted by P,
face-centred unit cell denoted by F,
body-centred unit cell denoted by I,
side-centred unit cell denoted by C.

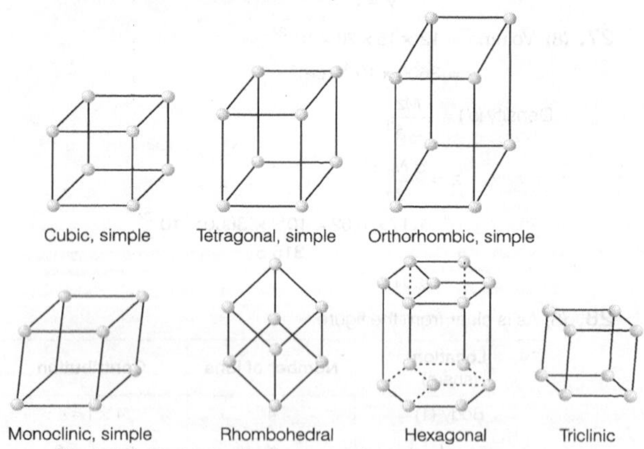

Simple (primitive) unit cells of seven crystal systems

31. (6) If there is small deviation from stoichiometry, then $[Ni^{2+}]$ and $[O^{2-}]$ can be taken as constant.

$$K_{eq} = \frac{[Ni^{3+}]^4\,[\Box_+]^2}{p_{O_2}}$$

Also, $$[\Box_+] = \frac{1}{2}[Ni^{3+}]$$

$\therefore$ $$K_{eq} = [Ni^{3+}]^6/p_{O_2}$$

$\therefore$ $$\frac{[Ni^{3+}]}{p_{O_2}^{1/6}} = K_{eq}$$

$\therefore$ $$[Ni^{3+}] \propto p_{O_2}^{1/6}$$

$\therefore$ $$n = 6$$

32. (4) Density of ice = 0.92 g/cm^3

Volume of unit cell $= a \times a \sin 60° \times c$

$$= a^2 \sin 60° \times c$$

$$= (4.53\times 10^{-8})^2\,\frac{\sqrt{3}}{2}\times 7.41\times 10^{-8}\text{ cm}^3$$

$$= 131.688\times 10^{-24}\text{ cm}^3$$

$$d = \frac{M}{a^3N_0}$$

$\therefore$ $$M\text{ (unit cell)} = da^3\,N_0$$

$$= 0.92\times 131.688\times 10^{-24}\times 6.02\times 10^{23}$$

$$= 71.35$$

Molar mass of H_2O = 18 g mol^{-1}

Thus, number of H_2O molecules in one unit cell $= \frac{71.35}{18} = 4$

33. (7) Ag crystallises in fcc unit cell with $z = 4$

$$d = \frac{Mz}{a^3N_0}$$

$$a^3 = \frac{Mz}{d\,N_0}$$

$$= \frac{108\times 4}{10.5\times 6.02\times 10^{23}\text{ cm}^3}$$

$$= 6.83\times 10^{-23}\text{ cm}^3$$

$\therefore$ $$a = 4.088\times 10^{-8}\text{ cm}$$

$\therefore$ $$\text{Area} = a^2 = 1.67\times 10^{-15}\text{ cm}^2$$

$$= 1.67\times 10^{-19}\text{ m}^2 \text{ containing one atom}$$

Hence, 10^{-12} m^2 has atoms $= \frac{10^{-12}}{1.67\times 10^{-19}}$

$$= 6.0\times 10^6 \text{ atoms}$$

Each surface has two Ag-atoms.

Thus, total Ag-atoms $= 6.0\times 10^6\times 2$

$$= 1.2\times 10^7$$

$$= y\times 10^x$$

Hence, $x = 7$

JEE Main & AIEEE Archive

1. (b) For bcc lattice,

$$r = \frac{\sqrt{3}}{4} a = \frac{\sqrt{3}}{4} \times 4.29 \text{ Å} = 1.86 \text{ Å}$$

2. (c) In CsCl, Cl^- lie at the corners of simple cube and Cs^+ at the body centre. Therefore, along the body diagonal, Cs^+ and Cl^- touch each other.

Hence, $\frac{\sqrt{3}}{2} a = r_{Cs^+} + r_{Cl^-}$

3. (b) Silicon exists as covalent crystal in solid state (Network like structure, like diamond).

4. (b) $M_{0.98}O$

Consider one mole of the oxide.

Moles of M = 0.98 mole of O^{2-} = 1

Let moles of $M^{3+} = x$

$$\text{Moles of } M^{2+} = 0.98 - x$$

On balancing charge,

$$(0.98 - x) \times 2 + 3x - 2 = 0$$

$$\Rightarrow \quad 1.96 - 2x + 3x - 2 = 0 \quad \Rightarrow \quad x = 0.04$$

$$\text{Percentage of } M^{3+} = \frac{0.04}{0.98} \times 100 = 4.08\%$$

5. (d) In body centred cubic structure,

Edge length, a = 351 pm, radius, r = ?

Then, $$r = \frac{\sqrt{3}a}{4} = \frac{\sqrt{3} \times 351}{4}$$

$$= 152 \text{ pm}$$

6. (d) Number of atoms (A) per unit cell

$$= 8 \times \frac{1}{8} = 1$$

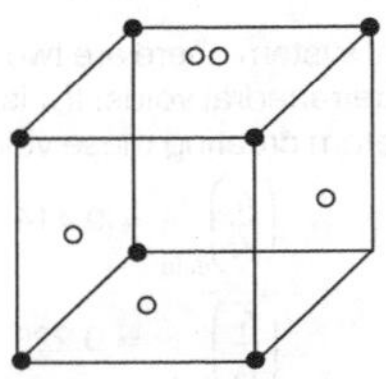

Number of atoms (B) per unit cell

$$= (6 - 1) \times \frac{1}{2}$$

$$= \frac{5}{2} \text{ (one atom } B \text{ is missing)}$$

Thus, formula is $A_1B_{5/2} = A_2B_5$

7. (c) Copper crystallises in fcc lattice.

If r = radius

a = edge length

then $r = \frac{a}{2\sqrt{2}} = \frac{361}{2\sqrt{2}}$ pm = 127.633 pm ≈ 128 pm

8. (d) For fcc arrangement,

$$2(r^+ + r^-) = \text{edge length}$$

$$2(110 + r^-) = 508$$

So, $$r^- = 144 \text{ pm}$$

9. (b) Packing fraction of ccp $= \frac{\pi}{3\sqrt{2}} = 0.74 = 74\%$

% free space in ccp = 26%

Packing fraction of bcc $= \frac{\pi\sqrt{3}}{8} = 0.68 = 68\%$

% free space in bcc = 32%

10. (a) Suppose atoms of element Y in ccp = 100

Number of tetrahedral voids = 2 × 100

Number of atoms of element

$$X = \frac{2}{3} \times 200 = \frac{400}{3}$$

$$\frac{X}{Y} = \frac{400}{300}$$

Formula = X_4Y_3

11. (d) Total number of atoms in face-centred cubic unit cell = 4

At corner $= 8 \times \frac{1}{8} = 1$

On faces $= 6 \times \frac{1}{2} = 3$

Total = 4

Total volume $= 4 \times \frac{4\pi r^3}{3} = \frac{16}{3}\pi r^3$

12. (b) Unit cell consists of A ions at the corners.

Thus, number of ions of the type $A = \frac{8}{8} = 1$

Unit cell consists of B ions at the centre of the six faces.

Thus, number of ions of the type $B = \frac{6}{2} = 3$

(Each corner is shared by 8 cubes and each face is shared by 2 faces)

Thus, formula is AB_3.

13. (b) When equal number of cations and anions (such that charges are equal) are missing ($1Na^+$, $1Cl^-$ / $1 Fe^{2+}$, $2Cl^-$) then, it is a case of Schottky defect.

14. (a) Mass of one unit cell (m) = Volume × Density

$$= a^3 \times d = a^3 \times \frac{Mz}{N_0 a^3} = \frac{Mz}{N_0}$$

$$m = \frac{58.5 \times 4}{6.02 \times 10^{23}} \text{ g}$$

$$\therefore \text{ Number of unit cells in 1 g} = \frac{1}{m} = \frac{6.02 \times 10^{23}}{58.5 \times 4}$$

$$= 2.57 \times 10^{21}$$

15. (c) In Na unit cell (body centred cubic),

$$8 \text{ (corner atoms)} \times \frac{1}{8} = 1$$

Atoms placed inside the unit cell = 1

Total atoms = 2

In Mg unit cell (face centred cubic),

$$8 \text{ (corner atoms)} \times \frac{1}{8} = 1$$

$$6 \text{ (face atoms)} \times \frac{1}{2} = 3$$

Total atoms = 4

JEE Advanced & IIT JEE Archive

1. (a) **Plan** Given arrangement represents octahedral void and for this,

$$\frac{r_+ \text{ (cation)}}{r_- \text{ (anion)}} = 0.414$$

$$\frac{r(A^+)}{r(X^-)} = 0.414$$

$$r(A^+) = 0.414 \times r(X^-)$$
$$= 0.414 \times 250 \text{ pm}$$
$$= 103.5 \text{ pm} \approx 104 \text{ pm}$$

2. (b) Contribution of atom from the edge centre is 1/4.

Therefore, number of M

$$= \frac{1}{4} \times 4 \text{ (from edge centre)} + 1 \text{ (from body centre)} = 2$$

$$\text{Number of } X = \frac{1}{8} \times 8 \text{ (from corners)} + \frac{1}{2} \times 6 \text{ (from face centre)} = 4$$

Hence, empirical formula $= M_2X_4 = MX_2$

3. (8) The truncated octahedron is the 14-faced Archimedean solid, with 14 total faces : 6 squares and 8 regular hexagons.

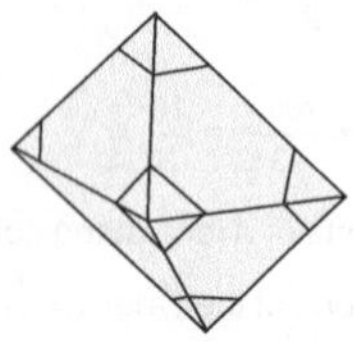

The truncated octahedron is formed by removing the six right square pyramids one from each point of a regular octahedron as

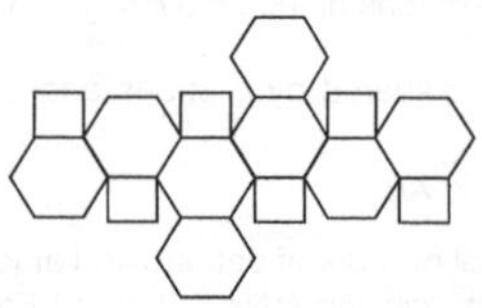

Truncated octahedron unfolded in two dimension.

4. (7) Ag crystallises in FCC unit cell with 4 atoms per unit cell.

$$\rho = \frac{4 \times 108}{6.023 \times 10^{23} \times a^3} = 10.5 \text{ g cm}^{-3}$$

$\Rightarrow$ a^3 (Volume of unit cell) $= 6.83 \times 10^{-23}$ cm^3

$\Rightarrow$ $a = 4 \times 10^{-8}$ cm $= 4 \times 10^{-10}$ m

$\Rightarrow$ Surface area of unit cell $= a^2 = 1.6 \times 10^{-19}$ m^2

$\Rightarrow$ Number of unit cells on 10^{-12} m^2 surface

$$= \frac{10^{-12}}{1.6 \times 10^{-19}} = 6.25 \times 10^6$$

$\because$ There are two atoms (effectively) on one face of unit cell number of atoms on 10^{-12} m^2 surface

$$= 2 \times \text{number of unit cell} = 1.25 \times 10^7$$

$\Rightarrow$ $x = 7$

5. (d) Contribution of circle from corner of square $= \frac{1}{4}$

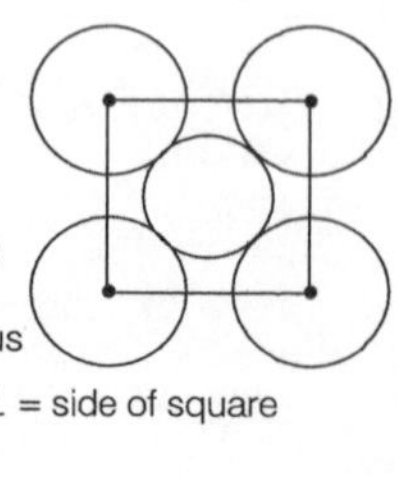

$\Rightarrow$ Effective number of circle per square

$$= \frac{1}{4} \times 4 + 1 \text{ (at centre)} = 2$$

$\Rightarrow$ Area occupied by circle $= 2\pi r^2$, r = radius

Also, diagonal of square $4r = \sqrt{2}\, L$, where L = side of square

$\Rightarrow$ Packing fraction $= \dfrac{\text{Area occupied by circles}}{\text{Area of square}}$

$$= \frac{2\pi r^2}{L^2} = \frac{2\pi r^2}{8r^2} = \frac{\pi}{4} = 0.785$$

$\Rightarrow$ % packing efficiency = 78.5%.

6. (a)

(i) Simple cubic and face-centred cubic both have cell parameters $a = b = c$ and $\alpha = \beta = \gamma = 90°$. Also both of them belongs to same, cubic, crystal system.

(ii) Both cubic and rhombohedral crystal systems have their cell parameters $a = b = c$ and $\alpha = \beta = \gamma$ but they belongs to different crystal systems.

(iii) Cubic and tetragonal are two different types of crystal systems having different cell parameters.

(iv) Hexagonal and monoclinic are two different crystal system and both have two of their crystallographic angles of 90°.

7. From the given information, the number of atoms per unit cell and therefore, type of unit cell can be known as

$$\rho = \frac{NM}{N_A a^3}$$

$$\Rightarrow \quad N = \frac{\rho N_A a^3}{M} = \frac{2 \times 6 \times 10^{23} \times (5 \times 10^{-8} \text{ cm})^3}{75}$$
$$= 2 \text{ (bcc)}$$

In bcc, $4r = \sqrt{3}a$

$$\Rightarrow \quad r = \frac{\sqrt{3}}{4}a$$
$$= \frac{\sqrt{3}}{4} \times 5 \times 10^{-10} \text{m}$$
$$= 2.17 \times 10^{-10} \text{ m}$$
$$= 217 \text{ pm}$$

8. In a cubic crystal system, there are two types of voids known as octahedral and tetrahedral voids. If r_1 is the radius of void and r_2 is the radius of atom creating these voids then

$$\left(\frac{r_1}{r_2}\right)_{\text{octa}} = 0.414$$

and

$$\left(\frac{r_1}{r_2}\right)_{\text{tetra}} = 0.225$$

The above radius ratio values indicate that octahedral void has larger radius hence for maximum diameter of atom to be present in interstitial space :

$$r_1 = 0.414\, r_2$$

Also in fcc, $4r_2 = \sqrt{2}a$

$\Rightarrow$ Diameter required $(2r_1) = (2r_2) \times 0.414$

$$= \frac{a}{\sqrt{2}} \times 0.414$$
$$= \frac{400 \times 0.414}{\sqrt{2}}$$
$$= 117 \text{ pm}$$

9. (b) In ZnS, S^{2-} (sulphide ions) are present at fcc positions giving four sulphide ions per unit cell. To comply with 1 : 1 stoichiometry, four Zn^{2+} ions must be present in four alternate tetrahedral voids out of eight tetrahedral voids present.

In NaCl, Na^+ ions are present in octahedral voids while in Na_2O, Na^+ ions are present in all its tetrahedral voids giving the desired 2 : 1 stoichiometry. In CaF_2, Ca^{2+} ions occupies fcc positions and all the tetrahedral voids are occupied by fluoride ions.

2. Solutions

DPP-1 Different Methods of Expressing Concentration of Solutions

1. **Plan** Molarity = $\frac{\text{Moles of solute}}{\text{Volume of solution (in L)}}$

Solution is 5% by mass of solution.

Thus, volume of solution = $\frac{\text{Mass}}{\text{Density of solution}} = \frac{100}{1.04}\text{ mL} = \frac{1}{10.4}\text{ L}$

$\text{Moles of solute} = \frac{5}{\text{Molar mass of solute}}$

Mass of solvent (H_2O) = 100 − 5 = 95 g = (95/1000) kg

***Step* I** Moles of solute ($Na_2S_2O_3$) = $\frac{5}{158}$ mol

Moles of solvent (H_2O) = $\frac{95}{18}$ mol

***Step* II** Total moles = $\left(\frac{5}{158} + \frac{95}{18}\right)$

(i) Molarity = $\frac{\text{Moles of solute}}{\text{Volume of solution (L)}} = \frac{5/158\text{ mol}}{1/10.4\text{ L}} = 0.329\text{ M}$

(ii) Molality = $\frac{\text{Moles of solute}}{\text{kg of solvent}} = \frac{5/158\text{ mol}}{95/1000\text{ kg}}$

= 0.333 molal (mol kg^{-1})

(iii) Total moles of solute and solvent = $\left(\frac{5}{158} + \frac{95}{18}\right)$

$\therefore$ Mole fraction of solute, $\chi_{\text{solute}} = \frac{5/158}{\left(\frac{5}{158} + \frac{95}{18}\right)} = 5.96 \times 10^{-3}$

2. (a) On changing temperature, solvent (H_2O) is vaporised or condensed. Thus, volume of solution and solvent changes. Thus, any concentration factor dependent on volume is affected by change in temperature.

(a) Molarity = $\frac{\text{Mole}}{\text{Litre}}$ (Volume dependent)

(b) Molality = $\frac{\text{Moles of solute}}{\text{kg of solvent}}$ (Weight dependent)

(c) Mole fraction = n_i/n_{total} (unitless)

(d) Weight fraction = w_i/w_{total} (unitless)

3. (b) Molarity (mol L^{-1}) = $\frac{\text{Moles of solute}}{\text{Volume of solution (in L)}}$

6.02×10^{23} molecules of urea = 1 mole

$\therefore$ 6.02×10^{20} molecules of urea = $\frac{6.02 \times 10^{20}}{6.02 \times 10^{23}} = 1 \times 10^{-3}$ mol

Volume = 100 mL = 0.10 L

Molarity = $\frac{1 \times 10^{-3}}{0.10} = 0.01$ M

4. (b) Glucose solution is one molal. Thus, 1000 g water (solvent) has glucose = one mole = 180 g

Total mass of solution = 1000 + 180 = 1180 g

1180 g of solution has glucose = 180 g

$\therefore$ 1000 g of solution has glucose = $\frac{180 \times 1000}{1180} = 152.5\text{ g} \approx 153\text{ g}$

5. (b) 1 dm^3 = 1000 cm^3 = 1 L = 1 kg

0.300 dm^3 of H_2O at STP = 0.300 kg of H_2O (density = 1.0 g cm^{-3})

Moles of urea = $\frac{0.0100}{60}$ mol

$\therefore$ Molality = $\frac{\text{Moles of solute (urea)}}{\text{kg of solvent}} = \frac{0.010/60\text{ mol}}{0.300\text{ kg}}$

= 5.55×10^{-4} mol kg^{-1}

6. (c) 20% ethanol by volume means

100 mL of (ethanol + H_2O) solution has ethanol = 20 mL

100 mL of solution = 100 × 0.977 g solution = 97.7 g solution

20 mL ethanol = 20 × 0.789 g ethanol = 15.78 g

Mass % = $\frac{15.78 \times 100}{97.7} = 16.15\%$

7. (b) 200 g H_2O at 60°C has 95 g NH_4Cl

Based on solubility graph, at 20°C,

100 g of H_2O can dissolve = 36 g NH_4Cl

hence, 200 g of H_2O can dissolve = 72g NH_4Cl

Thus, NH_4Cl separated (crystallised) = 95 − 72 = 23 g

Mass % = 36%

8. (a) This is a case of dilution.

$M_1V_1 = M_2V_2$

(15.6 M) 6.0 M

$V_1 \times 15.6 = 6.0 \times 10$

$\therefore$ $V_1 = \frac{60}{15.6} = 3.846$ L

Thus, 3.846 L of 15.6 M solution is diluted to 10.0 L by addition of 6.154 L of H_2O.

9. (c) Molality = $\frac{\text{Moles of (solute) } CH_3OH}{1\text{ kg of } H_2O} = \frac{5.2\text{ mol}}{1\text{ kg } H_2O}$

Thus, $n(CH_3OH) = 5.20$

$n(H_2O) = \frac{1000}{18} = 55.56$

$\therefore$ Total moles = 5.20 + 55.56 = 60.76 mol

$\therefore$ $\chi_{CH_3OH} = \frac{\text{Moles of } CH_3OH}{\text{Total moles}} = \frac{5.20}{60.76} = 0.086$

10. (a) 125 mL of 0.05 M $AgNO_3$ = 125 × 0.05 millimoles

$= \frac{125 \times 0.05}{1000}$ moles

$= \frac{125 \times 0.05 \times 170}{1000}$ g pure $AgNO_3$

$= \frac{125 \times 0.05 \times 170 \times 100}{1000 \times 85}$ g impure $AgNO_3$

= 1.25 g

11. (a) Hard water has hardness = 200 ppm

Thus, 10^6 mL(= 10^3 L) water has $CaCO_3$ = 200 g = 2 mol

Thus, molarity = $\frac{\text{Moles}}{\text{Litre}} = \frac{2}{10^3} = 2 \times 10^{-3}$ M

12. (b) 100 mL of 0.01 M $CaCl_2$ = 100 × 0.01 millimoles $CaCl_2$

= 0.001 mole $CaCl_2$ = 0.001 × 111g $CaCl_2$

= 0.111 g pure $CaCl_2$

But, residue = 0.15 g

Thus, impurity = 0.15 − 0.111 = 0.039 g

Percentage impurity = $\frac{0.039}{0.111} \times 100 = 35.14\%$

13. (b) 18 M H_2SO_4 means 18 mol L^{-1}.

1000 mL of H_2SO_4 solution has H_2SO_4 = 18 mol

= 18 × 98 g

Also, 1000 mL = 1000 × 1.8 g H_2SO_4 solution

Thus, (1000 × 1.8) g H_2SO_4 solution has H_2SO_4 = 18 × 98 g

100 g H_2SO_4 solution has H_2SO_4 = $\frac{18 \times 98}{1000 \times 1.8} \times 100 = 98$

Thus, mass percentage of H_2SO_4 = 98%

14. (a) 14.8 M $NH_3(aq)$ means 14.8 moles NH_3 in 1000 mL solution.

1000 mL solution = 1000 × 0.898 g solution

= 898 g solution (NH_3 + H_2O)

14.8 moles NH_3 = 14.8 × 17 g NH_3

= 251.6 g NH_3

Thus, H_2O in solution = 898.0 − 251.6 = 646.4 g

$= \frac{646.4}{18}$ moles = 35.91

$n(H_2O) + n(NH_3) = 35.91 + 14.8 = 50.71$

∴ $\chi_{NH_3} = \frac{14.8}{50.71} = 0.292$

15. (a) 100 g H_3PO_4 aqueous solution has 34 g H_3PO_4.

Thus, volume of solution $= \frac{100}{1.209}$ mL $= \frac{100}{1.209 \times 1000}$ L

$= \frac{1}{12.09}$ L

Mass of solvent = 100 − 34 = 66 g = 0.066 kg

Moles of solute $= \frac{34}{98}$ mol

I. Molarity $= \frac{\text{Moles of solute}}{\text{Volume (in L)}} = \frac{34/98}{1/12.09} = 4.19$ M

II. Molality $= \frac{\text{Moles of solute}}{\text{kg of solvent}} = \frac{34/98}{0.066} = 5.26$ m

III. Normality = Molarity × basicity

= 4.19 × 3 = 12.87 N

16. (c) 275 mL of 0.200 M KNO_3 has $NO_3^- = \frac{275 \times 0.20}{1000}$

325 mL of 0.400 M $Mg(NO_3)_2$ has $NO_3^- = \frac{325 \times 0.40 \times 2}{1000}$

Total moles of $NO_3^- = \frac{275 \times 0.20 + 325 \times 0.40 \times 2}{1000} = 0.315$ mol

Total volume of solution = 275 + 325 + 400 (H_2O)

= 1000 mL = 1 L

∴ $[NO_3^-] = \frac{0.315 \text{ mol}}{1 \text{ L}} = 0.315$ M

17. (b) 1.38 moles of CH_3OH in w kg water.

Molality $= \frac{1.38}{1} = 1.38$ molal

1.38 moles of CH_3OH in w kg water

so that molality = 1.0

∴ $1 = \frac{1.38}{w \text{ kg}} \Rightarrow w = 1.38$ kg = 1380 g

Thus, water added = 1380 − 1000 = 380 g

18. (d)

I. HCl = 3g $= \frac{3}{36.5}$ mol = 0.0822 mol

H_2O = 1 kg = 1000 g $= \frac{1000}{18}$ mol = 55.55 mol

$\chi_{HCl} = \frac{0.0822}{0.0822 + 55.55} = 1.5 \times 10^{-3}$

which is mole fraction of HCl in solution III.

Thus, I = III

II. V(HCl) = 2.0 L at room temperature (298 K)

$p = 1$ atm

$pV = nRT$

∴ $n(HCl) = \frac{pV}{RT} = \frac{1 \times 2}{0.0821 \times 298}$

= 0.0822 mol

which is same as in (I).

Thus, III = I.

19. (c)

(a) $Na_2SO_4 \rightleftharpoons 2Na^+ + SO_4^{2-}$

∴ $[Na^+] = 0.208 \times 2 = 0.416$ M

(b) 1.05 g NaCl/100 mL = 10.5 g NaCl/L $= \frac{10.5}{58.5}$ mol L^{-1}

= 0.1795 M

∴ $[Na^+] = 0.1795$ M

(c) 14.7 mg $Na^+ = 14.7 \times 10^{-3}$ g Na^+ mL^{-1}

= 14.7 g $Na^+L^{-1} = \frac{14.7}{23}$ mol Na^+L^{-1}

∴ $[Na^+] = \frac{14.7}{23} = 0.639$ M

Thus, (c) has the highest molar concentration of Na^+.

20. (a) $Ba(OH)_2 \rightleftharpoons Ba^{2+} + 2OH^-$

pH = 12

∴ $[H^+] = 10^{-12}$ M

Since, $[H^+][OH^-] = 1 \times 10^{-14} = K_w$

∴ $[OH^-] = \frac{1 \times 10^{-14}}{1 \times 10^{-12}} = 1 \times 10^{-2}$ M

∴ $[Ba(OH)_2] = \frac{1 \times 10^{-2}}{2}$ M (mol L^{-1}) $= \frac{1 \times 10^{-2}}{2 \times 10}$ mol in 100 mL

$= \frac{1 \times 10^{-2}}{20} \times 1000$ millimoles = 0.5 millimole

21. (c) This is a case of dilution.

NaOH solution has been diluted by NaCl or *vice-versa*.

Total volume = 35.0 mL

$M_1V_1 = M_2V_2$

***Case* I** 25 mL of 1.00 M NaOH diluted to 35 mL by addition of NaCl.

∴ $25 \times 1 = M_2$ (NaOH) × 35

∴ $M_2(NaOH) = \frac{25}{35} = 0.714$ M

∴ $[Na^+] = 0.714$ M ⇒ $[OH^-] = 0.714$ M

***Case* II** 10 mL of 0.50 M NaCl diluted to 35 mL using NaOH.

∴ $M_1V_1 = M_2V_2$

$10 \times 0.50 = M_2(NaCl) \times 35$

$M_2 = \frac{10 \times 0.50}{35} = 0.143$ M

∴ $M_2(NaCl) = 0.143$ M

∴ $[Na^+] = 0.143$ M

$[Cl^-] = 0.143$ M

Thus,	$[Na^+]$	$[OH^-]$	$[Cl^-]$
In NaOH	0.714 M	0.714 M	—
In NaCl	0.143 M	—	0.143 M
Total	0.857 M	0.714 M	0.143 M

22. (c) Volume can be additive or not depend on the nature of the solution.

$V_{mix} = V_1 + V_2$, if ideal

$V_{mix} \neq V_1 + V_2$, if non-ideal

Density $= \frac{\text{Mass}}{\text{Volume}}$

∴ $M(H_2O) = 50 \times 1 = 50$g

$M(C_2H_5OH) = 50 \times 0.7939 = 39.695$ g

M(mixture) = 89.695 g

$V(\text{mixture}) = \frac{\text{Mass}}{\text{Density}} = \frac{89.695}{0.9344} = 96.0$ mL

Actual volume = 100 mL > 96 mL

Thus, non-ideal

23. (b) $MgI_2 \rightleftharpoons Mg^{2+} + 2I^-$

Every mole of MgI_2 gives two moles of I^-.

$$MV = \text{mol L}^{-1} \times \frac{\text{mL}}{1000} = \text{mol}$$

$$250 \text{ mL of } 0.0876 \text{ M KI} = \frac{250 \times 0.0876}{1000} \text{ mol KI}$$

$$= 250 \times 0.0876 \times 10^{-3} \text{ mol } I^-$$

$$250 \text{ mL of } 0.10 \text{ M } I^- = \frac{250 \times 0.1}{1000} \text{ mol } I^-$$

Thus, I^- required $= (250 \times 0.1 \times 10^{-3} - 250 \times 0.0876 \times 10^{-3})$

$$= 250 \times 10^{-3} (0.10 - 0.0876)$$

$$= 250 \times 10^{-3} (0.0124) \text{ mol } I^-$$

$$= \frac{250 \times 10^{-3}(0.0124)}{2} \text{ mol } MgI_2$$

$$= 0.00155 \text{ mol } MgI_2$$

$$= 0.00155 \times 278 \text{ g} = 0.4309 \text{ g}$$

24. (d) Aqueous 6% methanol solution by mass means

100 g of aqueous solution has = 6g CH_3OH

Since, $\frac{\text{Mass}}{\text{Volume}} = \text{Density}$

Hence, $\text{Volume} = \frac{\text{Mass}}{\text{Density}}$

$\therefore$ $100 \text{ g} = \frac{100}{0.988} \text{mL solution}$

$$= \frac{100}{0.988} \times \frac{1}{1000} = \left(\frac{1}{9.88} \text{ L}\right)$$

$$6 \text{ g } CH_3OH = \frac{6}{32} \text{ mol}$$

$$\text{Molarity} = \frac{\text{mol}}{\text{L}} = \frac{6/32}{1/9.88} = \frac{6 \times 9.88}{32} = 1.8525 \text{ M}$$

25. (b) Volume of ethanol = 10 mL

Volume of solution = 100 mL

(i) Volume % ethanol $= \frac{10}{100} \times 100 = 10\%$

Thus, **(i) → (s)**

Since, $\text{Density} = \frac{\text{Mass}}{\text{Volume}}$

$$M(\text{ethanol}) = V \times d$$

$$= 10 \times 0.789 = 7.89 \text{ g}$$

$$100 \text{ mL solution} = 100 \times 0.982 = 98.2 \text{ g}$$

(ii) Mass % ethanol $= \frac{7.89}{98.2} \times 100 = 8.03\%$

Thus, **(ii) → (p)**

(iii) Mass/Volume % ethanol $= \frac{7.89}{100} \times 100 = 7.89\%$

Thus, **(iii) → (t)**

(iv) Mole % ethanol

Solution (ethanol + water) = 98.20 g

Ethanol = 7.89 g

Thus, water = 98.20 − 7.89 = 90.31 g

$$n(\text{ethanol}) = \frac{7.89}{46} = 0.171$$

$$n(H_2O) = \frac{90.31}{18} = 5.017$$

Total moles = 5.188

$$\chi_{\text{ethanol}} = \frac{0.171}{5.188} = 0.0329$$

Mole % = 3.29%

Thus, **(iv) → (q)**

(v) $\text{Molarity} = \frac{\text{Moles of solute (ethanol)}}{\text{Volume of solution (in L)}}$

$$= \frac{0.171}{0.100} = 1.71 \text{ mol L}^{-1}$$

Thus, **(v) → (u)**

(vi) $\text{Molality} = \frac{\text{Mole of solute (ethanol)}}{\text{kg of } H_2O}$

$$= \frac{0.171 \text{ mol}}{0.09031 \text{ kg}} = 1.89 \text{ mol kg}^{-1}$$

Thus, **(vi) → (r)**

26. (a,b,c) 1 kg aqueous solution = 1000 g aqueous solution

It has 0.4 kg NaOH = 0.4 × 1000 = 400 g NaOH

Thus, solvent (water) = 1000 − 400 = 600 g

(a) % by weight of solvent (water) $= \frac{400}{600} \times 100$

$$= 66.67\%$$

(b) % by weight of solution $= \frac{400}{1000} \times 100$

$$= 40\%$$

(c) $\text{Molality} = \frac{\text{Moles of NaOH}}{\text{kg of water}}$

$$= \frac{400/40}{600/1000} = \frac{10}{0.6} = 16.67$$

(d) Since, density is not given hence, molarity cannot be derived.

27. (a, b) $n_{(CH_3CH_2OH)} = \frac{n_{CH_3CH_2OH}}{n_{CH_3CH_2OH} + n_{H_2O}} = 0.25 = \frac{1}{4}$

Thus, $n_{CH_3CH_2OH} = 1 \text{ mole} = 46 \text{ g}$

then $n_{H_2O} = 3 \text{ moles} = 54 \text{ g}$

Total mass of solution = 46 + 54 = 100 g

Thus, CH_3CH_2OH by mass of solution $= \frac{46}{100} \times 100 = 46\%$

CH_3CH_2OH by mass of solvent $= \frac{46}{54} \times 100 = 85.2\%$

Explanations for Q. Nos. (28-29)

Molarity = 2.772 M = 2.772 mol L^{-1}

Thus, NaOH = 2.772 mol

$H_2O = 1.0 \text{ L} = 998 \text{ g} = 0.998 \text{ kg} = \frac{998}{18} \text{ mol}$

Total moles of H_2O and NaOH $= \left(\frac{998}{18} + 2.772\right)$

$\text{Molality of NaOH} = \frac{\text{Moles of solute (NaOH)}}{\text{kg of } H_2O}$

$$= \frac{2.772}{0.998} \text{ mol kg}^{-1}$$

$$= 2.778 \text{ molal}$$

$$\chi_{NaOH} = \frac{n(\text{NaOH})}{\text{Total moles}} = \frac{2.772}{\left(\frac{998}{18} + 2.772\right)}$$

$$= 4.76 \times 10^{-2}$$

28. (b) **29.** (c)

30. (8) 100 g brine solution has 3.87 g NaCl.

Thus, 0.32 g NaCl is in $= \frac{100 \times 0.32}{3.87}$ g brine solution.

Density = 1.036 g mL^{-1}

$\therefore$ $\text{Volume} = \frac{\text{Mass}}{\text{Density}}$

$$= \frac{100 \times 0.32}{3.87 \times 1.036} = 8.0 \text{ mL}$$

DPP-2 Henry's Law

1. Plan By Henry's law

$$p_{gas} = K_H \times \chi_{gas}$$

$$\chi_{N_2} = \frac{n_{N_2}}{n_{N_2} + n_{blood}}$$

Since, $n_{N_2} <<< n_{Blood}$

Hence, $\chi_{N_2} = \frac{n_{N_2}}{n_{blood}}$

$\therefore \quad p_{gas} = K_H \frac{n_{N_2}}{n_{blood}}$

$\therefore \quad n_{N_2} = \frac{p_{gas} n_{blood}}{K_H}$

5 L blood = 5000 mL = 5000 g

Also, partial pressure of $N_2 = p_{total}\chi_{N_2}$ in air $= p_{total}\ 0.80$

(i) At 1 bar, $n_{N_2} = 1 \times 0.80 \text{ bar} \times \left(\frac{5000 \text{ g}}{18}\right) \frac{\text{mol}}{9.04 \times 10^4 \text{bar}}$

$= 2.46 \times 10^{-3}$ mol

(ii) At 50 bar, $n_{N_2} = 50 \times 2.46 \times 10^{-3}$ mol $= 0.123$ mol

2. (b) Solubility $= K_H p_{CO_2}$

$\therefore \quad p_{CO_2} = \frac{\text{Solubility}}{K_H} = \frac{0.0506 \text{ mol kg}^{-1}}{0.034 \text{ mol kg}^{-1}\text{bar}^{-1}}$

$= 1.488$ bar

≈ 1.49 bar

Note *If K_H is in the unit of pressure, then,*

Concentration $\times K_H$ = pressure *If K_H is in the unit of (pressure)$^{-1}$, then*
Concentration $= K_H \times$ *pressure*

3. (a) The mole fraction of a gas in solution is related to partial pressure of the gas above the solution by Henry's law.

$$p_{CO} = \chi_{CO} K_H$$

$\therefore \quad \chi_{CO} = \frac{p_{CO}}{K_H} = \frac{0.01 \text{ atm}}{5.8 \times 10^4 \text{ atm}}$

$= 1.72 \times 10^{-7}$

4. (b) By Henry's law, $p = K_H\chi = K_H S$ (where, S = solubility)

$\therefore \quad K_H = \frac{p}{S} = \frac{1 \text{ atm}}{0.195 \text{ mol kg}^{-1}}$

$= 5.128$ atm kg mol^{-1}

5. (a) Partial pressure, $p = K_H\chi$

If temperature is lowered, solubility of gases increases at a given pressure. Thus, K_H increases with increase in temperature. It is due to this reason that aquatic species are more comfortable in cold water rather than in warm water.

6. (b)

Note *K_H (Henry's law constant) is in pressure unit, hence we use relation,*

Concentration $\times K_H$ = *Pressure*

$\therefore \quad \text{Concentration} = \frac{\text{Pressure of } CO_2}{K_H}$

$p_{CO_2} = 55 \text{ kPa} = 55 \times 10^3$ Pa

(Pa unit is to be converted into torr)

1 atm = 760 torr = 1.01325×10^5 Pa

$\therefore \quad 55 \times 10^3 \text{ Pa} = \frac{760 \times 55 \times 10^3}{1.01325 \times 10^5} \text{ torr} = 412.53$ torr

$\therefore \quad \chi_{CO_2} \text{(Mole fraction)} = \frac{412.53 \text{ torr}}{8.6 \times 10^4 \text{ torr}}$

$= 4.80 \times 10^{-3}$

7. (a) In this case, K_H has been given in the unit of mol/kg bar.
Thus, solubility of O_2 (or concentration) $= K_H\ p_{gas}$

χ_{O_2} (in air) $= 0.21$

$\therefore \quad p_{O_2} = p_{air}\chi_{O_2} = 0.21$ bar

$\therefore \quad \text{Solubility} = \frac{1.3 \times 10^{-3} \text{ mol}}{\text{kg bar}} \times 0.21 \text{ bar}$

$= 2.73 \times 10^{-4}$ mol/kg $= 2.73 \times 10^{-4} \times 32$ g/kg

$= 8.736 \times 10^{-3}$ g/kg

8. (d) By Henry's law at 273 K,

χ_{CO_2} (mole fraction of CO_2) $= \frac{p}{K_H} = \frac{0.10}{728}$

By Henry's law at 333 K, $\chi_{CO_2} = \frac{p}{K_H} = \frac{p_2}{3410}$

$\therefore \quad \frac{p_2}{3410} = \frac{0.10}{728}$

$p_2 = \frac{0.10 \times 3410}{728} = 0.468$ atm

9. (a) By Henry's law, $p = \chi K_H$

For solute, $p_2 = \chi_2 K_2 = \frac{n_2}{n_1 + n_2} K_2 = \frac{n_2 K_2}{n_t}$

Also, $p_2 = \frac{n_2 RT}{V}$

$\therefore \quad \frac{n_2 RT}{V} = \frac{n_2}{n_t} K_2$

$\therefore \quad \frac{RT}{V} = \frac{K_2}{n_t}$

$\therefore \quad V = \frac{n_t RT}{K_2}$

10. (a) Moles of O_2 at 0°C $= \frac{pV}{RT} = \frac{1.0 \text{ atm} \times \left(\frac{48.9}{1000}\right) \text{L}}{0.0821 \text{ L atm mol}^{-1} \text{K}^{-1} \times 273 \text{ K}}$

$= 2.182 \times 10^{-3}$ mol in 1 L H_2O

Thus, molarity of O_2 gas at 1 atm $= 2.182 \times 10^{-3}$ M

By Henry's law,

Concentration $= K_H p$

$C_1 = K_H p_1$

$C_2 = K_H p_2$

$\therefore \quad \frac{C_2}{C_1} = \frac{p_2}{p_1}$

$\therefore \quad C_2 = C_1 \frac{p_2}{p_1} = \left(2.182 \times 10^{-3} \times \frac{0.2095}{1}\right) \text{M} = 4.57 \times 10^{-4}$ M

11. (a) Henry's law constant is in the unit of pressure, hence we use relation

$$p = K_H \chi_{N_2}$$

$\chi_{N_2} = \frac{p}{K_H} = \frac{0.987 \text{ bar}}{76.48 \times 10^3 \text{ bar}} = 1.29 \times 10^{-5}$

1 L H_2O = 1000 mL H_2O = 1000 g H_2O (d = 1 g mL^{-3})

$= \frac{1000}{18} = 55.5$ moles

Let number of moles of nitrogen $= n$

then $\chi_{N_2} = \frac{n_{N_2}}{n_{N_2} + 55.5} \approx \frac{n_{N_2}}{55.5}$ (Since, n_{N_2} is very small)

$\therefore \quad n_{N_2} = 55.5\ \chi_{N_2} = 55.5 \times 1.29 \times 10^{-5}$

$= 7.16 \times 10^{-4}$ mol $= 7.16 \times 10^{-4} \times 1000$

$= 7.16 \times 10^{-2}$ millimole $= 0.0716$ millimole

12. (b) Volume of CO = 0.0354 mL = 0.0354×10^{-3} L

$p = 1$ atm, $T = 273$ K

$$\therefore \quad n(CO) = \frac{pV}{RT}$$

$$= \frac{1 \text{ atm} \times 0.0354 \times 10^{-3} \text{ L}}{0.0821 \text{ L atm mol}^{-1} \text{ K}^{-1} \times 273 \text{ K}}$$

$$= 1.5794 \times 10^{-6} \text{ mol mL}^{-1}$$

$$\therefore \quad M(CO) = \frac{1.5794 \times 10^{-6} \text{ mol}}{0.001 \text{ L}}$$

$$= 1.5794 \times 10^{-3} \text{ M}$$

By Henry's law, $C_1 = K_H p_1, C_2 = K_H p_2$

$$\therefore \quad \frac{C_2}{C_1} = \frac{p_2}{p_1}$$

$$p_2 = \frac{C_2}{C_1} p_1 = \frac{0.01 \text{ M} \times 1.0 \text{ atm}}{1.5794 \times 10^{-3} \text{ M}}$$

$$p_2 = 6.33 \text{ atm}$$

13. (a, b, c)

(a) $NH_3 + H_2O \rightleftharpoons NH_4OH \rightleftharpoons NH_4^+ + OH^-$

$CO_2 + H_2O \rightleftharpoons H_2CO_3 \rightleftharpoons 2H^+ + CO_3^{2-}$

$SO_2 + H_2O \rightleftharpoons H_2SO_3 \rightleftharpoons 2H^+ + SO_3^{2-}$

NH_3, CO_2, SO_3 ...... gases form ions in water, hence these gases are soluble in water. O_2, N_2 ...... do not form ions in water hence, insoluble thus correct.

(b) NH_3 and CO_2 are liquefied easily hence, soluble in water thus, correct.

(c) $p = K_H \chi$, $\chi \propto \frac{1}{K_H}$

Higher the value of K_H, smaller the solubility (χ) of the gas at a given pressure thus, correct.

14. (a,b,c)

(a) At constant temperature, different gases have different K_H values. Thus, K_H is a function of nature of gases.

Thus, correct.

(b) p (partial pressure) = $K_H \chi$

If p is constant and K_H value increases then χ (solubility) decreases. Thus, correct.

(c) K_H increases with increase in temperature thus, correct.

(d) K_H is independent of pressure thus, incorrect.

15. (b) Dissolution of gas can be considered as the reverse of vaporisation, i.e. condensation in which heat is evolved. It is thus, exothermic $\Delta H < 0$. By Le-Chatelier's principle, increase in temperature shifts the dissolution equilibrium in the direction of formation of $CO_2(g)$.

Thus, increase in temperature decreases solubility. Thus, both Statement I and Statement II are correct but Statement II is not the correct explanation of Statement I.

16. (b) At high altitudes, the partial pressure of oxygen is less than that at the ground level. This leads to the low concentration of O_2 in the blood and tissues of the people residing in hill areas. This produces weakness and loss of memory.

Symptoms of this conditions is known as ANOXIA. Thus, Statements I are correct but Statement II is not the correct explanation of Statement I.

17. (b) $CO_2(g) + H_2O(l) \rightleftharpoons H_2CO_3(l)$

(High vapour pressure) (Low vapour pressure)

When a gas dissolves in H_2O, vapour pressure decreases. Hence, increase in pressure of CO_2 leads to increase in solubility. Hence, soft drink bottles are sealed under high pressure of CO_2.

Thus, both Statements I are II are correct but Statement II is not the correct explanation of Statement I.

18. (a) When divers breathe under high pressure, solubility of atmospheric gases in blood increases. When divers come at the surface, pressure gradually decreases. Gases (N_2 and O_2) are released in the form of bubbles which blocks capillaries and creates a medical symptoms called **bends** which are painful and dangerous. To avoid bends as well as toxic effects of high concentration of nitrogen in the blood, the tanks used by divers are filled with air diluted with helium (11.7% He, 56.2% N_2 and 32.1% O_2). Thus, both Statements I and II are correct and Statement II is the correct explanation of Statement I.

19. (d) K_H is in pressure unit.

$$p = K_H \chi$$

$$\chi_{N_2} = \frac{p_{N_2}}{K_H} = \frac{10 \times 0.79 \times 760 \text{ mm}}{6.51 \times 10^7 \text{ mm}} = 9.22 \times 10^{-5}$$

Also, $n(H_2O) = \frac{1000}{18} = 55.55$

$$\chi_{N_2} = \frac{n_{N_2}}{n_{N_2} + n_{H_2O}} \approx \frac{n_{N_2}}{n_{H_2O}} \quad (n_{N_2} \text{ being very small})$$

$$\therefore \quad n_{N_2} = \chi_{N_2} \times n_{H_2O} = 9.22 \times 10^{-5} \times 55.55 = 5.117 \times 10^{-3}$$

Similarly, $\chi_{O_2} = \frac{p_{O_2}}{K_H} = \frac{10 \times 0.21 \times 760 \text{ mm}}{3.30 \times 10^7 \text{ mm}} = 4.84 \times 10^{-5}$

$\therefore \; n_{O_2} = 4.84 \times 10^{-5} \times n_{H_2O} = 4.84 \times 10^{-5} \times 55.5 = 2.686 \times 10^{-3}$

(i) $\chi_{N_2} = 9.22 \times 10^{-5}$ ...(r)

(ii) $\chi_{O_2} = 4.84 \times 10^{-5}$ ...(p)

(iii) Moles of $N_2 = 5.12 \times 10^{-3}$ ...(s)

(iv) Moles of $O_2 = 2.69 \times 10^{-3}$ ...(q)

Explanations for Q. Nos. (20-21)

Based on given figure, pressure of HCl $\propto \chi_{HCl}$

$$p_{HCl} = K_H \chi_{HCl}$$

$$\therefore \quad K_H = p_{HCl} / \chi_{HCl}$$

When $\chi_{HCl} = 0.01$, $p_{HCl} = 400$ torr

$$\therefore \quad K_H = \frac{400}{0.01} = 4 \times 10^4 \text{ torr}$$

At $p_{HCl} = 400$ torr, $\chi_{HCl} = 0.01$

At $p_{HCl} = 960$ torr, $\chi_{HCl} = 0.025$

$$\therefore \quad \frac{\chi_{HCl}(400 \text{ torr})}{\chi_{HCl}(960 \text{ torr})} = \frac{0.01}{0.025} = 0.4$$

20. (a) **21.** (b)

22. (8) By Raoult's law, $p_{H_2O} = p^{\circ}_{H_2O} \chi_{H_2O}$

$$= p^{\circ}_{H_2O} (1 - \chi_{NH_3}) = 258.02 \text{ torr}$$

Thus, difference = 266 − 258 = 8 torr

23. (2) By Henry's law, $p_i = \chi_i K_i$, $p_{O_2} = \chi_{O_2} K_{O_2}$

$\therefore \chi_{O_2}$ (solubility or mole fraction) $= \frac{p_{O_2}}{K_{O_2}}$

Similarly, $\chi_{N_2} = \frac{p_{N_2}}{K_{N_2}}$

$$\therefore \quad \frac{\chi_{N_2}}{\chi_{O_2}} = \frac{p_{N_2} K_{O_2}}{p_{O_2} K_{N_2}} = \frac{0.80 \times 6 \times 10^4}{0.20 \times 12 \times 10^4} = 2$$

24. (9) K_H is in pressure unit, hence $p_{O_2} = K_H \times$ Concentration

$[O_2] = 3.856 \text{ mg dm}^{-3} = 3.856 \times 10^{-3} \text{ g dm}^{-3}$

$$= \frac{3.856 \times 10^{-3}}{32} \text{ mol dm}^{-3}$$

$K_H = 74.68 \text{ kPa m}^3 \text{ mol}^{-1} = 74.68 \times 10^3 \text{ kPa dm}^3 \text{ mol}^{-1}$

$$\therefore \quad p_{O_2} = 74.68 \times 10^3 (\text{kPa dm}^3 \text{ mol}^{-1}) \times \frac{3.856 \times 10^{-3}}{32} (\text{mol dm}^{-3})$$

$$= 9.0 \text{ kPa}$$

DPP-3 Colligative Properties-I (Lowering of Vapour Pressure)

1. Plan For mixture of binary volatile liquids, benzene (b) and toluene (t), vapour pressures are $p_b = p_b^\circ \chi_b$

(p_b° = Vapour pressure of pure benzene with mole fraction χ_b)

$$p_t = p_t^\circ \chi_t$$

(p_t° = Vapour pressure of pure toluene with mole fraction χ_t)

$$p_{total} = p_b + p_t$$

Mole fraction of benzene in vapour phase above solution

$$\chi_b' = \frac{p_b}{p_{total}}$$

and that of toluene = $(1 - \chi_b')$

Moles of benzene in mixture, $n_b = \frac{11.7}{78} = 0.15$ mol

Moles of toluene in mixture, $n_t = \frac{4.6}{92} = 0.05$ mol

Total moles = $n_b + n_t = 0.20$

Mole fraction of benzene = $\frac{0.15}{0.20} = 0.75$

Mole fraction of toluene = $1 - 0.75 = 0.25$

$\therefore \quad p_b = p_b^\circ \chi_b = 3.6 \times 10^4 \times 0.75 = 2.7 \times 10^4$ Nm^{-2}

$p_t = p_t^\circ \chi_t = 1.12 \times 10^4 \times 0.25 = 2.8 \times 10^3$ Nm^{-2}

$p_{total} = p_b + p_t = 2.98 \times 10^4$ Nm^{-2}

X_b' = Mole fraction of benzene in vapour phase

$$= \frac{p_b}{p_{total}} = \frac{2.7 \times 10^4}{2.98 \times 10^4} = 0.906$$

Similarly, $\chi_t' = 1 - 0.906 = 0.094$

2. Plan Solution is a mixture of non-volatile solute (in solid state) in volatile solvent. We can use Raoult's law.

Based on Statement I For any solution, the partial vapour pressure of each volatile component in the solution is directly proportional to its mole fraction (volatile component).

Non-volatile solute (urea) = 6 g = $\frac{6}{60}$ = 0.1 mol

Volatile component (H_2O) = 90 g = $\frac{90}{18}$ = 5.0 mol

Total moles = 5.1

$$\chi_{H_2O} = \frac{5.0}{5.1} = \frac{50}{51}$$

$\therefore \quad p_{solution} \propto \chi_{H_2O}$

$p_{solution} = K\,\chi_{H_2O}$

When $\chi_{H_2O} = 1$,

$p_{solution} = p^\circ_{solvent}$

$K = p^\circ_{solvent}$

$\therefore \quad p_{solution} = \chi_{H_2O}\, p^\circ_{solvent} = \frac{50 \times 40}{51} = 39.22$ mm Hg

Based on Statement II When a non-volatile solute is added to volatile solvent, relating lowering of vapour pressure is equal to the mole fraction of solute.

p° = Vapour pressure of volatile component in pure state

p_s = Vapour pressure of the solution

$p^\circ - p_s$ = Lowering of vapour pressure

$\frac{p^\circ - p_s}{p^\circ}$ = Relative (wrt solvent) lowering of vapour pressure

χ_{solute} = Mole fraction of solute = $1 - \frac{50}{51} = \frac{1}{51}$

$\therefore \quad \frac{p^\circ - p_s}{p^\circ} = \chi_{solute} \Rightarrow 1 - \frac{p_s}{p^\circ} = \chi_{solute}$

$\frac{p_s}{p^\circ} = (1 - \chi_{solute}) = \chi_{solvent} = \frac{50}{51}$

$p_s = p^\circ \chi_{solvent}$ (as in I)

3. Plan For a binary mixture of two volatile liquids, toluene (t) and cresol (c),

$$p_t = p_t^\circ \chi_t$$

(p_t° = Vapour pressure of pure toluene with mole fraction χ_t)

$$p_c = p_c^\circ \chi_c$$

(p_c° = Vapour pressure of pure cresol with mole fraction χ_c)

$p_{total} = p_t^\circ \chi_t + p_c^\circ \chi_c$

$p_{total} = p_t^\circ \chi_t + p_c^\circ (1 - \chi_t)$, (since $\chi_t + \chi_c = 1$)

$= p_t^\circ \chi_t + p_c^\circ - p_c^\circ \chi_t$

$p_{total} = p_c^\circ + \chi_t (p_t^\circ - p_c^\circ)$

p_{total}, p_t° and p_c° are given, thus χ_t is determined.

0.50×760 torr = 150 torr + χ_t (400 − 150) torr

0.50×760 torr = 150 torr + $250\chi_t$ torr

$\therefore \quad \chi_t = 0.92$

Thus, toluene in the mixture is 92 mole%.

Mole fraction in vapour phase above solution = $\frac{p_t}{p_{total}} = \frac{p_t^\circ \chi_t}{p_{total}}$

$$\chi_t' = \frac{400 \times 0.92}{0.50 \times 760} = 0.968$$

Thus, toluene in vapour phase above solution is 96.8 mole%.

4. (d) Binary mixture contains non-volatile solute (B) in volatile solvent.

By Raoult's law, relative decrease in vapour pressure = χ_{solute}

$$\frac{\Delta p}{p^\circ} = \chi_{solute} \Rightarrow \frac{0.60}{0.80} = \chi_{solute}$$

$\therefore \quad \chi_{solute} = 0.75$

5. (c) By Raoult's law, for a mixture of non-volatile solute in volatile solvent.

$p_{solution} = p^\circ_{solvent}\chi_{solvent}$

(water) = 10 moles

(urea) = $\frac{6}{60}$ = 0.1 mol

$n_{(sucrose)} = \frac{68.4}{342} = 0.2$ mol

Total moles = 10.3

$\therefore \quad \chi_{H_2O} = \frac{10}{10.3} = \frac{100}{103}$

$\therefore \quad p_{solution} = 23.8 \times \frac{100}{103} = 23.1$ torr

6. (a) For immiscible solvents, composition of each component is based on

$$\frac{n_{(water)}}{n_{(nitrobenzene)}} = \frac{p^\circ_{(water)}}{p^\circ_{(nitrobenzene)}}$$

$$\frac{\frac{w}{m}\text{(water)}}{\frac{w}{m}\text{(nitrobenzene)}} = \frac{97.7}{3.6}$$

$$\therefore \quad \frac{w_{(water)}}{w_{(nitrobenzene)}} = \frac{97.7 \times 18}{3.6 \times 123} = \frac{3.97}{1.00}$$

Thus, total mass = 4.97 g

Percentage of nitrobenzene = $\frac{1.00}{4.97} \times 100$

= 20.12%

7. (d) Mixture boils at 1 atm and 350 K.

Thus, total vapour pressure is 1 atm = 760 torr

(h) used for hexane and (ch) for cyclohexane.

$$p_{\text{total}} = p_h^\circ \chi_h^\circ + p_{ch}^\circ \chi_{ch} \Rightarrow p_{\text{total}} = p_h^\circ \chi_h + p_{ch}^\circ (1 - \chi_h)$$

$$760 = 840\,\chi_h + 600\,(1 - \chi_h)$$

$$\therefore \quad \chi_h = 0.67$$

8. (c) By Raoult's law for a non-volatile solute in volatile solvent.

$$\frac{p^\circ - p_s}{p^\circ} = \chi_{\text{solute}} \Rightarrow \frac{23.76 - 22.98}{23.76} = \chi_{\text{solute}},\ \chi_{\text{solute}} = 0.033$$

Also, $\chi_{\text{solute}} = \dfrac{n_{\text{solute}}}{n_{\text{solvent}} + n_{\text{solute}}} \approx \dfrac{n_{\text{solute}}}{n_{\text{solvent}}} = \dfrac{0.033}{1}$

Thus, solute = 0.033 mole

Water (solvent) = 1 mole = 18 kg = 0.018 kg

$$\text{Molality} = \frac{\text{Moles of solute}}{\text{kg of solvent}} = \frac{0.033}{0.018} = 1.83$$

9. (b) Ethanol molecules are joined to each other by intermolecular H-bonding thus, they have high boiling point. When *n*-hexane is mixed into it, H-bonding breaks. Since, force of attraction between *n*-hexane and ethanol molecules is smaller than between the hexane-hexane and ethanol-ethanol. Thus, mixture is more volatile. This shows positive deviation from Raoult's law.

10. (d) At 100°C, vapour pressure of water = 1 atm. Thus, p° = 760 torr

Let vapour pressure of the solution = p_s

$$n_{\text{solute}}(C_6H_{12}O_6) = \frac{18}{180} = 0.1\ \text{mol}$$

$$n_{\text{solvent}}(H_2O) = \frac{178.2}{18} = 9.9\ \text{mol}$$

$$\chi_{\text{solute}} = \frac{0.1}{9.9 + 0.1} = \frac{1}{100}$$

By Raoult's law, $\dfrac{p^\circ - p_s}{p^\circ} = \chi_{\text{solute}} \Rightarrow \dfrac{760 - p_s}{760} = \dfrac{1}{100}$

$$\therefore \quad p_s = 760 - 7.60 = 752.4\ \text{torr}$$

11. (b)

(a) When HCl is mixed with H_2O, force of attraction between HCl and H_2O molecules is increased (due to dissolution), hence vaporisation decreased, a case of negative deviation.

(b) CH_3OH molecules are joined by intermolecular H-bonding which provides it a liquid state. On adding benzene, H-bonding breaks and force of attraction between CH_3OH and benzene molecules is less than between CH_3OH-CH_3OH and between benzene-benzene hence, more vaporisation of mixture, a case of positive deviation.

(c) As in (a), negative deviation.

(d) CH_3COCH_3 and $CHCl_3$ are volatile. On mixing, there is increase in intermolecular H-bonding due to the formation of chloretone (A).

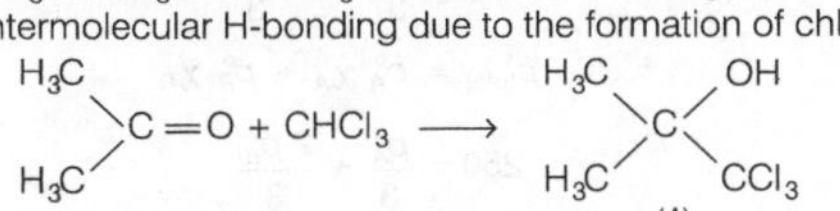

(A)

Thus, force of attraction between ($CHCl_3$ -CH_3COCH_3) molecules is larger than between ($CHCl_3$ - $CHCl_3$) and (CH_3COCH_3 - CH_3COCH_3) molecules. Thus, vaporisation decreases, a case of negative deviation.

12. (a) If force of attraction between *A* and *B* (solvents) is higher than that between *A* and *A* or *B* and *B*, then there is decrease in vaporisation. Thus, higher temperature (i.e. boiling point) is required to convert liquids into vapour state. This is a case of negative deviation from Raoult's law.

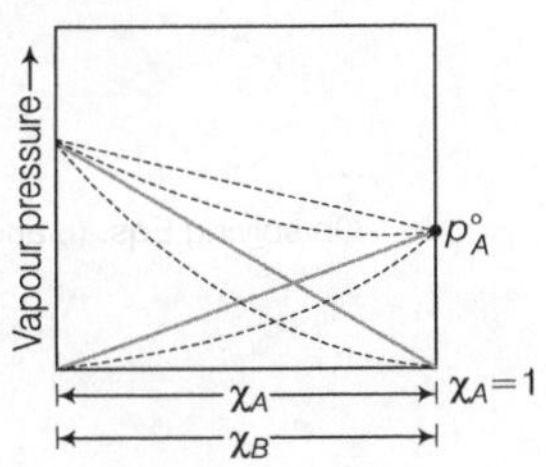

13. (c) Molality = 1 mol kg^{-1} (solvent)

Thus, solute = 1 mole

$$\text{Solvent} = 1\ \text{kg}\ H_2O = 1000\ \text{g}\ H_2O = \frac{1000}{18} = 55.55\ \text{mol}$$

$$\chi_{\text{solute}} = \frac{n_{\text{solute}}}{n_{\text{total}}} = \frac{1}{56.55}$$

By Raoult's law, $\dfrac{\Delta p}{p^\circ} = \chi_{\text{solute}} \Rightarrow \Delta p = \chi_{\text{solute}}\, p^\circ$

At 100°C, p° = 1 atm = 760 torr

$$\therefore \quad \Delta p = \frac{760}{56.55} = 13.44\ \text{torr}$$

14. (a) Density = 0.792 g/mL

Thus, 100 mL CH_3OH (methanol) = 79.2 g = $\dfrac{79.2}{32}$

= 2.475 moles

Urea in 100 mL CH_3OH saturated solution

$$= 17\ \text{g} = \frac{17}{60} = 0.283\ \text{g}$$

$$\therefore \quad \chi_{\text{solvent}}\ (CH_3OH) = \frac{2.475}{2.475 + 0.283} = 0.897$$

By Raoult's law, vapour pressure of solution

$$= p_{\text{solvent}}^\circ\, \chi_{\text{solvent}} = 96 \times 0.897 = 86.112\ \text{mm Hg}$$

15. (a) Liquid $\rightleftharpoons$ Vapour

(a) If vapour pressure is high, it means equilibrium is in forward side. This is attained when intermolecular forces holding the molecules are of weak nature. Low heat is required to convert liquid into vapour. Thus, (a) is correct.

(b) Thus, incorrect.

(c) When vapour pressure is low, it means intermolecular forces are strong hence (c) is incorrect.

(d) When vapour pressure is low, intermolecular forces are strong, hence large heat is required thus, (d) is incorrect.

16. (b,c,d) When binary mixture is ideal, then by Raoult's law

$\Delta V = 0$, $\Delta H_{\text{mix}} = 0$, no enthalpy change

Mixing is spontaneous, thus $\Delta G < 0$

By thermodynamic property,

$$\Delta G = \Delta H - T\Delta S$$

$$-\text{ve} = 0 - T\Delta S$$

$$\therefore \quad \Delta S > 0$$

Thus, on mixing, entropy increases

$$(\Delta S)_{\text{system}} + (\Delta S)_{\text{surroundings}} > 0$$

But $(\Delta S)_{\text{surroundings}} = 0$

Hence, $(\Delta S)_{\text{system}} > 0$

17. (a,c,d) p_A° = 100 mm Hg, n_A = 1 mol, $\chi_A = 0.25$

p_B° = 60 mm Hg, n_B = 3 mol, $\chi_B = 0.75$

$$\therefore \quad p_{\text{total}} = p_A^\circ \chi_A + p_B^\circ \chi_B = \frac{100 \times 1}{4} + \frac{60 \times 3}{4} = 70\ \text{mm Hg}$$

But, actual vapour pressure = 75 mm Hg. Vapour pressure is increased. Force of attraction between *A* and *B* is smaller than between *A* and *A* and between *B* and *B*. Thus, more vaporisation due to decrease in boiling point, a case of positive deviation.

18. (b) (A for acetone, C for carbon disulphide)

p_A° = 344 mm Hg, $\chi_A = 0.75$

p_C° = 512 mm Hg, $\chi_C = 0.25$

$$p_{\text{total}} = p_A^\circ \chi_A + p_C^\circ \chi_C$$

$$= 344 \times 0.75 + 512 \times 0.25 = 385\ \text{mm Hg}$$

But, actual p_{total} = 600 mm Hg

Thus, there is positive deviation. Boiling point is lowered.

Thus, ΔH = +ve (endothermic), i.e. heat is absorbed.

Explanations for Q. Nos. (19-20)

Let moles of $A = n_A$ and moles of $B = n_B$

$$\chi_A \text{ (mole fraction of } A) = \frac{n_A}{n_A + n_B}$$

$$p_{total} = p_A^\circ \chi_A + p_B^\circ \chi_B = p_A^\circ \chi_A + p_B^\circ (1 - \chi_A)$$

Mole fraction of A in vapour state

$$\chi'_A = \frac{p_A^\circ \chi_A}{p_A^\circ \chi_A + p_B^\circ \chi_B} \Rightarrow 0.35 = \frac{p_A^\circ \chi_A}{p_A^\circ \chi_A + p_B^\circ (1 - \chi_A)}$$

$$\therefore \quad \frac{p_A^\circ \chi_A + p_B^\circ (1 - \chi_A)}{p_A^\circ \chi_A} = \frac{100}{35}$$

$$1 + \frac{p_B^\circ (1 - \chi_A)}{p_A^\circ \chi_A} = \frac{100}{35} \Rightarrow \frac{p_B^\circ (1 - \chi_A)}{p_A^\circ \chi_A} = \frac{65}{35} = \frac{13}{7}$$

$$\therefore \quad \frac{1 - \chi_A}{\chi_A} = \frac{13}{7} \frac{p_A^\circ}{p_B^\circ} = \frac{13}{7} \times \frac{575}{390}$$

$$\frac{1}{\chi_A} - 1 = 2.738$$

$$\therefore \quad \chi_A = 0.27, \chi_B = 0.73$$

Thus, A is 27 mole %.

Total vapour pressure of binary mixture

$$p_{total} = p_A^\circ \chi_A + p_B^\circ \chi_B = 575 \times 0.27 + 390 \times 0.73 = 440 \text{ torr}$$

19. (a) **20.** (a)

Explanations for Q. Nos. (21-22)

Let vapour pressure of pure methanol = p_m°

and that of pure ethanol = p_e°

Mole fraction of methanol = χ_m

Mole fraction of ethanol, $\chi_e = (1 - \chi_m)$

$$\therefore \quad p_{total} = p_m^\circ \chi_m + p_e^\circ \chi_e$$

$$= p_m^\circ \chi_m + p_e^\circ (1 - \chi_m)$$

$$= p_m^\circ \chi_m + p_e^\circ - p_e^\circ \chi_m$$

$$= p_e^\circ + (p_m^\circ - p_e^\circ)\chi_m = 135 + 119\chi_m$$

Thus, $(p_m^\circ - p_e^\circ) = 119$ torr

$$p_e^\circ = 135 \text{ torr}$$

$$p_m^\circ = 135 + 119 = 254 \text{ torr}$$

Also, $\chi_e = 0.5, \chi_m = 0.5$

$$p_{total} = 135 + 119 \times 0.5 = 194.5$$

$\therefore \chi'_m$ (mole fraction of methanol in vapour phase)

$$= \frac{p_m^\circ \chi_m}{p_{total}} = \frac{254 \times 0.5}{194.5} = 0.65$$

21. (b) **22.** (c)

23. (c) Initially, $60 \text{ g } A = \frac{60}{60} = 1 \text{ mol}$

$$40 \text{ g } B = \frac{40}{40} = 1 \text{ mol}$$

Then, $\chi_A \text{ (in I)} = \frac{1}{2} = 0.5$

Total vapour pressure of mixture in I

$$p_I = p_A^\circ \chi_A + p_B^\circ \chi_B = \frac{24}{2} + \frac{12}{2} = 18 \text{ kPa}$$

$$\therefore \quad \chi_A \text{ (in II)} = \frac{p_A^\circ \chi_A}{p_{total}} = \frac{12}{18} = 0.67$$

III is liquefied form of II, thus,

$$\chi_A \text{ (in III)} = \chi_A \text{ (in II)} = 0.67$$

Vapour pressure of III = $p_A^\circ \chi_A\text{(III)} + p_B^\circ \chi_B\text{(III)}$

$$= 24 \times \frac{2}{3} + 12 \times \frac{1}{3} = 20 \text{ kPa}$$

Mole fraction of A above liquid in IV

$$\chi_A \text{ (in IV)} = \frac{p_A^\circ \chi_A\text{(III)}}{p_{total}} = \frac{16}{20} = 0.8$$

Thus, (i) → (q), (ii) → (r), (iii) → (r), (iv) → (s)

24. (4) Mixture of volatile solvents showing deviation from Raoult's law is an azeotropic mixture. In such cases, force of attraction between A and B in the mixture is higher/lower than that between A and A and between B and B.

S. No.	A	B	Force of attraction between A and B	Deviation
(i)	CCl_4	$CHCl_3$	Higher	Negative
(ii)	C_6H_{14}	C_7H_{16}	Same	No
(iii)	C_2H_5Br	C_2H_5Cl	Same	No
(iv)	C_6H_5Cl	C_6H_5Br	Same	No
(v)	C_2H_5OH	CH_3COCH_3	Lower	Positive
(vi)	$CHCl_3$	CH_3COCH_3	Higher	Negative
(vii)	C_6H_6	$C_6H_5CH_3$	Same	No
(viii)	*o*-nitrophenol	Benzene	Lower	Positive

Thus, (i), (v), (vi) and (viii) are azeotropic mixtures.

25. (1) By Raoult's law for a binary mixture of non-volatile solute (sucrose) in volatile solvent (water),

relative lowering of vapour pressure = mole fraction of solute

$$n_{water} = \frac{360}{18} = 20$$

$$\frac{\Delta p}{p^\circ} = \frac{p^\circ - p_s}{p^\circ} = \frac{n_{sucrose}}{n_{sucrose} + n_{water}}$$

$$\frac{1.19}{25} = \frac{n_{sucrose}}{(n_{sucrose} + 20)}$$

$$\frac{n_{sucrose} + 20}{n_{sucrose}} = \frac{25}{1.19} = 21$$

$$\therefore \quad n_{sucrose} = 1 \text{ mol}$$

26. (3)

Case I $\quad n_A = 1, n_B = 2$

$$\chi_A = \frac{1}{3}, \chi_B = \frac{2}{3}$$

$$\therefore \quad p_{total} = p_A^\circ \chi_A + p_B^\circ \chi_B$$

$$250 = \frac{p_A^\circ}{3} + \frac{2p_B^\circ}{3}$$

$$750 = p_A^\circ + 2p_B^\circ \quad \text{...(i)}$$

Case II $\quad n_A = 2, n_B = 2$

$$\chi_A = \frac{1}{2}, \chi_B = \frac{1}{2}$$

$$\therefore \quad 300 = \frac{p_A^\circ}{2} + \frac{p_B^\circ}{2}$$

$$\therefore \quad 600 = p_A^\circ + p_B^\circ \quad \text{...(ii)}$$

On solving Eqs. (i) and (ii), we get

$$p_A^\circ = 450 \text{ torr}$$

$$p_B^\circ = 150 \text{ torr}$$

$$\therefore \quad \frac{p_A^\circ}{p_B^\circ} = \frac{450}{150} = 3$$

DPP-4 **Colligative Properties-II** (Depression in Freezing Point)

1. (d) On adding $\left(\frac{w_1}{m_1}\right)$ mole of solute to w_2 g of solvent, depression in freezing point is

$$\Delta T_f = \frac{1000\, K_f w_1}{m_1 w_2}$$

$$(5.51 - 5.03) = \frac{1000 \times 5.12 \times 0.643}{m_1 \times (50 \times 0.879)}$$

$$\because \quad m_1 = 156 \text{ g mol}^{-1}$$

2. (a) Solute is ethylene glycol = 1 mL = 1.113 g (w_1)

Solvent (H_2O) = 1 g (w_2)

$$\Delta T_f = \frac{1000\, K_f w_1}{m_1 w_2} = \frac{1000 \times 1.86 \times 1.113}{62 \times 1}$$

$$= 33.39°$$

$\therefore$ Freezing point = 0 − 33.39° = − 33.39°

3. (a) ΔT_f (cane sugar) = 273.15 − 272.88 = 0.27°

ΔT_f (glucose) = ?

$$\frac{w_1}{w_2} = \frac{5}{100} \quad \text{(for both)}$$

$$\Delta T_f \text{ (glucose)} = \frac{1000 K_f w_1}{m_1 w_2} = \frac{1000\, K_f \times 5}{180 \times 100}$$

$$\Delta T_f \text{ (cane sugar)} = \frac{1000\, K_f \times 5}{342 \times 100}$$

$$\therefore \quad \frac{\Delta T_f \text{ (glucose)}}{\Delta T_f \text{ (cane sugar)}} = \frac{342}{180} = 1.9$$

$\therefore$ ΔT_f (glucose) = 1.9 × 0.27 = 0.513°

$\therefore$ Freezing point = 273.15 − 0.513 = 272.637 K

4. (b) Solute is CH_3OH = 30 mL = 30 × 0.7980 g

Solvent is H_2O = 70 mL = 70 × 0.9984 g

$$\Delta T_f = \frac{1000 K_f w_1}{m_1 w_2} = \frac{1000 \times 1.86 \times 30 \times 0.7980}{32 \times 70 \times 0.9984} = 19.91°$$

5. (d) −4°F is to be converted into °C.

$$\frac{F - 32}{180} = \frac{C}{100}$$

$$\frac{-4 - 32}{180} = \frac{C}{100}$$

$$\therefore \quad C = -20° \text{ C}$$

$$\therefore \quad \Delta T_f = -20°$$

1 L H_2O = 1000 g water

$$\Delta T_f = \frac{1000 K_f w_1}{m_1 w_2}$$

$$w_1 \text{ (ethyl alcohol)} = \frac{\Delta T_f m_1 w_2}{1000\, K_f} = \frac{20 \times 46 \times 1000}{1000 \times 1.86} = 495 \text{ g}$$

6. (a) CH_3OH solution is 10% by weight of solution.

Thus, 100 g solution has CH_3OH = 10 g

Water in solution = 100 − 10 = 90 g

$$\therefore \quad \Delta T_f = \frac{1000\, K_f w_1}{m_1 w_2} = \frac{1000 \times 1.86 \times 10}{32 \times 90} = 6.46°$$

$\therefore$ Freezing point = 0 − 6.46°

= − 6.46°C = − 6.46 + 273 = 266.5 K

7. (a) w_1 = solute = 50 g ethylene glycol

$$\frac{w_1}{m_1} = \frac{50}{62} \text{ mol}$$

Solvent required to maintain this amount = w_2 g

$$\Delta T_f = 9.3°$$

$$\therefore \quad \Delta T_f = \frac{1000\, w_1 K_f}{m_1 w_2}$$

$$\therefore \quad w_2 = \frac{1000 w_1\, K_f}{m_1 \Delta T_f} = \frac{1000 \times 50 \times 1.86}{62 \times 9.3} = 161.3 \text{ g}$$

Actual H_2O taken = 200 g

Thus, ice formed = 200 − 161.3 = 38.7 g

8. (a) For $\left(\frac{w_1}{m_1}\right)$ mole of non-volatile solute in w_2 g solvent, depression in freezing point is

$$\Delta T_f = \frac{1000 K_f w_1}{m_1 w_2}$$

$$0.30 = \frac{1000 \times 1.86\, w_1}{m_1 w_2}$$

$$\therefore \quad \frac{w_1}{m_1 w_2} = \frac{30}{1000 \times 186}$$

By Raoult's law,

$$\frac{p° - p_s}{p°} = \frac{n_1}{n_1 + n_2} \approx \frac{n_1}{n_2} = \frac{w_1}{m_1}\frac{m_2}{w_2}$$

$$1 - \frac{p_s}{p°} = \frac{30 \times 18}{1000 \times 186}$$

$$1 - \frac{p_s}{23.51} = 2.903 \times 10^{-3}$$

$$p_s = 23.44 \text{ torr}$$

9. (c) $\Delta T_f = \frac{1000 K_f w_1}{m_1 w_2}$

$\frac{w_1}{m_1}$ = moles of solute has not been given, but this amount has been obtained by condensation into benzene.

$$\frac{w_1}{m_1} = n_1 \text{ (moles of solute)} = \frac{pV}{RT}$$

$$= \frac{\left(\frac{748}{760}\right)(4)}{0.0821 \times 300} = 0.16 \text{ mol}$$

$$\therefore \quad \Delta T_f = \frac{1000 \times 5.12 \times 0.16}{58} = 14.12°$$

10. (a) T_0 = Freezing point of solvent

ΔH_f = Latent heat of fusion

$$K_f = \frac{RT_0^2}{1000\, \Delta H_f} = \frac{2 \times (280)^2}{1000 \times 30} = 5.2 \text{ K mol}^{-1} \text{ kg}$$

$$\Delta T_f = \text{molality} \times K_f = 0.1 \times 5.2 = 0.52°$$

11. (a) From the given graph, at point *A*

Liquid solvent $\rightleftharpoons$ Solid solvent

Thus, freezing point of solvent = 279 K

Freezing point of 1 molal solution of *X* at point *B* = 273 K

Thus, ΔT_f = 279 − 273 = 6 K

Also, $\Delta T_f = K_f \times$ molality $= K_f \times 1$

$$\therefore \quad K_f = 6 \text{ K mol}^{-1} \text{ kg}$$

For 10% solution of *X* in solvent,

w_1 = 10 g, w_2 = 100 g

m_1 = molar mass of solute *X*

At point *C*, freezing point of 10% solution = 268 K

$$\Delta T_f = 279 - 268 = 11 \text{ K}$$

$$\therefore \quad \Delta T_f = \frac{1000 K_f w_1}{w_2 m_1}$$

$$\therefore \quad m_1 = \frac{1000 K_f w_1}{w_2 \Delta T_f} = \frac{1000 \times 6 \times 10}{100 \times 11} = 54.54 \text{ g mol}^{-1}$$

12. (b) When a non-volatile solute is added to a solvent, temperature falls, hence vapour pressure also falls. Thus, freezing point is also lowered.

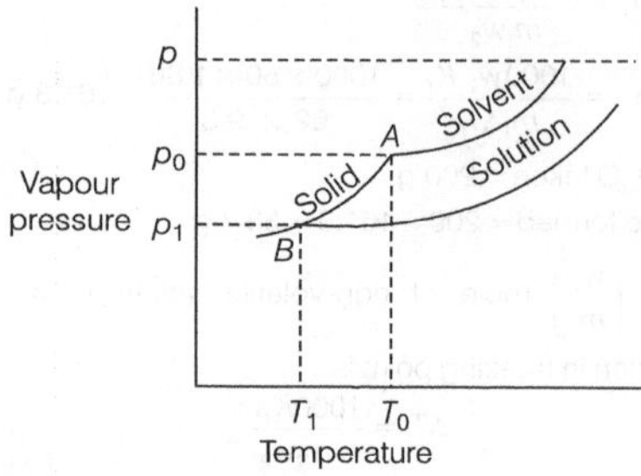

At A, vapour pressure = p_0

freezing point = T_0

At B, vapour pressure = $p_1 < p_0$ and freezing point = $T_1 < T_0$

Thus, freezing point is lowered. Hence, Statement I is correct.

At freezing point,

$$\text{liquid} \rightleftharpoons \text{solid}$$

Thus, $\Delta G = 0$

Also, $\Delta G = \Delta H - T\Delta S$

$\therefore$ $T = \dfrac{\Delta H}{\Delta S}$

ΔH is almost same for solvent and solution but entropy increases, $\Delta S > 0$. Hence, freezing point of solution is smaller than that of solvent.

Thus, Statement II is correct but not the correct explanation of Statement I.

13. (a, d) Liquid solvent $\rightleftharpoons$ Solid solvent

Freezing point is the temperature at which liquid solvent and solid remain at equilibrium. In the following figure,

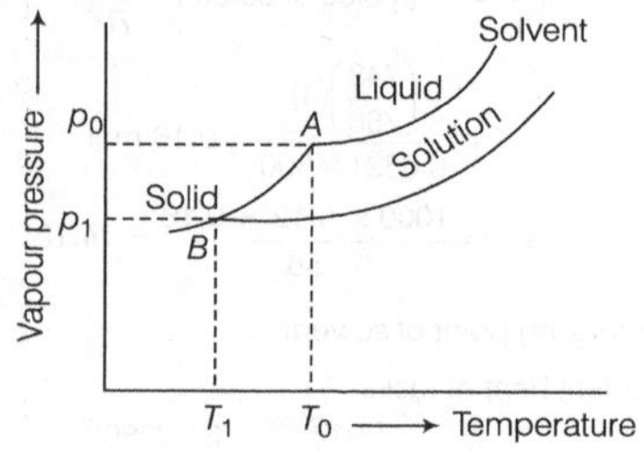

	Vapour pressure	Freezing point	At point
Solvent	p_0	T_0	A
Solution	p_1	T_1	B

On adding solute to solvent, vapour pressure falls $p_1 < p_0$ and $T_1 < T_0$.

Thus, (a) is correct.

Only solvent molecules solidify at the freezing point.

Thus, (d) is also correct.

14. (a,b,c,d) Solute, $w_1 = 3.5$ g

Solvent, $w_2 = 50.0$ g

Volume of solution = 52.5 mL

$\Delta T_f = 0.86°$

$K_f = 1.86° \text{ mol}^{-1}$ kg

$\Delta T_f = \text{molality} \times K_f$

(a) Molality $= \dfrac{\Delta T_f}{K_f} = \dfrac{0.86°}{1.86} = 0.46 \text{ mol kg}^{-1}$

Thus, (a) is correct.

(b) Total mass of solution = 50.0 + 3.5 = 53.5 g

Volume of solution = 52.5 mL

$\therefore$ Density $= \dfrac{53.5}{52.5} = 1.019 \text{ g mL}^{-1}$

50 g H_2O has X = 3.5 g

1000 g H_2O has $X = \dfrac{3.5}{50} \times 1000 = 70$ g X

Total mass of solution = 1000 + 70 = 1070 g

Volume of solution $= \dfrac{\text{Mass}}{\text{Density}} = \dfrac{1070 \text{ g}}{1.019 \text{ g mL}^{-1}}$

= 1050 mL = 1.050 L

$\therefore$ Molarity $= \dfrac{\text{mole}}{\text{Litre}} = \dfrac{0.46}{1.050} = 0.44$ M

Thus, (b) is correct.

(c) 70 g X = 0.46 mol (present in 1000 g H_2O)

$\therefore$ Molar mass of $X = \dfrac{70}{0.46} = 152.17 \text{ g mol}^{-1}$

Thus, (c) is correct.

(d) Solute = 0.46 mol kg^{-1} water

Solvent = 1000 g H_2O = 55.55 mol

$\therefore$ $\chi_{\text{solute}} = \dfrac{0.46}{55.55 + 0.46} = 0.0082$

Thus, (d) is correct.

15. (d) $\Delta T_f = \dfrac{1000 K_f w_1}{m_1 w_2}$

$\Delta T_f(\text{p}) = \dfrac{1000 \times 1.86 \times 6}{60 \times 100} = 1.86°$

Thus, **(i) → (q)**

$\Delta T_f(\text{q}) = \dfrac{1000 \times 1.86 \times 9.6}{32 \times 100} = 5.58°$

Thus, **(ii) → (r)**

$\Delta T_f(\text{r}) = \dfrac{1000 \times 1.86 \times 17.1}{342 \times 100} = 0.93°$

Thus, **(iii) → (s)**

$\Delta T_f(\text{s}) = \dfrac{1000 \times 1.86 \times 36.0}{180 \times 100} = 3.72°$

Thus, **(iv) → (p)**

Explanations for Q. Nos. (16-17)

Molality = 1 mol kg^{-1}, Solute (sucrose) = 1 mole

Solvent (H_2O) = 1 kg = 1000 g = 55.55 moles

$\chi_{\text{solute}} = \dfrac{1}{56.55}$

By Raoult's law for binary mixture of non-volatile solute in volatile solvent, relative lowering of vapour pressure = χ_{solute}

$$\frac{\Delta p}{p°} = \frac{1}{56.55} = 0.018$$

Sucrose solution is maintained at −3.534°C.

Thus, ΔT_f = 3.534°C

Total mass of 1 molal solution = 1000 g (solvent) + 342 g (solute) = 1342 g

1342 g solution contains sucrose = 342 g

1000 g solution contains sucrose $= \dfrac{342 \times 1000}{1342} = 255$ g

H_2O present in 1000 g solution = 1000 − 255 = 745 g

Solvent actually required to keep 255 g sucrose in solution at −3.534°C = w_2

$\Delta T_f = \dfrac{1000\, w_1 K_f}{w_2 m_1}$

$w_2 = \dfrac{1000\, w_1 K_f}{\Delta T_f m_1} = \dfrac{1000 \times 255 \times 1.86}{3.534 \times 342} = 392.0$ g (in liquid state)

Thus, ice formed = 745 − 392 = 353 g (in solid state)

16. (d) **17.** (c)

18. (6) Density of water vapour $= 6 \times 10^{-4}$ g cm^{-3}

Thus, weight of water vapour in 10 L steam $= 10 \times 1000 \times 6 \times 10^{-4} = 6$ g

Density of liquid water = 1 g cm^{-3}

Volume of liquid water in 6 g steam $= \dfrac{6\text{ g}}{1\text{ g cm}^{-3}} = 6\text{ cm}^3$

19. (4) $\dfrac{w_1}{m_1}$ = moles of solute, w_2 = mass of solvent

$$\Delta T_f = \frac{1000\,K_f w_1}{m_1 w_2}$$

$$(5.45 - 4.89) = \frac{1000 \times 5.12 \times 0.675}{m_1 \times 50}$$

$$\therefore \quad m_1 = 123.42$$

If molecular formula $= (P)_x$

$\therefore$ Molar mass $= 31x = 123.42$

$x = 4$

20. (6) Molality of urea solution = 1 mol kg^{-1}

$$\Delta T_f\text{ (urea)} = 0 - (-1.86°) = 1.86°$$

$$\Delta T_f = \left(\frac{1000\,w_1}{m_1 w_2}\right)K_f = \text{molality } K_f$$

$$\therefore \quad 1.86° = K_f \times 1\text{ mol kg}^{-1}$$

$$\therefore \quad K_f = 1.86°\text{ mol}^{-1}\text{ kg}$$

$$C_nH_{2n}O_n = 36\text{ g in }100\text{ g }H_2O$$

$$\Delta T_f(C_nH_{2n}O_n) = \frac{1000\,K_f\,w_1}{m_1 w_2}$$

$$\therefore \quad m_1(C_nH_{2n}O_n) = \frac{1000 K_f w_1}{\Delta T_f w_2} = \frac{1000 \times 1.86 \times 36}{3.72 \times 100}$$

$$m_1(C_nH_{2n}O_n) = 180\text{ g mol}^{-1}$$

$$12n + 2n + 16n = 180$$

$$\therefore \quad 30n = 180$$

$$n = 6$$

DPP-5 **Colligative Properties-III** (Elevation in Boiling Point)

1. (c) ΔT_b = Molality $\times K_b$

$$\Delta T_b = mK_b$$
$$\Delta T_f = mK_f$$
$$\therefore \quad \frac{\Delta T_f}{\Delta T_b} = \frac{K_f}{K_b}$$
$$\Delta T_f = \Delta T_b \frac{K_f}{K_b} = 1.04 \times \frac{1.86}{0.52} = 3.72°$$

2. (b) When (K_χ) mole fraction boiling point elevation constant is given, then

$$\Delta T_b = K_\chi \chi_{\text{solute}}$$

12.8 g of naphthalene $= \dfrac{12.8}{128} = 0.1$ mol

92 g of toluene $= \dfrac{92}{92} = 1.0$ mol

$$\therefore \quad \chi_{\text{solute}} = \frac{0.1}{0.1 + 1.0} = \frac{1}{11}$$

$$\therefore \quad \Delta T_b = \frac{1}{11} \times 36.1 = 3.28$$

Thus, boiling point of solution $= 110 + 3.28 = 113.28°$ C

3. (a) Molarity = 1 M = 1 mol L^{-1}

1000 mL of solution has glucose = 180 g

(1000×1.2) g of solution has glucose = 180 g

$\therefore$ Solute $(w_1) = 180$ g

Solution = 1200 g

Solvent, $H_2O(w_2) = 1200 - 180 = 1020$ g

$$\therefore \quad \Delta T_b = \frac{1000 K_b w_1}{m_1 w_2} = \frac{1000 \times K_b \times 180}{180 \times 1020} = 0.98\,K_b$$

4. (a) $\Delta T_b = \dfrac{1000\,K_b w_1}{m_1 w_2}$

$$2 = 1000 \times 0.76 \frac{w_1}{m_1 w_2}$$

$$\therefore \quad \frac{w_1}{m_1 w_2} = \frac{2}{1000 \times 0.76} = \frac{1}{380}$$

By Raoult's law,

$$\frac{p^\circ - p_s}{p^\circ} = \chi_{\text{solute}} = \frac{n_1}{n_1 + n_2}$$

But, $\quad n_2 >> n_1$

Thus, $\quad \dfrac{p^\circ - p_s}{p^\circ} = \dfrac{n_1}{n_2} = \dfrac{w_1 m_2}{m_1 w_2}$

$$\frac{p^\circ - p_s}{p^\circ} = \frac{18}{380} \quad (m_2\text{, molar mass of solvent} = 18)$$

$$p^\circ - p_s = \frac{18}{380} \times 760 = 36\text{ mm Hg}$$

$$\therefore \quad p_s = p^\circ - 36 = 760 - 36 = 724\text{ mm Hg}$$

5. (c) By Raoult's law,

$$\frac{\Delta p}{p^\circ} = 0.05 = \chi_{\text{solute}}$$

$$\Delta T_b = \frac{100\,K_b w_1}{m_1 w_2}$$

If solution is dilute,

$$\chi_{\text{solute}} = \frac{n_1}{n_1 + n_2} \approx \frac{n_1}{n_2} = \frac{w_1 m_2}{m_1 w_2}$$

$$\therefore \quad \frac{w_1 m_2}{m_1 w_2} = 0.05$$

$$\therefore \quad \frac{w_1}{m_1 w_2} = \frac{0.05}{18}$$

$$\therefore \quad \Delta T_b = \frac{1000 \times 0.52 \times 0.05}{18} = 1.44°$$

If solution is of moderate concentration,

then $\quad \dfrac{\Delta p}{p^\circ} = \dfrac{n_1}{n_1 + n_2} = \dfrac{0.05}{1}$

Thus, $\quad n_1 = 0.05$ mol

$$n_2 = 0.95\text{ mol} = 0.95 \times 18\text{ g}$$
$$= \frac{0.95 \times 18}{1000}\text{ kg} = 0.0171\text{ kg}$$

$$\therefore \quad \text{Molality} = \frac{0.05\text{ mol}}{0.017\text{ kg}} = 2.924\text{ mol kg}^{-1}$$

$$(\Delta T_b)' = K_b \times \text{molality}$$
$$= 0.52 \times 2.924 = 1.52°$$

$$\therefore \quad (\Delta T_b) - (\Delta T_b)' = 1.44 - 1.52 = -0.08°$$

6. (c) 6% urea solution, $\Delta T_b = \dfrac{1000\,K_b w_1}{m_1 w_2}$

$$\Delta T_b\text{ (urea)} = x_1 = \frac{1000 \times K_b \times 6}{60 \times 100} = K_b$$

$$\Delta T_b\text{ (glucose)} = x_2 = \frac{1000 \times K_b \times 18}{180 \times 100} = K_b$$

$$\Delta T_b \text{ (mixture)} = \frac{1000 \text{ (total moles of solute) } K_b}{\text{Total mass of solvent}}$$

$$= \frac{1000 \times K_b \times 0.2}{200}$$

$$x_3 = K_b$$

Thus, $x_1 = x_2 = x_3$

7. (a) Salt is sparingly soluble hence, can be considered as non-electrolyte.

$$K_{sp} = [A]^+[B^-] = 1 \times 10^{-8}$$

$\therefore$ Solubility $= \sqrt{K_{sp}} = 1 \times 10^{-4}$ M

Since, solubility is very low, molarity can be taken as molality.

$$\Delta T_b = K_b \times \text{molality} = 0.52 \times 10^{-4}$$

$$= 5.2 \times 10^{-5} \text{ K}$$

8. (b) In terms of mole fraction elevation constant

$$\Delta T_b = K_\chi \chi_{solute}$$

$$0.3 = 36.1\, \chi_{solute}$$

$$\frac{0.3}{36.1} = \chi_{solute} = \frac{n_1}{n_1 + n_2}$$

Thus, solute = 0.3 mole

$$n_1 + n_2 = 36.1 \text{ moles}$$

$\therefore$ $n_2 = 35.8$ moles $= 35.8 \times 92$ g

$$= \frac{35.8 \times 92}{1000} \text{ kg} = 3.2936 \text{ kg}$$

$\therefore$ Molality $= 0.3/3.2936 = 0.091$ mol kg^{-1}

9. (b) $\Delta T_b = \frac{1000\, K_b w_1}{m_1 w_2}$

Boiling point of solution = $(373 + \Delta T_b)$

Larger the value of ΔT_b, higher the boiling point.

	w_1	w_2	m_1	ΔT_b	Boiling point
(a)	18 g	82 g	180	0.63	373.63 K
(b)	34.2 g	65.8 g	342	0.79	373.79 K
(c)	6.0 g	94.0 g	60	0.55	373.55 K

10. (a) From I and II, $(\Delta T_b)_X = \frac{1000\, K_b w_1}{m_1 w_2}$

$$5° = K_b \text{ (molality)}$$

$\therefore$ K_b (solvent) $= 5°$ mol^{-1} kg

From I and III, $(\Delta T_b)_Y = 10°$

$\therefore$ $(\Delta T_b)_Y = \frac{1000 K_b w_1}{m_1 w_2}$

$\therefore$ $m_1(Y) = \frac{1000\, K_b w_1}{(\Delta T_b)_Y w_2} = \frac{1000 \times 5 \times 10}{10 \times 100}$

$$= 50 \text{ g mol}^{-1}$$

11. (b) $H_2O(l)$ changes to steam $H_2O(g)$ at 373 K. Thus, this represents latent heat of vaporisation. K_b (molal elevation constant) is related to $\Delta H°$ and boiling point by equation,

$$K_b = \frac{RT_0^2}{1000\, \Delta H°}$$

Here, $\Delta H°$ is in energy unit per gram of solvent.

$$\Delta H° = 9.72 \text{ kcal mol}^{-1}$$

$$= \frac{9.72}{18} \text{ kcal g}^{-1}$$

$\therefore$ $K_b = \frac{0.002 \times (373)^2}{1000 \times \left(\frac{9.72}{18}\right)} = 0.52° \text{ mol}^{-1} \text{ kg}$

ΔT_b (sucrose solution) = molality $\times K_b = 0.1 \times 0.52 = 0.052°$

$\therefore$ Boiling point of solution $= 373 + 0.052° = 373.052$ K

12. (d) Water boils at 1 atmospheric pressure.

Thus, boiling point of H_2O = 100°C

1 molal urea solution boils at 100.52°C

$\therefore$ $\Delta T_b = 0.52°$C

$$\Delta T_b(\text{urea}) = K_b \times \text{molality}$$

$$0.52° = K_b \times 1$$

$\therefore$ $K_b(H_2O) = 0.52°$ mol^{-1} kg

Glucose solution is 20% by weight of solution.

Thus, w_1(glucose) = 20 g

w_2 (solvent H_2O) = 80 g

$$\Delta T_b = \frac{1000\, K_b w_1}{m_1 w_2}$$

$$= \frac{1000 \times 0.52 \times 20}{180 \times 80} = 0.72°$$

$\therefore$ Boiling point of glucose solution = 100.72°C

13. (d) K_b (molal elevation constant) is related to latent heat of vaporisation (ΔH_v) and boiling point (T_b) of solvent by relation.

$$K_b = \frac{RT_b^2}{1000\, \Delta H_v}$$

Similarly, K_f (molal depression constant) is related to latent heat of fusion (ΔH_f) and boiling point T_f of solvent by relation.

$$K_f = \frac{RT_f^2}{1000\, \Delta H_f}$$

Thus, $\frac{K_b}{K_f} = \left(\frac{T_b}{T_f}\right)^2 \left(\frac{\Delta H_f}{\Delta H_v}\right) = \left(\frac{3}{2}\right)^2 \left(\frac{1}{6}\right) = \frac{3}{8}$

$$\Delta T_b = m\, K_b, \quad \Delta T_f = m\, K_f$$

$\therefore$ $\frac{\Delta T_b}{\Delta T_f} = \frac{K_b}{K_f} = \frac{3}{8}$

14. (a,c,d) Any property which is dependent on the number of the particles of solute in a solution, is called colligative property.

(a) Osmotic pressure, $\pi = \frac{n}{V} RT$

$$\pi \propto n$$

(b) Boiling point is not a colligative property.

(c) ΔT_f (depression in freezing point) $= 1000 \frac{w_1 K_f}{m_1 w_2}$

$$\Delta T_f \propto \frac{w_1}{m_1} \propto n_1 \text{ (solute)}$$

(d) $\frac{\Delta p}{p°}$ (relative lowering of vapour pressure) $= \frac{n_1}{n_1 + n_2}$

$$\frac{\Delta p}{p°} \propto n_1 \text{ (solute)}$$

If there is change in number of moles of solute, the following properties also change:

(i) If solute is non-electrolyte, equal moles will have same property.

(ii) There is ionisation, n changes hence, properties change.

(iii) There is polymerisation, n changes hence, properties change.

15. (a,b,c,d)

$$\Delta T_b = \frac{100 K_b w_1}{m_1 w_2}$$

Given, $\Delta T_b = x \Rightarrow w_1 = 10$ g, $m_1 = m$, $w_2 = 100$ g

$\therefore$ $x = \frac{1000 \times K_b \times 10}{100\, m} = 10^2\left(\frac{K_b}{m}\right)$

$$\log x = \log 10^2 + \log\left(\frac{K_b}{m}\right)$$

$$\log x = 2 + \log\left(\frac{K_b}{m}\right)$$

This represents linear equation.

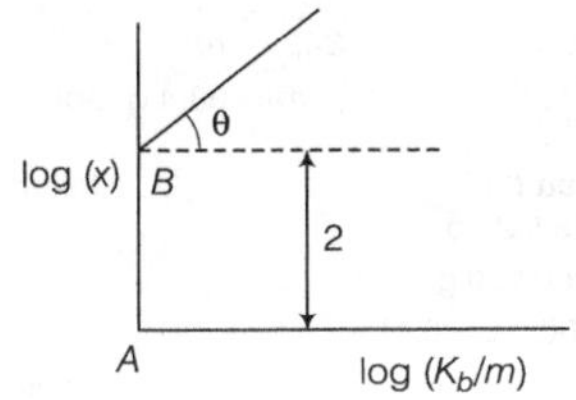

With slope $=1=\tan\theta$ ($\theta = 45°$)

Intercept, $AB = 2$

As $\log(x)$ increases, $\log\left(\frac{K_b}{m}\right)$ also increases. Hence, x is dependent on $\left(\frac{K_b}{m}\right)$.

When boiling point of solution $= 200°C$.

Then, $\Delta T_b = 200 - 100 = 100C$

$\therefore \quad x = 100$

$\therefore \quad \log x = 2 + \log\left(\frac{K_b}{m}\right)$

$\Rightarrow \quad \log 100 = 2 + \log\left(\frac{K_b}{m}\right)$

$2 = 2 + \log\left(\frac{K_b}{m}\right)$

$\Rightarrow \log\left(\frac{K_b}{m}\right) \to 0$

Thus, (a), (b), (c) and (d) are correct.

Explanations for Q. Nos.(16-17)

For aqueous solution of urea,

$$\Delta T_b = \text{molality} \times K_b$$

$\therefore \quad 0.052 = 0.1 \times K_b$

$\therefore \quad K_b = 0.52° \text{ mol}^{-1}\text{ kg}$

Thus, $K_b(D_2O) = 0.52 \times \frac{110}{100}$ 10% higher $= 0.572 \text{ K mol}^{-1}\text{ kg}$

Also, $K_b = \frac{RT_0^2}{1000\,\Delta H_v}$

$\therefore \quad \Delta H_v = \frac{RT_0^2}{1000\,K_b}$

$$\Delta H_v = \frac{0.002 \text{ kcal mol}^{-1}\text{K}^{-1} \times (374.42)^2\text{ K}^2}{1000 \text{ g kg}^{-1} \times 0.572 \text{ K mol}^{-1}\text{kg}} = 0.490 \text{ kcal g}^{-1}$$

$$= 0.490 \times 20 \text{ kcal mol}^{-1} = 9.8 \text{ kcal mol}^{-1}$$

16. (a) **17.** (b)

18. (c) ΔT_b (sucrose) $= K_b \times$ molality

$0.52° = K_b \times 1$

$\therefore \quad K_b = 0.52° \text{ mol}^{-1}\text{ kg}$

$\Delta T_b(AB_2) = \frac{1000\, w_1 K_b}{m_1 w_2}$

$0.26 = \frac{1000 \times 5 \times 0.52}{m(AB_2) \times 100}$

$\therefore \quad m(AB_2) = 100$

$\Delta T_b(A_2B) = \frac{1000\, w_1 K_b}{m_1 w_2}$

$0.52° = \frac{1000 \times 9 \times 0.52}{m(A_2B) \times 100}$

$\therefore \quad m(A_2B) = 90$

$\therefore \quad A + 2B = 100$ ...(i)

$2A + B = 90$ ...(ii)

On solving Eqs. (i) and (ii), we get

$A = 40$

$B = 10$

Hence, (i) → (s), (ii) → (r), (iii) → (q), (iv) → (p)

19. (8) $\Delta T_b = 319.67 - 319.30 = 0.37$ K

$\Delta T_b = \frac{1000\, K_b w_1}{m_1 w_2}$

$\therefore \quad m_1(S_x) = \frac{1000\, K_b w_1}{w_2(\Delta T_b)}$

$= \frac{1000 \times 2.406 \times 3.975}{100 \times 0.37} = 258.5$

Molar mass of $(S)_x = 32x = 258.5$

$\therefore \quad x = 8$

20. (4) $\Delta T_b = (101.24 + 273) - 373 = 1.24°$

$\Delta T_b = \frac{1000 K_b w_1}{m_1 w_2}$

$\therefore \quad m_1(C_nH_{2n}O_n) = \frac{1000 K_b w_1}{\Delta T_b w_2} = \frac{1000 \times 0.52 \times 28.4}{1.24 \times 100}$

$= 119.1 \text{ g mol}^{-1}$

$C_nH_{2n}O_n = 12n + 2n + 16n$

$\therefore \quad 30n = 119.1$

$\Rightarrow \quad n = 4$

DPP-6 **Colligative Properties-IV** (Osmotic Pressure)

1. (c) I. Excretion of urine.

II. Reverse osmosis is used in desalination of sea water.

IV. Interchange of nutrients and waste products between tissue cells and their surroundings.

These are the processes involving osmosis.

2. (c) Pressure (due to insulin) = 0.0072 atm

$= 0.0072 \times 76$ cm of Hg

$= 0.0072 \times 76 \times 13.6$ cm of Hg column

Height of water column $= h$ cm

$\therefore \quad p = h \times 1$ cm of water column

(Density of water $= 1 \text{ g mL}^{-1}$)

$\therefore \quad h \times 1 = 0.0072 \times 76 \times 13.6 = 7.44$ cm

3. (d) $\pi = MRT$

(a) π (urea) $= RT$

(b) π (glucose) $= RT$

(c) M (mixture) $= \frac{M_1V_1 + M_2V_2}{V_1 + V_2} = \frac{100 \times 1 + 300 \times 1}{100 + 300} = 1$ M

$\therefore \quad \pi$ (mixture) $= RT$

Thus, equal osmotic pressure. Thus, isotonic.

4. (d) Side X (0.1 M) (Concentrated) Side Y (0.01 M) (Dilute)

There is osmosis of solvent (H_2O) from dilute solution (Y) to concentrated solution (X). There is no osmosis of ions, hence no blue colour formation.

5. (d) Osmotic pressure, $p = \frac{n}{V}RT = \frac{w}{mV}RT$

$$p_1\text{(glucose)} = \frac{10}{180 \times 0.2}RT = 0.28\,RT$$

$$p_2\text{(urea)} = \frac{10}{60 \times 0.1}RT = 1.67\,RT$$

$$p_3\text{(sucrose)} = \frac{10}{342 \times 0.05}RT = 0.585\,RT$$

Thus, $p_2 > p_3 > p_1$

6. (d)

HCHO	$C_6H_{12}O_6$
$w_1 = x$g	$w_1 = y$g
Volume = V L	Volume = VL
$m_1 = 30$ g mol^{-1}	$m_1 = 180$ g mol^{-1}

$\pi(\text{HCHO}) = \frac{w_1}{m_1 V}RT$ $\quad \pi(C_6H_{12}O_6) = \frac{w_1}{m_1 V}RT$

$\pi(\text{HCHO}) = \pi\,(C_6H_{12}O_6)$

$$\frac{x}{30 \times V}RT = \frac{y}{180 \times V}RT$$

$\therefore$ $\frac{x}{y} = 1:6$

7. (a) $\pi = \frac{w}{mV}RT$

Given, $\pi = 66.66$ kPa

$R = 8.3143$ JK^{-1}mol^{-1} = 8.3143 kPa dm^{3}K^{-1}mol^{-1}

$w = 0.4$g, $m = ?$, $V = 0.1$ dm^3, $T = 300$ K

$$\therefore \quad m = \frac{wRT}{\pi V} = \frac{0.4 \times 8.3143 \times 300}{66.66 \times 0.1}$$

$= 149.67 \approx 150.0$ g mol^{-1}

8. (c) At a column of height h,

Osmotic pressure $\pi = \rho g h$ = density $\times g \times h$

Density, $\rho = 1$g cm^{-3}

$$\rho = \left(\frac{1}{1000}\text{kg}\right)10^6\text{m}^{-3}$$

$$\rho = 10^3 \text{ kg m}^{-3}$$

$$g = 10.0 \text{ ms}^{-2}$$

$$h = 0.17 \text{ m}$$

$\therefore$ $\pi = 10^3 \times 10 \times 0.17$ kg m^{-1}s^{-2} = 1.7×10^3Pa

$$\pi = \frac{1.7 \times 10^3 \text{Pa}}{1.013 \times 10^5 \text{ Pa atm}^{-1}} = 0.0168 \text{ atm}$$

Thus, (a) and (b) both are correct.

9. (c) $NaCl \rightleftharpoons \underbrace{Na^+ + Cl^-}_{y=2}$

van't Hoff factor $(i) = [1 + (y-1)x] = 1 + x = 2$

$\pi = MRTi$

$= 0.7 \times 0.0821 \times 298 \times 2$ for 100% recovery

$= \frac{0.7 \times 0.0821 \times 298 \times 2 \times 100}{50}$ for 50% recovery = 68.5 atm

10. (c)

	Cane sugar	Unknown solute
Amount	$5\text{g} = \frac{5}{342}$mol	$1\text{ g} = \frac{1}{m}$mol
Volume	100 mL = 0.1 L	100 mL = 0.1 L
Temperature	300 K	300 K
Osmotic pressure	π_1	π_2

$$\pi_1 = \frac{n}{V}RT \qquad \pi_2 = \frac{n}{V}RT$$

$$= \frac{5 \times R \times 300}{342 \times 0.1} \qquad = \frac{1 \times R \times 300}{m \times 0.1}$$

Solutions are isotonic, thus $\pi_1 = \pi_2$

$$\frac{5}{342} = \frac{1}{m}$$

Thus, $m = 68.4$ g mol^{-1}

11. (d)

	Urea (1)	Unknown solute (2)
	Solute 5.25 g	21.5 g
	Solution 100 g	100 g
	Volume 100 mL = 0.1 L	100 mL = 0.1 L (density=1g cm^{-3})
Concentration	$M_1 = \left(\frac{5.25}{60}\right)/0.10$ mol L^{-1}	$M_2 = \left(\frac{21.5}{m}\right)/0.10$ mol L^{-1}
Temperature	300 K	300 K
Osmotic pressure	$\pi_1 = M_1RT$	$\pi_2 = M_2RT$

Solutions are isotonic, hence

$$\pi_1 = \pi_2$$

$$M_1RT = M_2RT$$

$$M_1 = M_2$$

$\therefore$ $\frac{5.25}{60}/0.10 = \frac{21.5}{m}/0.10$

$\therefore$ $m = 246$ g mol^{-1}

12. (b) ΔT_f = molality $\times K_f$

$0.56°$ = molality $\times$ 1.86° mol^{-1}kg

$\therefore$ Molarity = $\frac{0.56}{1.86}$ mol kg^{-1}

or molarity = 0.301 mol L^{-1}

Also, $\pi = MRT = 0.301 \times 0.0821 \times 310 = 7.66$ atm

13. (a) Osmotic pressure, $\pi = (M_1 + M_2)RT$

$$M_1\text{(urea)} = \frac{0.4\text{ g}}{60\text{ g mol}^{-1} \times 0.1\text{ L}} = \frac{1}{15}\text{ mol L}^{-1}$$

$$M_2(X) = \frac{0.9\text{ g}}{m\text{ g mol}^{-1} \times 0.1\text{ L}} = \frac{9}{m}\text{ mol L}^{-1}$$

$\pi = 2.85$ atm (given)

$\therefore$ $2.85 = \left(\frac{1}{15} + \frac{9}{m}\right) \times 0.0821 \times 298$

$$\left(\frac{1}{15} + \frac{9}{m}\right) = 0.1165$$

$$\frac{9}{m} = \left(0.1165 - \frac{1}{15}\right)$$

$$\frac{9}{m} = 0.0498$$

$\therefore$ $m = 180$ g mol^{-1}

14. (d) $n_1\text{(glucose)} = \frac{1}{180}$, $n_2\text{(sucrose)} = \frac{1}{342}$

$$\pi\text{(mixture)} = \pi(X)$$

$$\left(\frac{n_1 + n_2}{V}\right)RT = \frac{n_X}{V}RT$$

$$\frac{\left(\frac{w_1}{m_1} + \frac{w_2}{m_2}\right)RT}{V} = \frac{w_X}{m_X}\frac{RT}{V}$$

$$\left(\frac{1}{180} + \frac{1}{342}\right) = \frac{2}{m_X}$$

$\therefore$ $m_X = 235.86 \approx 236$ g mol^{-1}

15. (b) Osmotic pressure, $\pi = MRT = \frac{w_1}{m_1V}RT$

(V = volume of solution in L)

$$\frac{\text{Mass}}{\text{Volume}} = \text{Density}$$

$\therefore$ $$\text{Volume} = \frac{\text{Mass}}{\text{Density}} = \frac{w_2\ \text{g}}{8.149 \times 10^2\ \text{g dm}^{-3}} = \frac{w_2}{8.149 \times 10^2}\text{dm}^3$$

$\therefore$ $$\pi = \frac{w_1 \times 8.149 \times 10^2}{m_1 w_2} RT$$

By Raoult's law, for dilute solute

$$\frac{p^\circ - p}{p^\circ} = \chi_{\text{solute}} = \frac{n_1}{n_1 + n_2} \approx \frac{n_1}{n_2} = \frac{w_1\, m_2}{m_1\, w_2}$$

$$\frac{w_1}{m_1 w_2} = \frac{p^\circ - p}{p^\circ m_2}$$

$\therefore$ $$\pi = \frac{p^\circ - p}{p^\circ m_2} \times 8.149 \times 10^2 \times 8.314 \times 353$$

$$= \frac{(10.0268 - 9.8960)\, 10^4 \times 8.149 \times 10^2 \times 8.314 \times 353}{10.0268 \times 10^4 \times 0.078\ \text{kg}}$$

$$= 4.00 \times 10^5\ \text{Nm}^{-2}$$

16. (a) $[Mg^{2+}] = 2.0$ milliequivalent L^{-1}

$$= 2.0 \times 10^{-3} \text{equivalent L}^{-1}$$

$$= \frac{2.0 \times 10^{-3}}{2} \text{mol L}^{-1} = 1 \times 10^{-3}\text{M}$$

$[PO_4^{3-}] = 3.0$ milliequivalent $L^{-1} = 3.0 \times 10^{-3}$ equivalent L^{-1}

$$= \frac{3.0 \times 10^{-3}}{3} \text{mol L}^{-1} = 1 \times 10^{-3}\text{M}$$

$\therefore$ $$\pi = M(Mg^{2+} + PO_4^{3-})RT = 2 \times 10^{-3} \times 0.0821 \times 300$$

$$= 0.04926\ \text{atm}$$

17. (a) $$\pi = \frac{n_1\,(\text{solute})}{V(\text{L})} = \frac{w_1}{m_1 V} RT$$

$$3.632 = \frac{3}{m_1 \times 0.1} \times 0.0821 \times 298$$

$\therefore$ $$m_1 = \frac{3 \times 0.0821 \times 298}{0.1 \times 3.632} = 202.08\ \text{g mol}^{-1}$$

By Raoult's law for dilute solution,

$$\frac{\Delta p}{p^\circ} = \chi_{\text{solute}} = \frac{n_1}{n_1 + n_2} \approx \frac{n_1}{n_2}$$

$$= \frac{w_1}{m_1} \cdot \frac{m_2}{w_2}$$

$$= \frac{3}{202.08} \times \frac{18}{100} = 2.67 \times 10^{-3}$$

$\therefore$ $$\Delta p = 2.67 \times 10^{-3} \times 23.7\ \text{mm} = 0.063\ \text{mm}$$

$\therefore$ $$p^\circ - p_s = 0.063\ \text{mm}$$

$\therefore$ $$p_s = p^\circ - 0.063 = 23.7 - 0.063 = 23.64\ \text{mm}$$

18. (a) $$\pi = MRT = \frac{n}{V} RT = \frac{w}{mV} RT$$

Here, $w = 34.2$ g

$$m(\text{sucrose}) = 342.0\ \text{g mol}^{-1}$$

$$V = 100\ \text{mL} = 0.1\ \text{L}$$

$\therefore$ $$\pi = \frac{34.2}{342 \times 0.1} \times 0.0821 \times 300 = 24.63\ \text{atm}$$

But, experimental value is 30.0 atm. This value is much higher than the theoretical value.

It is because ($\pi = mRT$) is valid for dilute solution.

Thus, both Statements I and II are correct and Statement II is the correct explanation of Statement I.

19. (d) As per definition

Thus,

(i) $\to$ (r), (ii) $\to$ (s), (iii) $\to$ (q), (iv) $\to$ (t), (v) $\to$ (p)

Explanations for Q. Nos. (20-21)

$$\text{Pressure} = \frac{\text{Force}}{\text{Area}} = h\rho g$$

$$h = \text{height} = 77.8\ \text{mm} = 0.0778\ \text{m}$$

$$\rho = \text{density} = 1 \times 10^3\ \text{kg m}^{-3}$$

$$g = 9.81\ \text{ms}^{-2}$$

$\therefore$ $$\pi\ (\text{Osmotic pressure}) = 0.0778\ \text{m} \times 1 \times 10^3\ \text{kg m}^{-3} \times 9.81\ \text{m s}^{-2}$$

$$= 763.2\ \text{kg m}^{-1}\ \text{s}^{-2} = 763.2\ \text{N m}^{-2}$$

Also, $$\pi = \frac{n}{V} RT = \frac{w}{mV} RT$$

$\therefore$ $$m = \frac{wRT}{\pi V}$$

$$w = 20\ \text{g} = 20 \times 10^{-3}\ \text{kg}$$

$$V = 1\ \text{dm}^3 = 1 \times 10^{-3}\ \text{m}^3$$

$$R = 8.314\ \text{J mol}^{-1}\ \text{K}^{-1}, T = 298\ \text{K}$$

$$m\ (\text{molar mass}) = \frac{20 \times 10^{-3}\ \text{kg} \times 8.314\ \text{J mol}^{-1}\ \text{K}^{-1} \times 298\ \text{K}}{763.2\ \text{Nm}^{-2} \times 1 \times 10^{-3}\ \text{m}^3}$$

$$= 64.92\ \text{kg mol}^{-1}$$

20. (a) **21.** (b)

22. (8) Osmotic pressure, $\pi = MRT = \frac{w}{mV} RT$

$\therefore$ $$m(\text{molar mass of } C_nH_{2n}O_n) = \frac{wRT}{\pi V}$$

$$m = \frac{6 \times 0.0821 \times 300}{24.63 \times 0.1} = 60\ \text{g mol}^{-1}$$

Molar mass of $C_nH_{2n}O_n = 12n + 2n + 16n = 30n$

$\therefore$ $$30n = 60$$

$$n = 2$$

Thus, solute is $C_2H_4O_2$.

Total atoms = 2 + 4 + 2 = 8

23. (6) $\Delta T_f = K_f$ molality

$$1.86 = K_f \times 1$$

$\therefore$ $$K_f = 1.86\ \text{mol}^{-1}\ \text{kg}$$

$$\Delta T_f = \frac{1000\, K_f w_1}{m_1 w_2}$$

$$0.93^\circ = \frac{1000 \times 1.86 \times 9}{m_1 \times 100}$$

$\therefore$ m_1 (molar mass of A) = 180 g mol^{-1}

$$\pi\ (\text{osmotic pressure}) = \frac{n}{V} RT = \frac{w}{mV} RT$$

$$= \frac{4.41 \times 0.0821 \times 298}{180 \times 0.1} = 6.00\ \text{atm}$$

24. (8) π (osmotic pressure) $= \frac{n}{V} RT = \frac{w}{mV} RT$

$$\pi = \frac{CRT}{m}$$

$$\frac{\pi}{C} = \frac{RT}{m}$$

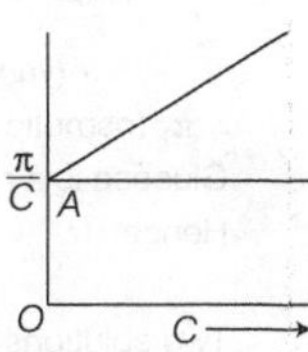

At $C = 0$, $$\frac{\pi}{C} = \text{Constant} = \frac{RT}{M} = OA$$

$\therefore$ $$OA = \frac{0.0821\ \text{L atm mol}^{-1}\text{K}^{-1} \times 300\ \text{K}}{308\ \text{g mol}^{-1}}$$

$$= 8 \times 10^{-2} = x \times 10^{-y}\ \text{L atm g}^{-1}$$

Thus, $x = 8$

DPP-7 van't Hoff Factor

1. **Plan** Osmotic pressure is a colligative property.

$$NaCl \rightleftharpoons Na^+ + Cl^-, (y = 2)$$

$$\text{van't Hoff factor } (i) = 1 + (y-1)x = 1 + x$$
$$= 1 + 0.50 = 1.50$$

$$\text{Osmotic pressure, } \pi = mRTi = \frac{n}{V}RTi = \frac{w}{mV}RTi$$

$$w = 5.85\text{ g}, m = 58.5\text{ g mol}^{-1}$$
$$V = 100\text{ mL} = 0.1\text{ L}$$

$$\therefore \quad \pi = \frac{5.85 \times 0.0821 \times 300 \times 1.50}{58.5 \times 0.1}$$
$$= 36.945\text{ atm}$$

2. **Plan** Depression in freezing point is a colligative property.

$$HA \rightleftharpoons H^+ + A^-$$

	HA	H^+	A^-
Initial	C	0	0
After ionisation	$C(1-\alpha)$	$C\alpha$	$C\alpha$

$$\Delta T_f = K_f \times \text{molality} \times i$$
$$0.18786 = 1.86 \times 0.1 \times i$$
$$\therefore \quad i = 1.01$$

(i) van't Hoff factor $(i) = 1.01$

(ii) $HA \rightleftharpoons H^+ + A^-, \quad y = 2$

$$i = 1 + (y-1)\alpha$$
$$i = 1 + \alpha$$
$$\therefore \quad 1.01 = 1 + \alpha$$
$$\therefore \quad \alpha = 0.01$$

Degree of ionisation = 0.01

(iii) Due to ionisation, $[H^+] = C\alpha$

Take molality = molarity

$$\therefore \quad [H^+] = 0.1 \times 0.01 = 1 \times 10^{-3}$$
$$\therefore \quad pH = -\log(1 \times 10^{-3}) = 3$$

(iv) Ionisation constant (by Ostwald's dilution law)

$$K_a = \frac{C\alpha^2}{(1-\alpha)} = \frac{0.1\,(0.01)^2}{(1-0.01)} \approx 1 \times 10^{-5} = 1.0 \times 10^{-5}$$

3. (a) van't Hoff factor $(i) = [1 + (n-1)\alpha]$

where, n = number of ions from one mole of solute

α = degree of ionisation/association

$$A_xB_y \rightleftharpoons xA + yB$$

Total ions = $(x + y)$

$$\therefore \quad i = 1 + (x + y - 1)\alpha$$
$$\therefore \quad \alpha = \frac{i-1}{(x+y-1)}$$

4. (a) $\underset{1\text{ mol}}{Na_2SO_4} \rightleftharpoons \underbrace{2Na^+ + SO_4^{2-}}_{3\text{mol}}$

$$i \text{ (van't Hoff factor)} = 1 + (y-1)x = (1 + 2x)$$

$\therefore \pi_1$ (osmotic pressure) $= MRTi = 0.004 \times R \times 300 \times (1 + 2x)$

Glucose is a non-electrolyte.

Hence, $\quad i = 1$

$$\pi_2 = mRTi = 0.01 \times R \times 300 \times 1$$

Two solutions are isotonic, hence

$$\pi_1 = \pi_2$$
$$\therefore \quad 0.004 \times R \times 300 \times (1 + 2x) = 0.01 \times R \times 300 \times 1$$
$$\therefore \quad (1 + 2x) = \frac{0.01}{0.004} = 2.5$$
$$x = 0.75$$

Thus, 75% dissociation.

5. (a) $\Delta T_f = \dfrac{1000\,K_f\,w_1\text{(solute)}}{m_1\text{(solute)}\,w_2}(i)$

$$w_1 = 20.0\text{ g}$$
$$m_1 = 11 \times 12 + 8 \times 1 + 2 \times 16 = 172\text{ g mol}^{-1}$$
$$w_2 = 50.0\text{ g}$$
$$\therefore \quad \Delta T_f = 2 = \frac{1000 \times 1.72 \times 20 \times i}{172 \times 50}$$
$$i = 0.5$$

6. (b) $\Delta T_b = 101.08 - 100 = 1.08°$

$$\Delta T_f = 0 - (-1.80°) = 1.80°$$

$AB \rightleftharpoons A^+ + B^-, i = 2$ at boiling point

$$\Delta T_b = \text{molality} \times K_b \times i = m \times K_b \times 2$$
$$\Delta T_f = m \times K_f \times i$$

(van't Hoff factor of AB at freezing point)

$$\therefore \quad \frac{\Delta T_b}{T_f} = \frac{m \times K_b \times 2}{m \times K_f \times i}$$
$$\frac{1.08}{1.80} = \frac{0.3 \times 2}{i}$$
$$\therefore \quad i = \frac{0.3 \times 2 \times 1.8}{1.08} = 1$$

Thus, AB behaves as a non-electrolyte at the freezing point of solution (remains non-ionised).

7. (a) Osmotic pressure is a colligative property, hence osmolarity is the molarity of the solution involving van't Hoff factor.

$$K_4[Fe(CN)_6] \rightleftharpoons \underbrace{4K^+ + [Fe(CN)_6]^{4-}}_{y=5}$$

$$i = [1 + (y-1)x] = 1 + 4x = 5 \qquad (x = 1)$$

Thus, osmolarity $= 0.02 \times i = 0.02 \times 5 = 0.10$ osmol

8. (d) For $AB, i = 1 + (y-1)\alpha$

y = number of particles from one mole of solute = 2

$$i = 1 + (2-1)\alpha = (1 + \alpha)$$

For CD, $\quad i = \left(1 + \left(\frac{1}{2} - 1\right)\beta\right) = \left(1 - \frac{\beta}{2}\right) = \left(\frac{2-\beta}{2}\right)$

$$\therefore \quad \Delta T_b(AB) = K_b m\text{(molality)}i = K_b m(1 + \alpha)$$
$$\Delta T_b(CD) = K_b m i = K_b m\left(\frac{2-\beta}{2}\right)$$

Given, $\quad \dfrac{\Delta T_b(AB)}{\Delta T_b(CD)} = 2$

$$\therefore \quad \frac{K_b m(1+\alpha)}{K_b m\left(\frac{2-\beta}{2}\right)} = 2$$
$$\frac{1+\alpha}{(2-\beta)} = 1$$
$$1 + \alpha = 2 - \beta$$
$$\therefore \quad \beta = 1 - \alpha = 1 - 0.80 = 0.20$$

9. (d) T_b (boiling point) $= T_0$(solvent) $+ \Delta T_b$

$$\Delta T_b = m \times K_b \times i$$
$$m\text{(same)} = 0.1\text{ molal}$$
$$\Delta T_b \propto i$$

Greater the value of i, larger the boiling point of solution.

		y	$i = 1 + (y-1)x$
(a) $Na_2SO_4 \rightleftharpoons 2Na^+ + SO_4^{2-}$,		3	$= (1 + 2x)$
(b) $C_6H_{12}O_6$ (non-electrolyte),		1	1
(c) $MgCl_2 \rightleftharpoons Mg^{2+} + 2Cl^-$,		3	$(1 + 2x)$
(d) $Al(NO_3)_3 \rightleftharpoons Al^{3+} + 3NO_3^-$,		4	$(1 + 3x)$

10. (b,c,d) $\pi_I = MRT = RT$

$$NaCl \rightleftharpoons \underbrace{Na^+ + Cl^-}_{y=2}, x = 1$$

$$i = 1 + (y-1)x = 1 + x = 2$$

$\therefore$ $\pi_{II} = MRTi = 2RT$

$$(NH_4)_3PO_4 \rightleftharpoons \underbrace{3NH_4^+ + PO_4^{3-}}_{y=4}, x = 1$$

$$i = 1 + (y-1)x = 1 + 3x = 4$$

$\therefore$ $\pi_{III} = MRTi = 4RT$

In benzene, benzoic acid forms dimer by intermolecular H-bonding.

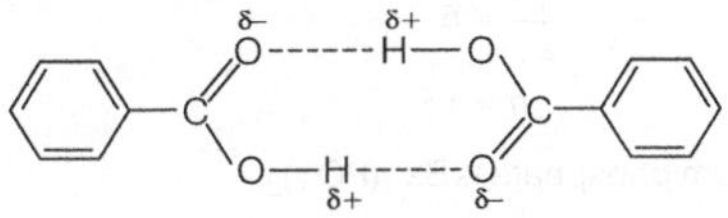

$$A \rightleftharpoons \frac{1}{2}(A_2), x = 1$$

$$y = \frac{1}{2}$$

$$i = [1 + (y-1)x] = \left(\frac{1}{2}\right)$$

$\therefore$ $\pi_{IV} = MRTi = \left(\frac{RT}{2}\right)$

Thus, osmotic pressure

$$\pi_{III} > \pi_{II} > \pi_I > \pi_{IV}$$

Thus, solutions are not isotonic, (a) is incorrect.

III has maximum osmotic pressure, thus hypertonic of I, II and IV, (b) is correct.

IV has minimum osmotic pressure, thus hypotonic of I, II and III. Thus, (c) is correct.

II is hypotonic of III but hypertonic of I and IV, thus (d) is correct.

11. (a,c)

(a) Vapour pressure of a component in a solution, $p_A = p_A^\circ \chi_A$.

by Raoult's law, thus correct.

(b) $\pi = i\,MRT$

For NaBr,

$$y = 2, i = 1 + (y-1)x = 2$$
$$M = 1, T = (273 + t)$$

$\therefore$ $\pi = 2R\,(273 + t)$ thus, incorrect.

(c) $\pi = i\,MRT$

$\pi \propto i$ for equimolar solution

$i = 1$ for sucrose

$i = (1 + x)$ for CH_3COOH $(x << 1)$

for CH_3COOH being weak acid

$i = 2$ for KCl

$i = 3$ for K_2SO_4

Thus, π (osmotic pressure) of 0.01M solutions of

sucrose < CH_3COOH < KCl < K_2SO_4

Thus, correct.

(d) $\Delta T_f = K_f \times \text{molality} \times i$

Urea is non-electrolyte, hence

$$i = 1$$
$$\Delta T_f = K_f \times \text{molality}$$
$$\Delta T_f (\text{in water}) = K_f$$
$$\Delta T_f (\text{in } CH_3OH) = K_f^1$$

K_f^1 and K_f are different, hence ΔT_f will be different, thus incorrect.

12. (b)

	Number of ions
(a) $PtCl_4 \rightleftharpoons Pt^{4+} + 4Cl^-$	5
$PtCl_4 + 2KCl \rightleftharpoons K_2[PtCl_6] \rightleftharpoons 2K^+ + [PtCl_6]^{2-}$	3
(b) $ZnCl_2 \rightleftharpoons Zn^{2+} + 2Cl^-$	3
$ZnCl_2 + 4NH_3 \rightleftharpoons [Zn(NH_3)_4]Cl_2 \rightleftharpoons [Zn(NH_3)_4]^{2+} + 2Cl^-$	3
(c) $FeCl_3 \rightleftharpoons Fe^{3+} + 3Cl^-$	3
$4FeCl_3 + K_4[Fe(CN)_6] \rightleftharpoons Fe_4[Fe(CN)_6]_3 + 12KCl$; $Fe_4[Fe(CN)_6]_3 \rightleftharpoons 4Fe^{3+} + 3[Fe(CN)_6^{4-}]$	7
(d) $HgCl_2 \rightleftharpoons Hg^{2+} + 2Cl^-$	2
$HgCl_2 + 4KI \rightleftharpoons K_2HgI_4 + 2KCl \rightleftharpoons 2K^+ + [HgI_4]^{2-} + 2K^+ + 2Cl^-$	7

Except in (b), number of ions changes. Hence, van't Hoff factor also changes in (a), (c) and (d). In (b), no change.

13. (a) $\Delta T_b = K_b m(i) = K_b i$ for molality, $m = 1$

$\therefore$ $i = [1 + (y-1)x]$

		y	i	ΔT_b
(i)	Urea	1	1	K_b
(ii)	$KCl \rightleftharpoons K^+ + Cl^-$	2	$(1 + x) = 2$	$2K_b$
(iii)	$K_2SO_4 \rightleftharpoons 2K^+ + SO_4^{2-}$	3	$(1 + 2x) = 1 + 2 \times 0.75 = 2.5$	$2.5K_b$
(iv)	$K_4[Fe(CN)_6] \rightleftharpoons 4K^+ + [Fe(CN)_6^{4-}]$	5	$(1 + 4x) = (1 + 4 \times 0.5) = 3$	$3.0\,K_b$
(v)	$Fe_2(SO_4)_3 \rightleftharpoons 2Fe^{3+} + 3SO_4^{2-}$	5	$1 + 4x = 1 + 4 \times 0.25 = 2$	$2.0\,K_b$

Thus, (i) $\rightarrow$ (p), (ii) $\rightarrow$ (r), (iii) $\rightarrow$ (s), (iv) $\rightarrow$ (t), (v) $\rightarrow$ (r)

14. (d) Depression in freezing point is a colligative property.

$$\Delta T_f = K_f \times \text{molality} \times i$$

$$HX \rightleftharpoons \underbrace{H^+ + X^-}_{y=2}$$

$\therefore$ $i = [1 + (y-1)x] = (1 + x)$

$$0.383 = 1.86 \times 0.2 \times i$$

van't Hoff factor, $i = \dfrac{0.383}{1.86 \times 0.2} = 1.03$

$$(1 + x) = 1.03$$

$\therefore$ $x = 0.03$ (degree of ionisation)

	HX	$\rightleftharpoons H^+$	$+ X^-$
Initial	C	0	0
At equilibrium	$(C - Cx)$	Cx	Cx

$$K_a = \frac{[H^+][X^-]}{[HX]} \quad \text{(By Ostwald's dilution law)}$$

$$= \frac{Cx^2}{(1-x)} = \frac{0.2\,(0.03)^2}{0.97} = 1.86 \times 10^{-4}$$

$$pK_a = -\log K_a = 3.73$$
$$[H^+] = Cx = 0.2 \times 0.03 = 6 \times 10^{-3}$$
$$pH = -\log [H^+] = -\log 6 \times 10^{-3} = 2.22$$

Thus, (i) $\rightarrow$ (r) (ii) $\rightarrow$ (p) (iii) $\rightarrow$ (s) (iv) $\rightarrow$ (q)

Explanations for Q. Nos. (15-16)

$$2KI + HgI_2 \rightleftharpoons K_2HgI_4$$

$$\underbrace{2(K^+ + I^-)}_{i=4} \quad \underbrace{2K^+ + HgI_4^{2-}}_{i=3}$$

Thus, van't Hoff factor decreases by 1 unit.

$$\Delta T_f = m \times K_f \times i$$
$$\Delta T_f(\text{KI}) = m \times K_f \times 4 = 4mK_f$$

$\therefore$ Freezing point $= 0 - \Delta T_f = 0 - 4mK_f$

$$\Delta T_f\ (K_2HgI_4) = 3mK_f$$

Freezing point $= 0 - 3mK_f = -3mK_f$

Freezing point of K_2HgI_4 $(-3mK_f)$ > Freezing point of KI $(-4mK_f)$

15. (c) **16.** (c)

Explanations for Q. Nos. (17-18)

$$H_2A \rightleftharpoons 2H^+ + A^{2-}$$
$$C(1-x) \quad 2Cx \quad Cx$$

$y = 3 =$ number of ions from one unit

$i =$ van't Hoff factor $= 1 + (y-1)x = (1 + 2x)$

Also, 10 mL of dibasic acid ≡ 12 mL of 0.1 N NaOH

N_1 (dibasic acid) $= 0.12$ g equiv L^{-1}

$0.12 \times$ equivalent weight $=$ concentration $= 10\text{ g L}^{-1}$

$$\text{Equivalent weight} = \frac{10}{0.12} = 83.33$$

Molar mass of acid $= 83.33 \times 2 = 166.66$

[Acid] $= 10/166.6$ mol L^{-1}

$$\Delta T_f = 0 - (-0.15°) = 0.15°$$
$$\Delta T_f = \frac{1000\ K_f w_1 i}{m_1\ w_2}$$
$$0.15° = \frac{1000 \times 1.86 \times 10 \times i}{166.66 \times 1000}$$

$\therefore$ $i = 1.34$

Also, $i = (1 + 2x) = 1.34$

$\therefore$ $x = 0.17$

$\therefore$ $[H^+] = 2Cx = \dfrac{2 \times 10}{166.66} \times 0.17 = 0.02$ M

$\therefore$ $pH = -\log[H^+] = 1.70$

17. (a) **18.** (b)

Explanations for Q. Nos. (19-20)

From ΔT_f of cane sugar, K_f can be determined.

$$\Delta T_f\ (\text{cane sugar}) = 0 - (-0.037°) = 0.037°$$
$$\Delta T_f = \frac{1000\ w_1\ (\text{solute}) K_f}{m_1\ (\text{solute}) w_2\ (\text{solvent})}$$
$$0.037 = \frac{1000 \times 0.684 \times K_f}{342 \times 100}$$
$$K_f = 1.85°\ \text{mol}^{-1}\ \text{kJ}$$

For NaCl solution,

$$\Delta T_f = 0 - (-0.342°) = 0.342°$$
$$NaCl \rightleftharpoons Na^+ + Cl^-$$
$$y = 2$$

i (van't Hoff factor) $= [1 + (y-1)x] = (1 + x)$

$$\Delta T_f = \frac{1000 K_f\ w_1\ (i)}{m_1\ w_2}$$
$$0.342° = \frac{1000 \times 1.85 \times 0.585 \times i}{58.5 \times 100}$$
$$i = 1.85$$

Also, $i = \dfrac{\text{Normal molecular weight}}{\text{Apparent molecular weight}}$

$\therefore$ Apparent molecular weight $= \dfrac{\text{Normal molecular weight}}{i}$

$$= \frac{58.5}{1.85} = 31.6$$

19. (b) **20.** (c)

21. (1) By Raoult's law,

$$\frac{\Delta p}{p^o} = \chi_{\text{solute}} = \frac{n_1}{n_1 + n_2}$$

If n_1 (NaCl) $= n$

then due to ionisation of NaCl,

$$n_1 = n \times i = 2n$$

$\therefore$ $n_2(\text{water}) = \dfrac{180}{18} = 10$

$\therefore$ $0.167 = \dfrac{2n}{2n + 10}$

$$\frac{2n + 10}{2n} = \frac{1}{0.167} \Rightarrow 1 + \frac{10}{2n} = 6$$
$$\frac{10}{2n} = 5$$
$$n = 1$$

22. (8) Barium phosphate is $Ba_3(PO_4)_2$.

$$Ba_3(PO_4)_2 \rightleftharpoons \underbrace{3Ba^{2+} + 2PO_4^{3-}}_{y = 5}$$
$$i = 1 + (y-1)x \quad (x = 1, 100\%)$$
$$= 1 + 4x = 1 + 4 \times 1 = 5$$

$\therefore$ $\Delta T_f =$ molality $\times K_f \times i$

$$\frac{\Delta T_f}{K_f} = 0.40(\text{given}) = \text{molality} \times 5$$

$\therefore$ Molality $= \dfrac{0.40}{5} = 0.08 = 8 \times 10^{-2}$ molal

Thus, $x = 8$

23. (2) $\Delta T_f = 0 - (-0.00732°) = 0.00732°$

$$\Delta T_f = \text{Molality} \times K_f \times i$$
$$0.00732 = 0.002 \times 1.86 \times i$$
$$i = 1.97 \approx 2.0$$

Two moles of ions are formed.

$$[Co(NH_3)_5NO_2]\,Cl \rightleftharpoons \underbrace{[Co(NH_3)_5NO_2]^+ + Cl^-}_{y=2}$$

24. (2) Let molar mass of $X = m$

Then molar mass of X in $A = 2m$ (dimer)

and that in $B = 3m$ (trimer)

$$\Delta T_f(A) = \text{molality } K_f(A) = \frac{1000\ K_f\ w_1}{m_1\ w_2}$$
$$\alpha = \frac{1000 \times 1.86 \times w_1}{2m \times w_2}$$

Similarly, $\beta = \dfrac{1000 \times 2.72 \times w_1}{3m \times w_2}$

$$\frac{\alpha}{\beta} = \frac{1.86}{2} \times \frac{3}{2.72} = \frac{1}{1}$$

Thus, $\alpha + \beta = 2$

25. (3) Let number of ions from one unit of substance $= y$

then, $i = 1 + (y-1)\alpha$

$\alpha = 1$ (given)

$\therefore$ $i = 1 + (y-1) = y$

$$1.28 = \Delta T_f\ (\text{in benzene}) = \frac{1000\ K_f w_1}{m_1\ w_2}$$
$$1.40 = \Delta T_f\ (\text{in water}) = \frac{1000\ K_f'\ w_1}{m_1\ w_2}(i)$$
$$\frac{1.28}{1.40} = \frac{K_f\ (\text{benzene})}{K_f'\ (\text{water})\, i}$$

$\therefore$ $i = \dfrac{5.12 \times 1.40}{1.86 \times 1.28} = 3.01 \approx 3$

Revisal Problems (JEE Main)

1. (c) By Henry's law,

$$p = K_H S \text{ (solubility of ethane)}$$

$\therefore$ $p_1 = K_H S_1$

$p_2 = K_H S_2$

$\therefore$ $\frac{p_2}{p_1} = \frac{S_2}{S_1}$

$\therefore$ $p_2 = \frac{S_2}{S_1} p_1 = \frac{5.0 \times 10^{-2} \times 1 \text{ bar}}{6.56 \times 10^{-3}} = 0.762 \text{ bar}$

2. (a) HNO_3 is 63 % by mass, thus

100 g of HNO_3 solution has HNO_3 = 63 g

Since, $\frac{m}{V} = d$

Hence, $V = \frac{\text{Mass}}{\text{Density}}$

Hence, $\frac{100}{1.5}$ mL of HNO_3 solution has HNO_3 = 63 g

$= \frac{63}{63} = 1 \text{ mol}$

Molarity $= \frac{\text{mole}}{\text{litre}}$, Volume $= \frac{100}{1.5}$ mL $= \frac{1}{15}$ L

Molarity (HNO_3) $= \frac{1}{1/15} = 15$ M

10 mL of this solution is diluted to 100 mL.

$$M_1 V_1 = M_2 V_2$$

$$15 \times 10 = M_2 \times 100$$

$\therefore$ $M_2 = 1.50$ M

3. (d) $\frac{w_1}{m_1}$ = moles of solute in w_2 g solvent

$$\Delta T_f = \frac{1000 K_f w_1}{m_1 w_2}$$

$\therefore$ $= m_1(AB_2) = \frac{1000 \times 5.5 \times 1}{2.3 \times 20} = 120$

$\therefore$ $m_1(AB_4) = \frac{1000 \times 5.5 \times 1}{1.3 \times 20} = 212$

$\therefore$ $A + 2B = 120$ ...(i)

$A + 4B = 212$ ...(ii)

On solving Eqs. (i) and (iii), we get

$\therefore$ $B = 46$

$A = 28$

4. (d) 100 cc of solution has glucose = 20 g

100 × 0.9971 g of solution has glucose = 20 g

99.71 g of solution has glucose = 20 g

Thus, water (solvent) = 99.71 − 20 = 79.71 g $= \frac{79.71}{18}$ mol

Glucose (solute) $= \frac{20}{180}$ mol

$$p_{\text{solution}} = p^{\circ}_{H_2O} \times \chi_{\text{solvent}}$$

$$\chi_{\text{solvent}} = \frac{n(H_2O)}{n(H_2O) + n(\text{glucose})}$$

$$= \frac{\frac{79.71}{18}}{\frac{79.71}{18} + \frac{20}{180}} = \frac{\frac{797.1}{180}}{\frac{797.1}{180} + \frac{20}{180}}$$

$$= \frac{797.1}{817.1} = 0.9755$$

$\therefore$ $p_{\text{solution}} = 23.8 \times 0.9755 = 23.217 \text{ mm}$

5. (a) By Raoult's law,

$$\frac{p^{\circ} - p}{p^{\circ}} = \frac{\text{Loss in weight in solvent}}{\left(\begin{array}{l}\text{Loss in weight of solvent} \\ + \text{Loss in weight of solution}\end{array}\right)} = \chi_{\text{solute}}$$

$$\chi_{\text{solute}} = \frac{\frac{2.25}{m}}{\left(\frac{150}{78} + \frac{2.25}{m}\right)}$$

$\therefore$ $\frac{2.25/m}{\left(\frac{150}{78} + \frac{2.25}{m}\right)} = \frac{0.0160}{2.154 + 0.0160} = 7.373 \times 10^{-3}$

$$\frac{\left(\frac{150}{78} + \frac{2.25}{m}\right)}{\left(\frac{2.25}{m}\right)} = 135.625$$

$$\frac{\left(\frac{150}{78}\right)}{\left(\frac{2.25}{m}\right)} + 1 = 135.625$$

$$\frac{150}{78} \times \frac{m}{2.25} = 134.625$$

$\therefore$ $m = 157.51 \approx 158$

6. (d) $\Delta T_f = \frac{1000 K_f w_1 \text{(solute)}}{w_2 \text{(solvent)} m_1 \text{(solute)}}$

$$m_1 \text{ (solute, } CH_3OH) = \frac{\Delta T_f \times w_2 \times m_1}{1000 K_f} = \frac{9.3 \times 100 \times 32}{1000 \times 1.86} = 16 \text{ g}$$

7. (b) For carbohydrate,

w_1 (solute) = 18 g

Moles $= \frac{18}{m_1}$

Volume = 100 mL = 0.1 L

Molarity $= \frac{18}{m_1 \times 0.1} = \frac{180}{m_1}$ mol L^{-1}

i = 1 (non-electrolyte)

$$\pi \text{ (carbohydrate)} = mRT = \frac{180}{m_1} RT$$

For NaCl,

$$NaCl \rightleftharpoons Na^+ + Cl^-$$

w_1 (solute) = 2.925 g

Moles $= \frac{2.925}{58.5}$

Volume = 100 mL = 0.1L

Molarity $= \frac{2.925}{58.5 \times 0.1} = 0.5$ mol L^{-1}

$i = 2$

$$\pi(NaCl) = mRTi = 0.5 \times RT \times 2 = RT$$

Solutions are isotonic, hence

$$\frac{180}{m_1} RT = RT$$

$\therefore$ $m_1 = 180$ g mol^{-1}

8. (b)

Solution	Volume	Solute
I(25%)	300 g	$\frac{300 \times 25}{100} = 75$ g
II(40%)	400 g	$\frac{400 \times 40}{100} = 160$ g
Mixture	700 g	75 + 160 = 235 g

$\therefore$ Mass percentage $= \frac{235}{700} \times 100 = 33.57\%$

9. (a) Let, $Na_2CO_3 = x\,g = \frac{x}{106}$ mol

then, $NaHCO_3 = (1 - x)g = \frac{1-x}{84}$ mol

$$\underset{1\text{ mol}}{Na_2CO_3} + \underset{2\text{ mol}}{2HCl} \longrightarrow 2NaCl + H_2O + CO_2$$

$$\frac{x}{106}\text{mol} \left(\frac{2x}{106}\text{ mol}\right)$$

$$\underset{1\text{ mol}}{NaHCO_3} + \underset{1\text{ mol}}{HCl} \longrightarrow NaCl + H_2O + CO_2$$

$$\left(\frac{1-x}{84}\right)\text{mol} \left(\frac{1-x}{84}\right)\text{mol}$$

Total HCl required $= \left(\frac{2x}{106}\right) + \left(\frac{1-x}{84}\right)$

Actual HCl required $= \frac{V \times 0.1}{1000}$ mol

$\therefore \quad \frac{2x}{106} + \frac{1-x}{84} = \frac{V \times 0.1}{1000}$

Also, $\quad \frac{x}{106} = \frac{1-x}{84}$ (equimolar mixture)

$\therefore \quad x = 0.558$

$$\frac{2 \times 0.558}{106} + \frac{0.442}{84} = \frac{V \times 0.1}{1000}$$

$$V = 158 \text{ mL}$$

10. (a) $V_{total} = n_1\bar{V}_1 + n_2\bar{V}_2$

where, $\bar{V}_1$ and $\bar{V}_2$ are molar volumes and n_1 and n_2 are number of moles.

Let propanone $= wg = \frac{w}{58}$ mol

then butanone $= (1000 - w)\,g = \left(\frac{1000-w}{72}\right)$ mol

$$\chi_{propanone} = \frac{w/58}{\frac{w}{58} + \left(\frac{1000-w}{72}\right)} = 0.4 \text{(given)}$$

$w = 350$ g

$\therefore$ Propanone $= 350\text{ g} = \frac{350}{58}$ moles

Butanone $= 650\text{ g} = \frac{650}{72}$ moles

Thus, $\quad V_{total} = \frac{350}{58} \times 75 + \frac{650}{72} \times 80 = 1175\text{ cm}^3$

11. (a) $\underset{1\text{ mol}}{CaCl_2(aq)} \rightleftharpoons \underbrace{Ca^{2+}(aq) + 2Cl^-(aq)}_{y = 3\text{ mole}}$

i (van't Hoff factor) $= 1 + (y + 1)x$

$i = (1 + 2x)$

By Raoult's law, relative lowering of vapour pressure $= \chi_{solute}$

$$\frac{p^\circ - p_s}{p^\circ} = \frac{n_1(i)}{n_1(i) + n_2} \approx \frac{n_1 i}{n_2} \quad \text{(solute being an electrolyte)}$$

p° (vapour pressure of water) $= 0.03126$ atm

p_s (vapour pressure of solution) $= 0.02970$ atm

$$\frac{0.03126 - 0.02970}{0.03126} = \frac{w_1\, m_2}{m_1\, w_2}(i)$$

$$\frac{w_1 m_2 i}{m_1 w_2} = 0.05$$

$$\therefore \quad \frac{w_1 i}{m_1 w_2} = \frac{0.05}{18}$$

Depression in freezing point

$$\Delta T_f = \frac{1000 K_f\, w_1(i)}{m_1\, w_2}$$

$$= \frac{1000 \times 1.86 \times 0.05}{18} = 5.17^\circ$$

12. (c) $\Delta T_b = \text{molality} \times K_b \times i$

Solutions are equimolar in same solvent, thus $\Delta T_b \propto i$

Also, boiling point of the solution = (boiling point of $H_2O + \Delta T_b$)

Thus, greater the value of i, larger the boiling point.

x = degree of ionisation.

Solution	***y* (number of ions)**	**$=[1 + (y-1)x]$**
I (Na_2SO_4)	3	$1 + 2x = 3$
II (Na_3PO_4)	4	$1 + 3x = 4$
III NaBr	2	$1 + x = 2$
IV (glucose) (non-electrolyte)	1	$1 + 0 = 1$

Thus, van't Hoff factor is in the order, IV < III < I < II and boiling point is in the order, IV < III < I < II.

13. (b) $Na_2SO_4(aq) \rightleftharpoons 2Na^+(aq) + SO_4^{2-}(aq)$

Number of ions from 1 mol of $Na_2SO_4 = y = 3$

$\therefore \quad x = 1$ (100% ionisation)

i (van't Hoff factor) $= 1 + (y - 1)x = 1 + 2 = 3$

$\frac{w_1}{m_1}$ = moles of solute = 0.01 mol

Solvent (H_2O) = 1 kg

$\therefore \quad$ Molality $= \frac{0.01}{1} = 0.01$ mole kg^{-1}

$\Delta T_f = \text{molality} \times K_f \times i = 0.01 \times 1.86 \times 3 = 0.0558^\circ$

14. (c) There is dimer formation of benzoic acid in benzene due to intermolecular H-bonding.

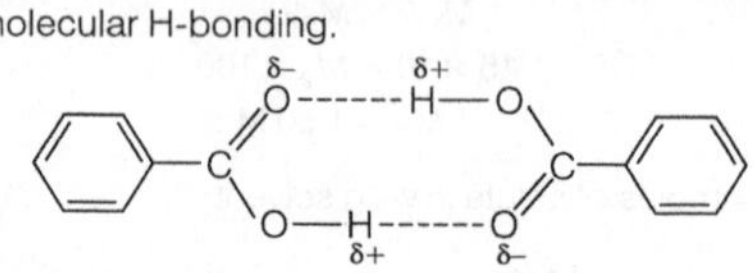

Thus, molar mass of dimer = 244.0 g mol^{-1}

$$\chi_{solvent} = \frac{n_{solvent}}{n_{total}}$$

By Raoult's law,

$$p_{solution} = p^\circ \chi_{solvent} = 66.6 \times \frac{\left(\frac{1000}{78}\right)}{\left(\frac{1000}{78} + \frac{122}{244}\right)} = 64.10 \text{ cm of Hg}$$

15. (c) Potassium ferrocyanide is $K_4[Fe(CN)_6]$.

$$K_4[Fe(CN)_6] \rightleftharpoons \underbrace{4K^+ + [Fe(CN)_6]^{4-}}_{\text{Number of ions, } y = 5}$$

i = van't Hoff factor $= 1 + (y - 1)x$

$= 1 + 4x = 1 + 4 \times 0.4 = 2.6$

$\pi = MRT(i)$

$R = 8.3143$ J mol^{-1}K^{-1}

$M = 0.1$ mol dm^{-3} $= 0.1 \times 10^3$ mol m^{-3}

$\pi = 0.1 \times 10^3 \times 8.3143 \times 300 \times 2.6$

$= 6.485 \times 10^5$ Nm^{-2} = 6.5 bar

Note 1×10^5 Nm^{-2} = 1 bar

16. (a) Let, glycerol to be added = V mL

$= V \times 1.26\text{ g} = 1.26V\text{g} = \frac{1.26}{92} V$ mole

Water $= 1000\text{ g} = \frac{1000}{18} = 55.55$

Total moles $= \left(55.55 + \frac{1.26V}{92}\right)$

$$\text{Mole \% of glycerol } \frac{\left(\frac{1.26V}{92}\right) \times 100}{\left(55.55 + \frac{1.26V}{92}\right)} = 4.85 \text{ (given)}$$

Thus, $\quad V = 207$ mL

17. (a) Water vapour will be transferred from the more dilute solution to the more concentrated solution until both solutions reach at the same concentration.

Moles of H_2O transferred = n

Moles of H_2O in first (dilute solution) = $\left(\frac{100}{18} - n\right)$

Moles of H_2O in second (concentrated solution) = $\left(\frac{100}{18} + n\right)$

Mole fractions are equal at equilibrium.

$$\chi_{C_{12}H_{22}O_{11}} = \frac{0.01}{\left(0.01 + \frac{100}{18} - n\right)} = \frac{0.02}{\left(0.02 + \frac{100}{18} + n\right)}$$

$\therefore \quad n = 1.852$

$\therefore \quad \chi_{C_{12}H_{22}O_{11}} = 2.693 \times 10^{-3} \approx 0.002693$

18. (a) By Raoult's law, $\frac{p^\circ - p_s}{p^\circ} = \chi_{solute} = \frac{n_{solute}}{n_{solute} + n_{solvent}}$

Given, $p^\circ = 640$ mm Hg, $p_s = 600$ mm Hg

$$n_{solute} = \frac{2.175}{m_1} (m_1 = \text{molecular weight of solute})$$

$$n_{solvent} = \frac{39}{78} = 0.5$$

$$\therefore \quad \frac{40}{640} = \left(\frac{2.175/m_1}{\frac{2.175}{m_1} + 0.5}\right) \Rightarrow \frac{(2.175/m_1) + 0.5}{(2.175/m_1)} = 16$$

$$\frac{0.5}{\left(\frac{2.175}{m_1}\right)} = 15 \Rightarrow \frac{2.175}{m_1} = \frac{0.5}{15}$$

$\therefore \quad m_1 = 65.25 \text{ g mol}^{-1}$

19. (a) ΔT_b (elevation in boiling point) = $\frac{1000 K_b w_1}{m_1 w_2}$

$$\frac{w_1}{m_1} = \text{moles of solute}$$

$$w_2 = \text{amount of solvent (in g)}$$

$$2 = \frac{1000 \times 0.76 \times w_1}{m_1 w_2}$$

$$\therefore \quad \frac{w_1}{m_1 w_2} = \frac{2}{760} = \frac{1}{380}$$

By Raoult's law for dilute solution,

$$(p^\circ = 1 \text{atm} = 760 \text{ mm Hg})$$

$$\frac{\Delta p}{p^\circ} = \frac{n_1}{n_1 + n_2} \approx \frac{n_1}{n_2} = \frac{w_1 m_2}{m_1 w_2}$$

$$\frac{\Delta p}{760} = \frac{18}{380}$$

$$\therefore \quad \Delta p = 760 \times \frac{18}{380} = 36 \text{ mmHg}$$

$$\therefore \quad p_s = p^\circ - \Delta p = 760 - 36 = 724 \text{ mmHg}$$

20. (d) By Henry's law, $\quad p = K_H \chi$

$$\therefore \quad \chi = \frac{p}{K_H}$$

$p = 2.5 \text{ atm} = 2.5 \times 1.01325 \times 10^5$ Pa, $K_H = 1.67 \times 10^8$ Pa

$$\therefore \chi \text{ (mole fraction of } CO_2) = \frac{2.5 \times 1.01325 \times 10^5}{1.67 \times 10^8} = 1.517 \times 10^{-3}$$

Let $CO_2 = n_1$ mole, $\quad H_2O = 500 \text{ mL} = 500 \text{ g} = \frac{500}{18}$ mol

$$\therefore \chi_{CO_2} = \frac{n_1}{n_1 + \left(\frac{500}{18}\right)} \approx \frac{n_1}{\left(\frac{500}{18}\right)} \quad \text{(since, } \chi_{CO_2} \text{ is very small)}$$

$$\therefore n_1 (CO_2) = \chi_{CO_2} \cdot \left(\frac{500}{18}\right) = 1.517 \times 10^{-3} \times \frac{500}{18} = 0.042 \text{ mol}$$

21. (a) Salts have been assumed as strong electrolytes.

van't Hoff factor, $i = [1 + (y - 1)x]$

y = number of ions from one mole of salt

x = degree of ionisation = 1

C_2H_5OH is a non-electrolyte hence, $y = 1$

$$\pi = \frac{n}{V} RTi = MRTi$$

Thus, $\pi \propto Mi$, RT being constant. Greater the value of Mi, greater the osmotic pressure.

Solute	Concentration	y	$i = 1 + (y-1)x$	Mi
C_2H_5OH	0.500 M	1	1	0.500
$Mg_3(PO_4)_2$	0.100 M	5	5	0.500
KBr	0.250 M	2	2	0.500
Na_3PO_4	0.125 M	4	4	0.500

Thus, $\pi \propto Mi$, have equal values.

22. (c)

	HA	$\rightleftharpoons$	H^+	+	A^-
Initial	C		0		0
After equilibrium	$(C - Cx)$		Cx		Cx

$$pK_a = -\log K_a = 4$$

$$\therefore \quad K_a = 10^{-4} = \frac{Cx^2}{(1 - x)} \approx Cx^2$$

$$\therefore \quad 10^{-4} = (0.01) x^2$$

$$\therefore \quad x^2 = 0.01$$

$$x = 0.1$$

van't Hoff factor = $i = (1 + x) = 1.1$

23. (c) **Urea** (NH_2CONH_2)

Solute = 6 g

Molar mass = 60 g mol^{-1}

Solvent = 100 mL = 0.1 L

Molarity = mol L^{-1} = $\frac{6/60}{0.1}$ = 1 mol L^{-1}

Acetic acid (CH_3COOH)

Solute = 6 g molar mass = 60 g mol^{-1} solvent = 100 mL = 0.1 L

$$\text{Molarity} = \frac{6/60}{0.1} = 1 \text{ mol L}^{-1}$$

Thus, solutions are equimolar. Thus, Statement I is correct.

$$\pi \text{(urea)} = MRTi$$

$$i = 1 \text{ (non-electrolyte)}$$

$$\pi = RT$$

$$\pi \text{(acetic acid)} = MRTi$$

$$i = (1 + \alpha)$$

$$\pi' = RT(1 + \alpha)$$

$$\pi'(CH_3COOH) > \pi \text{(urea)}$$

Thus, solutions are not isotonic (they have different osmotic pressure. Thus, Statement II is incorrect.

24. (b) $\Delta T_f = \frac{1000 \, w_1 \text{(solute)} \, K_f \, (i)}{m_1 \text{(solute)} \times w_2 \text{(solvent)}}$

$$1.62 = \frac{1000 \times 2.0 \times 4.86}{122 \times 25} (i)$$

$$i = 0.5$$

In benzene, benzoic acid is dimerised, thus freezing point depression is smaller than expected ($i = 0.5$).

ΔT_f depends on the number of ions hence, it is a colligative property.

Thus, both statements I and II are correct but Statement II is not the correct explanation of Statement I.

Revisal Problems (JEE Advanced)

1. (d) ΔT_f (urea) $= 1.86° =$ molality $\times K_f$

$$1.86° = 1 \times K_f$$

$$\therefore \quad K_f = 1.86° \text{ mol}^{-1}\text{kg}$$

$$\chi_{\text{solute}} = 0.07 = \frac{n_1}{n_1 + n_2}$$

Thus, if $\quad n_1(\text{solute}) = \frac{w_1}{m_1} = 0.07$

Then, $\quad n_2(\text{solvent}) = \frac{w_2}{m_2} = 0.93$

$$w_2 = 0.93 \times 18 \text{ g}$$

$$\Delta T_f = \frac{1000\, K_f\, w_1}{m_1\, w_2} = \frac{1000 \times 1.86 \times 0.07}{0.93 \times 18} = 7.78°$$

2. (b) $\Delta T_b =$ molality $\times K_b \times i$

$$2.08 = 1 \times 0.52 \times i$$

$$i = 4 = [1 + (y-1)x] = 1 + (y-1)1 = y$$

where, y = number of ions of the complex

x = degree of ionisation = 1

$$\therefore \quad y = 4$$

Four ions are formed from

$$K_3[Fe(CN)_6] \rightleftharpoons 3K^+ + [Fe(CN)_6]^{3-}$$

3. (b)

(a) ΔT_b (NaCl) = molality $K_b(i) = 1 \times K_b \times 2 = 2K_b$

$$\therefore \quad T_b(\text{NaCl}) = (100 + 2K_b)$$

$$\Delta T_b(\text{sucrose}) = 1 \times K_b \times 1 = K_b$$

$$T_b(\text{sucrose}) = (100 + K_b) \neq \frac{T_b(\text{NaCl})}{2}$$

Thus, (a) is incorrect.

(b) Based on (a), $\Delta T_b(\text{glucose}) = \frac{\Delta T_b(K_a)}{2}$. Thus, (b) is correct.

(c) Elevation in boiling point (ΔT_b) is a colligative property. Boiling point is not a colligative property. Thus, (c) is incorrect.

4. (b) Let sucrose be $= w\text{g} = \frac{w}{342}$ mole

$$\text{Water} = 100\text{g} = \frac{100}{18} \text{ moles}$$

$$\chi_{\text{sucrose}} = \frac{w/342}{\left(\frac{w}{342} + \frac{100}{18}\right)} = 0.124 \text{ (given)}$$

$$\therefore \quad w = 269\text{g}$$

5. (d) $K_3[Fe(CN)_6] \rightleftharpoons \underbrace{3K^+ + [Fe(CN)_6]^{3-}}_{y=4}$

$$i \text{ (van't Hoff factor)} = [1 + (y-1)x] = 1 + 3x$$

$$= 1 + 3 \times 0.8 = 3.4$$

$$\Delta T_f \text{ (depression in freezing point)} = \frac{1000 K_f\, w_1(i)}{m_1\, w_2}$$

$$= \frac{1000 \times 1.86 \times 0.1 \times 3.4}{329 \times 100} = 0.0192°$$

$\therefore$ Freezing point $= 0 - 0.0192 = -1.92 \times 10^{-2}$°C

6. (a)

(I) Entropy change between solution and vapour is smaller than the entropy change between pure solvent and vapour, thus correct.

(II) Heats of vaporisation for a pure solvent and for a solution are similar because similar intermolecular forces between solvent molecules must be overcome in both cases, thus correct.

(III) When a non-volatile solute is added to a solvent, vapour pressure falls hence, boiling point is elevated, thus correct.

7. (a) Solute (urea) = 60 g

Solution (urea + water) = 1100 g

$\therefore$ Solvent (water) = 1100 − 60 = 1040 g

$$\Delta T_f = \frac{1000 K_f\, w_1}{m_1\, w_2}$$

$$\frac{\Delta T_f}{K_f} = \frac{1000 w_1}{m_1\, w_2} = 1 \text{ mol kg}^{-1} \text{ (given)}$$

$$\frac{w_1}{m_1 w_2} = 60 \text{ g urea in 1000 g water}$$

Thus, water separated in the form of ice = 1040 − 1000 = 40g

8. (b) When a solute is added to a solvent, freezing point is lowered. Thus, freezing point of solution is less than that of the solvent. Let freezing point of solvent $= T_1$ and freezing point of solution $= T_2$.

$$T_2 < T_1$$

On lowering temperature, first T_1 is attained hence, acetic acid is converted into solid (crystal).

9. (a) $[Cu(NH_3)_3Cl]Cl \rightleftharpoons \underbrace{[Cu(NH_3)Cl]^+ + Cl^-}_{y=2}$

$$i \text{ (van't Hoff factor)} = 1 + (y-1)x = 1 + x$$

By Raoult's law, n_1 = solute, n_2 = solvent

$$\frac{\Delta p}{p°} = \frac{n_1 i}{n_1 i + n_2}$$

$$0.5 = \frac{2(1+x)}{2(1+x)+3}$$

$$0.5[2(1+x)+3] = 2(1+x)$$

$$2(1+x) + 3 = 4(1+x)$$

$$2(1+x) = 3$$

$$1 + x = 1.5$$

$$x = 0.5 \text{ (50\% ionised)}$$

$$[Cu(NH_3)_3Cl]Cl \rightleftharpoons [Cu(NH_3)Cl]^+ + Cl^-$$

$$Cl^- + Ag^+ \rightleftharpoons AgCl$$

Cl^- from 2 moles complex $= 2 \times 0.5 = 1.0$

$\therefore$ AgCl formed = 1 mol

10. (c) 46 g ethanol $= \frac{46}{46} = 1.0$ mol

72 g water $= \frac{72}{18} = 4.0$ mol

Total moles = 5.0 mol

$$\chi_{\text{ethanol}} = \frac{1}{5} = 0.2,\ \chi_{\text{water}} = \frac{4}{5} = 0.8$$

From the graph, partial molar volume of ethanol when

$$(\chi = 0.2) = 55\text{cm}^3\text{mol}^{-1}$$

and that of water when $(\chi = 0.8) = 14$ cm^{3}mol^{-1}

Hence, total volume $= \bar{V}_1 n_1 + \bar{V}_2 n_2 = 55 \times 1 + 14 \times 4 = 111.0$ cm^3

11. (d) H_2A (dibasic acid) $\rightleftharpoons \underbrace{\underset{2Cx}{2H^+} + \underset{Cx}{A^{2-}}}_{y=3}$

$$i = 1 + (y-1)x = 1 + 2x$$

$$pH = 3$$

$$\therefore \quad [H^+] = 10^{-3}\text{M} = 2Cx = 2 \times 0.1 \times x$$

$$\therefore \quad x = \frac{10^{-3}}{0.2} = 5 \times 10^{-3}$$

$$\therefore \quad i = 1 + 2x = 1 + 2 \times 5 \times 10^{-3} = 1.01$$

$$\pi = M\,i\,RT = 0.1 \times 1.01\, RT$$

$$= 0.101RT$$

12. (d) $RH_2 + Ca^{2+} \longrightarrow RCa + 2H^+$

Every Ca^{2+} replaced furnishes two H^+

Thus, $2[Ca^{2+}] = [H^+]$

$\therefore \quad [Ca^{2+}] = \frac{[H^+]}{2}$

$pH = 2$

Thus, $[H^+] = 10^{-2}$ M

$\therefore \quad [Ca^{2+}] = \frac{10^{-2}}{2} M = 0.005 \text{ mol L}^{-1} = 0.005 \times 40 \text{ gL}^{-1}$

10^3 mL has = 0.20 gCa^{2+}

10^6 mL(ppm) has = 200 ppm

13. (a) $CHCl_3$ is 15 ppm by mass, it means

10^6 g of $CHCl_3$ solution has = 15 g of $CHCl_3$

100 g of $CHCl_3$ solution has = $\frac{15 \times 100}{10^6}$ g $CHCl_3 = 15 \times 10^{-4}$%

10^6 g of $CHCl_3$ solution has = 15 g of $CHCl_3$

$(10^6 - 15)$g $H_2O \approx 10^6$g H_2O has = 15 g of $CHCl_3$

10^3 g of H_2O has = 15×10^{-3}g$CHCl_3 = \frac{15 \times 10^{-3}}{119.5}$ mole

$\therefore$ Molality = 1.255×10^{-4} mol kg^{-1}

14. (d) Figure shows ideal behaviour of solution.

At point A, $\chi_A = 0, \chi_B = 1$, thus

AD = vapour pressure of pure B = 135 units.

At point B, $\chi_B = 0, \chi_A = 1$, thus

BC = vapour pressure of pure A = 85 units

As A is added into B, vapour pressure of A decreases and that of B increases.

Thus, AC shows Raoult's law for a solution of B in A (solvent) and BD shows Raoult's law for a solution of A in B (solvent).

Thus, (I) is correct.

Total vapour pressure at any composition is given by line CD, thus (II) is correct.

$$p_B^\circ - p_A^\circ = 135 - 85 = 50 \text{ units}$$

Thus, (III) is correct.

When, $\chi_B = 0.4, \chi_A = 0.6$,

vapour pressure of the mixture is given by CD.

$$p_{\text{total}} = p_A^\circ \chi_A + p_B^\circ \chi_B$$

$$= 85 \times 0.6 + 135 \times 0.4$$

$$= 105$$

It is same as given by graph.

Thus, (IV) is also correct.

15. (b) $Ag^+ + Cl^- \longrightarrow AgCl$

When titration is complete, CrO_4^{2-} gives chocolate coloured precipitate with Ag^+.

$$2Ag^+ + CrO_4^{2-} \longrightarrow Ag_2CrO_4 \downarrow$$

This end point is detected.

$$M_1V_1 \equiv M_2V_2$$

$$M_1 \times 25 \text{ (KCl)} = 1.0 \times 20.0 \text{ (AgNO}_3)$$

$\therefore \quad M_1 \text{(KCl)} = \frac{1.0 \times 20.0}{25}$

$= 0.8$ mol L^{-1}

$= 0.8$ mol kg^{-1} (since, molarity = molality)

$KCl \rightleftharpoons K^+ + Cl^-$ $\quad (y = 2)$

$i = [1 + (y-1)x] = 1 + x = 1 + 1 = 2$ $\quad (x = 1)$

ΔT_f = molality $\times K_f \times i = 0.8 \times 2.0 \times 2 = 3.2°$

16. (c) Let total number of ions due to ionisation of the complex = y

$\therefore$ van't Hoff factor $(i) = 1 + (y-1)x$

$x = 1$ (100% ionisation)

$i = y$

ΔT_f = molality $\times K_f \times i$

$0.0054 = 0.001 \times 1.86 \times y$

$y = \frac{0.0054}{0.001 \times 1.86} = 2.90 \approx 3.0$

NH_3 is always a ligand, hence Cl can be ligand or ionisable ions.

Thus, complex is $[Pt(NH_3)_4Cl_2]Cl_2$

$$[Pt(NH_3)_4Cl_2]Cl_2 \rightleftharpoons \underbrace{[Pt(NH_3)_4Cl_2]^{2+} + 2Cl^-}_{y=3}$$

17. (d) Carbon $\xrightarrow{[O]} CO_2$

12 g $\qquad$ 44 g

$$C\% = \frac{12 \times 100 \times \text{mass of } CO_2}{44 \times \text{mass of substance}} = \frac{12 \times 100 \times 5.46}{44 \times 2.40} = 62.05\ \%$$

$2H \xrightarrow{[O]} H_2O$

2 g $\qquad$ 18 g

$$H\% = \frac{2 \times \text{mass of } H_2O \times 100}{18 \times \text{mass of substance}}$$

$$= \frac{2 \times 2.23 \times 100}{18 \times 2.40} = 10.32\%$$

Oxygen (by difference) = 100 − (62.05 + 10.32) = 27.63%

Element	%	Moles	Ratio
C	62.05	5.17	3
H	10.32	10.32	6
O	27.63	1.73	1

Empirical formula = C_3H_6O

$$\Delta T_f = \frac{1000\, K_f w_1}{m_1 w_2}$$

$\therefore \quad m_1 = \frac{1000\, K_f w_1}{w_2\, \Delta T_f}$

$$m_1 \text{ (solute)} = \frac{1000 \times 1.86 \times 8.69}{281 \times 0.97} = 59.3$$

$$m_1 (C_3H_6O) = 58$$

Thus, solute is C_3H_6O.

18. (a, c)

10% ethanol by mass means

100 g (ethanol + water) solution has = 10 g ethanol

90 g H_2O has ethanol = 10 g ethanol

$$\text{Molality} = \frac{10/46}{0.09 \text{ kg}} = 2.415 \text{ mol kg}^{-1}$$

Molality does not involve density, hence same at 15°C and 25°C.

Thus, (c) is correct.

$$\frac{\text{Mass}}{\text{Volume}} = \text{Density}$$

$\therefore \quad \text{Volume} = \frac{\text{Mass}}{\text{Density}}$

$$V_{15} = \text{Volume at 15°C} = \frac{100}{0.9831} \text{ mL} = 0.1017 \text{ L}$$

$$V_{25} = \text{Volume at 25°C} = \frac{100}{0.9804} \text{ mL} = 0.1020 \text{ L}$$

Molarity = $\frac{\text{mol}}{\text{L}}$, thus molarity $\propto \frac{1}{V}$,

$$V_{15} < V_{25}$$

Hence, molarity at 15°C > molarity at 25°C

Thus, (a) is also correct.

19. (b,d)

Glucose solution is = 1 mol kg^{-1} water

Solute (glucose) = 180 g

Solvent (water) = 1000 g

Total mass of solution = 1180 g

$$= \frac{1180}{1.180} \text{ mL} = 1000 \text{ mL}$$

Thus, molarity = 1 mol L^{-1},

$$\frac{\Delta T_b}{\Delta T_f} = \frac{K_b}{K_f} = \frac{1}{3} = 0.33$$

Osmotic pressure, $\pi = mRT = 1 \times 0.0821 \times 300 = 24.63$ atm

20. (a,b,c,d)

Let χ_A = mole fraction of component A

χ_B = mole fraction of component B

p_A° = vapour pressure of pure A

p_B° = vapour pressure of pure B

p_A = vapour pressure of A in binary mixture

$= p_A^\circ \chi_A$ by Raoult's law

Similarly, $p_B = p_B^\circ \chi_B$

$$p_{total} = p_A + p_B = p_A^\circ \chi_A + p_B^\circ \chi_B$$
$$= p_A^\circ \chi_A + p_B^\circ (1 - \chi_A)$$
$$= p_A^\circ \chi_A + p_B^\circ - p_B^\circ \chi_A$$
$$= p_B^\circ + (p_A^\circ - p_B^\circ)\chi_A \text{ (in terms of mole fraction of } A)$$

Thus, (a) and (d) are correct.

Variation of p_A with χ_A, p_B with χ_B and p_{total} with χ_A and χ_B both can be represented as

At start $\chi_A = 1$ thus, p_A° is the vapour pressure of pure A. As B is added into it, χ_B increases. Hence, p_A decreases and that of p_B increases. When $\chi_B = 1$, vapour pressure of $B = p_B^\circ$ (pure).

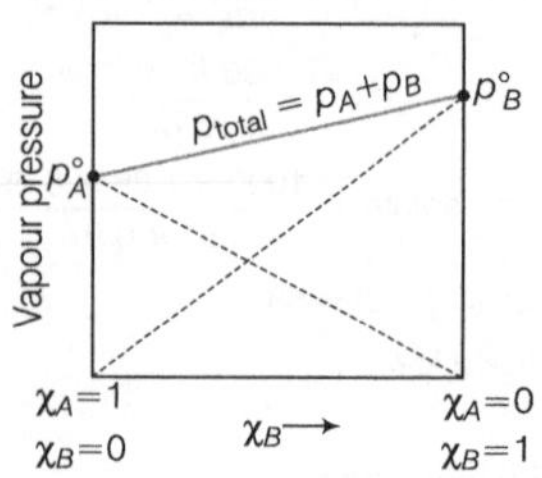

Thus, vapour pressure of B increase as χ_B increases and decrease as χ_A decreases.

Line joining p_A° and p_B° represent total vapour pressure.

Thus, (b) and (c) are also correct.

Note *Vapour pressure of $A(p_A^\circ)$ is less than vapour pressure of $B(p_B^\circ)$.*

Hence, A is less volatile than B.

(d) p_{total} can also be expressed as

$$p_{total} = p_A^\circ \chi_A + p_B^\circ \chi_B$$
$$= p_B^\circ + (p_A^\circ - p_B^\circ)\chi_A$$

21. (a, c)

Two liquids (A, B) form an ideal solution, when force of attraction between A and B (in mixture) is same as between A and A and between B and B. In such case, $\Delta H_{mix} = 0$.

Since, disorder increases, $\Delta S_{mix} > 0$. Also, process is spontaneous at a given temperature,

$$\Delta G = \Delta H - T\Delta S = 0 - T\Delta S$$

$\therefore$ $\Delta G_{mix} < 0$

Explanations for Q. Nos. (22-23)

Variation of vapour pressure with temperature is given by Clapeyron-Clausius equation.

$$\Delta H = \frac{2.303\, RT_1T_2}{(T_2 - T_1)} \log \frac{p_2}{p_1}$$

$$\log \frac{p_2}{p_1} = \frac{\Delta H (T_2 - T_1)}{2.303\, RT_1\, T_2}$$

$$= \frac{10 \text{ kcal mol}^{-1} (373 - 313) \text{ K}}{2.303 \times 0.002 \text{ kcal mol}^{-1}\text{K}^{-1} \times 373 \text{ K} \times 313 \text{ K}} = 1.116$$

$p_2 = 1$ atm at 373 K

$$\log \left(\frac{1}{p_1}\right) = 1.116$$

$$\frac{1}{p_1} = 13.0617$$

$$p_1 = 0.07656 \text{ atm}$$

$$p_1 = 0.07656 \times 760 \text{ torr} = 58.19 \text{ torr}$$

Vapour pressure at 40°C(313 K) = 58.19 torr

By Raoult's law, $\dfrac{\Delta p}{p^\circ} = \chi_{solute}$

Urea = 0.2 molal = 0.2 mole urea per kg of water

$\therefore$ Urea = 0.2 mol

$$\text{Water} = 1000 \text{ g} = \frac{1000}{18} \text{ mol} = 55.56 \text{ mol}$$

$$\chi_{urea} = \frac{0.2}{0.2 + 55.56} = 3.587 \times 10^{-3}$$

$\therefore$ $\Delta p = p^\circ \times \chi_{solute}$

$= 58.19 \times 3.587 \times 10^{-3}$

$= 0.2087$ torr

22. (b) **23.** (a)

Explanations for Q. Nos. (24-25)

This is a case of negative deviation, thus force of attraction between ($A \cdots B$) is larger than average of ($A \cdots A$) and ($B \cdots B$). Thus, vapour pressure of mixture is decreased, boiling point is increased, hence

$$\Delta H < 0, \Delta V = \text{negative}$$

Diethylether and chloroform show negative deviation.

24. (b) **25.** (d)

Explanations for Q. Nos. (26-27)

By Raoult's law,

$$\frac{p_{H_2O}^\circ - p_s^\circ}{p_{H_2O}^\circ} = \chi_{solute}$$

***Case* I** Solute $= \dfrac{6}{m}$, Water = 10 mol

$$\frac{p_{H_2O}^\circ - 20}{p_{H_2O}^\circ} = \frac{6/m}{\frac{6}{m} + 10} = \frac{6}{6 + 10m} \quad \ldots(i)$$

***Case* II** Solute $= \dfrac{6}{m}$, Water = 11 mol

$$\therefore \quad \frac{p_{H_2O}^\circ - 20.02}{p_{H_2O}^\circ} = \frac{6}{6 + 11m} \quad \ldots(ii)$$

$$20.22 = p_{H_2O}^\circ \left(\frac{11}{11 + \frac{6}{m}}\right)$$

$$20.22 = p_{H_2O}^\circ \left(\frac{11m}{11m + 6}\right)$$

From Eqs. (i) and (ii), we get

$$\frac{20.02}{20.00} = \left(\frac{10m+6}{11m+6}\right)\frac{11}{10} \Rightarrow \frac{10m+6}{11m+6} = \frac{200.2}{220.0}$$

$\therefore \quad m = 54.0$ g mol and $p^{\circ}_{H_2O} = 20.22$ torr

26. (d) **27.** (a)

28. (4) Molarity = 4.58 mol L^{-1}

1000 mL has 4.58 mol Na^+ ions

$$\frac{m}{V} = \text{Density}$$

$\therefore$ Mass of solution $= V \times d = 1000 \times 1.25$ g $= 1250$ g

Na^+ ions $= 4.58$ mol $= 4.58 \times 23$ g $= 105.34$ g

$\therefore H_2O$ in lake water $= 1250 - 105.34 = 1144.66$ g $= 1.145$ kg

$$\therefore \quad \text{Molality} = \frac{4.58 \text{ mol}}{1.145 \text{ kg}} = 4$$

29. (2) p_{O_3} (partial pressure of ozone) $= p_{\text{total}}\, \chi_{O_3}$

$$\therefore \quad \chi_{O_3} = \frac{p_{O_3}}{p_{\text{total}}} = \frac{1.2 \times 10^{-5} \text{ mm}}{10 \text{ mm}} = 1.2 \times 10^{-6}$$

$$\therefore \quad \chi_{\text{air}} + \chi_{O_3} = 1$$

$$\therefore \quad \chi_{\text{air}} = 1 - \chi_{O_3} = 1 - 1.2 \times 10^{-6} \approx 1$$

$$\therefore \quad \frac{\chi_{O_3}}{\chi_{\text{air}}} = \frac{1.2 \times 10^{-6}}{1} = \frac{n_{O_3}}{n_{\text{air}}} = \frac{w(O_3)/m(O_3)}{w(\text{air})/m(\text{air})}$$

$$m(\text{air}) = \frac{m(N_2)\,\chi(N_2) + m(O_2)\,\chi\, O_2}{\chi(N_2) + \chi(O_2)}$$

$$m(\text{air}) = \frac{28 \times 4 + 32 \times 1}{4+1} = 28.8$$

$$\frac{w(O_3)}{48} \times \frac{28.8}{w(\text{air})} = 1.2 \times 10^{-6}$$

$$\therefore \quad \frac{w(O_3)}{w(\text{air})} = \frac{1.2 \times 10^{-6} \times 48}{28.8}$$

28.8 g air has ozone $= 1.2 \times 10^{-6} \times 48$ g

$$\therefore 10^6 \text{g (ppm) has ozone} = \frac{1.2 \times 10^{-6} \times 48 \times 10^6}{28.8} = 2 \text{ ppm}$$

30. (2) By Raoult's law,

$$\frac{\Delta p}{p^{\circ}} = \chi_{\text{solute}} = (1 - \chi_{\text{solvent}})$$

Case I $\quad \dfrac{(\Delta p_1)}{p^{\circ}} = 0.2,\ (\chi_{\text{solvent}})_1 = 0.8$

Case II $\quad (\chi_{\text{solvent}})_2 = 0.8 - \dfrac{0.8 \times 25}{100} = 0.8 - 0.2 = 0.6$

$$\therefore \quad (\chi_{\text{solute}}) = 0.4$$

$$\therefore \quad \frac{(\Delta p_2)}{p^{\circ}} = \chi_{\text{solute}} = 0.4$$

$$(\Delta p_2)/(\Delta p_1) = \frac{0.4}{0.2} = 2$$

31. (3) $K_x[Fe(CN)_6] \rightleftharpoons \underbrace{xK^+ + [Fe(CN)_6]^{x-}}_{y\,=\,\text{number of ions}} = (x+1)$

$\therefore \quad i =$ van't Hoff factor $= 1 + (y-1)\alpha$

where, α = degree of ionisation $= 1 = 1 + (x + 1 - 1)\,1 = (x+1)$

$$\Delta T_f = \text{molality} \times K_f \times i$$

$$7 \times 10^{-3} = 1 \times 10^{-3} \times 1.86 \times (x+1)$$

$$x + 1 = 3.76 \approx 4$$

$$\therefore \quad x = 3$$

32. (d) w_1 (solute) $= 17.1$ g $= \dfrac{17.1}{342} = 0.05$ mole (n_1)

w_2(solvent) $= 1000$ g $= \dfrac{1000}{18} = 55.55$ moles (n_2)

(i) $$\frac{\Delta p}{p^{\circ}} = \chi_{\text{solute}} = \frac{n_1}{n_1 + n_2} \approx \frac{n_1}{n_2}$$

$$\therefore \quad \Delta p = p^{\circ}\frac{n_1}{n_2} = 26.50 \times \frac{0.05}{55.55} = 0.024 \text{ mm}$$

(ii) $\Delta T_f = \dfrac{1000\, K_f\, w_1}{m_1 w_2} = \dfrac{1000 \times 1.86^{\circ} \times 17.1}{342 \times 1000} = 0.093^{\circ}$

(iii) $\Delta T_b = \dfrac{1000\, K_b\, w_1}{m_1 w_2} = \dfrac{1000 \times 0.52 \times 17.1}{342 \times 1000} = 0.026^{\circ}$

(iv) $\pi = MRT = \dfrac{w_1}{m_1 V} RT = \dfrac{0.05 \times 0.0821 \times 300}{1} = 1.2315$

Hence, (i) $\to$ (s), (ii) $\to$ (r), (iii) $\to$ (q), (iv) $\to$ (p)

JEE Main & AIEEE Archive

1. (c) p°(acetone) = 185 torr, Solute = 1.2 g $= \dfrac{1.2}{m}$ mol

acetone = 100 g $= \dfrac{100}{58}$ mol, $p_{\text{solution}} = 183$ torr

By Raoult's law

$$p_{\text{solution}} = p^{\circ}\chi_{\text{solvent}} = 185\left(\frac{\frac{100}{58}}{\frac{1.2}{m} + \frac{100}{58}}\right)$$

$$183 = 185\left[\frac{\frac{100}{58}}{\frac{1.2}{m} + \frac{100}{58}}\right] \Rightarrow \frac{\frac{1.2}{m} + \frac{100}{58}}{\frac{100}{58}} = \frac{185}{183}$$

$$\frac{1.2}{m} \times \frac{58}{100} = \frac{2}{183}$$

$$\therefore \quad M = \frac{1.2 \times 58}{100} \times \frac{183}{2} = 63.68$$

$$\approx 64.0 \text{ g mol}^{-1}$$

Alternate Method (For dilute solution)

$$\frac{\Delta p}{p^{\circ}} = \frac{w_1}{m_1} \times \frac{m_2}{w_2} \Rightarrow \frac{2}{185} = \frac{1.2 \times 58}{m_1 \times 100}$$

$$m_1 = 64.38 \approx 64.0 \text{ g mol}^{-1}$$

2. (a) Salts have been assumed as strong electrolytes.

van't Hoff factor, $i = [1 + (y-1)x]$

y = number of ions from one mole of salt,

x = degree of ionisation = 1

C_2H_5OH is a non-electrolyte hence, $y = 1$

$$\pi = \frac{n}{V}RTi = MRTi$$

Thus, $\pi \propto Mi$, RT being constant. Greater the value of Mi, greater the osmotic pressure.

Solute	Concentration	y	$i = 1 + (y-1)x$	Mi
C_2H_5OH	0.500 M	1	1	0.500
$Mg_3(PO_4)_2$	0.100 M	5	5	0.500
KBr	0.250 M	2	2	0.500
Na_3PO_4	0.125 M	4	4	0.500

Thus, $\pi \propto Mi$, have equal values.

3. (b) Coolant is glycol ($C_2H_6O_2$) and it is non-electrolyte.

$$\Delta T_f = 2.8°C$$

$$\Delta T_f = \frac{1000 K_f w_1}{m_1 w_2} \Rightarrow 2.8 = \frac{1000 \times 1.86 \times w_1}{62 \times 1000}$$

$$\therefore \quad w_1 = 93.33 \text{ g}$$

4. (c) 5.2 molal aqueous solution of CH_3OH suggests that 5.2 moles of CH_3OH are dissolved in 1000 g of water.

$$n_1(CH_3OH) = 5.2 \text{ mol}$$

$$n_2(H_2O) = \frac{1000}{18} = 55.56 \text{ mol}$$

$\therefore$ $$n_1 + n_2 = 5.20 + 55.56 = 60.76 \text{ mol}$$

$\therefore$ $$\chi_{CH_3OH} = \text{Mole fraction of } CH_3OH$$

$$= \frac{n_1}{n_1 + n_2} = \frac{5.2}{60.76} = 0.086$$

5. (a) ΔT_f = Freezing point of H_2O – freezing point of ethylene glycol solution

$$= 0 - (-6°) = 6°$$

$K_f = 1.86°$ K kg mol^{-1}

w_1 = Mass of ethylene glycol in grams

w_2 = Mass of solvent (H_2O) in grams = 4000 g

m_1 = Molar mass of ethylene glycol = 62 g mol^{-1}

$$\Delta T_f = \frac{1000\, K_f\, w_1}{m_1\, w_2}$$

$\therefore$ $$6 = \frac{1000 \times 1.86 \times w_1}{62 \times 4000}$$

$$w_1 = 800 \text{ g}$$

6. (a) $A_xB_y \rightleftharpoons xA + yB$

Total ions = $(x + y)$

van't Hoff factor $(i) = 1 + [(x + y) - 1]\alpha$

$\therefore$ $$\alpha = \frac{i - 1}{(x + y - 1)}$$

7. (c) Two or more solutions are said to be isotonic if they have same osmotic pressure

π_1 (cone sugar) $= M_1RT_1$, π_2 (unknown) $= M_2RT_2$

If $\pi_1 = \pi_2$

then $M_1RT_1 = M_2RT_2$

Also, $T_1 = T_2$

Hence, $M_1 = M_2$

$$\frac{5}{0.1 \times 342} = \frac{1}{0.1 \times m_2}$$

$\therefore$ $$m_2 = \frac{342 \times 1}{5} = 68.4 \text{ g mol}^{-1}$$

8. (b) Molality $= \dfrac{\text{moles of solute}}{\text{kg of water}}$, Moles of urea $= \dfrac{0.010}{60}$ mol

Water at STP (d = 1g/cm^3 = 1 kg/dm^3) = 0.3 dm^3 = 0.3 kg

$\therefore$ $$\text{Molality} = \frac{0.010}{60 \times 0.3} = 5.55 \times 10^{-4} \text{ m}$$

9. (b) $$Na_2SO_4 \longrightarrow 2Na^+ + SO_4^{2-}$$

van't Hoff factor (i) for $Na_2SO_4 = 1 + (y - 1)x = 3$

$$\Delta T_f = i \times K_f \times m$$

$$= 3 \times 1.86 \times 0.01 \quad \left[\because m = \frac{0.01 \text{ mol}}{1 \text{ kg}} = 0.01 \text{ mol kg}^{-1}\right]$$

$$= 0.0558 \text{ K}$$

10. (a) $$p_T = \chi_H \cdot p_H^\circ + \chi_O \cdot p_O^\circ$$

$$\chi_H = \frac{\frac{25}{100}}{\frac{25}{100} + \frac{35}{114}} = 0.45$$

$\therefore$ $$\chi_O = 0.55$$

$$p_T = 0.45 \times 105 + 0.55 \times 45$$

$$= 72.0 \text{ k Pa}$$

11. (c) $$p_T = p_A^\circ\chi_A + p_B^\circ\chi_B$$

$$550 = p_A^\circ \times \frac{1}{4} + p_B^\circ \times \frac{3}{4}$$

Thus, $p_A^\circ + 3p_B^\circ = 2200$ ...(i)

When 1 mole of Y is further added to the solution, vapour pressure of a solution becomes 560 mm Hg.

Thus, $$560 = p_A^\circ \times \frac{1}{5} + p_B^\circ \times \frac{4}{5}$$

Thus, $p_A^\circ + 4p_B^\circ = 2800$ ...(ii)

On subtracting Eq. (ii) by Eq. (i), we get

$$p_B^\circ = 2800 - 2200$$

$\Rightarrow$ $$p_B^\circ = 600$$

Putting the value of p_B° in Eq. (i)

$$p_A^\circ + 3 \times 600 = 2200$$

$$p_A^\circ = 2200 - 1800 = 400$$

12. (b) n-heptane and ethanol forms non-ideal solution. As a result, n-heptane-ethanol molecular interaction is very poor in comparison to ethanol-ethanol or n-heptane so the resulting gives positive deviation from Raoult's law.

13. (d) $$p_T = p_A^\circ\chi_A + p_B^\circ\chi_B$$

Mixture solution boils at 1 atm = 760 mm = total pressure

$$760 = 520\chi_A + 1000(1 - \chi_A)$$

$$\chi_A = 0.5, \text{ mol \% of } A = 50\%$$

14. (d) This problem is related to the relative lowering of vapour pressure as glucose is a non-volatile solute. Thus, according to Raoult's law,

$$\frac{p^0 - p_{\text{solute}}}{p^0} = \chi_{\text{solute}}$$

Moles of glucose $= \dfrac{18}{180} = 0.1$, Moles of $H_2O = \dfrac{178.2}{18} = 9.9$

$$\chi_{\text{glucose}} = \frac{0.1}{9.9 + 0.1} = \frac{0.1}{10}$$

Thus, $$\frac{17.5 - p_s}{17.5} = \frac{0.1}{10}$$

So, $$p_s = 17.325 \text{ mm Hg}$$

15. (c) 100 g H_2SO_4 has = 29 g H_2SO_4

$\dfrac{100}{d}$ mL H_2SO_4 has = 29 g H_2SO_4

$$\therefore \text{Molarity} = 3.6 = \frac{\text{Moles}}{\text{Volume in litres}} = \frac{29/98}{(100/d)/1000}$$

$\therefore$ $$\frac{29 \times 1000 \times d}{98 \times 100} = 3.6$$

$$d = 1.22 \text{ g mL}^{-1}$$

16. (a) According to Raoult's law,

$$p = p_A^\circ\chi_A + p_B^\circ\chi_B$$

$$290 = 200 \times 0.4 + p_B^\circ \times 0.6$$

$$p_B^\circ = 350 \text{ mm}$$

17. (d) Osmotic pressure of isotonic solutions are equal.

Therefore, $$\pi_1 = \pi_2$$

$\Rightarrow$ $$C_1RT = C_2RT \quad \Rightarrow \quad C_1 = C_2$$

If densities of both the solutions are equal, then 5.25 g of a substance and 1.5 g of urea are dissolved in the same volume of the solvent, thus

$$C_1 = \frac{5.25/M}{V}$$

[$\because M$ = molecular weight of unknown substance]

$$C_2 = \frac{5.25/60}{V}$$

$\therefore$ $$\frac{5.25/M}{V} = \frac{1.5/60}{V} \quad \Rightarrow \quad M = 210.0 \text{ g mol}^{-1}$$

18. (c) Molality $(m) = \dfrac{M}{1000\,d - MM_1} \times 100$

where, M = Molarity, M_1 = Molecular mass and d = Density

$$= \frac{2.05}{(1000 \times 1.02) - (2.05 \times 60)} \times 1000$$

$$= 2.28 \text{ mol kg}^{-1}$$

19. (d) According to Raoult's law, relative lowering of vapour pressure,

$$\frac{p^\circ - p_s}{p^\circ} = \chi_{\text{solute}}$$

$$\frac{760 - p_s}{760} = \frac{18/180}{18/180 + 178.2/18}$$

$$\frac{760 - p_s}{760} = \frac{0.1}{10}$$

$$\Rightarrow \quad p_s = 760 - 7.6 = 752.40 \text{ torr}$$

20. (b) The molarity of the resulting solution is given by

$$\text{Total molarity} = \frac{M_1V_1 + M_2V_2}{V_1 + V_2} = \frac{1.5 \times 480 + 1.2 \times 520}{480 + 520} = 1.34 \text{ M}$$

21. (b) Boiling point and freezing point depend on K_b (molal elevation constant) and K_f (molal depression constant) of the solvent. Thus, equimolar solution (of the non-electrolyte), will have same boiling point and also same freezing point.

$$\Delta T_f = K_f \times \text{molality}$$

$$\Delta T_b = K_b \times \text{molality}$$

Note *In this question, nature of solute has not been mentioned. Hence, we have assumed that solute is non-electrolyte.*

22. (d) Mixture contains 78 g benzene = 1 mol benzene and 46 g toluene = 0.5mol toluene

Total moles of benzene and toluene = 1.5 mol

$$\text{Mole fraction of benzene in mixture} = \frac{1}{1.5} = \frac{2}{3}$$

Vapour pressure of benzene $p_b^\circ = 75$ torr

$$\therefore \text{ Partial vapour pressure of benzene} = p^\circ_b\, \chi_b = 75 \times \frac{2}{3} = 50 \text{ torr}$$

23. (b) $$Na_2SO_4 \rightleftharpoons 2Na^+ + SO_4^{2-}$$

van't Hoff factor, $i = [1 + (y - 1)\,\alpha]$

where y is the number of ions from one mole solute, (in this case = 3), α is the degree of dissociation, $i = (1 + 2\,\alpha)$

24. (b) Avogadro's number, N_A

$= 6.02 \times 10^{23}$ molecules = 1 mol

$\therefore$ 6.02×10^{20} molecules = 0.001 mol in 100 mL or (0.1 L) solution

$$\therefore \text{ Molar concentration} = \frac{\text{mol}}{\text{volume in litre}} = \frac{0.001}{0.1} = 0.01 \text{ M}$$

25. (c) H_3PO_3 is a dibasic acid (i.e. contains two ionisable protons attached directly to O).

$$H_3PO_3 \rightleftharpoons 2H^+ + HPO_3^{2-}$$

$\therefore$ $0.1 \text{ M } H_3PO_3 = 0.2 \text{ N } H_3PO_3$

and $0.1 \text{ M KOH} = 0.1 \text{ N KOH}$

$$\underset{(KOH)}{N_1V_1} = \underset{(H_3PO_3)}{N_2V_2}$$

$$0.1\,V_1 = 0.2 \times 20 \quad \Rightarrow \quad V_1 = 40 \text{ mL}$$

26. (b) Water and hydrochloric acid; and water and nitric acid form miscible solutions. They show negative deviation.

In case of CH_3COCH_3 and $CHCl_3$, there is interaction between them, thus force of attraction between $CH_3COCH_3....CHCl_3$ is larger than between $CHCl_3....CHCl_3$ or $CH_3COCH_3....CH_3COCH_3$ and thus, vapour pressure is less than expected—a negative deviation.

In case of CH_3OH, there is association by intermolecular H-bonding. When benzene is added to CH_3OH, H-bonding breaks and thus, force of attraction between CH_3OH and benzene molecule is smaller than between CH_3OH or benzene molecules (in pure state). Vapour pressure of mixture is greater than expected—a positive deviation.

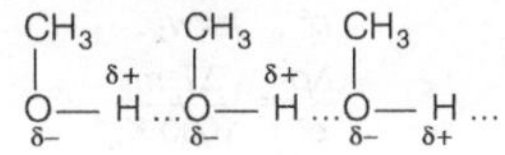

27. (a) The colligative property will be given as

$$\Delta T_b = i \times K_b \times m$$

Where, ΔT_b = Elevation in boiling point

i = van't Hoff factor

K_b = molal elevation constant

m = molality of the solution

From the above equation, it is clear that,

$$\Delta T_b \propto i$$

This suggests that higher the value of i, larger is the boiling point of the solution.

For 100% dissociation, i = number of ions produced by one mole of solute

Thus, for

(a) Na_2SO_4, $\quad Na_2SO_4 \rightleftharpoons 2Na^+ + SO_4^{2-}$ [3 ions]

$\therefore$ $i = 3$

(b) KNO_3, $\quad KNO_3 \rightleftharpoons K^+ + NO_3^-$ [2 ions]

$\therefore$ $i = 2$

(c) Urea $[(NH_2CONH_2)$ and

(d) glucose are non-electrolytes therefore, $i = 1$.

Hence, 0.01 M Na_2SO_4 has highest boiling point.

28. (d)

(a) $p_A = \chi_A p_A^\circ$ true

(b) $\pi = iMRT = MRT$

true [If van't Hoff factor, $i = 1$]

(c) $\pi \propto i$

Greater the value of i, larger is the value of π.

i for $BaCl_2$ (strong electrolyte) is 3.

$[BaCl_2 \rightleftharpoons Ba^{2+} + Cl^-$ (2 ions)]

i for KCl is 2. $[KCl \rightleftharpoons K^+ + Cl^-]$

i for CH_3COOH is

2 $[CH_3COOH \rightleftharpoons CH_3COO^- + H^+]$

But CH_3COOH is weak electrolyte than KCl.

i for sucrose is 1 as it is a non-electrocyte.

Thus, i for $BaCl_2 > KCl > CH_3COOH >$ sucrose

Thus, (c) is also true.

(d) $\Delta T_f = K_f m$.

K_f is dependent on solvent.

Thus, freezing points = [T (solvent) – ΔT_f] are different.

Thus, (d) is false.

29. (a) For ideal solution, $\Delta H = 0$, $\Delta V = 0$

$$F_{A-A} = F_{B-B} = F_{A-B}$$

30. (a) $$HX \rightleftharpoons H^+ + X^{-1}$$

$$\alpha = 0.3$$

$$i = 1 + \alpha = 1.3$$

$$\Delta T_f = \text{molality} \times K_f \times i$$

$$= 0.2 \times 1.85 \times 1.3 = 0.481^\circ$$

$\therefore$ Freezing point = $-\,0.481^\circ$C

31. (a) Molarity gets affected as it is the number of moles per unit volume (volume increases with increase in temperature).

JEE Advanced & IIT JEE Archive

1. (a) $\Delta T_b = \dfrac{1000\, K_b w_1}{m_1 W_2}$

Also, $\dfrac{\Delta p}{p^\circ} = \dfrac{w_1 m_2}{m_1 W_2}$

$\therefore \quad \dfrac{\Delta p}{p^\circ} = \dfrac{\Delta T_b m_2}{1000\, K_b}$

$\dfrac{\Delta p}{p^\circ} = \dfrac{2 \times 18}{1000 \times 0.76}$

$\dfrac{\Delta p}{p^\circ} = 0.04737$

$p^\circ = 1\text{ atm} = 760\text{ mm Hg}$

$\Delta p = 0.04737 \times 760 = 36\text{ mm Hg}$

$\therefore \quad p_{\text{solution}} = 760 - 36 = 724\text{ mm Hg}$

2. (a) $K_3[Fe(CN)_6] \rightleftharpoons \underbrace{3K^+ + [Fe(CN)_6]^{3-}}_{y=4}$

$i = 1 + (y-1)x = 1 + 3x = 4 \quad (\text{if } x = 1)$

$\Delta T_f = \dfrac{1000\, K_f w_1(i)}{m_1 W_2} = \dfrac{1000 \times 1.86 \times 0.1 \times 4}{329 \times 100}$

$= 2.26 \times 10^{-2}$°

Thus, freezing point of solution $= T_b - \Delta T_f = 0 - 2.26 \times 10^{-2}$

$= -2.26 \times 10^{-2}\ ^\circ\text{C}$

3. (a) Total pressure = 5 atm

$\therefore \quad p_{N_2} = \chi_{N_2} p = 0.8 \times 5 = 4.0\text{ atm}$

$K_H \chi_{N_2} (\text{in } H_2O) = p$

$\therefore \quad \chi_{N_2} (\text{in } H_2O) = \dfrac{p}{K_H} = \dfrac{4.0}{1 \times 10^5}$

$= 4.0 \times 10^{-5}$

Water = 10 mol

$N_2 = n$ mol

$\therefore \quad \chi_{N_2} = \dfrac{n}{10 + n} \approx \dfrac{n}{10}$

$\therefore \quad n(N_2) = 4.0 \times 10^{-5} \times 100 = 4.0 \times 10^{-4}$

4. (d) Water is solute and ethanol is solvent

$\chi_w = 0.1$

$\chi_e = 0.9$

$\dfrac{\chi_w}{\chi_e} = \dfrac{1}{9} = \dfrac{n_w}{n_e}$

$= \dfrac{w_1 (H_2O)\, m_2 (\text{ethanol})}{m_1 (H_2O)\, W_2 (\text{ethanol})}$

$\dfrac{w_1}{w_1 W_2} = \dfrac{1}{9 \times 46}$

$\therefore \quad \Delta T_f (\text{water as solute}) = \dfrac{1000\, K_f w_1}{m_1 W_2}$

$= \dfrac{1000 \times 2.0}{9 \times 46} = 4.83^\circ$

$\therefore$ Freezing point = 155.70 − 4.83 = 155.86 K

5. (a) $p_{\text{total}} = \chi_{H_2O} p^\circ_{H_2O} + \chi_{\text{ethanol}}\, p^\circ_{\text{ethanol}}$

$= 0.1 \times 32.8 + 0.9 \times 40.0$

$= 39.28$ mmHg

6. (b) In this case H_2O is solvent

$\dfrac{\chi_e}{\chi_w} = \dfrac{1}{9} = \dfrac{n_e}{n_w} = \dfrac{w_1 \times 18}{m_1 W_2}$

$\dfrac{w_1}{m_1 W_2} = \dfrac{1}{9 \times 18}$

$\therefore \quad \Delta T_b = 1000 \times 0.52 \times \dfrac{1}{9 \times 18} = 3.20^\circ$

$\therefore$ Boiling point of solution as water solvent = 373.0 + 3.2

= 376.2 K

7. (a) $M_{\text{Naphthoic acid}} = 11 \times 12 + 8 + 2 \times 16$

$= 172\text{ g mol}^{-1}$

$\Delta T_f = \dfrac{1000\, K_f\, w_1(i)}{m_1 W_2}$

$2 = \dfrac{1000 \times 1.72 \times 2(i)}{172 \times 50}$

$\therefore \quad i = 0.5$

8. Phenol forms dimer due to H − bonding

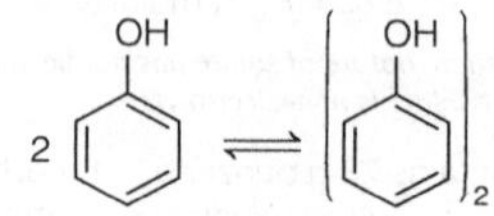

$i = [1 + (y-1)x] = \left(1 - \dfrac{x}{2}\right) \qquad \left(y = \dfrac{1}{2}\right)$

$\Delta T_f = \dfrac{1000\, K_f w_1}{m_1 W_2}(i)$

$7 = \dfrac{1000 \times 14 \times 75.2}{94 \times 800}(i)$

$\therefore \quad i = 0.5$

$\therefore \quad \left(1 - \dfrac{x}{2}\right) = 0.5$

$\therefore \quad x = 1.0$

100% dimer formation

9. (c) $\Delta T_b = \dfrac{1000 \times K_b w_1(i)}{m_1 W_2}$

For $\quad CuCl_2 \rightleftharpoons \underbrace{Cu^{2+} + 2Cl^-}_{y=3}$

$i = 1 + (y-1)x$

$= 1 + 2x$

$= 1 + 2$

$= 3$

$\therefore \quad \Delta T_b = \dfrac{1000 \times 0.52 \times 13.44 \times 3}{134.4 \times 1000} = 0.156^\circ$

3. Electrochemistry

DPP-1 Conduction

1. Plan (i) Cell constant $= \dfrac{\text{Distance between two electrodes}}{\text{Area between two electrodes}}$

$$= \frac{0.020\text{m}}{0.0006\text{m}^2} = 33.33\ \text{m}^{-1}$$

(ii) Conductance $= \dfrac{1}{\text{Resistance}} = \dfrac{1}{40\Omega} = 0.025\ \Omega^{-1}$

Ω^{-1} is also written as S(Siemen)

$\therefore$ Conductance = 0.025 S

(iii) Specific conductance = Conductance × Cell constant

$$= 0.025\ \text{S} \times 33.33\ \text{m}^{-1}$$
$$= 0.0833\ \text{Sm}^{-1}$$

2. Plan Specific conductance (KCl) = Conductance × Cell constant

$$= \frac{1}{\text{Resistance}} \times \text{cell constant}$$

$\therefore$ Cell constant = Specific constant × Resistance

$$= 0.2768 \times 82.4\text{m}^{-1}$$

Specific conductance $(K_2SO_4) = \dfrac{1}{\text{Resistance}} \times$ Cell constant

$$= \frac{1}{300} \times 0.2768 \times 82.4$$
$$= 0.076\ \text{Sm}^{-1}$$

3. Plan κ (specific conductance) = Conductance × Cell constant

$$1\text{cm}^{-1} = \frac{1}{\text{cm}} = \frac{100}{\text{m}} = 100\ \text{m}^{-1}$$

$$\text{Molarity} = 1\ \text{molL}^{-1} = 1\ \text{moldm}^{-3}$$

$$= \frac{1\text{mol}}{\text{dm}^3} = \frac{1\text{mol} \times 1000}{\text{m}^3} = 1000\ \text{mol m}^{-3}$$

$$\frac{1}{\text{ohm}} = \text{ohm}^{-1} = \text{S(Siemen)}$$

(i) Λ_m = Molar conductance $= \dfrac{\text{Specific conductance (Sm}^{-1})}{\text{Molarity (in mol m}^{-3})}$

$$= \cdots\ \text{Sm}^2\text{mol}^{-1}$$

Specific conductance of KOH $= \dfrac{0.3792\,\text{cm}^{-1}}{31.6\,\Omega}$

$$= 0.012\ \Omega^{-1}\text{cm}^{-1} = 0.012\ \text{Scm}^{-1}$$
$$= 0.012 \times 100\text{Sm}^{-1}$$

$\therefore$ Molar conductance $= \dfrac{0.012 \times 100\ \text{Sm}^{-1}}{.05 \times 1000\ \text{mol m}^{-3}} = 0.024\ \text{Sm}^2\text{mol}^{-1}$

Also, $\Lambda_m = \dfrac{1000 \times \text{Specific conductance}}{\text{Molarity (mol L}^{-1})}\ (\text{Scm}^{-1})$

$$= \frac{1000 \times 0.012}{0.05} = 240\ \text{Scm}^2\text{mol}^{-1}$$

(ii) Equivalent mass of KOH = molar mass of KOH

$\therefore$ 0.05M = 0.05N

Equivalent conductance $\Lambda_{eq} = \Lambda_m = 0.024\ \text{Sm}^2\text{equiv}^{-1}$

$$= 240\ \text{Scm}^2\text{equiv}^{-1}$$

4. Plan $0.075\ \text{Sm}^{-1} = \dfrac{0.075\text{S}}{\text{m}} = \dfrac{0.075\text{S}}{100\text{cm}} = 75 \times 10^{-4}\ \text{Scm}^{-1}$

$$0.0025\ \text{M} = 0.0025 \times 2\text{N}$$

$$\left(\frac{\text{Molar mass of } K_2SO_4}{2} = \text{equivalent mass}\right)$$

$$0.0025\text{M} = 0.0025\ \text{mol L}^{-1}$$
$$= 0.0025\ \text{mol dm}^{-3}$$
$$= 0.0025 \times 1000\ \text{mol m}^{-3}$$

(i) Λ_m = Molar conductance $= \dfrac{\text{Specific conductance (Sm}^{-1})}{\text{Molarity (molm}^{-3})}$

$$= \frac{0.075\ \text{Sm}^{-1}}{0.0025 \times 1000\ \text{mol m}^{-3}}$$
$$= 0.03\ \text{Sm}^2\text{mol}^{-1}$$
$$= 0.03\text{S} \times 10000\ \text{cm}^2\text{mol}^{-1}$$
$$= 300\ \text{Scm}^2\text{mol}^{-1}$$

(ii) Λ_{eq} = Equivalent conductance $= \dfrac{0.075\ \text{Sm}^{-1}}{0.0050 \times 1000\ \text{mol m}^{-3}}$

$$= 0.015\ \text{Sm}^2\text{equiv}^{-1}$$
$$= 0.015\ \text{S} \times 1000\ \text{cm}^2\text{equiv}^{-1}$$
$$= 150\ \text{Scm}^2\text{equiv}^{-1}$$

5. Plan $CH_3COOH \rightleftharpoons CH_3COO^- + H^+$

	CH_3COOH	CH_3COO^-	H^+
Initial	C	0	0
At equli	$(C - Cx)$	Cx	Cx

By Ostwald's dilution law,

K_a (ionisation constant) $= K_a = \dfrac{[CH_3COO^-][H^+]}{[CH_3COOH]}$

$$= \frac{Cx \cdot Cx}{C(1-x)} = \frac{Cx^2}{(1-x)},$$

$$x = \frac{\Lambda_C}{\Lambda_0}$$

$$\Lambda_C = \frac{1000 \times \text{Specific conductance (conductivity)}}{\text{molarity (mol cm}^{-3})}$$

$$= \frac{1000 \times 7.896 \times 10^{-5}\ \text{Scm}^{-1}}{0.00241\ \text{mol cm}^{-3}} = 32.76$$

Note $0.00241\ \text{mol dm}^{-3} = \dfrac{0.00241}{1000\text{cm}^3}$

$\therefore$ $x = \dfrac{32.76}{390.5} = 0.084$

$\therefore$ $K_a = \dfrac{Cx^2}{(1-x)} = \dfrac{0.00241\,(0.084)^2}{(1-0.084)} = 1.856 \times 10^{-5}$

6. (b) $\Lambda_C = \dfrac{1000\kappa\ (\text{Specific conductance) in Scm}^{-1})}{\text{Concentration (equiv dm}^{-3})}$

$$= 1000\,\kappa \left(\frac{\text{S}}{\text{cm}}\right) \frac{\text{dm}^3}{\text{equiv}}$$

$$= 1000\kappa \left(\frac{\text{S} \times 100}{\text{m}}\right) \frac{\text{m}^3}{1000 \times \text{equiv}} = \left(\frac{100\kappa}{C}\right) \text{Sm}^2\text{equiv}^{-1}$$

7. (b) Molar conductance $= \dfrac{\text{Specific conductance}}{\text{Molarity}}$

On dilution, molarity decreases and molar conductance increases. Hence, specific conductance decreases.

8. (b) Number of ions remains constant but due to dilution, ionic mobility increases (as ionic attraction decreases). Hence, equivalent conductance increases.

9. (a) Specific conductance = Conductance × Cell constant

$\therefore$ $x = \kappa\ (0.01\ \text{N KCl}) = y \times \text{cell constant}$

$\kappa(0.01\text{N NaCl}) = z \times \text{cell constant}$

$\therefore$ $\dfrac{\kappa(0.01\text{N NaCl})}{x} = \dfrac{z}{y}$

$\therefore$ $\kappa(0.01\ \text{N NaCl}) = \dfrac{zx}{y}$

10. (b) For a weak acid,

	$HA \rightleftharpoons$	H^+ +	A^-
Initial	C	0	0
At equilibrium	$(C - Cx)$	Cx	Cx

$$\therefore \quad K_a = \frac{[H^+][A^-]}{[HA]} = \frac{Cx^2}{(1-x)}$$

By Kohlrausch's law, $x = \frac{\Lambda_m}{\Lambda_m^\circ}$

$$\therefore \quad K_a = \frac{C\left(\frac{\Lambda_m}{\Lambda_m^\circ}\right)^2}{\left(1 - \frac{\Lambda_m}{\Lambda_m^\circ}\right)} = \frac{C\Lambda_m^2}{\Lambda_m^\circ(\Lambda_m^\circ - \Lambda_m)}$$

11. (b) $\lambda_H^\infty = 350 \text{ Scm}^2\text{equiv}^{-1}$

$\lambda_{SO_4^{2-}} = 80 \text{ Scm}^2 \text{ equiv}^{-1}$

$\therefore \Lambda_{H_2SO_4}^\infty$ (equivalent conductance) $= 350 + 80$

$= 430 \text{Scm}^2\text{equiv}^{-1}$

Since, equivalent mass of $H_2SO_4 = \frac{1}{2}$ molar mass

Hence, $\Lambda_{H_2SO_4}^\infty = 430 \times 2 = 860 \text{ S cm}^2\text{mol}^{-1}$

12. (d) Molarity (KCl) $= \frac{\text{Mole}}{\text{Litre}} = \frac{0.100}{0.500} = 0.2 \text{ molL}^{-1}$

Molar conductance $= \frac{1000 \times \text{Specific conductance}}{\text{Molarity}}$

$= \frac{1000 \times x}{0.2} = 5000\, x \text{ Scm}^2\text{mol}^{-1}$

13. (a) Resistance = (Resistivity) × Cell constant

$\therefore$ Resistance = Resistivity

when cell constant = 1cm^{-1}

14. (d) Cell constant $= \frac{\text{Distance between two electrodes}}{\text{Area between two electrodes}}$

(a) $= \frac{100\text{cm}}{100\text{cm}^2} = 1\text{cm}^{-1}$

(b) $= \frac{10\text{cm}}{10\text{cm}^2} = 1\text{ cm}^{-1}$

(c) $= \frac{1\text{cm}}{1 \text{ cm}^2} = 1\text{cm}^{-1}$

Thus, equal.

15. (c) When temperature increases, conductance increases. Thus, resistivity and conductivity both are temperature dependent. By Nernst equation,

$$E = E^\circ - \frac{2.303\,RT}{nF}\log K$$

Thus, Emf of the cell is dependent on temperature.

Cell constant $= \frac{\text{Distance between two electrodes}}{\text{Area between them}}$

Thus, cell constant is temperature independent.

16. (c) $\Lambda_m = \frac{1000 \text{ specific conductance}}{\text{Molarity } (M)}$

If solution is diluted to x times then new molarity will be

$$VM = VxM'$$

$$\therefore \quad M' = \left(\frac{M}{x}\right)$$

Then new molar conductance will be

$$\Lambda_m' = \frac{1000 \text{ specific conductance}}{M/x} = x\Lambda_m$$

0.1 diluted to 0.01 M

Thus, $x = \frac{0.10}{0.01} = 10$

$\therefore$ New molar conductance will be 10 times.

17. (c) $S \propto$ concentration in mol dm^{-3}

or $S \propto \text{mol m}^{-3}$

$S \propto \text{area (m}^2)$

$\propto \frac{1}{\text{Length (m)}}$

$\therefore \quad S \propto \frac{\text{mol m}^{-3} \times \text{m}^2}{\text{m}}$

$S = \kappa$(Specific conductivity) mol m^{-2}

$\therefore \quad \kappa = \text{Sm}^2\text{mol}^{-1}$

18. (d) $\Lambda_\infty(CH_3COOH) = \Lambda_\infty(CH_3COONa) + \Lambda_\infty(HCl) - \Lambda_\infty(NaCl)$

$= 91.0 + 426.2 - 126.5$

$= 517.2 - 126.5$

$= 390.7 \text{ Scm}^2\text{mol}^{-1}$

19. (a) $\Lambda^\circ_{CH_3COOH} = \Lambda^\circ_{CH_3COO^-} + \Lambda^\circ_{H^+}$

$= \Lambda^\circ_{CH_3COONa} + \Lambda^\circ_{HCl} - \Lambda^\circ_{NaCl}$

Thus, Λ°_{NaCl} is required.

20. (b) Specific conductance (κ) of 0.2 M solution

= Conductance × Cell constant

$= \frac{1}{\text{Resistance}} \times \left(\frac{\rho}{a}\right)$

$1.3\text{Sm}^{-1} = \left(\frac{1}{50}\right)\text{S} \times \frac{\rho}{a}$

$\therefore \quad \left(\frac{\rho}{a}\right) = 50 \times 1.3 \text{ m}^{-1}$

Specific conductance of 0.4M solution $= \frac{1}{\text{Resistance}} \times \left(\frac{\rho}{a}\right)$

$\kappa \text{ (0.4 m)} = \left(\frac{1}{260} \times 50 \times 1.3\right)$

$= 0.25 \text{ Sm}^{-1}$

Molar conductance $= \frac{\text{Specific conductance}}{\text{Concentration}}$

$0.4 \text{ M} = 0.4 \text{ mol dm}^{-3}$

$= \frac{0.4 \text{ mol}}{\text{dm}^3} = \frac{0.4 \text{ mol} \times 1000}{\text{m}^3}$

$\therefore$ Molar conductance, $\Lambda_m = \frac{0.25 \text{ Sm}^{-1}}{0.4 \times 1000 \text{ mol m}^{-3}}$

$= 6.25 \times 10^{-4} \text{ Sm}^2\text{mol}^{-1}$

21. (c) By Debye-Huckel Onsager equation

$$\lambda_C = \lambda_\infty - B\sqrt{C}$$

when, λ_C = equivalent conductivity at concentration C

and λ_∞ = limiting equivalent conductivity at infinite dilution

Limit $C \to 0, \lambda_C = \lambda_\infty$

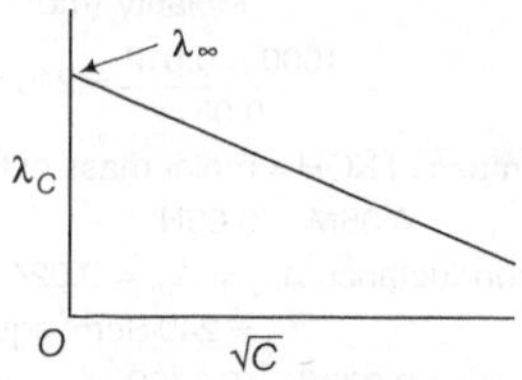

22. (c) Water has also conductance, hence to obtain pure value of conductance of AgCl, conductance of water should be substracted from the conductance of AgCl given.

Pure specific conductance of AgCl $= (3.41 \times 10^{-6} - 1.60 \times 10^{-6})$

$= 1.81 \times 10^{-6}\Omega^{-1}\text{cm}^{-1}$

AgCl is sparingly soluble, hence its solubility may be taken its concentration.

Also, $\Lambda_C = \Lambda_0$ in saturated solution sparingly soluble salt

$$\Lambda_0 = 138.3 = \frac{1000 \times \text{specific conductance}}{\text{Solubility or concentration}}$$

$$= \frac{1000 \times 1.81 \times 10^{-6}}{S(\text{g equiv L}^{-1})}$$

$$\therefore \quad S = \frac{1000 \times 1.81 \times 10^{-6}}{138.3}$$

$$= 1.31 \times 10^{-5} \text{g equiv L}^{-1}$$

$$S = 1.31 \times 10^{-5} \text{ mol L}^{-1}$$

(for AgCl, equivalent mass = molar mass)

$$\therefore \quad AgCl\,(s) \rightleftharpoons Ag^+ + Cl^-$$

$$\therefore \quad K_{sp} = [Ag^+][Cl^-] = S^2$$

$$= (1.31 \times 10^{-5})^2$$

$$= 1.72 \times 10^{-10} \text{ mol}^2\text{L}^{-2}$$

23. (d) $KCl + AgNO_3 \longrightarrow AgCl(s) + KNO_3$

$$K^+ + Cl^+ + Ag^+ + NO_3^- \longrightarrow AgCl(s) + K^+ + NO_3^-$$

Net $\quad Cl^- + Ag^+ \longrightarrow AgCl(s)$

When $AgNO_3$ is added to KCl solution, insoluble AgCl(*s*) is formed. Thus, conductance *a* remains constant. After KCl has been completely precipitated as AgCl, further addition of $AgNO_3$ causes increases in conductance.

24. (a) Specific conductance (0.2 M) = Conductance × Cell constant

$$1.4\,\text{Sm}^{-1} = \kappa = \frac{1}{50\,\Omega} \times \text{Cell constant}$$

$$= \frac{1\text{S}}{50} \times \text{Cell constant} \quad \left(\frac{1}{\Omega} = \text{S(Siemen)}\right)$$

$$\therefore \quad \left(\frac{\rho}{a}\right) = \text{Cell constant} = 1.4 \times 50 \text{ m}^{-1}$$

Specific conductance (0.5M)

$$= \text{Conductance} \times \text{Cell constant}$$

$$= \frac{1}{R} \times \frac{l}{a} = \frac{1}{280\Omega} \times 1.4 \times 50 \text{ m}^{-1}$$

$$\kappa = 0.25 \text{ S m}^{-1}$$

Molar conductance

$$\Lambda_m = \frac{\text{Specific conduction}}{\text{Molarity}}$$

$$= \frac{0.25 \text{ Sm}^{-1}}{0.5 \text{ mol dm}^{-3}}$$

$$= \frac{0.25 \text{ Sm}^{-1}}{0.5 \times 10^3 \text{ mol m}^{-3}}$$

$$= 5 \times 10^{-4} \text{ Sm}^2 \text{ mol}^{-1}$$

Note $1 \text{ mol L}^{-1} = 1 \text{mol dm}^{-3} = 10^3 \text{ mol m}^{-3}$

25. (a,c) $126.36 \text{ Scm}^2 \text{ equiv}^{-1} = 126.36 \text{ S}\left(\frac{\text{m}^2}{10^4}\right)\text{equiv}^{-1}$

$$= 126.36 \times 10^{-4} \text{ Sm}^2 \text{ equiv}^{-1}$$

$$CaCl_2 \rightleftharpoons Ca^{2+} + 2Cl^-$$

$$1 \text{ mol } CaCl_2 = 2 \text{ equivalent } CaCl_2$$

$$1 \text{ mol L}^{-1} = 2 \text{ equivalent L}^{-1}$$

$$\therefore \quad 0.01\text{N } CaCl_2 = 0.005 \text{ M } CaCl_2$$

$$\Lambda_C \text{ (molar conductance)} = \frac{\text{Specific conductivity}}{\text{Molarity}}$$

$$\Lambda_C \text{(equivalent conductance)} = \frac{\text{Specific conductivity}}{\text{Normality}}$$

$$\therefore \quad \Lambda_C \text{ (molar conductance)} = 126.36 \times 2 \text{ Scm}^2\text{mol}^{-1}$$

$$= 252.72 \times 10^{-4} \text{ Sm}^2\text{mol}^{-1}$$

26. (a,c,d) $x = \dfrac{\Lambda_C \text{ (H}_2\text{O, given concentration)}}{\Lambda_0 (\text{H}_2\text{O - limiting})}$

$$\therefore \quad \Lambda_C = x\,\Lambda_0$$

$$\Lambda_0(H_2O) = \lambda_0(H^+) + \lambda_0(OH^-)$$

$$= (350 + 200) \text{ Scm}^2\text{equiv}^{-1}$$

$$= 550 \text{ Scm}^2\text{equiv}^{-1}$$

$$\Lambda_C = 1.9 \times 10^{-9} \times 550$$

$$= 1.045 \times 10^{-6} \text{ Scm}^2\text{equiv}^{-1}$$

$$= 1.045 \times 10^{-6} \text{ S}\left(\frac{\text{m}^2}{10^4}\right)\text{equiv}^{-1}$$

$$= 1.045 \times 10^{-10} \text{ Sm}^2\text{equiv}^{-1}$$

Since, $\quad H^+ \equiv OH^-$ in H_2O

Hence, $\quad$ 1 equivl ≡ 1mol

$$\therefore \quad \Lambda_C = 1.045 \times 10^{-10} \text{ Sm}^2\text{mol}^{-1}$$

27. (a,c,d)

(a) $Al_2(SO_4)_3 \rightleftharpoons 2Al^{3+} + 3SO_4^{2-}$

$$\therefore \quad \Lambda_m^\infty = 2\lambda^\infty_{Al^{3+}} + 3\lambda^\infty_{SO_4^{2-}}$$

Thus, (a) is true.

(b)

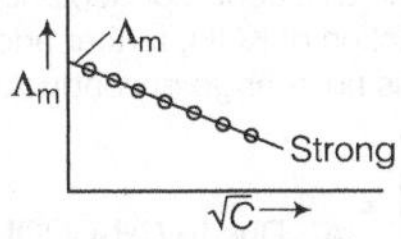

For strong electrolyte, variation in graph is uniform, hence Λ_m can be extrapolated to $\sqrt{C} \longrightarrow 0$, to obtain Λ_m^∞.

In case of weak electrolyte, variation of Λ_m with $\sqrt{C}$ is rapid in lower concentration range (higher dilution), hence graph is not uniform and Λ_m^∞ cannot be obtained by extrapolation to $\sqrt{C} \longrightarrow 0$.

Thus, (b) is incorrect.

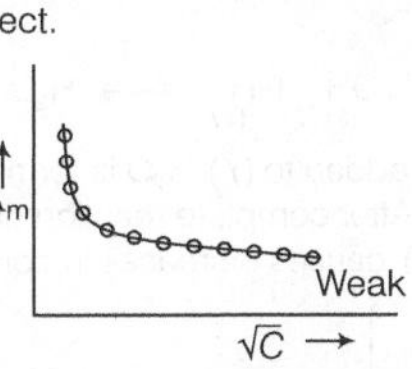

(c) $\Lambda^\infty_{AgCl} = \Lambda^\infty_{AgNO_3} + \Lambda^\infty_{HCl} - \Lambda^\infty_{HNO_3} = \Lambda^\infty_{Ag^+} + \Lambda^\infty_{Cl^-}$

Thus, correct based on Kohlrausch's law.

(d) Also, $\Lambda^\circ_{BaSO_4} = \Lambda^\circ_{BaCl_2} + \Lambda^\infty_{H_2SO_4} - 2\Lambda^\infty_{HCl}$

Thus, it is also correct based on Kohlrausch's law.

28. (a)

(i) *X* and *Y* both are weak electrolytes, hence have minimum conductance. When *X* $[(C_2H_5)_3N]$ is added to *Y* (CH_3COOH), ions are formed hence conductance increases.

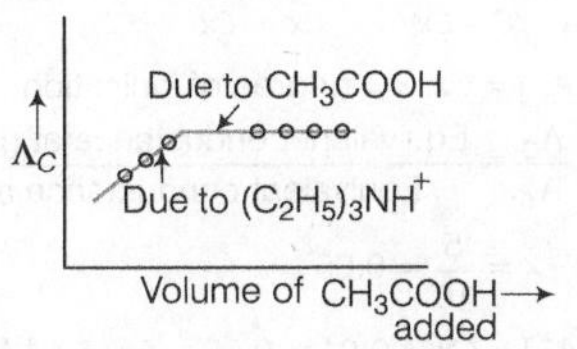

$$\underset{(Y)}{CH_3COOH} + \underset{(X)}{(C_2H_5)_3N} \longrightarrow CH_3COO^- + (C_2H_5)_3NH^+$$

After complete reaction further addition of *Y*, makes no change in conductance.

Thus, **(i) → (r)**

(ii) Net reaction is

$$\underset{(X)}{I^-} + \underset{(Y)}{Ag^+} \longrightarrow \underset{\text{Insoluble}}{AgI(s)}$$

where $X(I^-)$ is added to $Y(Ag^+)$ insoluble AgI is formed. Hence, there is no change in conductance. After complete reaction of Ag^+, further addition of I^- causes increase in conductance.

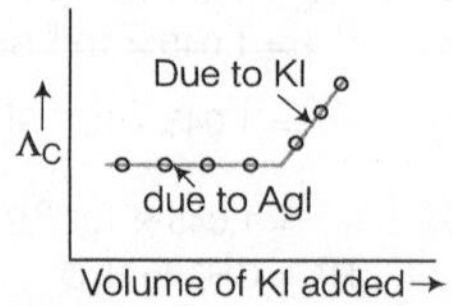

Thus, **(ii)** → **(s)**

(iii) KOH is a strong electrolyte. As $CH_3COOH(X)$ is added, OH^- is neutralised by CH_3COOH.

$$\underset{(Y)}{OH^-} + \underset{(X)}{CH_3COOH} \longrightarrow CH_3COO^- + H_2O$$

CH_3COO^- is hydrolysed at the same time.

$$CH_3COO^- + H_2O \longrightarrow CH_3COOH + OH^-$$

But being in small extent, conductance decreases. After complete reaction of KOH, further addition of CH_3COOH, (weak) there is no change in conductance.

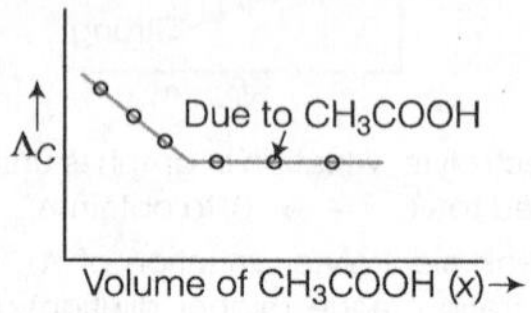

Thus, **(iii)** → **(q)**

(iv) NaOH(X) and HI(Y) both are strong electrolytes. Thus, net reaction is

$$\underset{(X)}{OH^-} + \underset{(Y)}{H^+} \longrightarrow H_2O$$

When (X) is added to (Y), H_2O is formed hence conductance decreases. After complete reaction of OH^-, further addition of NaOH (X), causes increases in conductance.

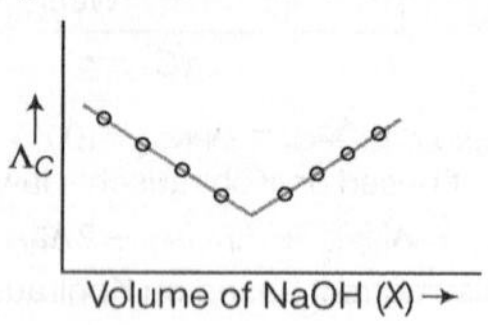

Thus, **(iv)** → **(p)**

29. (d) $HA \rightleftharpoons H^+ + A^-$

	HA	H^+	A^-
Initial	C	0	0
At equilibrium	$(C - Cx)$	Cx	Cx

$\therefore \quad [A^-] = Cx$, x = degree of ionisation

$$x = \frac{\Lambda_C}{\Lambda_0} = \frac{\text{Equivalent conductance at given concentration}}{\text{Equivalent conductance at infinite diution}}$$

$$x = \frac{5}{10} = 0.05$$

$$[A^-] = Cx = 0.01 \times 0.05 = 5 \times 10^{-4}\ \text{M}$$

$$[H^+] = Cx = 5 \times 10^{-4}\ \text{M}$$

$$\therefore \quad \text{pH} = -\log[H^+] = 4 \cdot \log 5 = 3.3$$

By Ostwald's dilution law

$$K_a = \frac{[H^+][A^-]}{[HA]} = \frac{Cx^2}{(1-x)} = \frac{0.01\,(0.05)^2}{(1-0.05)} = 2.63 \times 10^{-5}$$

$$\therefore \quad pK_a = -\log K_a = 4.58$$

Thus, (i) → (r), (ii) → (s), (iii) → (p), (iv) → (q).

Explanations for Q. Nos. (30-31)

For 0.1 M solution, $\kappa = 1.30\ \text{Sm}^{-1}$

$$\text{Conductivity}\ (\kappa) = \text{Conductance} \times \text{Cell constant}$$

$$1.30 = \frac{1}{100} \times \text{Cell constant}$$

For 0.2 M solution, $\kappa' = \dfrac{1}{520} \times \text{Cell constant}$

$$\therefore \quad \frac{\kappa'}{1.30} = \frac{100}{520}$$

$$\therefore \quad \kappa' = \frac{100}{520} \times 1.30 = 0.25\ \text{Sm}^{-1}$$

$$0.2\ \text{M} = 0.2\ \text{mol dm}^{-3} = 0.2\ \text{mol} \times 1000\ \text{m}^{-3}$$

$$\therefore \quad \text{Molar conductance} = \frac{\text{Conductivity}\ (\kappa)}{\text{Molarity (mol m}^{-3})} = \frac{0.25\ \text{Sm}^{-1}}{0.2 \times 1000\ \text{mol m}^{-3}} = 1.25 \times 10^{-3}\ \text{Sm}^2\ \text{mol}^{-1}$$

30. (b) **31.** (a)

Explanations for Q. Nos. (32-33)

By Kohlrausch's law,

$$\Lambda_0(NH_4OH) = \Lambda_0(NH_4Cl) + \Lambda_0(NaOH) - \Lambda_0(NaCl)$$
$$= 1.50 \times 10^{-2} + 2.48 \times 10^{-2} - 1.26 \times 10^{-2}$$
$$= 2.72 \times 10^{-2}\ \text{Sm}^2\text{mol}^{-1}$$

$\Lambda_C(NH_4OH)$ at 0.001 M $= 3.00 \times 10^{-3}\ \text{Sm}^2\text{mol}^{-1}$

$$x\ (\text{degree of ionisation}) = \frac{\Lambda_c}{\Lambda_0}$$

$$x = \frac{3.00 \times 10^{-3}}{2.72 \times 10^{-2}} = 0.11$$

Also, $NH_4OH \rightleftharpoons NH_4^+ + OH^-$

	NH_4OH	NH_4^+	OH^-
Initial	C	0	0
At equilibrium	$(C - Cx)$	Cx	Cx

$$K_b\ (\text{ionisation constant}) = \frac{[NH_4^+][OH^-]}{[NH_4OH]} = \frac{Cx^2}{(1-x)}$$

$$K_b = \frac{0.001\,(0.11)^2}{(1-0.11)} = 1.36 \times 10^{-5}$$

$$pK_b = -\log K_b = -\log 1.36 \times 10^{-5} = 4.87$$

32. (a) **33.** (d)

34. (5) Dibasic acid is $H_2A \rightleftharpoons 2H^+ + A^{2-}$

$$\therefore \quad 0.01\text{M} = 0.02\text{N}$$

Equivalent conductance at 0.02 N

$$\Lambda_{eq} = \frac{\text{Specific conductance}}{\text{Normality}} = \frac{\kappa}{0.02}$$

Equivalent conductance at 0.004 N

$$\Lambda'_{eq} = \frac{k}{0.004}$$

$$\frac{\Lambda'_{eq}}{\Lambda_{eq}} = x = \frac{\kappa}{0.004} \times \frac{0.02}{\kappa} = 5$$

35. (7) Reciprocal of resistance is Siemen (Conductance Ω^{-1}).

DPP-2 Electrochemical Series

1. **Plan** If $E_{cell} > 0$, then reducing agent with less positive value of ($E°_{red}$) is the anode and the oxidising agent with more positive value ($E°_{red}$) is the cathode.

Also $E°_{oxi} = -E°_{red}$

and $E°_{cell} = E°_{oxi} + E°_{red}$

but be careful : take values according to the reaction

$E°_{Cu^{2+}/Cu} = 0.34\text{V}$, hence is the anode

Oxidation at anode,

$$Cu(s) \longrightarrow Cu^{2+} + 2e^-,$$

$$E°_{Cu/Cu^{2+}} = -E°_{Cu^{2+}/Cu} = -0.34\text{ V}$$

Reduction at cathode,

$$2Ag^+ + 2e^- \longrightarrow 2Ag, \quad E°_{Ag^+/Ag} = 0.80\text{ V}$$

$$Cu + 2Ag^+ \longrightarrow Cu^{2+} + 2Ag \quad E°_{cell} = 0.46\text{ V}$$

Oxidation half-cell is $Cu \mid Cu^{2+}$ (1 M) $\ominus$

Reduction half-cell is Ag^+ (1M) $\mid$ Ag $\oplus$

Two half-cells are joined by salt-bridge, thus cell is

$$\underset{\ominus}{Cu} \mid Cu^{2+}(1\text{ M}) \underset{\text{Salt bridge}}{||} Ag^+(1\text{M}) \mid \underset{\oplus}{Ag}$$

(ii) For the cell reaction,

$$Q \text{ (reaction quotient)} = \frac{[Cu^{2+}]}{[Ag^+]^2} = \frac{(0.01)}{[Ag^+]^2}$$

$$E_{cell} = 0.00\text{ V}$$

$$E°_{cell} = 0.46\text{V}$$

$$\therefore \quad E_{cell} = E°_{cell} - \frac{2.303 \times RT}{nF}\log Q$$

At 298 K, $\frac{2.303RT}{F} = 0.0591$

$$\therefore \quad 0 = 0.46 - \frac{0.0591}{2}\log\frac{0.01}{[Ag^+]^2}$$

$$\log\frac{0.01}{[Ag^+]^2} = \frac{0.92}{0.0591}$$

$$\log[[Ag^+]^2 \times 100] = -15.57$$

$$[Ag^+]^2 \times 100 = 2.71 \times 10^{-16}$$

$$[Ag^+]^2 = 2.71 \times 10^{-18}$$

$$[Ag^+] = 1.65 \times 10^{-9}\text{M}$$

2. (a) Based on electrochemical series, oxidising power of

$$F_2 > Cl_2 > Br_2 > I_2$$

F_2 oxidises Cl^- to Cl_2

Cl_2 oxidises Br^- to Br_2

Br_2 oxidises I^- to I_2

On passing Cl_2 into a solution containing KF, KI and KBr,

$$2KBr + Cl_2 \longrightarrow 2KCl + Br_2 \text{ (orange)}$$

$$2KI + Cl_2 \longrightarrow 2KCl + I_2 \text{ (violet)}$$

$$2KI + Br_2 \longrightarrow 2KBr + I_2 \text{ (violet)}$$

Br_2 formed also oxidises KI to I_2 (violet)

3. (d) $E°_{cell} = 1.10$ V

(a) Hence, when $E_{ext} < 1.10$ V, reaction continues till the E_{ext} reaches 1.10 V and

$$Zn(s) + Cu^{2+}(aq) \longrightarrow Cu(s) + Zn^{2+}(aq)$$

thus, zinc dissolves at anode and Cu deposits at cathode.

(b) When $E_{ext} = 1.10$V, no reaction.

(c) When $E_{ext} > 1.10$V, reverse reaction takes place,

$$Zn^{2+}(aq) + Cu(s) \rightleftharpoons Cu^{2+}(aq) + Zn(s)$$

It now acts as electrolytic cell.

Zinc deposits at zinc plate which is now the cathode (+ve plate) and Cu dissolves as Cu^{2+} at copper plate which is now the anode (−ve plate)

4. (c) Based on electro chemical series, reactivity series is

$$\text{Mg} \quad \text{Cu} \quad \text{Hg} \quad \text{Ag}$$

$\longleftarrow$ Reducing behaviour of metals

Oxidising behaviour of metals $\longrightarrow$

(a) $2Ag^+ + Cu \longrightarrow Cu^{2+} + 2Ag$

Cu is a better reducing agent than Ag hence Ag^+ is reduced by Cu. Thus $AgNO_3$ cannot be stored in copper vessel.

(b) $Cu^{2+} + Mg \longrightarrow Mg^{2+} + Cu$

Mg will reduce Cu^{2+} to Cu thus cannot be stored.

(c) $Cu^{2+} + 2Ag \longrightarrow Cu + 2Ag^+$

Cu is better reducing agent hence $CuCl_2$ can be stored in silver vessel.

(d) $Hg^{2+} + Cu \longrightarrow Cu^{2+} + Hg$

Cu is better reducing agent. Hence, $HgCl_2$ cannot be stored in copper vessel.

5. (b) $Ag_2CO_3(s) \longrightarrow Ag_2O(s) + CO_2(g)$

$$Ag_2O(s) \longrightarrow 2Ag(s) + \frac{1}{2}O_2(g)$$

Oxides of the metals are decomposed to metals if in electro chemical series, $E°_{red} > 0.79$ V

For Ag, $E°_{Ag^+/Ag} = 0.80$V hence Ag_2O is decomposed to metals. Products are Ag, CO_2 and O_2. Since reaction takes place in an open vessel, CO_2 and O_2 escape from mixture, residue being 2 moles of silver.

6. (d) Blue colour fades or changes to colourless if Cu^{2+} is reduced to Cu.

$$\underset{\text{Blue}}{Cu^{2+}} + Zn \longrightarrow \underset{\text{Colourless}}{Zn^{2+}} + Cu$$

$$Cu^{2+} + 2Ag \longrightarrow \underset{\text{Colourless}}{2Ag^+} + Cu$$

$$Cu^{2+} + Fe \longrightarrow \underset{\text{Very light green}}{Fe^{2+}} + Cu$$

In electrochemical series (ECS), reactivity order is

Zn, Fe, Cu, Ag

Thus, Zn and Fe reduce Cu^{2+} to Cu but Ag does not reduce Cu^{2+}.

Thus, in I and III, $CuSO_4$ changes to colourless or fades by zinc and iron plates.

7. (a) $3Mn^{2+}(aq) + 2MnO_4^-(aq) + 2H_2O(l) \longrightarrow 5MnO_2(s) + 4H^+$

$$E°_{cell} = 0.47\text{ V}$$

Since $E°_{cell} > 0$, hence spontaneous.

Thus, Mn^{2+} is oxidised to MnO_2 by MnO_4^- in acidic medium.

8. (c) $E°_{cell}$ remains unchanged on division, thus intensive property.

$\Delta_r G$ is dependent on the division, thus extensive property.

9. (b)

$$Sn^{2+} \longrightarrow Sn^{4+} + 2e^-, \quad E°_{Sn^{2+}/Sn^{4+}} = -0.13\text{V}$$

$$Sn^{2+} + 2e^- \longrightarrow Sn, \quad E°_{Sn^{2+}/Sn} = -0.14\text{V}$$

$$2Sn^{2+} \longrightarrow Sn^{4+} + Sn \quad E°_{cell} = -0.27\text{V}$$

Since $E°_{cell} < 0$, disproportionation of Sn^{2+} to Sn^{4+} (by oxidation) and Sn (by reduction) is not spontaneous.

10. (a) $\underset{1\text{ bar}}{H_2} \longrightarrow \underset{1\text{ M}}{2H^+} + 2e^-$

It is oxidation half-cell

$$\underset{1\text{ bar}}{Pt(H_2)} \mid \underset{\ominus}{H^+(1M)}$$

H_2 (reduced part) is at 1 bar. H^+ (oxidised part) is at 1 M.

$$\underset{1\text{ M}}{2H^+} + 2e^- \longrightarrow \underset{1\text{ bar}}{H_2(g)}$$

It is reduction half-cell

$$\underset{\oplus}{H^+(1M)} \mid \underset{1\text{ bar}}{Pt(H_2)}$$

In this case also, oxidised and reduced parts are at unity.

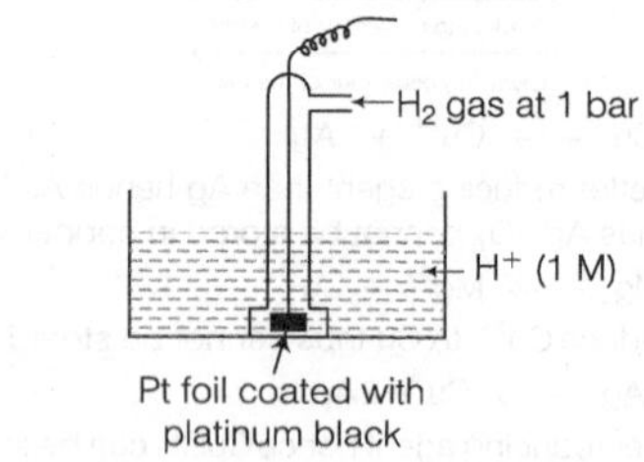

Then (I), (II) and (III) are correct facts.

11. (d) Most positive $E°_{red}$/value means most oxidising agent

Most negative $E°_{red}$ value means most reducing agent

$$Cr^{3+} > Cr_2O_7^{2-} > Cl^- > MnO_4^-$$

$\longleftarrow$ Reducing power

Oxidising power $\longrightarrow$

12. (d) $E°_{Z/Z^-} < E°_{Y/Y^-} < E°_{X/X^-}$

Oxidising power $\longrightarrow$

$\longleftarrow$ Reducing power

In ECS, pair with more negative values of $E°_{red}$ reducing agent is above oxidising agent.

Thus, Z/Z^- is the best reducing agent

$$Z^- \longrightarrow Z + e^-$$
$$Y + e^- \longrightarrow Y^-$$
$$X + e^- \longrightarrow X^-$$

Thus, Z^- will reduce both X and Y and itself will be oxidised to Z.

13. (a) $E°_X = -0.52$ V

$E°_Y = -3.03$ V

$E°_Z = -1.18$ V

Their placements in ECS is in order $Y > Z > X$.

Thus, reducing nature is also in same order

$$Y > Z > X$$

14. (c) These metals based on $E°_{red}$ values are placed in the following order:

	$E°$
$Mg^{2+} + 2e^- \longrightarrow Mg,$	-2.37 V
$Cu^{2+} + 2e^- \longrightarrow Cu,$	$+0.34$ V
$Hg_2^{2+} + 2e^- \longrightarrow 2Hg,$	$+0.79$ V
$Ag^+ + e^- \longrightarrow Ag,$	$+0.80$ V

Reducing power ↑ Oxidising power ↓

Mg will reduce Cu^{2+}, Hg_2^{2+} and Ag^+ to metals Cu, Hg, Ag.

Cu will reduce Hg_2^{2+} and Ag^+ to Hg, Ag,

Hg will reduce Ag^+ to Ag.

Ag^+ is reduced by all metals thus Ag first. Hg^{2+} is reduced by Cu, Mg thus Hg. Cu^{2+} is reduced by Mg only, thus Cu.

Thus, deposition : Ag, Hg, Cu $\longrightarrow$

15. (a) In electrochemical series (ECS), elements have been arranged in the increasing standard reduction potential starting from most negative to most positive value w.r.t. SHE ($E°_{SHE} = 0.00$ V).

Most negative standard reduction potential means most easily oxidised and thus is the best reducing agent.

OR Reducing agent in (ECS) is always above oxidising agent.

Thus, Zn(s) is the best reducing agent in given set.

16. (a)

	$E°$
$Cr^{3+} + e^- \longrightarrow Cr^{2+}$	-0.41 V
$Mn^{3+} + e^- \longrightarrow Mn^{2+}$	$+1.57$ V
$Fe^{3+} + e^- \longrightarrow Fe^{2+}$	$+0.77$ V
$Co^{3+} + e^- \longrightarrow Co^{2+}$	$+1.97$ V

Reducing agent ↑ Oxidising agent ↓

Cr^{3+}/Cr^{2+} with most negative $E°_{red}$ is the best reducing agent.

$$Cr^{2+} \longrightarrow Cr^{3+} + e^-, \quad E° = 0.41 \text{ V}$$

Thus, is oxidised from Cr^{2+} to Cr^{3+} most easily.

17. (b) Most negative $E°_{red}$ means the M^{n+}/M is at the top of ECS and is the best reducing agent.

18. (d) $E°_{Zn^{2+}/Zn} = -0.76$ V

$E°_{Cu^{2+}/Cu} = +0.34$ V

In electrochemical series, zinc is above copper and thus zinc is a better reducing agent than copper. When $CuSO_4$ is placed in zinc vessel, copper is displaced

$$Zn + CuSO_4 \longrightarrow ZnSO_4 + Cu\downarrow$$

Thus, $CuSO_4$ cannot be stored in a vessel made of zinc.

Thus, statement I is incorrect and statement II is correct.

19. (a) $Cu + \underset{\text{Colourless}}{2AgNO_3} \longrightarrow \underset{\text{Blue}}{Cu(NO_3)_2} + 2Ag$

In electrochemical series, copper is above silver thus is a better reducing agent. When $AgNO_3$ solution is stirred with copper spoon, Ag is displaced and copper is oxidised to Cu^{2+} (blue).

Thus, statement I and II are correct and statement II is the correct explanation of statement I.

20. (b, d)

$$Cu^{2+} + H_2 \longrightarrow Cu^+ + 2H^+$$

$Cu^{2+} + 2e^- \longrightarrow Cu,$	$E°_{Cu^{2+}/Cu} = 0.34$ V
$H_2 \longrightarrow 2H^+ + 2e^-,$	$E°_{SHE} = 0.00$ V
$Cu^{2+} + H_2 \longrightarrow Cu + 2H^+$	$E°_{cell} = 0.34$ V

$E°_{cell} > 0$ thus spontaneous

Cu^{2+}/Cu couple is thus a stronger oxidising agent. Reverse reaction is non-spontaneous.

$$Cu + 2H^+ \longrightarrow Cu^{2+} + H_2, \; E°_{cell} = -0.34 \text{ V}$$

Thus, copper cannot displace H_2 from acid.

21. (a,c)

$$\underset{\text{Dil}}{H_2SO_4(aq)} \rightleftharpoons 2H^+ + SO_4^{2-}$$

$$SO_4^{2-} + H_2O \rightleftharpoons H_2SO_4 + O^{2-}$$

At cathode

$$2H^+ + 2e^- \longrightarrow H^2$$

Thus, (a) is correct.

At anode

$$2O^{2-} \longrightarrow O_2 + 4e^-$$

Thus, (c) is correct.

Concentrated H_2SO_4 ionises as H^+ and HSO_4^- and then water is not oxidised in concentrated H_2SO_4. Thus, (b) is incorrect.

SO_4^{2-} is not oxidised to $S_2O_8^{2-}$ in dilute H_2SO_4. Thus, (d) is incorrect.

(d) In dilute sulphuric acid solution, SO_4^{2-} ions will be oxidised to tetrathionate ion ($S_2O_8^{2-}$) at anode.

22. (b,c,d)

If dil. H_2SO_4 is used, then H^+ is the reacting species.

If conc. H_2SO_4 is used, then SO_4^{2-} is the reacting species.

(a) $2H^+ + 2e^- \longrightarrow H_2, \quad E°_{SHE} = 0.00$ V

$Cu \longrightarrow Cu^{2+} + 2e^-, \quad E°_{Cu/Cu^{2+}} = -0.34$ V

$Cu + 2H^+ \longrightarrow Cu^{2+} + H_2 \quad E°_{cell} = -0.34$ V

$E°_{cell} < 0$

Thus, copper does not react with H^+ (dil. H_2SO_4) forming H_2.

(b) $SO_4^{2-} + 4H^+ + 2e^- \longrightarrow SO_2 + 2H_2O, \quad E° = 1.17$ V

$Cu \longrightarrow Cu^{2+} + 2e^-, \quad E° = -0.34$ V

$Cu + SO_4^{2-} + 4H^+ \longrightarrow Cu^{2+} + SO_2 + 2H_2O,$

$E°_{cell} = 0.83$ V

Since $E°_{cell} > 0$, Thus, copper reacts with conc. H_2SO_4 forming SO_2.

(c) $Zn + SO_4^{2-} + 4H^+ \longrightarrow Zn^{2+} + SO_2 + 2H_2O$

$E°_{cell} = 1.93$ V, $E°_{cell} > 0$, thus zinc also reacts with conc. H_2SO_4 forming SO_2.

(d) $Zn + 2H^+ \longrightarrow Zn^{2+} + H_2, \quad E°_{cell} = 0.76$ V

$E°_{cell} > 0$, thus zinc reacts with dil. H_2SO_4 forming H_2.

23. (a)

(i)

(C) $O_2 + 4H^+ + 4e^- \longrightarrow 2H_2O,$

$n_1 = 4, E°_1 = 1.229$ V

(D) $H_2O_2(aq) + 2H^+ + 2e^- \longrightarrow 2H_2O,$

$n_2 = 2, E°_2 = 1.763$ V

(C) – (D) gives

$O_2 + 2H^+ + 2e^- \longrightarrow H_2O_2(aq),$

$n_3 = 2, E°_3 = ?$

$$\Delta G_3 = \Delta G_1 - \Delta G_2$$

$$-n_3FE°_3 = -n_1FE°_1 + n_2FE°_2$$

$$E°_3 = \frac{n_1E°_1 - n_2E°_2}{n_3}$$

$$= \frac{4 \times 1.229 - 2 \times 1.763}{2}$$

$$= 0.695 \text{ V}$$

Thus, **(i) → (p)**

(ii)

(B) $O_2(g) + 2H_2O + 4e^- \longrightarrow 4OH^- \quad n_1 = 4, E_1 = 0.401$ V

(C) $O_2(g) + 4H^+ + 4e^- \longrightarrow 2H_2O, \quad n_2 = 4, E_2 = 1.229$ V

(G) $4H^+ + 4e^- \longrightarrow 2H_2, \quad n_3 = 4, E_3 = 0.00$ V

(B) + (G) – (C) gives

$4H_2O + 4e^- \longrightarrow 2H_2 + 4OH^-$

$2H_2O + 2e^- \longrightarrow H_2 + 2OH^-$

$\therefore \quad E_4 = 0.401 + 0.00 - 1.229$

$= -0.83$ V

Thus, **(ii) → (s)**

(iii) $O_3 + 2H^+ + 2e^- \longrightarrow O_2(g) + H_2$

(A) or $2O_3 + 4H^+ + 4e^- \longrightarrow 2O_2 + 2H_2O, \quad E°_1 = 2.075$ V

(G) $2H_2 \longrightarrow 4H^+ + 4e^-, \quad E°_2 = 0.00$

(H) $4H_2O + 4e^- \longrightarrow 2H_2 + 4OH^-,$

$E°_3 = -0.83$ V

(A) + (G) + (H) gives

$2O_3 + 2H_2O + 4e^- \longrightarrow 2O_2 + 4OH^-$

$O_3 + H_2O + 2e^- \longrightarrow O_2 + 2OH^-$

$E°_{cell} = 1.245$ V

Thus, **(iii) → (r)**

(iv)

(E) $Cr^{2+} + 3e^- \longrightarrow Cr, n_1 = 3, E°_1 = -0.74$ V

(F) $Cr^{2+} + 2e^- \longrightarrow Cr, n_2 = 2, E°_2 = 0.91$ V

(iv) $Cr^{3+} + e^- \longrightarrow Cr^{2+}, \quad n_3 = 1, E°_3 = ?$

(E) – (F) = (iv)

$$\therefore \quad \Delta G_3 = \Delta G°_1 - \Delta G°_2$$

$$-n_3F E°_3 = -n_1F E°_1 + n_2F E°_2$$

$$\therefore \quad E°_3 = \frac{n_1E°_1 - n_2E°_2}{n_3}$$

$$= \frac{3(-0.74) - 2 \times 0.91}{1}$$

$$= -2.24 + 1.82$$

$$= -0.42 \text{ V}$$

Thus, **(iv) → (s)**

Explanations for Q. Nos. (24-25)

I. $Cr_2O_7^{2} + 14H^+ + 6e^- \longrightarrow 2Cr^{3+}, E° = 1.33$ V

↑ Reduce species

II. $MnO_4^- + 8H^+ + 5e^- \longrightarrow Mn^{2+}, E° = 1.51$ V

↑ Reduced species

III. $Cr \longrightarrow Cr^{3+} + 3e^-, \quad E° = +0.74$ V

↑ Oxidised species

For (I) and (II), $E° > 0$, but for reduced species.

For (III), $E° > 0$, thus Cr^{3+} is the most stable.

More negative value of standard electrode reduction potential, better the reducing agent. Thus,

$$Mn^{2+} < Cl^- < Cr^{3+} < Cr$$

24. (a) **25.** (b)

Explanations for Q. Nos. (26-27)

$4[Al \longrightarrow Al^{3+} + 3e^-,$		$E° = 1.66$ V
$3O_2 + 12H^+ + 12e^- \longrightarrow 6H_2O,$		$E° = 1.23$ V
$4Al + 3O_2 + 12H^+ \longrightarrow 4Al^{3+} + 6H_2O,$		$E° = 2.89$ V

$E°_{cell} > 0$, thus above cell reaction is spontaneous

26. (b) **27.** (a)

Explanations for Q. Nos. (28-29)

CH_3CH_2OH is oxidised to CH_3COOH and $Cr_2O_7^{2-}$ is reduced to Cr^{3+}.

$$\therefore \quad E°_{cell} = E°_{CH_3CH_2OH/CH_3COOH} + E°_{Cr_2O_7^{2-}/Cr^{3+}}$$

$$= -0.06 + 1.33$$

$$= 1.27 \text{ V}$$

Reaction quotient

$$(Q) = \frac{[CH_3COOH]^3 [Cr^{3+}]^4 [H_2O]^{11}}{[CH_3CH_2OH^3] [Cr_2O_7^{2-}]^4 [H^+]^{16}}$$

$$(Q) = \frac{1}{[H^+]^{16}} = \frac{1}{(10^{-4})^{16}} = 10^{64}$$

$$\therefore \quad E_{cell} = E°_{cell} - \frac{0.0591}{12} \log 10^{64}$$

$$= 1.27 - \frac{0.0591}{12} \times 64 \log 10$$

$$= 1.27 - \frac{0.0591}{12} \times 64 = 0.095 \text{ V}$$

28. (b) **29.** (d)

30. (4) Fe, Al, Zn and Mg can be used to protect Ni from being oxidised to Ni^{2+} since, $E°_{cell} > 0$ and $\Delta G° < 0$.

DPP-3 Reference Half-Cell and Nernst Equations

1. Plan

(i) Oxidation takes place at anode (LHS) and reduction takes place at cathode (RHS)

$$Zn(s) \longrightarrow Zn^{2+}(aq) + 2e^-$$

$$E^\circ_{Zn/Zn^{2+}} = -E^\circ_{Zn^{2+}/Zn} = +0.76\text{ V}$$

$$Cu^{2+}(aq) + 2e^- \longrightarrow Cu(s)$$

$$E^\circ_{Cu^{2+}/Cu} = +0.34\text{ V}$$

Net $Zn(s) + Cu^{2+}(aq) \longrightarrow Cu(s) + Zn^{2+}(aq)$

$$E^\circ_{cell} = 1.10\text{ V}$$

Note *Take values according to reaction.*

We have been given E°_{red}. ($E^\circ_{Zn^{2+}/Zn} = -0.76$ V)

and $E^\circ_{Cu^{2+}/Cu} = 0.34$ V. But at anode oxidation takes place, hence we take

$$E^\circ_{Zn/Zn^{2+}} = E^\circ_{oxid} = -E^\circ_{redu} = -E^\circ_{Zn^{2+}/Zn} = +0.76\text{V}$$

and $E^\circ_{Cu^{2+}/Cu} = E^\circ_{red} = 0.34$ V for reaction.

(ii) Q (reaction quotient) $= \dfrac{[Cu][Zn^{2+}]}{[Zn][Cu^{2+}]} = \dfrac{[Zn^{2+}]}{[Cu^{2+}]} = \dfrac{1.0}{0.1} = 10$

We use Nernst equation to calculate EMF of the cell

$$E_{cell} = E^\circ_{cell} - \frac{2.303\,RT}{nF}\log Q$$

At 298 K

$$\frac{2.303\,RT}{F} = \frac{2.303 \times 8.314\text{ J} \times 298\text{ K}}{96500\text{ C}}$$

$$= 0.0591\text{ JKC}^{-1} = 0.0591\text{ V}$$

n = number of electrons exchanged = 2

$$\therefore \quad E_{cell} = 1.10 - \frac{0.0591}{2}\log 10$$

$$= 1.10 - 0.0295 = 1.07\text{ V}$$

(iii) If concentrations are exchanged,

$$[Cu^{2+}] = 1.0\text{ M}$$
$$[Zn^{2+}] = 0.1\text{ M}$$

Reaction quotient $(Q) = \dfrac{[Zn^{2+}]}{[Cu^{2+}]} = \dfrac{0.1}{1.0} = \dfrac{1}{10}$

$$\therefore \quad E_{cell} = 1.10 - \frac{0.0591}{2}\log\left(\frac{1}{10}\right)$$

$$= 1.10 + \frac{0.0591}{2}\log 10$$

$$= 1.10 + 0.0295 = 1.13\text{ V}$$

Since, 1.13 V > 1.07V, hence emf increases.

(iv) In both cases, $E_{cell} > 0$, thus cell reaction is spontaneous.

2. Plan Oxidation takes place at anode and reduction takes place at cathode, electron flows from anode to cathode and thus direction of the current is from cathode to anode

if $E_{cell} > 0$, spontaneous

$E_{cell} < 0$, non-spontaneous

(a) $Cu|Cu^{2+}$ (1.0 M) || Fe^{2+} (1.0 M) | Fe

(i) Net reaction $Cu + \underset{1.0\text{M}}{Fe^{2+}} \longrightarrow Fe + \underset{1.0\text{M}}{Cu^{2+}}$

$$E^\circ_{cell} = E^\circ_{Cu/Cu^{2+}} + E^\circ_{Fe^{2+}/Fe}$$

$$= -0.34 - 0.44$$

$$= -0.78\text{ V}$$

(ii) Reaction quotient $(Q) = \dfrac{[Cu^{2+}]}{[Fe^{2+}]} = \dfrac{1.0}{1.0} = 1$

$$\therefore \quad E_{cell} = E^\circ_{cell} = -0.78\text{ V}$$

$E_{cell} < 0$, hence electrons flow from B to A and for which $E^\circ_{cell} = 0.78$ V

(b) Pt | Sn^{2+} (1.0 M), Sn^{4+} (1.0 M) || Ag^+ (1.0 M) | Ag

(i) Net reaction

$$Sn^{2+} + 2Ag^+ \longrightarrow 2Ag + Sn^{4+}$$

$$E^\circ_{cell} = E^\circ_{Sn^{2+}/Sn^{4+}} + E^\circ_{Ag^+/Ag}$$

$$= -E^\circ_{Sn^{4+}/Sn^{2+}} + E^\circ_{Ag^+/Ag}$$

$$= -0.15 + 0.80$$

$$= 0.65\text{ V}$$

(ii) Reaction quotient $(Q) = \dfrac{[Sn^{4+}]}{[Ag^+]^2[Sn^{2+}]} = 1.0$

$$E_{cell} = E^\circ_{cell} - \frac{0.0591}{2}\log Q$$

$$= E^\circ_{cell} - \frac{0.0591}{2}\log(1)$$

$$= E^\circ_{cell}$$

$$= 0.65\text{ V}$$

$E_{cell} > 0$, thus spontaneous reaction, electrons flow A to B.

(c) $Zn|Zn^{2+}$ (0.10M) || Fe^{2+} (0.001 M) | Fe

(i) Net reaction

$$Zn + Fe^{2+} \longrightarrow Zn^{2+} + Fe$$

$$E^\circ_{cell} = E^\circ_{Zn/Zn^{2+}} + E^\circ_{Fe^{2+}/Fe}$$

$$E^\circ_{cell} = -E^\circ_{Zn^{2+}/Zn} + E^\circ_{Fe^{2+}/Fe}$$

$$= -0.76 - 0.44$$

$$= -1.20\text{ V}$$

(ii) Reaction quotient $(Q) = \dfrac{[Zn^{2+}]}{[Fe^{2+}]} = \dfrac{0.1}{0.001} = 100$

$$\therefore \quad E_{cell} = E^\circ_{cell} - \frac{0.0591}{2}\log 100$$

$$= -1.20 - 0.0591$$

$$= -1.26\text{ V}$$

$E_{cell} < 0$, thus, electrons flow from B to A.

3. Plan

(i) If Ag^+| Ag electrode is used

$$Ag^+ + e^- \longrightarrow Ag, \; E^\circ_{Ag^+/Ag} = 0.799\text{ V}$$

$$[Ag^+] = \frac{K_{sp}(AgCl)}{[Cl^-]} = \frac{1.82 \times 10^{-10}}{0.05}$$

$$= 3.64 \times 10^{-9}\text{M}$$

$$E_{Ag^+/Ag} = E^\circ_{Ag^+/Ag} - 0.0591\log\left(\frac{1}{3.64 \times 10^{-9}}\right)$$

$$= 0.799 - 0.500$$

$$= 0.299\text{ V}$$

(ii) If AgCl | Ag electrode is used

$$AgCl + e^- \longrightarrow Cl^- + Ag$$

$$E = E^\circ - 0.0591\log(Cl^-)$$

$$= 0.222 - 0.0591\log(0.05)$$

$$= 0.222 + 0.077$$

$$= 0.299\text{ V}$$

Thus, electrode potential is same in both cases.

4. **Plan** Reduction potential in a half-cell can be calculated using Nernst equation.

$$E = E^\circ - \frac{2.303\,RT}{nF}\log Q$$

At 298 K, $\frac{2.303\,RT}{F} = \frac{2.303 \times 8.314 \times 298}{96500} = 0.0591$ V

Reaction quotient $(Q) = \frac{[NO_2][H_2O]}{[NO_3^-][H^+]^2}$

$$Q = \frac{1}{[H^+]^2}$$

Since, $[NO_2] = [H_2O] = [NO_3^-] = 1$ M

(i) $[H^+] = 8$ M

$\therefore\ Q = \frac{1}{(8)^2} = \left(\frac{1}{64}\right)$

$\because\ E = 0.78 - \frac{0.0591}{1}\log\left(\frac{1}{64}\right)$

$= 0.78 + \frac{0.0591}{1}\log(64) = 0.78 + \frac{0.0591}{1}\log 2^6$

$= 0.78 + \frac{0.0591}{1} \times 6\log 2 = 0.89$ V

(ii) In neutral solution,

$[H^+] = 1 \times 10^{-7}$M

$\therefore\ Q = \frac{1}{[H^+]^2} = \frac{1}{(10^{-7})^2} = 10^{14}$

$\therefore\ E = E^\circ - \frac{0.0591}{1}\log Q$

$= 0.78 - \frac{0.0591}{1}\log 10^{14}$

$= 0.78 - \frac{0.0591 \times 14}{1}\log 10$

$= -0.05$ V

5. (c) Standard electrode potential is the potential difference under standard state, i.e.

When $[M^{n+}] = 1$M

It can be measured by coupling it with SHE electrode

$Pt(s)\,|\,H_2(g, 1\text{ bar})\,|\,H^+\ (1.0\text{ M})$

6. (b) For SHE $E^\circ_{SHE} = 0.00$ V

Oxidation at anode (left)

$\underset{p_1}{H_2(g)} \longrightarrow \underset{1\text{M}}{2H^+} + 2e^-$ $\quad E^\circ_{ox} = 0.00$ V

Reduction at cathode (right)

$\underset{1\text{M}}{2H^+} + 2e^- \longrightarrow \underset{p_2}{H_2(g)}$ $\quad E^\circ_{red} = 0.00$ V

Net $\underset{p_1}{H_2(g)} \longrightarrow \underset{p_2}{H_2(g)}$ $\quad E^\circ_{cell} = 0.00$ V

This is the type of the cell in which electrodes at different pressures are dipped in same electrolyte and connectivity is made by a salt-bridge.

Reaction quotient $(Q) = \frac{p_{H_2}\text{ (right)}}{p_{H_2}\text{ (left)}} = \frac{p_2}{p_1}$

$\because\ E_{cell} = E^\circ_{cell} - \frac{RT}{nF}\log_e Q$

$= 0 - \frac{RT}{2F}\log_e\frac{p_2}{p_1} = \frac{RT}{2F}\log_e\frac{p_1}{p_2}$

7. (c) This is a type of concentration cell using hydrogen electrode as anode and cathode.

Anode (left)

$\underset{x\text{ atm (L)}}{H_2(g)} \longrightarrow \underset{0.01\text{M(L)}}{2H^+(aq)} + \underset{\text{(SHE)}}{2e^-}$ $\quad E^\circ_{ox} = 0.00$ V

Cathode (right) $\underset{0.1\text{ M (R)}}{2H^+(aq)} + 2e^- \longrightarrow \underset{y\text{ atm (R)}}{H_2(g)}$; $E^\circ_{red} = 0.00$V

$\underset{x\text{ atm}}{H_2(g)} + \underset{0.1\text{ M}}{2H^+(aq)} \longrightarrow \underset{y\text{ atm}}{H_2(g)} + \underset{0.01\text{ M}}{2H^+(aq)},\ E^\circ_{cell} = 0.00$

Reaction quotient, $Q = \frac{(p_{H_2})_R\,([H^+]_L^2)}{(p_{H_2})_L\,([H^+]_R^2)}$

$Q = \left(\frac{y}{x}\right)\left(\frac{0.01}{0.10}\right)^2 = \frac{y}{100\,x}$

$E_{cell} = E^\circ_{cell} - \frac{2.303\,RT}{nF}\log Q$

$0 = 0 - \frac{0.0591}{2F}\log\left(\frac{y}{100\,x}\right)$

$= \frac{0.0591}{2F}\log\left(\frac{y}{100\,x}\right) = 0$

$\log\left(\frac{y}{100\,x}\right) = 0$

$\frac{y}{100x} = 1$

$\left(\frac{y}{x}\right) = 100$

$\therefore\ \left(\frac{x}{y}\right) = \frac{1}{100} = 0.01$

8. (c) When Fe^{2+} is 50% titrated

$Fe^{2+} \longrightarrow Fe^{3+} + e^-$

when $[Fe^{2+}] = [Fe^{3+}]$

Thus, reaction quotient$(Q) = \frac{[Fe^{3+}]}{[Fe^{2+}]} = 1$

$\because\ E_{Pt/Fe^{2+}, Fe^{3+}} = E^\circ_{Pt/Fe^{2+}, Fe^{3}} - \frac{0.0591}{1}\log Q$

$= -0.77 - \frac{0.0591}{1}\log 1$

$= -0.77$ V

When Fe^{2+} is 90% titrated

$[Fe^{2+}] = 10,\ [Fe^{3+}] = 90$

Thus, $Q = \frac{[Fe^{3+}]}{[Fe^{2+}]} = \frac{90}{10} = 9$

$\therefore\ E_{Pt/Fe^{2+}, Fe^{3+}} = E^\circ_{Pt/Fe^{2+}, Fe^{3+}} - \frac{0.0591}{1}\log Q$

$= -0.77 - 0.0591\log 9 = -0.826$ V

9. (b,a) K_{sp} (AgCl) $= 2.8 \times 10^{-10}$

$[Ag^+][Cl^-] = 2.8 \times 10^{-10}$

$\therefore\ [Ag^+]_{left} = \frac{2.8 \times 10^{-10}}{0.2} = 1.4 \times 10^{-9}$ M

K_{sp} (AgBr) $= 3.3 \times 10^{-12}$

$\therefore\ [Ag^+][Br^-] = 3.3 \times 10^{-13}$

$\therefore\ [Ag^+]_{right} = \frac{3.3 \times 10^{-13}}{0.001} = 3.3 \times 10^{-10}$ M

Anodic $Ag \longrightarrow [Ag^+]_L + e^-$

Cathodic $[Ag^+]_R + e^- \longrightarrow Ag$

Net $[Ag^+]_R \longrightarrow [Ag^+]_L,\ E^\circ_{cell} = 0$

$E_{cell} = E^\circ_{cell} - \frac{0.0591}{1}\log\frac{[Ag^+]_L}{[Ag^+]_R}$

$= 0 - \frac{0.0591}{1}\log\frac{(1.4 \times 10^{-9})}{(3.3 \times 10^{-10})} = -0.037$ V

Thus, cell reaction is non-spontaneous

(b) Cell is reversed of (a), thus spontaneous.

10. (b) $InHA \rightleftharpoons H^+ + A^-$

For weak acid by Ostwald's dilution law

$$[H^+] = \sqrt{K_a C}$$

$$(pH) = \frac{1}{2}[pK_a - \log C]$$

$$\because \quad (pH)_{HA} = \frac{1}{2}[(3) - \log 1] = 1.5$$

$$(pH)_{HB} = \frac{1}{2}[5 - \log 1] = 2.5$$

At anode $\frac{1}{2}H_2 \longrightarrow [H^+]_L + e^-$

At cathode $[H^+]_R + e^- \longrightarrow \frac{1}{2}H_2$

$$[H^+)_R \longrightarrow [H^+]_L$$

$$E^\circ_{cell} = 0.00\text{ V}$$

$$\text{Reaction quotient } (Q) = \frac{[H^+]_L\ (HA)}{[H^+]_R\ (HB)}$$

$$E_{cell} = E^\circ_{cell} - \frac{0.0591}{1}\log\frac{[H^+]_L}{[H^+]_R}$$

$$= 0 - 0.0591\,[\log [H^+]_L - \log [H^+]_R]$$

$$= 0.0591\,[-\log (H^+)_L - (-\log (H^+)_R]$$

$$= 0.0591\,[(pH)_{HA} - (pH)_{HB}]$$

$$= +\ 0.0591\,[1.5 - 2.5] = -\ 0.0591$$

11. (a) This is a type of concentration cell with gas electrodes at the same pressure (1 bar) but dipped in aqueous solution of different concentration. Hence, a potential difference is set up.

At anode $2\underset{L}{Cl^-}\ (C_1) \longrightarrow \underset{1\text{ bar}}{Cl_2} + 2e^-$ $\quad E^\circ_{ox} = x$V

At cathode $\underset{1\text{ bar}}{Cl_2} + 2e^- \longrightarrow 2\underset{R}{Cl^-}\ (C_2)$ $\quad E^\circ_{red} = -\ x$V

$$2\underset{L}{Cl^-}\ (C_1) \longrightarrow 2\underset{R}{Cl^-}\ (C_2)$$

$$E^\circ_{cell} = 0.00\text{ V}$$

$$n = 2$$

$$\text{Reaction quotient } (Q) = \frac{[Cl^-]_R^2}{[Cl^-]_L^2} = \frac{(C_2)^2}{(C_1)^2}$$

$$E_{cell} = E^\circ_{cell} - \frac{2.303\,RT}{nF}\log Q$$

$$E_{cell} = 0 - \frac{0.0591}{2}\log\left(\frac{C_2}{C_1}\right)^2 = -\frac{0.0591}{2} \times 2\log\left(\frac{C_2}{C_1}\right)$$

$$E_{cell} = 0.0591\log\left(\frac{C_1}{C_2}\right)$$

To make cell reaction spontaneous,

$$E_{cell} > 0, \text{ hence } C_1 > C_2$$

12. (d) $2H^+ + 2e^- \longrightarrow H_2$

This represents reduction half-cell reaction

$$Q\text{ (reaction quotient)} = \frac{p_{H_2}}{[H^+]^2} = \frac{1}{[H^+]^2}$$

$$E = E^\circ_{SHE} - \frac{2.303\,RT}{2F}\log Q$$

$$= 0.00 - \frac{0.0591}{2}\log\frac{1}{[H^+]^2}$$

$$= 0.0591\log [H^+]$$

(a) $[H^+] = 1$ M, $E = 0$

(b) pH = 4, $[H^+] = 10^{-4}$M

$$\because \quad E = 0.0591\log 10^{-4}$$

$$= -\ 4 \times 0.0591 = -\ 0.2364\text{ V}$$

(c) Pure water, $[H^+] = 10^{-7}$M

$$\because \quad E = 0.0591\log 10^{-7}$$

$$= -\ 7 \times 0{,}0591 = -\ 0.4137\text{ V}$$

(d) 1.0 M NaOH

$$[OH^-] = 1\text{ M}$$

$$\therefore \quad [H^+] = K_w/[OH^-] = \frac{1 \times 10^{-14}}{1}$$

$$[H^+] = 1 \times 10^{-14}\text{M}$$

$$\because \quad E = 0.0591\log 1 \times 10^{-14}$$

$$= -\ 14 \times 0.0591$$

$$= -\ 0.8274\text{ V (maximum)}$$

13. (d) For the cell, $\underset{\ominus}{Zn}|Zn^{2+}(1M)||H^+\ (1M)|\underset{\oplus}{Pt}\ \underset{1\text{bar}}{(H_2)}$

$$E^\circ_{cell} = E^\circ_{Zn/Zn^{2+}} + E^\circ_{SHE}$$

$$0.76 = E^\circ_{Zn/Zn^{2+}} + 0.00$$

$$\therefore \quad E^\circ_{Zn/Zn^{2+}} = 0.76\text{ V}$$

For the cell $E^\circ_{Zn^{2+}/Zn} = -\ 0.76$ V

$Pt(H_2)\,|\,H^+\ (1M)\,||\,Ag^+\ (1M)\,|\,Ag$

$$E^\circ_{cell} = E^\circ_{SHE} + E^\circ_{Ag^+/Ag}$$

$$0.80 = 0.00 + E^\circ_{Ag^+/Ag}$$

$$\therefore \quad E^\circ_{Ag^+/Ag} = 0.80\text{ V}$$

$$\therefore \quad E^\circ_{Ag/Ag^+} = -\ 0.80\text{ V}$$

For the cell

$$\underset{\ominus}{Ag}\,|\,Ag^+(0.1M)\,||\,Zn^{2+}\ (0.1M)\,|\,\underset{\oplus}{Zn}$$

Oxidation at anode

$$2Ag \longrightarrow \underset{0.1\text{M}}{2Ag^+} + 2e^-$$

$$E^\circ_{Ag/Ag^+} = -\ 0.80\text{V}$$

Reduction at cathode

$$\underset{0.1\text{M}}{Zn^{2+}} + 2e^- \longrightarrow Zn,$$

$$E^\circ_{Zn^{2+}/Zn} = -\ 0.76\text{ V}$$

$$\underset{0.1\text{M}}{Zn^{2+}} + 2Ag \longrightarrow \underset{0.1\text{M}}{2Ag^+} + Zn$$

$$E^\circ_{cell} = -\ 1.56\text{ V}$$

$$\text{Reaction quotient } (Q) = \frac{[Ag^+]^2}{[Zn^{2+}]} = \frac{(0.1)^2}{(0.1)} = \frac{1}{10}$$

$$\because \quad E_{cell} = E^\circ_{cell} - \frac{2.303\,RT}{nF}\log Q$$

$$\text{At 298 K,} \quad \frac{2.303\,RT}{F} = 0.0591\text{ V}$$

$$\therefore \quad E_{cell} = -\ 1.56 - \frac{0.0591}{2}\log\left(\frac{1}{10}\right)$$

$$= -\ 1.56 + \frac{0.0591}{2}\log 10$$

$$= -\ 1.56 + 0.0295 = -\ 1.53\text{ V}$$

$E_{cell} < 0$, hence reaction is non-spontaneous.

14. (d) pH = 3, $[H^+] = 10^{-3}$M

Also at 298 K, $\frac{2.303\,RT}{F} = 0.0591$ V

Electrons involved $n = 4$

$$\text{Reaction quotient, } Q = \frac{[Fe^{2+}]^2}{[H^+]^4\ p_{O_2}}$$

$$= \frac{(1 \times 10^{-3})^2}{(1 \times 10^{-3})^4\ (0.1)} = 10^7$$

By Nernst equation,

$$\therefore \quad E_{cell} = E^\circ_{cell} - \frac{0.0591}{n}\log Q$$

$$E^\circ = 1.67 - \frac{0.0591}{4}\log 10^7$$

$$= 1.67 - 0.10 = 1.57\text{ V}$$

15. (c) This is a type of concentration cell in which two electrodes are at different pressure but dipped in same electrolyte. Salt-bridge is used to make connectivity and liquid junction potential is minimised and

$$E^\circ_{cell} = 0.00\ V$$

At anode (left) $\quad 2Cl^- \underset{1\,M}{(aq)} \longrightarrow \underset{p_1}{Cl_2 (g)} + 2e^-$

At cathode (right)

$$\underset{p_2}{Cl_2 (g)} + 2e^- \longrightarrow 2Cl^-\ (1M)$$

$$\underset{p_2}{Cl_2(g)} \longrightarrow \underset{p_1}{Cl_2(g)}$$

Reaction quotient, $Q = \dfrac{p_{Cl_2}(L)}{p_{Cl_2}(R)} = \dfrac{p_1}{p_2}$

By Nernst equation

$$E_{cell} = E^\circ_{cell} - \frac{2.303\,RT}{nF}\log Q$$

$$= 0 - \frac{0.0591}{2}\log\left(\frac{p_1}{p_2}\right)$$

$$= \frac{0.0591}{2}\log\left(\frac{p_2}{p_1}\right)$$

To make cell reaction spontaneous

$$E_{cell} > 0$$

Thus, $\quad p_2 > p_1$

16. (c) $Cl^- / Hg_2Cl_2, Hg\ (l)$

This is reduction half-cell

$$Hg_2Cl_2(s) + 2e^- \longrightarrow 2Hg\ (l) + 2Cl^-$$

Reaction quotient $(Q) = [Cl^-]^2$

$$E_{calomel} = E^\circ_{calomel} - \frac{0.0591}{n}\log Q$$

$$= 0.28 - \frac{0.0591}{2}\log [Cl^-]^2$$

$$= 0.28 - 0.0591\log [Cl^-]$$

Thus, larger the value of $[Cl^-]$, smaller the value of $E_{calomel}$.

(a) $E_{calomel} = E^\circ_{calomel} - 0.0591\ [\log 1] = 0.28\ V$

(b) $E_{calomel} = 0.28 - \dfrac{0.0591}{1}\log 0.1$

$= 0.28 + 0.0591 = 0.3391\ V$

(c) $E_{calomel} = 0.28 - 0.0591\log 0.01$

$= 0.28 + 0.0591 \times 2$

$= 0.3982\ V$

(d) $E_{calomel} = 0.28 - 0.0591 = 0.2209\ V$

17. (b) **At anode** (calomel)

$$2Hg + \underset{0.1\,M}{2Cl^-} \longrightarrow Hg_2Cl_2 + 2e^-$$

$$E^\circ = -0.268\ V$$

At cathode (quinhydrone)

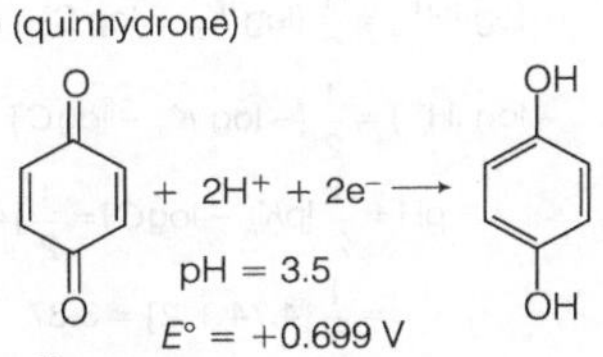

$pH = 3.5$

$E^\circ = +0.699\ V$

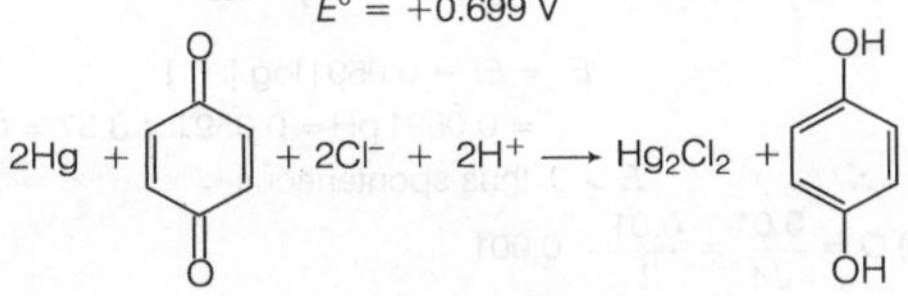

$$E^\circ_{cell} = 0.431\ V$$

$$pH = 3.5$$

$\because \quad [H^+] = 10^{-pH} = 10^{-3.5}$

$= 10^{-0.5}\ 10^{-4}$

$= 3.16 \times 10^{-4}$

Reaction quotient $(Q) = \dfrac{1}{[Cl^-]^2[H^+]^2} = \dfrac{1}{[0.1]^2\ (3.16 \times 10^{-4})^2}$

$Q = 1 \times 10^9$

Note *In quinhydrone electrode*

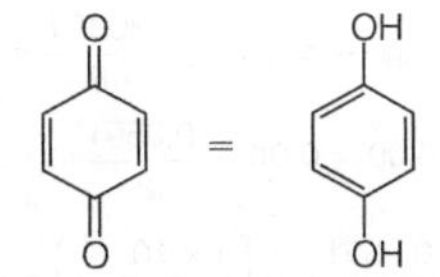

and [solid] or [element]= 1

$\because \quad E_{cell} = E^\circ_{cell} - \dfrac{0.0591}{2}\log Q$

$= 0.431 - \dfrac{0.0591}{2}\log 1 \times 10^9$

$= 0.431 - \dfrac{0.0591 \times 9}{2}\log 10$

$= 0.431 - 0.27\ V$

$= 0.16\ V$

18. (b) $\Delta G^\circ = -nFE^\circ_{cell}$

Larger the value of E°_{cell} (> 0), larger the value ΔG (< 0)

(a) $E^\circ_{cell} = E^\circ_{Zn/Zn^{2+}} + E^\circ_{Mg^{2+}/Mg}$

$= -E^\circ_{Zn^{2+}/Zn} + E^\circ_{Mg^{2+}/Mg}$

$= +0.76 - 2.37$

$= -1.61\ V$

(b) $E^\circ_{cell} = E^\circ_{Zn/Zn^{2+}} + E^\circ_{Ag^+/Ag}$

$= -E^\circ_{Zn^{2+}/Zn} + E^\circ_{Ag^+/Ag}$

$= -(-0.76) + 0.80$

$= 1.56\ V$

(c) $E^\circ_{cell} = E^\circ_{Cu/Cu^{2+}} + E^\circ_{Ag^+/Ag}$

$= -E^\circ_{Cu^{2+}/Cu} + E^\circ_{Ag^+/Ag}$

$= 0.34 + 0.80$

$= 1.56\ V$

(d) $E^\circ_{cell} = E^\circ_{Ag/Ag^+} + E^\circ_{Mg^{2+}/Mg}$

$= -E^\circ_{Ag^+/Ag} + E^\circ_{Mg^{2+}/Mg}$

$= -0.80 - 2.37$

$= -3.17\ V$

In (b) E°_{cell} is most positive, then ΔG° is most negative.

19. (a) **Oxidation** $\quad C^{3-} \longrightarrow C^- + 2e^- \qquad E^\circ = 1.25\ V$

Reduction $\quad 2B^- + 2e^- \longrightarrow 2B^{2-} \qquad E^\circ = 1.25\ V$

$2B^- + C^{3-} \longrightarrow C^- + 2B^{2-} \qquad E^\circ_{cell} = 2.50\ V$

$$\underset{\ominus}{Pt \mid C^{3-}, C^-} \parallel \underset{\oplus}{B^-, B^{2-} \mid Pt}$$

20. (b) $\quad Ag \longrightarrow Ag^+ + e^- \qquad E^\circ = -0.800\ V$

$V^{3+} + Ag^+ + H_2O \longrightarrow VO^{2+} + 2H^+ + Ag \qquad E^\circ = +0.439\ V$

$\therefore \quad V^{3+} + H_2O \longrightarrow VO^{2+} + 2H^+ + e^- \qquad E^\circ = -0.361\ V$

$V^{2+} + VO^{2+} + 2H^+ \longrightarrow 2V^3 + H_2O \qquad E^\circ = 0.616\ V$

Adding $\quad V^{2+} \longrightarrow V^{3+} + e^- \qquad E^\circ = +0.255\ V$

$V^{3+} + e^- \longrightarrow V^{2+} \qquad E^\circ = -0.255\ V$

21. (b) $Pt(H_2)|\ H_2O$

This is oxidation half-cell. Instead of

$$[H^+] = 1\text{ M and } p_{H_2} = 1\text{ bar}$$

H_2O has been taken as a source of $[H^+]$

$$H_2(g) \rightleftharpoons 2H^+(aq) + 2e^-$$

$$\text{Reaction quotient } (Q) = \frac{[H^+]^2}{p_{H2}} = \frac{(1\times 10^{-7})^2}{p_{H2}}$$

$$E_{SHE} = E°_{SHE} - \frac{2.303RT}{nF}\log Q$$

$$0.00 = 0.00 - \frac{0.0591}{2}\log\frac{1\times 10^{-14}}{p_{H2}}$$

$$\because \quad \frac{0.0591}{2}\log\left(\frac{1\times 10^{-14}}{p_{H2}}\right) = 0$$

$$\log\left(\frac{1\times 10^{-14}}{p_{H_2}}\right) = 0$$

$$\frac{1\times 10^{-14}}{p_{H_2}} = 1$$

$$\therefore \quad p_{H_2} = 1\times 10^{-14}\text{ bar}$$

22. (c) Half-cell reaction is $Hg + 2OH^- \longrightarrow HgO + H_2O + 2e^-$

$$\text{Reaction quotient } (Q) = \frac{1}{[OH^-]^2}$$

$$\therefore \quad E = E° - \frac{2.303\,RT}{2F}\log Q$$

$$= E° - \frac{2.303\,RT}{2F}\log\frac{1}{[OH^-]^2}$$

$$= E° - \frac{2.303\,RT}{2F}\log[OH^-]^2$$

$$= E° + \frac{2.303\,RT}{F}\text{pOH}$$

$$= E° + \frac{2.303\,RT}{F}[14 - \text{pH}]$$

Thus, electrode potential is dependent on

(i) pH, (ii) temperature.

23. (a,b,c)

(a) If salt-bridge is removed, anodic and cathodic solution intermix and potential difference falls to zero — correct.

(b)

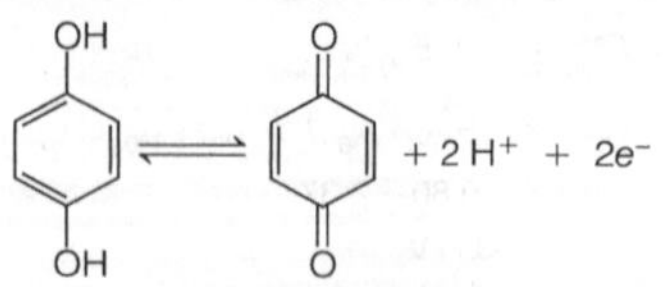

(A) (B)

Quinhydrone is a mixture of A and B in (1 : 1) molar ratio joined by H-bonding. Above equilibrium is set up in aqueous solution.

$$= [H^+]^2\frac{[B]}{[A]} = [H^+]^2$$

$$E_{QH} = E°_{QH} - \frac{0.0591}{3}\log K$$

$$E_{QH} = E°_{QH} - \frac{0.0591}{2}\log[H^+]^2$$

$$E_{QH} = E°_{QH} - 0.0591\log[H^+]$$

If $[H^+]$ changes, E_{QH} also changes. Thus, quinhydrone electrode is reversible to H^+ ion correct

(c) Liquid junction potential is minimised by use of concentration cell—correct.

(d) Calomel electrode is $Hg(l), Hg_2Cl_2(s)|\ Cl^-$.
Thus, (d) is incorrect.

24. (a,b,c) $Zn + 2H^+ \longrightarrow Zn^{2+} + H_2$

$$\text{Reaction quotient } (Q) = \frac{[Zn^{2+}]\,p_{H_2}}{[H^+]^2}$$

$$E = E°_{cell} - \frac{0.0591}{2}\log Q$$

(a) $Q = 1,\ E = E°_{cell}$

(b) $Q = \dfrac{(0.01)(1)}{(0.1)^2} = \dfrac{0.01}{0.01} = 1$

$\therefore\ E_{cell} = E°_{cell}$

(c) $Q = \dfrac{(1)(0.01)}{(0.1)^2} = 1$

$\therefore\ E_{cell} = E°_{cell}$

(d) $Q = \dfrac{(0.1)(1)}{(0.1)^2} = 10$

$\therefore\ E_{cell} = E°_{cell} - \dfrac{0.0591}{2}\log 10 = E°_{cell} - \dfrac{0.0591}{2}$

25. (c,d) Reaction taking place in a Daniell cell is

$$Zn(s) + Cu^{2+}(1\text{ M}) \rightleftharpoons Cu(s) + Zn^{2+}(1\text{ M}),\quad E°_{cell} = 1.10\text{ V}$$

$$E_{cell} = E°_{cell} - \frac{2.303\,RT}{2F}\log_{10}Q$$

At equilibrium, $E_{cell} = 0.00$ V and $Q = K_c$

$$\because \quad E°_{cell} = \frac{2.303\,RT}{2\,F}\log_{10}K_c$$

$$1.1 = \frac{2.303\,RT}{2\,F}\log_{10}K_c$$

Thus, (d) is correct.

$$2.2 = \frac{2.303\,RT}{F}\log_{10}K_c$$

$$2.2 = 0.0591\log_{10}K_c$$

$$\therefore \quad \log_{10}K_c = \frac{2.2}{0.0591}$$

Thus, (c) is correct.

26. (a,b,c,d) $\frac{1}{2}H_2 \longrightarrow H^+ + e^-$

(a) Reaction quotient $(Q) = \dfrac{[H^{-1}]}{\sqrt{p_{H2}}} = 1\times 10^{-5}$

$$E = E° - \frac{0.0591}{1}\log[H^+] = 0 + 0.0591\text{ pH}$$

$$= 0.0591\times 5 = 0.2955\text{ V}$$

$E > 0.2955$ V thus spontaneous

(b) $Q = \dfrac{0.01}{\sqrt{0.50}} = 0.0141$

$$E = 0 - 0.0591(\log 0.0141) = 0.11\text{ V}$$

$E > 0$, thus spontaneous

(c) $[H^+] = \sqrt{K_aC}$

$$\log[H^+] = \frac{1}{2}(\log K_a + \log C)$$

$$-\log[H^+] = \frac{1}{2}[-\log K_a - \log C]$$

$$\text{pH} = \frac{1}{2}[pK_a - \log C] = \frac{1}{2}[4.74 - \log 0.01]$$

$$= \frac{1}{2}[4.74 + 2] = 3.37$$

$$E = E° - 0.0591\log[H^+]$$

$$= 0.0591\text{ pH} = 0.0591\times 3.37 = 0.2\text{ V}$$

$\therefore$ $E > 0$, thus spontaneous

(d) $Q = \dfrac{0.01}{\sqrt{4}} = \dfrac{0.01}{2} = 0.001$

$$E = 0 - \frac{0.0591}{1}\log(0.005) = 0.14\text{ V}$$

$E > 0$, thus spontaneous.

Explanations for Q. Nos. (27-28)

Cell reaction

$$\frac{1}{2}H_2 \longrightarrow H^+ + e^-, \quad E^\circ_{SHE} = 0.00\text{ V}$$

$$Ag^+ + e^- \longrightarrow Ag, \quad E^\circ_{Ag^+/Ag} = 0.80\text{ V}$$

$$Ag^+ + \frac{1}{2}H_2 \longrightarrow Ag + H^+, \quad E^\circ_{cell} = 0.80\text{V}$$

$$\text{Reaction quotient (Q)} = \frac{[H^+]}{[Ag^+]} = \frac{1\text{ M}}{x}$$

$$E_{cell} = E^\circ_{cell} - \frac{0.0591}{1}\log\left(\frac{1}{x}\right)$$

$$0.500 = 0.800 + 0.0591 \log x$$

$$= \frac{-0.300}{0.0591} = \log x$$

$$\therefore \quad x = 8.4 \times 10^{-6}\text{ M}$$

$$[Ag^+] = 8.4 \times 10^{-6}\text{ mol L}^{-1}$$

$$= 8.4 \times 10^{-6} \times 0.350 \text{ mol in 350 mL}$$

$$= 8.4 \times 10^{-6} \times 0.350 \times 108 \text{ g in 350 mL}$$

$$= 3.1752 \times 10^{-4} \text{ g in 1.05 g sample}$$

$$\therefore \quad \text{Percentage} = \frac{3.175 \times 10^{-4}}{1.05} \times 100 = 0.03\%$$

27. (a) **28.** (a)

Explanations for Q. Nos. (29-30)

Anodic $\frac{1}{2}\underset{1\text{ bar}}{H_2} \longrightarrow \underset{\text{left}}{H^+(aq)} + e^-, \quad E^\circ_{SHE} = 0.00\text{ V}$

Cathodic $\underset{\text{right}}{H^+(aq)} + e^- \longrightarrow \frac{1}{2}\underset{1\text{bar}}{H_2}, \quad E^\circ_{SHE} = 0.00\text{ V}$

$$\underset{\text{right}}{H^+(aq)} \longrightarrow \underset{\text{left}}{H^+(aq)}, \quad E^\circ_{cell} = 0.00\text{ V}$$

$$\text{Reaction quotient }(Q) = \frac{[H^+]_L}{[H^+]_R} = \frac{[H^+]_L}{(0.01)} = 100\,[H^+]_L$$

$$E_{cell} = E^\circ_{cell} - \frac{0.0591}{1}\log[Q]$$

$$0.19 = 0 - \frac{0.0591}{1}\log 100\,[H^+]_L$$

$$\frac{0.19}{0.0591} = -\log[100] - \log[H^+]_L$$

$$\frac{0.19}{0.0591} = -2 + \text{pH}$$

$$\text{pH} = 2 + \frac{0.190}{0.0591}$$

$$\text{pH} = 5.21$$

$$HA \rightleftharpoons H^+ + A^-$$

$$\because \quad [H^+] = \sqrt{K_a C}$$

$$\log[H^+] = \frac{1}{2}\log(K_a C) = \frac{1}{2}[\log K_a + \log C]$$

$$-\log[H^+] = \frac{1}{2}[-\log K_a - \log C]$$

$$2\text{pH} = \text{p}K_a - \log C$$

$$\therefore \quad \text{p}K_a = 2\text{pH} + \log C = 2 \times 5.21 + \log(0.02)$$

$$= 10.42 - 1.70$$

$$\text{p}K_a = 8.72$$

29. (a) **30.** (d)

31. (2) **Anode** $X \longrightarrow \underset{0.05\text{M}}{X^{n+}} + ne^-, \quad E^\circ = \text{V}$

Cathode $\underset{0.05\text{M}}{X^{n+}} + ne^- \longrightarrow X, \quad E^\circ = -\text{V}$

$$X^{n+}(0.05\text{ M}) \longrightarrow X^{n+}(0.50\text{ M})$$

$$E^\circ_{cell} = 0.00\text{ V}$$

$$\text{Reaction quotient }(Q) = \frac{0.50}{0.05} = 10$$

At 290 K

$$\frac{2.303\,RT}{F} = 0.0575$$

$$E_{cell} = E^\circ_{cell} - \frac{0.0575}{n}\log Q$$

$$-0.028 = 0 - \frac{0.0575}{n}\log 10$$

$$= -\frac{0.0575}{n}$$

$$\therefore \quad n = \frac{0.0575}{0.028} \approx 2$$

32. (3) This question relates colligative properties to emf of the cell

$$HA \rightleftharpoons H^+ + A^-$$

van't Hoff factor $(i) = (1 + x)$

$$\therefore \quad \text{Osomotic pressure } (\pi) = MRTi$$

$$2.5 \times 10^{-2} = 0.001 \times 0.0821 \times 300\,(1 + x)$$

$$\therefore \quad (1 + x) = 1.01$$

$$x = 0.01$$

$$\therefore \quad [H^+]_{HA} = Cx = 0.001 \times 0.01 = 1 \times 10^{-5}\text{ M}$$

$$\frac{1}{2}H_2 \longrightarrow H^+ + e^-$$

$$\text{Reaction quotient (Q)} = \frac{[H^+]}{p_{H_2}} = \frac{1 \times 10^{-5}}{1}\text{ M}$$

$$E = E^\circ - \frac{0.0591}{1}\log 1 \times 10^{-5}$$

$$= 0 + 0.0591 \times 5\text{ V}$$

$$= 0.30\text{ V}$$

$$= 0.3 \times 10 \text{ deci volt}$$

$$= 3.0 \text{ decivolt}$$

33. (1) $Pt(H_2)\,|\,\text{pH} = x\,||\,Ni^{2+}(1\text{ M})\,|\,Ni$

Anode $H_2(g) \longrightarrow 2H^+(aq) + 2e^-, \quad E^\circ_{SHE} = 0.00\text{ V}$

Cathode $Ni^{2+}(aq)\,2e^- \longrightarrow Ni(s), \quad E^\circ_{Ni^{2+}/Ni} = E^\circ$

$$H_2(g) + Ni^{2+}(aq) \longrightarrow Ni(s) + 2H^+(aq), \quad E^\circ_{cell} = E^\circ$$

$$\text{Reaction quotient }(Q) = \frac{[H^+]^2}{[Ni^{2+}]} = \frac{[H^+]^2}{1}$$

$$E_{cell} = E^\circ_{cell} - \frac{2.303\,RT}{nF}\log Q$$

$$(E_{cell} - E^\circ_{cell}) = -\frac{2.303\,RT}{nF}\log Q$$

$$0.0591 \text{ (given)} = -\frac{0.0591}{2}\log[H^+]^2$$

$$= -\frac{0.0591}{2} \times 2\log[H^+]$$

$$0.0591 = 0.0591\,\text{pH}$$

$$\therefore \quad \text{pH} = 1\;[-\log(H^+) = \text{pH}]$$

DPP-4 Thermodynamics of Electrochemistry

1. Plan $2Cr(s) \longrightarrow 2Cr^{3+}(aq) + 6e^-$,

$E^\circ_{Cr/Cr^{3+}} = -E^\circ_{Cr^{3+}/Cr} = 0.74\,V$

$3Cd^{2+}(aq) + 6e^- \longrightarrow 3Cd(s)$, $E^\circ_{Cd^{2+}/Cd} = -0.40\,V$

$\therefore\ 2Cr(s) + 3Cd^{2+}(aq) \longrightarrow 2Cr^{3+}(aq) + 3Cd(s)$

and $E^\circ_{cell} = +0.74 - 0.40 = 0.34\,V$

Note *If* $E^\circ_{X/X^+} = a$

Then $E^\circ_{X^+/X} = -a$

Take value according to reaction.

Net EMF is the sum of EMF of stepwise reaction. It will not be confusing and always rewarding.

$$\Delta G^\circ = -nFE^\circ_{cell} \qquad (n = 6)$$
$$= -6 \times 96500 \times 0.34$$
$$= -196860\,J$$
$$= -196.860\,kJ$$
$$\Delta G^\circ = -2.303RT\log K$$

or
$$E^\circ_{cell} = \frac{2.303RT\log K_{eq}}{nF}$$

$$\therefore \log K_{eq} = \frac{E^\circ_{cell} \times nF}{2.303RT}$$
$$= \frac{0.34 \times 6 \times 96500}{2.303 \times 8.314 \times 298}$$
$$\log K_{eq} = 34.50$$
$$\therefore K_{eq} = 3.17 \times 10^{34}$$

2. (a) By Gibbs-Helmholtz equation

$$\Delta G = \Delta H + T \cdot \left[\frac{d(\Delta G)}{dT}\right]_p$$

Also,
$$\Delta G = \Delta H - T\Delta S$$
$$\therefore \Delta S = -\left[\frac{d(\Delta G)}{dT}\right]_p$$

But
$$\Delta G = -nFE_{cell}$$
$$\Delta S = -\left[\frac{d(-nFE_{cell})}{dT}\right]_p$$
$$= nF\left[\frac{dE_{cell}}{dT}\right]_p$$
$$= \left[\frac{dE_{cell}}{dT}\right]_p = \frac{\Delta S}{nF}$$

$\left[\frac{dE_{cell}}{dT}\right]_p$ is called temperature coefficient of EMF of a cell. Thus,

temperature coefficient of EMF $= \frac{\Delta S}{nF}$.

3. (a) $H_2O + e^- \rightleftharpoons \frac{1}{2}H_2 + OH^-$, $E^\circ_{red} = -0.8277\,V$

$\frac{1}{2}H_2 \rightleftharpoons H^+ + e^-$, $E^\circ_{ox} = 0.00\,V$

$H^+ + H_2O \rightleftharpoons H_3O^+$

$H_2O \rightleftharpoons H_3O^+ + OH^-$ $E^\circ_{cell} = -0.8277\,V$

$$\therefore K_{eq} = K_w = [H_3O^+][OH^-]$$

Equilibrium constant is ionic-product of water (K_w)

$$\therefore E_{cell} = E^\circ_{cell} - \frac{0.0591}{1}\log K_w$$
$$0 = E^\circ_{cell} - \frac{0.0591}{1}\log K_w$$
$$\therefore \log K_w = \frac{E^\circ_{cell}}{0.0591} = \frac{-0.8277}{0.0591} = -14$$
$$\therefore K_w = 1.0 \times 10^{-14}$$

4. (c) $2 \times [Fe^{3+} + e^- \longrightarrow Fe^{2+}$, $E^\circ_{red} = 0.77\,V$

$2I^- \longrightarrow I_2 + 2e^-$, $E_{ox} = -0.54\,V$

$I_2 + I^- \rightleftharpoons I_3^-$

$2Fe^{3+} + 3I^- \rightleftharpoons 2Fe^{2+} + I_3^-$, $E^\circ_{cell} = 0.23\,V$

$$\therefore E_{cell} = E^\circ_{cell} - \frac{0.0591}{2}\log K_{eq}$$
$$E_{cell} = 0.00\,V \text{ at equilibrium}$$
$$K_{eq} = \frac{[Fe^{2+}]^2[I_3^-]}{[Fe^{3+}]^2[I^-]^3}$$
$$\therefore E^\circ_{cell} = \frac{0.0591}{2}\log K_{eq}$$
$$\frac{0.23 \times 2}{0.0591} = \log K_{eq}$$
$$\therefore \log K_{eq} = 7.78$$

5. (d) $Cu^{2+} + e^- \longrightarrow Cu^+$, $E^\circ_{red} = 0.15\,V$

$In^{2+} \longrightarrow In^{3+} + e^-$, $E^\circ_{ox} = ? = x$

$\therefore\ Cu^{2+} + In^{2+} \rightleftharpoons In^{3+} + Cu^+$ $E^\circ_{cell} = 0.15 + x$

But I. $In^+ \longrightarrow In^{2+} + e^-$, $E^\circ_{ox} = 0.40\,V, n = 1$

II. $In^+ \longrightarrow In^{3+} + 2e^-$, $E^\circ_{ox} = 0.42\,V, n = 2$

(II – I) gives $In^{2+} \longrightarrow In^{3+} + e^-$ $E^\circ_{ox} = x, n = 1$

Since, different number of electron are involved, hence

$$\therefore \Delta G^\circ_3 = \Delta G^\circ_2 - \Delta G^\circ_1$$
$$-n_3FE^\circ_3 = -n_2FE^\circ_2 + n_1FE^\circ_1$$
$$\therefore E^\circ_3 = x = \frac{n_2E^\circ_2 - n_1E^\circ_1}{n_3}$$
$$= \frac{0.84 - 0.40}{1}$$
$$= 0.44\,V$$
$$\therefore E^\circ_{cell} = 0.15 + 0.44 = 0.59\,V$$
$$\therefore E^\circ_{cell} = \frac{0.0591}{1}\log K_{eq}$$
$$\therefore 0.59 = 0.0591\log K_{eq}$$
$$\therefore \log K_{eq} = 10$$

6. (a) When equilibrium is set up

$$E_{cell} = 0$$

$2Hg(l) \longrightarrow Hg_2^{2+}(aq) + 2e^-$, $E^\circ_{Hg/Hg_2^{2+}} = x\,V$

$2Fe^{3+}(aq) + 2e^- \longrightarrow 2Fe^{2+}(aq)$, $E^\circ_{Fe^{3+}/Fe^{2+}} = 0.77\,V$

$2Hg(l) + 2Fe^{3+}(aq) \longrightarrow Hg_2^{2+}(s) + 2Fe^{2+}$,

$E^\circ_{cell} = (x + 0.77)\,V$

	$2Fe^{3+}$	Hg_2^{2+}	$2Fe^{2+}$
Initial	$(1 \times 10^{-3})\,M$	0	0
At equilibrium	$(1 \times 10^{-3} \times 0.05)$	$\left(\frac{1 \times 10^{-3} \times 0.95}{2}\right)M$,	$(1 \times 10^{-3} \times 0.95)\,M$

$$K_{eq} = \frac{[Hg^{2+}][Fe^{2+}]^2}{[Fe^{3+}]^2}$$
$$K_{eq} = \frac{\left(\frac{1 \times 10^{-3} \times 0.95}{2}\right) \times (1 \times 10^{-3} \times 0.95)^2}{(1 \times 10^{-3} \times 0.05)^2} = 0.17$$
$$\therefore E^\circ_{cell} = E^\circ_{cell} - \frac{0.0591}{2}\log K_{eq}$$
$$\therefore E^\circ_{cell} = \frac{0.0591}{2}\log 0.17$$
$$= -0.023\,V$$
$$\therefore (x + 0.77) = -0.023\,V$$
$$x = -0.793\,V$$

7. (b) **Oxidation** $Zn \longrightarrow \underset{0.1M}{Zn^{2+}(aq)} + 2e^-$

Reduction $\underset{0.01\,M}{Fe^{2+}} + 2e^- \longrightarrow Fe$

Reaction $\underset{0.01\,M}{Fe^{2+}} + Zn \longrightarrow Fe + \underset{0.1\,M}{Zn^{2+}}$

Reaction quotient $(Q) = \dfrac{[Zn^{2+}]}{[Fe^{2+}]} = \dfrac{0.10}{0.01} = 10$

$$\therefore \quad E = E^\circ - \frac{0.0591}{2}\log Q$$

$$E^\circ = E + \frac{0.0591}{2}\log Q = E + 0.0295\log Q$$

$$= 0.2905 + 0.03\log 10 = 0.32\text{ V}$$

$$E^\circ = \frac{2.303RT}{n}\log K_{eq} = 0.0295\log K_{eq}$$

$$\therefore \quad \log K_{eq} = \frac{E^\circ}{0.0295} = \frac{0.32}{0.0295}$$

$$\therefore \quad K_{eq} = 10^{0.32/0.0295}$$

8. (b) In rusting of iron, Fe^{2+} is formed. Thus

$$Fe(s) \longrightarrow Fe^{2+} + 2e^-, \qquad E^\circ = +0.44\text{V}$$

$$2H^+ + 2e^- + \frac{1}{2}O_2 \longrightarrow H_2O(l), \qquad E^\circ = +1.23\text{V}$$

$$Fe + 2H^+ + \frac{1}{2}O_2 \longrightarrow Fe^{2+} + H_2O, \qquad E^\circ_{cell} = 1.67\text{ V}$$

$$\therefore \quad \Delta G^\circ = -nFE^\circ_{cell} \qquad (n = 2)$$

$$= -2 \times 96500 \times 1.67\text{ J} \qquad (1\text{FV} = 1\text{J})$$

$$= -322310\text{ J}$$

$$= -322.3\text{ kJ}$$

9. (c)

$$6O^{2-} \longrightarrow 3O_2 + 12e^-$$

$$4Al^{3+} + 12e^- \longrightarrow 4Al$$

$$\Delta G = 2898\text{kJ} = 2898 \times 10^3\text{ J}$$

n = electrons exchanged = 12 in the given reaction

$$\therefore \quad \Delta G = -nFE_{cell}$$

$$2898 \times 10^3\text{ J} = -12 \times 96500 \times E_{cell}$$

$$\therefore \quad E_{cell} = -\frac{2898 \times 10^3}{12 \times 96500}\text{V} = -2.5\text{ V}$$

10. (c)

E°/V	Reaction	Electrons involved	Free energy change
$E^\circ_1 = E^\circ_{Fe^{2+}/Fe} = -0.439$ V	$Fe^{2+} + 2e^- \longrightarrow Fe$	$n_1 = 2$	$\Delta G_1 = -n_1FE_1$
$E^\circ_2 = E^\circ_{Fe^{3+}/Fe} = -0.036$ V	$Fe^{3+} + 3e^- \longrightarrow Fe$	$n_2 = 3$	$\Delta G_2 = -n_2FE_2$
$E^\circ_3 = E^\circ_{Fe^{3+}/Fe^{2+}} = ?$	$Fe^{3+} + e^- \longrightarrow Fe^{2+}$	$n_3 = 1$	$\Delta G_3 = -n_3FE_3$

Since, different number of electrons are involved hence

$$E^\circ_3 \neq (E^\circ_2 - E^\circ_1)$$

But

$$\Delta G^\circ_3 = \Delta G^\circ_2 - \Delta G^\circ_1$$

$$-n_3FE^\circ_3 = -n_2FE^\circ_2 - (-n_1FE^\circ_1)$$

$$E^\circ_3 = \frac{n_2E^\circ_2 - n_1E^\circ_1}{n_3}$$

$$= \frac{3 \times (-0.036) - 2 \times (-0.439)}{1}$$

$$= 0.77\text{ V}$$

11. (a) $\Delta G^\circ = -nFE^\circ_{cell} = \Delta H^\circ - T\Delta S^\circ$

$$\therefore \quad nFE^\circ_{cell} = -\Delta H^\circ + T\Delta S^\circ$$

$$\therefore \quad T\Delta S^\circ = nFE^\circ_{cell} + \Delta H^\circ$$

$$\therefore \quad \Delta S^\circ = \frac{nFE^\circ_{cell} + \Delta H^\circ}{T}$$

$n = 4$ (four electrons are involved) as

$$2H_2 \longrightarrow 4H^+ + 4e^-$$

$$O_2 + 4e^- \longrightarrow 2O^{2-}$$

$$4H^+ + 2O^{2-} \longrightarrow 2H_2O$$

$$F = 96500\text{ C mol}^{-1}$$

$$\Delta H^\circ = 2 \times \Delta H^\circ_f(H_2O) \text{ for two moles of } H_2O$$

$$= 2 \times (-285.8)\text{ kJ}$$

$$= 2 \times (-285.8 \times 10^3)\text{J}$$

$$\therefore \quad \Delta S^\circ = \frac{(4 \times 96500 \times 1.23)\text{J} + 2(-285.8 \times 10^3)\text{J}}{298\text{K}}$$

$$= -324.9\text{JK}^{-1}$$

$$= -0.325\text{ kJ K}^{-1}$$

12. (a) Electron involved are different, hence

$$E^\circ_{net} \neq E^\circ_1 + E^\circ_2$$

But $\quad \Delta G^\circ_{net} = \Delta G^\circ_1 \pm \Delta G^\circ_2$

Signs are taken as per required reaction

	Reaction	Electrons	E°
I.	$MnO_4^- \xrightarrow{(+5e^-)} Mn^{2+}$	$n_1 = 5$	$E^\circ_1 = 1.51$V
II.	$MnO_2 \xrightarrow{(+2e^-)} Mn^{2+}$	$n_2 = 2$	$E^\circ_2 = 1.23$V
III.	$MnO_4^- \xrightarrow{(+3e^-)} MnO_2$	$n_3 = 3$	$E^\circ_3 = ?$

III is obtained as (I) – (II)

$$\Delta G^\circ_3 = \Delta G^\circ_1 - \Delta G^0_2$$

$$-n_3FE^\circ_3 = -n_1FE^\circ_1 - (-n_2FE^\circ_2)$$

$$\therefore \quad E^\circ_3 = \frac{n_1E^\circ_1 - n_2E^\circ_2}{n_3}$$

$$= \frac{5 \times 1.51 - 2 \times 1.23}{3} = 1.70\text{V}$$

13. (c) When a chemical reaction is multiplied/divided EMF of the changed equation remains constant but equilibrium constant is raised to power of that change.

Thus, $\quad x = y$

$$A + B \rightleftharpoons C + D,$$

$$K_1 = \frac{[C][D]}{[A][B]}$$

$$2A + 2B \rightleftharpoons 2C + 2D,$$

$$K_2 = \frac{[C]^2[D]^2}{[A]^2[B]^2} = K_1^2$$

14. (d) For a spontaneous reaction,

$$\Delta G < 0$$

Also, $\quad \Delta G = -nFE_{cell}$

$$\therefore \quad -\text{ve} = -nFE_{cell}$$

$$\therefore \quad E_{cell} > 0$$

Also, $\quad E_{cell} = E_{cell} - \dfrac{2.303RT}{nF}\log Q$

If equilibrium is reached, $E_{cell} = 0$

and $\quad Q = K_{eq}$

$$E^\circ_{cell} = \frac{2.303RT}{nF}\log K_{eq}$$

$$\therefore \quad E_{cell} = \frac{2.303RT}{nF}\log K_{eq} - \frac{2.303T}{nF}\log Q$$

$$= \frac{2.303RT}{nF}\left[\log\frac{K_{eq}}{Q}\right]$$

To make $\quad E_{cell} > 0$

$$K_{eq} > Q$$

15. (d) ΔG°_f (element, as Zn, Ag) = 0

Thus, ΔG° (button cell)

$$= \Delta G^\circ_f(ZnO) - \Delta G^\circ_f(Ag_2O)$$
$$= -318.30 - (-11.21)$$
$$= -307.09 \text{ kJ mol}^{-1}$$
$$\Delta G^\circ = -nFE^\circ_{cell} \qquad (n = 2)$$
$$\therefore \quad E^\circ_{cell} = -\frac{\Delta G^\circ}{nF}$$
$$= -\frac{(-307.09 \times 1000 \text{ J})}{2 \times 96500} = 1.591 \text{ V}$$

16. (c) For the given reaction,

$$E^\circ_{cell} = 2.73 \text{V}$$
$$n(\text{electrons exchanged}) = 12$$
$$(4\text{Al} \longrightarrow \text{Al}^{3+} + 12e^-)$$
$$F = 96500 \text{ C mol}^{-1}$$
$$\Delta G^\circ_f(\text{element}) = 0$$
$$\therefore \quad \Delta G^\circ(\text{reaction}) = -nFE^\circ_{cell}$$
$$= -12 \times 96500 \times 2.73$$
$$= 3161340 \text{J} = 3161.34 \text{ kJ}$$

Thus,

$$-3161.34 = \Delta G_f[Al(OH)_4^-] - [6\Delta G_f(H_2O) + 4\Delta G_f(OH^-)]$$
$$= \Delta G_f[Al(OH)_4^-] - [6 \times (-237.2) + 4(-57.0)]$$
$$\Delta G_f[Al(OH)_4^-] = -3161.34 + 6(-237.2) + 4(-57.0)$$
$$= -4812.54 \text{ kJ mol}^{-1}$$
$$= -4.812 \times 10^3 \text{ kJ mol}^{-1}$$

17. (d) In all cases, we use

$$\Delta G^\circ_3 = \Delta G^\circ_1 + \Delta G^\circ_2$$

with proper sign and not

$$E^\circ_3 = E^\circ_1 + E^\circ_2$$

(i)

	n (Electron)	E°
$Fe^{3+} + e^- \longrightarrow Fe^{2+}$	$n_1 = 1$	$E^\circ_1 = 0.77$ V
$Fe^{2+} + 2e^- \longrightarrow Fe$	$n_2 = 2$	$E^\circ_2 = -0.44$ V
$Fe^{3+} + 3e^- \longrightarrow Fe$	$n_3 = 3$	$E^\circ_3 = ?$

$$\therefore \quad \Delta G^\circ_3 = \Delta G^\circ_1 + \Delta G^\circ_2$$
$$-n_3FE^\circ_3 = -n_1FE^\circ_1 - n_2FE^\circ_2$$
$$\therefore \quad E^\circ_3 = \frac{n_1E^\circ_1 + n_2E^\circ_2}{n_3}$$
$$E^\circ_3 = \frac{0.77 \times 1 + 2(-0.44)}{3}$$
$$= \frac{0.77 - 0.88}{3}$$
$$= -\frac{0.11}{3} = -0.04 \text{V}$$

Thus, **(i) → (r)**

(ii)

	n	E°
$O_2 + 4H^+ + 4e^- \longrightarrow 2H_2O,$	$n_1 = 4,$	$E_1 = 1.23$ V
$O_2 + 2H_2O + 4e^- \longrightarrow 4OH,$	$n_2 = 4,$	$E^\circ_2 = +0.40$V
$4H_2O \rightleftharpoons 4H^+ + 4OH^-,$		$E^\circ_3 = E_2 - E_1$

(Equal number of electrons are involved)

$$= (0.40 - 1.23)\text{V}$$
$$= -0.83\text{V}$$

Thus, **(ii) → (s)**

(iii)

	n	E°
$Cu^{2+} + 2e^- \longrightarrow Cu,$	$n_1 = 2$	$E^\circ_1 = 0.34$V
$2 \times (Cu \longrightarrow Cu^+ + e^-,$	$n_2 = 2$	$E^\circ_2 = -0.52$ V
$\therefore Cu^{2+} + Cu \longrightarrow 2Cu^+,$		$E^\circ_3 = E^\circ_1 + E^\circ_2 = -0.18$V

Equal number of electrons are involved.

Thus, **(iii) → (p)**

(iv)

	n	E°
$Cr^{3+} + 3e^- \longrightarrow Cr,$	$n_1 = 3,$	$E^\circ_1 = -0.74$ V
$Cr^{2+} + 2e^- \longrightarrow Cr,$	$n_2 = 2,$	$E^\circ_2 = +0.91$ V
$Cr^{3+} + e^- \longrightarrow Cr^{2+}$	$n_3 = 1,$	$E^\circ_3 = ?$

$$\Delta G^\circ_3 = \Delta G^\circ_1 - \Delta G^\circ_2$$
$$-n_3FE^\circ_3 = -n_1FE^\circ_1 - (-n_2FE^\circ_2)$$
$$E^\circ_3 = \frac{n_1E^\circ_1 - n_2E^\circ_2}{n_3}$$
$$= \frac{3 \times (-0.74) + 2(0.91)}{1}$$
$$= \frac{-2.22 + 1.82}{1}$$
$$= -0.40 \text{ V}$$

Thus, **(iv) → (q)**

18. (c) (i) Temperature coefficient of EMF

$$= \left(\frac{dE_{cell}}{dT}\right) = \frac{\text{Change in EMF}}{\text{Change in temperature}}$$
$$= \frac{E^\circ_{T_2} - E^\circ_{T_1}}{(T_2 - T_1)}$$
$$= \frac{(0.21 - 0.23)\text{Volt}}{(308 - 288)\text{K}}$$
$$= \frac{-0.02 \text{ Volt}}{20\text{K}}$$
$$= -1 \times 10^{-3} \text{ VK}^{-1}$$
$$= -1.0 \text{ mV K}^{-1}$$

Thus, **(i) → (s)**

(ii) $\Delta G^\circ = \Delta H^\circ - T\Delta S^\circ$

By Gibb's Helmholtz equation,

$$\Delta G^\circ = \Delta H^\circ + T \cdot \left[\frac{d\,\Delta G^\circ}{dT}\right]_p$$
$$\therefore \quad \Delta S^\circ = -\left(\frac{d(\Delta G^\circ)}{dT}\right)_p$$
$$= -d\left(\frac{-n(dE^\circ_{cell})}{dT}\right)_p$$
$$= +\frac{nF(dE^\circ_{cell})}{dT}$$
$$= nF\left(\frac{dE^\circ_{cell}}{dT}\right)$$
$$= -1 \times 96500 \times 1.0 \times 10^{-3}$$
$$= -96.5 \text{ JK}^{-1}$$

Thus, **(ii) → (q)**

(iii)

$$\Delta G^\circ = \Delta H^\circ - T\Delta S^\circ$$
$$-nFE^\circ_{cell} = \Delta H^\circ - T\Delta S^\circ$$

At 288 K, $E^\circ_{cell} = 0.23$ V

$$-1 \times 96500 \times (0.23) = \Delta H^\circ - 288(-96.5)$$
$$\therefore \quad \Delta H^\circ = [-96500 \times 0.23 - 288 \times 96.5]\text{J}$$
$$= -50000 \text{ J} = -50 \text{ kJ}$$

At 308 K, $E^\circ_{cell} = 0.21$ V

$$-1 \times 96500 \times 0.21 = \Delta H^\circ - 308(-96.5)$$
$$\Delta H^\circ = -50000 \text{ J} = -50\text{kJ}$$

Thus, **(iii) → (r)**

At 298 K,

$$-1 \times 96500\, E^\circ_{cell} = -50{,}000 - 298 \times (-96.5)$$
$$E^\circ_{cell} = \frac{500000 - 298 \times 96.5}{96500} = 0.22 \text{ V}$$

$$\begin{array}{lll} AgCl(s)+e^- \longrightarrow Ag + Cl^-, & E^\circ = 0.22\text{ V} \\ Ag \longrightarrow Ag^+ + e^-, & E^\circ = -0.80\text{V} \\ \hline AgCl(s) \longrightarrow Ag^+ + Cl^- & E^\circ = -0.58\text{V} \end{array}$$

$$K = K_{sp} = [Ag^+][Cl^-]$$

$\therefore$ $$E^\circ = \frac{0.0591}{1}\log K_{sp}$$

$\therefore$ $$\log K_{sp} = -\frac{0.58}{0.0591} = -9.81$$

Thus **(iv)** $\rightarrow$ **(p)**

Explanations for Q. Nos. (19-20)

At anode $M(s) \longrightarrow \underset{0.5\text{M}}{M^+(aq)} + e^-, \quad E^\circ_{ox} = x\text{V}$

At cathode $\underset{1.0\text{M}}{M^+(aq)} + e^- \longrightarrow M(s), \quad E^\circ_{red} = -x\text{V}$

$$\underset{1.0\text{ M}}{M^+(aq)} \longrightarrow \underset{0.05\text{ M}}{M^+(aq)}, \quad E^\circ_{cell} = 0.00\text{V}$$

$$E_{cell} = E^\circ_{cell} - \frac{0.0591}{1}\log\frac{[M^+]_L}{[M^+]_R}$$

$$= 0 - \frac{0.0591}{1}\log\left(\frac{0.05}{1.0}\right)$$

$$E_{cell} = 0 + 0.077\text{ V} = 77\text{ mV}$$

$\therefore$ $E_{cell} > 0$, $\Delta G = -nFE^\circ_{cell}$

and $\Delta G < 0$

When $[M^+]_L$ changes to 0.0025 M, then

$$Q = \frac{0.0025}{1}$$

$\therefore$ $$E_{cell} = 0 - \frac{0.0591}{1}\log(0.0025) = 0.154\text{ V} = 154\text{ mV}$$

19. (b) **20.** (c)

Explanations for Q. Nos. (21-22)

ΔG°_f (net reaction)

$$= \Delta G^\circ_f(AgCl) - (\Delta G^\circ_f(Ag^+) + \Delta G_f(Cl^-))$$
$$= -109 - [77 - 129]$$
$$= -109 - 77 + 129$$
$$= -57\text{kJ}$$
$$= -57 \times 10^3\text{ J}$$

$$\Delta G^\circ = -nFE^\circ_{cell}$$

$$= E^\circ_{cell} = -\frac{\Delta G^\circ}{nF} = -\frac{(-57\times10^3)}{1\times 96500}$$

$$E^\circ_{cell} = 0.59\text{ V}$$

For equilibrium, $AgCl(s) \rightleftharpoons Ag^+ + Cl^-, \quad E^\circ = -0.59\text{V}$

$$K = [A^+][Cl^-] = K_{sp}$$

$\therefore$ $$E^\circ_{cell} = \frac{0.0591}{1}\log K_{sp}$$

$$-0.59 = 0.0591\log K_{sp}$$

$\therefore$ $\log K_{sp} = -10$

$\therefore$ $K_{sp} = 10^{-10}$

21. (a) **22.** (b)

23. (6)

$$\begin{array}{lll} Fe^{3+} + e^- \longrightarrow Fe^{2+} & E^\circ_{red} = x \\ \frac{1}{2}H_2 \longrightarrow H^+ + e^- & E^\circ_{ox} = 0.00\text{ (SHE)} \\ \hline Fe^{3+} + \frac{1}{2}H_2 \longrightarrow Fe^{2+} + H^+ & E^\circ_{cell} = x \end{array}$$

$$K = \frac{[Fe^{2+}][H^+]}{[Fe^{3+}]} = \left(\frac{0.1}{0.1}\right)(1) = 1$$

$\therefore$ $$E_{cell} = E^\circ_{cell} - \frac{0.0591}{1}\log K$$
$$= x - 0.0591\log 1$$
$$E_{cell} = x$$

In case II, $Fe^{3+} = 1.0\text{M}$

$\therefore$ $$K' = \left(\frac{0.1}{1.0}\right)^1 = \frac{1}{10}$$

$\therefore$ $$E'_{cell} = E^\circ_{cell} - \frac{0.0591}{1}\log\frac{1}{10}$$
$$= x + 0.06\log 10$$
$$E'_{cell} = x + 0.06$$
$$(E'_{cell} - E_{cell}) = 0.06\text{V}$$
$$= 0.06 \times 100\text{ centivolt}$$
$$= 6\text{ centivolt}$$

24. (3)

$$\underset{(+12)}{Cr_2O_7^{2}} \longrightarrow \underset{\uparrow 2x}{2Cr^{x+}} \quad \text{(of two } Cr_2\text{-atoms)}$$

reduced product

Change in oxidation number = $(12 - 2x)$

$$\Delta G^\circ = -xFE^\circ_{red}$$

$$-777.07 \times 10^3\text{mol}^{-1} = -(12 - 2x) \times 96500 \times 1.33$$

$$(12 - 2x) = 6.00$$

$\therefore$ $2x = 6$

$x = 3$

25. (4)

At anode

$$Cu \longrightarrow Cu^{2+} + 2e^-,$$
$$E^\circ_{Cu/Cu^{2+}} = -E^\circ_{Cu^{2+}/Cu} = -0.3385$$

At cathode

$$Fe^{3+} + e^- \longrightarrow Fe^{2+}]\times 2,$$
$$E^\circ_{Fe^{3+}/Fe^{2+}} = +0.77\text{ V}$$

$$Cu + 2Fe^{3+} \rightleftharpoons 2Fe^{2+} + Cu^{2+}$$

$$K_{eq} = \frac{[Fe^{2+}]^2[Cu^{2+}]}{[Fe^{3+}]^2}$$

$$E^\circ_{cell} = 0.4315\text{ V}$$

$$E^\circ_{cell} = \frac{0.0591}{2}\log K_{eq}$$

$$\frac{0.4315\times 2}{0.0591} = \log K_{eq}$$

$$K_{eq} = 4\times10^{14}$$

Thus, $y = 4$

DPP-5 Applications of EMF-Measurements

1. Plan

(i) Oxidation takes place at anode.

$Zn(s)$ is oxidised to $Zn^{2+}(aq)$

$$Zn \longrightarrow Zn^{2+} + 2e^-$$

Thus, it is anode, $\underset{\ominus}{Zn} | Zn^{2+}$ (0.1M)

Reduction takes place at cathode. $H^+(aq)$ is reduced to H_2

$$2H^+ + 2e^- \longrightarrow H_2(g)$$

Thus, it is cathode, $H^+ (pH = ?) | \underset{\oplus}{Pt(H_2)}$

(ii) Thus, electrochemical cell is formed by joining two half-cell using a salt-bridge.

$$\underset{\ominus}{Zn} \mid Zn^{2+}(0.1M) \underset{\text{Salt-bridge}}{||} H^+(pH = ?) \mid \underset{\oplus}{Pt} \underset{1\,atm}{(H_2)}$$

(iii) $E°_{cell} = E°_{Zn/Zn^{2+}} + E°_{SHE} = 0.76 + 0.00 = 0.76V$

For the electrochemical reaction,

$$Zn(s) + 2H^+(aq) \longrightarrow Zn^{2+}(aq) + \underset{1atm}{H_2(g)}$$

$$\text{Reaction quotient } (Q) = \frac{[Zn^{2+}](p_{H_2})}{[H^+]^2} = \frac{(0.1)\times 1}{[H^+]^2}$$

$$\therefore \quad E_{cell} = E°_{cell} - \frac{0.0591}{2}\log Q$$

$$0.28 = 0.76 - \frac{0.0591}{2}\log\frac{(0.1)}{[H^+]^2}$$

$$-0.48 = -\frac{0.0591}{2}\left[\log\left(\frac{1}{10}\right) + \log\frac{1}{[H^+]^2}\right]$$

$$= -\frac{0.0591}{2}\{-1 - 2\log[H^+]\}$$

$$= -\frac{0.0591}{2}[-1 + 2pH]$$

[since, $-\log[H^+] = pH$]

$$pH = \frac{1}{2}\left(\frac{0.48\times 2}{0.0591} + 1\right) = 8.6$$

2. Plan When $AgI(s)$ is in equilibrium with $Ag^+(aq)$ and $I^-(aq)$

$$E_{cell} = 0.00\text{ V}$$

and reaction quotient $(Q) = K_{sp}$

$$Ag | Ag^+ || I^- | AgI, Ag$$

Anode $Ag(s) \longrightarrow Ag^+(aq) + e^-$,

$$E°_{Ag/Ag^+} = -E°_{Ag^+/Ag} = -0.80V$$

Cathode $AgI(s) + e^- \longrightarrow Ag(s) + I^-(aq)$

$$E°_{I^-/AgI, Ag} = -0.15\text{ V}$$

$$AgI(s) \longrightarrow Ag^+(aq) + I^-(aq),\ E°_{cell} = -0.95\text{ V}$$

$\therefore$ $Q = [Ag^+][I^-] = K_{sp}$

$\therefore$ $E_{cell} = E°_{cell} - \frac{0.0591}{1}\log K_{sp}$

$$0 = E°_{cell} - 0.0591\log K_{sp}$$

$$E°_{cell} = 0.0591\log K_{sp}$$

$$-0.95 = 0.0591\log K_{sp}$$

$\therefore$ $\log K_{sp} = -\frac{0.95}{0.0591} = -16.07$

3. (b) It is an anode and thus oxidation takes place:

$$\underset{1\,bar}{H_2} \longrightarrow 2H^+ + 2e^-$$

Source of $[H^+]$ is H_2O in which at 298 K

$$[H^+] = 1\times 10^{-7}M$$

$\therefore$ $E_{Pt(H_2)/H^+} = E°_{Pt(H_2)/H^+} - \frac{0.0591}{2}\log[H^+]^2$

$$= 0 - 0.0591\log[H^+]$$

$$= +0.0591\,pH = 0.0591\times 7 = 0.4137\text{ V}$$

4. (d) $Cu^{2+} + 2e^- \longrightarrow Cu$

$$\text{Reaction quotient } (Q) = \frac{[Cu]}{[Cu^{2+}]} = \frac{1}{[Cu^{2+}]}$$

$$E_{Cu^{2+}/Cu} = E°_{Cu^{2+}/Cu} - \frac{2.303\,RT}{2F}\log\frac{1}{[Cu^{2+}]}$$

$$E_{Cu^{2+}/Cu} = E°_{Cu^{2+}/Cu} + \frac{0.0591}{2}\log[Cu^{2+}]$$

This represents a straight line of the type

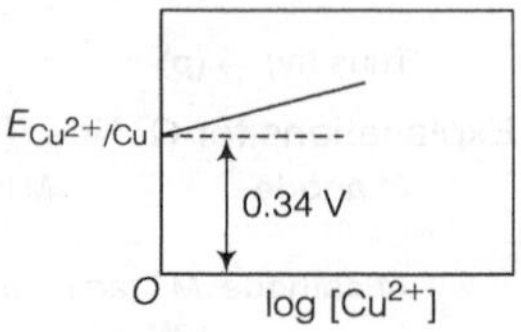

$$y = C + mX$$

Intercept at $(E_{Cu^{2+}/Cu})$ axis is

when $\log[Cu^{2+}] = 0$,

or $[Cu^{2+}] = 1M$

Thus, $E°_{Cu^{2+}/Cu} = 0.34$ V

Given half-cell is oxidation half-cell (anode)

$\therefore$ $Cu(s) \longrightarrow Cu^{2+}(aq) + 2e^-$

$$Q = [Cu^{2+}] = 0.1M$$

Since, $E°_{Cu^{2+}/Cu} = 0.34$ V

Hence, $E°_{Cu/Cu^{2+}} = -0.34$ V

$\therefore$ $E_{Cu/Cu^{2+}} = E°_{Cu/Cu^{2+}} - \frac{0.0591}{2}\log Q$

$$= -0.34 - \frac{0.0591}{2}\log(0.1) = \left(-0.34 + \frac{0.0591}{2}\right)V$$

5. (c) pH = 12

$$pOH = 2,\ [OH^-] = 10^{-2}M$$

$$Cu(OH)_2(s) \rightleftharpoons Cu^{2+}(aq) + 2OH^-(aq)$$

$\therefore$ $K_{sp} = [Cu^{2+}][OH^-]^2$

$$1\times 10^{-19} = [Cu^{2+}][10^{2-}]^2$$

$\therefore$ $[Cu^{2+}] = 1\times 10^{-15}M$

$$Cu^{2+} + 2e^- \longrightarrow Cu$$

$$\text{Reaction quotient } (Q) = \frac{[Cu]}{[Cu^{2+}]} = \frac{1}{[Cu^{2+}]} = \frac{1}{10^{-15}} = 10^{15}M$$

$$E = E° - \frac{0.0591}{2}\log[Q]$$

$$= 0.34 - \frac{0.0591}{2}\log 10^{15}$$

$$= \left(0.34 - \frac{0.0591}{2}\times 15\right)$$

$$= 0.34 - 0.44 = -0.10\text{ V}$$

6. (a) **Hydrogen electrode** has been coupled to **calomel electrode.**

Anode $H_2(g) \longrightarrow 2H^+(aq) + 2e^-$, $E°_{SHE} = 0.00$

Cathode

$$Hg_2Cl_2(s) + 2e^- \longrightarrow 2Hg(e) + 2Cl^-(aq),\quad E°_{cal} = 0.28\text{ V}$$

$$Hg_2Cl_2(s) + H_2(g) \longrightarrow 2Hg(l) + 2H^+(aq) + 2Cl^-(aq)$$

$$E°_{cell} = 0.28\text{ V}$$

$$\text{Reaction quotient } (Q) = \frac{[H^+]^2[Cl^-]^2}{pH_2} = \frac{[H^+]^2[1]^2}{1} = [H^+]^2$$

Note [*Solid*] = [*Element*] = 1molL^-

$\therefore$ $E_{cell} = E°_{cell} - \frac{2.303RT}{nF}\log Q$

$$0.67\text{ V} = 0.28\text{ V} - \frac{0.0591}{2}\log[H^+]^2$$

$$0.39\text{ V} = \frac{0.0591\times 2}{2}(pH)$$

$$0.39\text{ V} = 0.591\,pH$$

$\therefore$ $pH = \frac{0.39}{0.0591} = 6.6$

7. (b) For A^{3+}, $E_1 = E_1^\circ - \dfrac{0.0591}{1}\log\dfrac{[A^{2+}]}{[A^{3+}]}$

For B^{4+}, $E_2 = E_2^\circ - \dfrac{0.0591}{2}\log\dfrac{[B^{2+}]}{[B^{4+}]}$

or $2E_2 = 2E_2^\circ - 0.0591\log\dfrac{[B^{2+}]}{[B^{4+}]}$

At equivalence point

$$E_1 = E_2 = E_{eq}$$

$$\therefore \quad E_1 + 2E_2 = 3E_{eq}$$

$$3E_{eq} = (E^\circ_1 + 2E^\circ_2) - 0.0591\log\frac{[A^{2+}][B^{2+}]}{[A^{3+}][B^{4+}]}$$

Also, $\dfrac{[A^{2+}][B^{2+}]}{[A^{3+}][B^{4+}]} = 1$ at equivalence point

$$\therefore \quad E_{eq} = \left(\frac{E^\circ_1 + 2E^\circ_2}{3}\right) - 0.0591\log 1 = \frac{E^\circ_1 + 2E^\circ_2}{3}$$

$$= \frac{1.42 + 0.4 \times 2}{3} = 0.74\,\text{V}$$

8. (b) $Zn(s) + Cu^{2+}(aq) \rightleftharpoons Cu(s) + Zn^{2+}(aq)$

$$Q = \frac{[Zn^{2+}]}{[Cu^{2+}]}, n = 2$$

$$\therefore \quad E_{cell} = E^\circ_{cell} - \frac{2.303RT}{nF}\log Q$$

$$E_{cell} = E^\circ_{cell} - \frac{0.0591}{2}\log Q$$

This equation represents a straight line

$$y = c - mx$$

At $Q = 1$, $E_{cell} = E^\circ_{cell}$

Thus, $y = c$ = intercept = $OA = E^\circ_{cell}$

Thus, $E^\circ_{cell} = 1.10\,\text{V}$

$$\therefore \quad 1.1591 = 1.10 - \frac{0.0591}{2}\log\frac{[Zn^{2+}]}{[Cu^{2+}]}$$

$$\therefore \quad +0.0591 = -\frac{0.0591}{2}\log\frac{[Zn^{2+}]}{[Cu^{2+}]}$$

$$\therefore \quad -2 = \log\frac{[Zn^{2+}]}{[Cu^{2+}]}$$

$$\frac{[Zn^{2+}]}{[Cu^{2+}]} = 0.01$$

9. (d) $Ag(s) \longrightarrow Ag^+(aq) + e^-$, $E^\circ_{Ag/Ag^+} = -E^\circ_{Ag^+/Ag} = -0.80\,\text{V}$

$Ag(NH_3)_2^+ + e^- \longrightarrow Ag + 2NH_3(aq)$, $E^\circ = x\text{V}$

$Ag(NH_3)_2^+ \rightleftharpoons Ag^+(aq) + 2NH_3(aq)$, $E^\circ_{cell} = (x - 0.80)$

Reaction quotient $(Q) = K = \dfrac{[Ag^+][NH_3]^2}{[Ag(NH_3)_2^+]} = 6 \times 10^{-18}$

$$\therefore \quad E_{cell} = E^\circ_{cell} - \frac{2.303RT}{nF}\log K$$

$$0 = (x - 0.80) - \frac{0.0591}{1}\log 6 \times 10^{-8}$$

$$(x - 0.80) = \frac{0.0591}{1}\log(6 \times 10^{-8}) = -0.43$$

$$x = 0.80 - 0.43 = 0.37\text{ V}$$

10. (a) In this case Ag^+ ion collects as solid AgCl on the electrode itself. Some Ag^+ exists in equilibrium with $AgCl(s)$ in solution.

$$AgCl(s) \rightleftharpoons Ag^+(aq) + Cl^-(aq)$$

$$K_{sp} = [Ag^+][Cl^-]$$

Given, $[Cl^-] = 1\text{M}$

$$\therefore \quad [Ag^+] = \frac{K_{sp}}{[Cl^-]} = \frac{1 \times 10^{-10}}{1} = 1 \times 10^{-10}\text{M}$$

$$Ag^+ + e^- \longrightarrow Ag$$

$$E_{Ag^+/Ag} = E^\circ_{Ag^+/Ag} - \frac{2.303RT}{F}\log\frac{1}{[Ag^+]}$$

$$= 0.80 - 0.0591\log\frac{1}{(1 \times 10^{-10})}$$

$$= 0.80 - 0.0591\log 10^{10}$$

$$= 0.80 - 0.0591 \times 10$$

$$= (0.80 - 0.591) = 0.209\text{ V}$$

11. (b) **Anode** $H_2(g) \longrightarrow 2H^+(aq) + 2e^-$

Cathode $2\,AgCl(s) + 2e^- \longrightarrow 2Ag(s) + 2Cl^-$

$$n = 2$$

By Gibbs Helmholtz equation

$$\Delta G = \Delta H + T\left[\frac{d(\Delta G)}{dT}\right]_p$$

$$-nFE = \Delta H + T\left[\frac{d(-nFE)}{dT}\right]_p$$

$$-nFE = \Delta H - nFT\left(\frac{dE}{dT}\right)_p$$

$$\therefore \quad \Delta H = nFT\left(\frac{dE}{dT}\right)_p - nFE$$

$$\Delta H = nF\left[T\left(\frac{dE}{dT}\right) - E\right]$$

$$E_{cell} = 0.2650\text{V at 298 K}$$

$$E_{cell} = 0.2595\text{ V at 308 K}$$

$$\therefore \quad dE_{cell} = (0.2595 - 0.2650)\text{ V}$$

$$dT = 308 - 298 = 10\text{ K}$$

$$\therefore \quad \left(\frac{dE_{cell}}{dT}\right) = -\frac{55 \times 10^{-2}}{10}\text{ VK}^{-1}$$

$$= -5.5 \times 10^{-4}\text{ VK}^{-1}$$

$\therefore$ Heat of reaction

$$\Delta H = 2 \times 96500\,(298(-5.5 \times 10^{-4}) - 0.2650)]$$

$$= -82777.7\text{ J} = -82.8\text{ kJ}$$

12. (c)

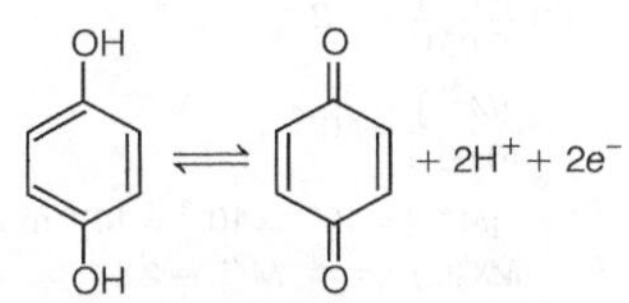

A B

Reaction quotient $(Q) = \dfrac{[B][H^+]^2}{[A]} = [H^+]^2$

$$\therefore \quad E = E^\circ - \frac{0.0591}{2}\log[H^+]^2$$

$$= E^\circ - 0.0591\log[H^+]$$

$$= E^\circ + 0.0591\,\text{pH}$$

Thus, E_{cell} is dependent on pH

As (NaOH) is added to HCl,

$$H^+ + \underset{\text{Being added}}{OH^-} \rightleftharpoons H_2O$$

Initially in acid (HCl), pH < 7 at neutral point pH = 7 and after HCl is neutralised, pH increases due to further addition of NaOH. Hence, E_{cell} also increases.

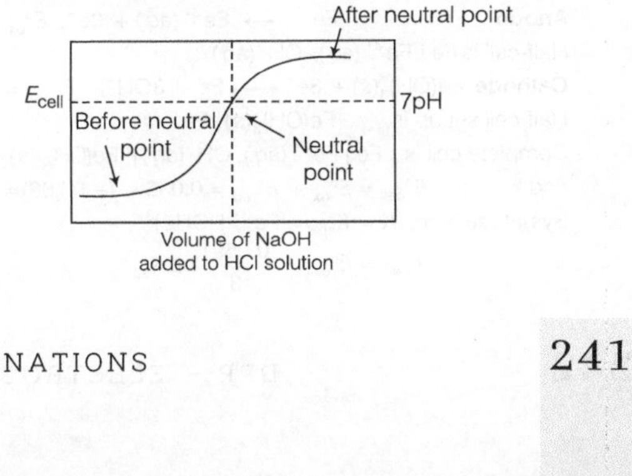

13. (a) When equivalence point is reached

$[Fe^{3+}] = [Mn^{2+}]$, $[Fe^{2+}] = [MnO_4^-]$ and $\frac{[Fe^{2+}][Mn^{2+}]}{[MnO_4^-][Fe^{3+}]} = 1$

and $E_{eq} = E_1 = E_2$

$$E_1 = E°_1 - \frac{0.0591}{1}\log\frac{[Fe^{2+}]}{[Fe^{3+}]}$$

$$E_2 = E°_2 - \frac{0.0591}{5}\log\frac{[Mn^{2+}]}{[MnO_4^-][H^+]^8}$$

$$5E_2 = 5E°_2 - 0.0591\log\frac{[Mn^{2+}]}{[MnO_4^-][H^+]^8}$$

$$(E_1 + 5E_2) = 6E_{eq} = (E°_1 + 5E°_2) - 0.0591\log\frac{[Fe^{2+}][Mn^{2+}]}{[MnO_4^-][Fe^{3+}][H^+]^8}$$

$$\therefore E_{eq} = \left(\frac{E°_1 + 5E°_2}{6}\right) - \frac{0.0591}{6}\log\frac{1}{[H^+]^8}$$

$$= \left(\frac{E°_1 + 5E°_2}{6}\right) - \frac{0.0591 \times 8}{6}\log\frac{1}{[H^+]}$$

$$= \left(\frac{E°_1 + 5E°_2}{6}\right) - 0.08\,pH$$

Explanations for Q. Nos. (14-15)

$M\text{ (LHS)} \rightleftharpoons \underset{(\text{in } MX_2)}{M^{2+}} + 2e^-$, $\quad E°_{ox} = xV$

$\underset{(0.001\text{ M})}{M^{2+}}\text{ (RHS)} + 2e^- \longrightarrow M$ $\quad E°_{red} = -XV$

$M^{2+}\text{ (0.001M)} \longrightarrow M^{2+}\text{ (in } MX_2)$, $\quad E°_{cell} = 0.00\text{ V}$

$$\therefore \quad Q = \frac{[M^{2+}](\text{in } MX_2)}{[M^{2+}](0.001\text{M})}$$

By Nernst equation, $E_{cell} = E°_{cell} - \frac{2.303RT}{nF}\log Q$

$$0.059 = 0 - \frac{0.059}{2}\log\frac{[M^{2+}]}{(0.001)}$$

$$\log\frac{[M^{2+}]}{(0.001)} = -2$$

$$\frac{[M^{2+}]}{(0.001)} = 10^{-2}$$

$\therefore \quad [M^{2+}] = 10^{-2} \times 10^{-3} = 10^{-5}\text{ mol dm}^{-3}$

$MX_2(s) \rightleftharpoons M^{2+} + 2X^-$

$\therefore \quad [X^-] = 2 \times 10^{-5}\text{ mol dm}^{-9}$

$K_{sp} = [M^{2+}][X^-]^2 = (10^{-5})(2 \times 10^{-5})^2$

$K_{sp} = 4 \times 10^{-15}\text{ mol}^3\text{dm}^{-9}$

Also, $\Delta G = -nFE_{cell}$

$n = 2$

$F = 96500\text{ C mol}^{-1}$

$E_{cell} = 0.059\text{ V}$

$\therefore \quad \Delta G = -2 \times 96500 \times 0.059\text{ J} = -11387\text{ J} = -11.4\text{ kJ}$

14. (b) **15.** (d)

Explanations for Q. Nos. (16-17)

For $Fe(OH)_3(s)$, $Fe(OH)_3(s) \rightleftharpoons Fe^{3+}(aq) + 3\,OH^-(aq)$

Above reaction is the net reaction in a cell for determination of K_{sp}. Thus,

Anode $Fe \longrightarrow Fe^{3+}(aq) + 3e^-$, $E°_{ox} = 0.036\text{V}$

Half-cell is $Fe \mid Fe^{3+}(aq) \cdot OH^-(aq)$

Cathode $Fe(OH)_3(s) + 3e^- \longrightarrow Fe + 3OH^-$, $E°_{red} = -0.786\text{V}$

Half-cell set up is $Fe(OH)_3(s) \mid Fe$

Complete cell is $Fe \mid Fe^{3+}(aq), OH^-(aq) \parallel Fe(OH)_3(s) \mid Fe$

and $E°_{cell} = E°_{ox} + E°_{red} = 0.036 + (-0.786) = -0.750\text{V}$

By net reaction, $K = K_{sp} = [Fe^{3+}][OH^-]^3$

$$\therefore \quad E_{cell} = E°_{cell} - \frac{0.0591}{3}\log K_{sp}$$

$E_{cell} = 0.00$ at equilibrium

$$\therefore \quad E°_{cell} = \frac{0.0591}{3}\log K_{sp}$$

$$\therefore \quad \log_{10} K_{sp} = \frac{3 \times E°_{cell}}{0.0591} = \frac{3(-0.75)}{0.0591}$$

$\therefore \quad \log_e K_{sp} = -87.7$

16. (b) **17.** (d)

Explanations for Q. Nos. (18-19)

$$2\,Ag^+ + 2e^- \longrightarrow 2Ag$$

$$C_6H_{12}O_6 + H_2O \longrightarrow C_6H_{12}O_7 + 2H^+ + 2e^-$$

$E_r^\circ = 0.80\text{ V}$ $\quad E° = -0.05\text{ V}$

$$2\,Ag^+ + C_6H_{12}O_6 + H_2O \longrightarrow 2Ag + C_6H_{12}O_7 + 2H^+$$

$E°_{cell} = 0.75\text{ V}$

At equilibrium, $E_{cell} = 0$

$$\therefore \quad E°_{cell} = \frac{0.0591}{2}\log_{10} K$$

$$\therefore \quad \log_{10} K = \frac{0.75 \times 2}{0.0591} = 25.38$$

$\log_e K = 2.303\log_{10} K = 2.303 \times 25.38 = 58.45$

When NH_3 is added, pH increases to 11

$\therefore \quad [H^+] = 10^{-11}\text{M}$

$$\therefore E_{ox} = E°_{ox} - \frac{0.0591}{2}\log[H^+]^2 = E°_{ox} - \frac{0.0591}{2} \times 2\log(10^{-11})$$

$= E°_{ox} + 0.0591 \times 11 = E°_{ox} + 0.65$

Thus E_{ox} increases by 0.65 V from $E°_{ox}$

18. (b) **19.** (a)

20. (4) $\underset{1\text{ bar}}{Pt(H_2)} \mid \underset{\ominus}{H^+}(pH = x)$

It is oxidation half-cell, and anode reaction is

$$H_2 \longrightarrow 2H^+ + 2e^-$$

$$\therefore \quad E_{ox} = E°_{ox} - \frac{2.303RT}{nF}\log Q$$

Reaction quotient $(Q) = \frac{[H^+]^2}{p_{H_2}} = [H^+]^2$

$n = 2$ $\quad \frac{2.303RT}{F} = 0.0591\text{V at 298 K}$

For SHE, $E°_{ox} = E°_{Pt(H_2)/H^+} = 0.00\text{V}$

$$\therefore \quad 0.2364 = 0 - \frac{0.0591}{2}\log[H^+]^2$$

$0.2364 = -0.0591\log[H^+] = 0.0591\,pH$

$$\therefore \quad pH = \frac{0.2364}{0.0591} = 4$$

Note $-\log[H^+] = pH$

21. (1) $Cu^{2+} + 2e^- \longrightarrow Cu$, $E° = 0.34\text{V}$

$$Q_{(\text{reaction quotient})} = \frac{1}{[Cu^{2+}]}$$

$$E = E° - \frac{0.0591}{2}\log(Q)$$

$$0.0455 = 0.34 - \frac{0.0591}{2}\log\frac{1}{[Cu^{2+}]} = 0.34 + \frac{0.0591}{2}\log[Cu^{2+}]$$

$$0.0455 - 0.34 = \frac{0.0591}{2}\log[Cu^{2+}]$$

$\log[Cu^{2+}] = -10$

$\therefore \quad [Cu^{2+}] = 10^{-10}\text{ M}$

$pH = 12$

$\therefore \quad pOH = 14 - 12 = 2$

$\therefore \quad [OH^-] = 10^{-2}\text{ M}$

$K_{sp} = [Cu^{2+}][OH^-]^2 = 10^{-10} \times 10^{-4} = 1 \times 10^{-14}$

Thus, $x = 1$

DPP-6 Battery and Corrosion

1. (d)

(a) Fe $\longrightarrow$ Fe^{2+} + $2e^-$ at anode–correct.

$O_2 + 4H^+ + 4e^- \longrightarrow 2H_2O$ at cathode –correct.

(b) Electrons flow from anode to cathode–correct.

(c) $4Fe^{2+} + O_2 + 4H_2O \longrightarrow Fe_2O_3(s) + 8H^+$ –correct.

2. (d)

(a) A typical 12V battery consists of six cells connected in series each cell providing a potential of about 2 V –correct

(b) No separation of cathodic and anodic compartment, species being in solid state –correct

(c) During recharging, $E° < 0$, thus $\Delta G° > 0$–correct.

3. (b) In lead-acid storage battery, specific gravity of H_2SO_4 changes and water is to be added periodically. In maintenance free battery calcium-containing lead alloy that prevents the electrolytic decomposition of H_2O. These batteries are sealed and no maintenance is required.

4. (a) I. Mercury battery has anode made of zinc and cathode is a steel vessel in contact with HgO in alkaline medium of KOH and $Zn(OH)_2$.

Anode $Zn(s) + 2OH^-(aq) \longrightarrow ZnO(s) + H_2O(l) + 2e^-$

Cathode $HgO(s) + H_2O(l) + 2e^- \longrightarrow Hg(l) + 2OH^-(aq)$

Net $Zn(s) + HgO(s) \longrightarrow ZnO(s) + Hg(l)$

Net reaction does not involve OH^-, hence emf is independent of $[OH^-]$.

II. In nickel, cadmium battery, net reaction is

$Cd(s) + 2Ni(OH)(s) + 2H_2O(l) \longrightarrow Cd(OH)_2(s) + 2Ni(OH)_2(s)$

Since, (OH^-) species are in solid state, hence emf is independent of (OH^-).

III. Alkaline dry cell is a modified version of the Leclanche cell. Net reaction is

$$Zn(s) + 2MnO_2(s) \longrightarrow ZnO(s) + Mn_2O_3(s)$$

Thus, emf is independent of (OH^-).

5. (d) I. **Primary** Leclanche cell.

Secondary Lead-storage

Flow battery Fuel-cell

II. True III. True

6. (a) $2Cu + H_2O + \underbrace{SO_3 + O_2}_{\text{Smog}} \longrightarrow Cu_2(OH)_2 \cdot SO_4$

$$2Cu + H_2O + CO_2 + O_2 \longrightarrow Cu_2(OH)_2CO_3$$

7. (b) In **fuel-cell**, reactions are not contained within the cell but are supplied from external source. (It is used as a source of electric power in space vehicles).

At anode

$$2H_2 \longrightarrow 4H^+ + 4e^-$$
$$4H^+ + 4OH^- \rightleftharpoons 4H_2O$$
$$2H_2 + 4OH^- \longrightarrow 4H_2O + 4e^-$$

At cathode

$$O_2 + 2H_2O + 4e^- \longrightarrow 4OH^-$$

Net $2H_2 + O_2 \longrightarrow 2H_2O$

8. (a) Pipes are made of Iron. Iron can be oxidised to Fe^{2+}.

$$Fe \longrightarrow Fe^{2+} + 2e^-$$

but when connected to Mg, which is more electropositive that Fe, Mg is oxidised to Mg^{2+} and in turn reduces Fe^{2+} (rust) to Fe

Anode $Mg \longrightarrow Mg^{2+} + 2e^-$

Cathode $Fe^{2+} + 2e^- \longrightarrow Fe$

It is called **cathodic protection** but at the cost of Mg (sacrified anode).

9. (b) In presence of moisture, rate of corrosion of metal is enhanced.

$$2Fe + O_2 + 2H_2O \longrightarrow 2Fe(OH)_2$$

10. (b) More noble metal is cathode and is protected from rusting by active metal which is anode.

Anode $Zn \longrightarrow Zn^{2+} + 2e^-$

Cathode $Fe^{2+} + 2e^- \longrightarrow Fe$

11. (a)

I. Temperature affects corrosion. However, solubility of dissolved gases decreases with rise in temperature.

II. In acidic medium, pH < 7 (acid rain), enhances corrosion

$$Fe + 2H^+ \longrightarrow Fe^{2+} + H_2$$

In basic medium, pH > 7

Fe^{2+} changes to $Fe(OH_2)$ and then to $Fe(OH_3)(s)$

(Fe_2O_3 - rust)

III. Impure metals set up their own electrochemical cells and thus, corrosion is enhanced.

IV. Conc. HNO_3 forms an oxide protective layer on the surface of a metal. When metal (as Al) reacts with conc. HNO_3, there is reaction but it ceases due to formation of an oxide (Al_2O_3) layer. Thus, HNO_3 can be stored in Al vessel.

12. (d) During discharge, sulphuric acid is consumed as :

$$Pb + PbO_2 + 2H_2SO_4 \longrightarrow 2PbSO_4 + 2H_2O$$

Note *Pb is oxidised to Pb^{2+} (as $PbSO_4$) $PbO_2(+4)$ is reduced to Pb^{2+} (as $PbSO_4$) and specific gravity of H_2SO_4 decreases.*

13. (d)

(a) In oxidised state, shine of the metal is furnished, thus, corrosion is the oxidative deterioration of a metal.

(b) A mini-cell is set up at the surface at which corrosion takes place.

Anode

$$Fe \longrightarrow Fe^{2+} + 2e^- \qquad E° = 0.44V$$

Cathode

$$O_2 + 4H^+ + 4e^- \longrightarrow 2H_2O \qquad E° = 1.23V$$

Thus, $E°_{cell} = 1.67$ V

Reaction is spontaneous-correct.

(c) Impurities also react with active metals by making own cells. Thus, correct.

14. (c)

(a) $O_2 + 2H_2O + 4e^- \longrightarrow 4OH^-$

$$Fe \longrightarrow Fe^{2+} + 2e^-$$
$$Fe^{2+} + 2OH^- \longrightarrow Fe(OH)_2(s)$$
$$4Fe(OH)_2(s) + O_2 + 2H_2O \longrightarrow 4Fe(OH)_3(s)$$
$$4Fe(OH)_3(s) \longrightarrow \underbrace{Fe_2O_3 \cdot H_2O}_{\text{Rust}} + 2H_2O$$

(b) CO_2 (from industries) + $H_2O \longrightarrow H_2CO_3$ (acid rain)

SO_2 (from industries) + $H_2O \longrightarrow H_2SO_3$ (acid rain)

$$H_2CO_3 \rightleftharpoons 2H^+ + CO_3^{2-}$$
$$H_2SO_3 \rightleftharpoons 2H^+ + SO_3^{2-}$$

Acid rain causes corrosion of metals. Even Cu, Ag and Pb are corroded.

15. (b) In the anodic reaction, the metal that is more easily oxidised loses electrons to produce metal ions. In case (a), zinc (more electropositive than Fe) is oxidised. In case (b), iron (more electropositive than Cu is oxidised).

In the cathodic reaction, oxygen gas which is dissolved in a thin film of water on the metal is reduced to OH^-.

Thus, rusting of iron does not take place in (a) but it is in (b).

Fe^{2+} and OH^- ions from the half-reactions initiate these further reactions

$$Fe^{2+} + 2OH^- \longrightarrow Fe(OH)_2(s)$$
$$4Fe(OH)_2(s) + O_2 + 2H_2O \longrightarrow 4Fe(OH)_3(s)$$
$$2Fe(OH)_3(s) \longrightarrow \underbrace{Fe_2O_3 \cdot H_2O}_{\text{Rust}} + 2H_2O$$

16. (d) **Oxidation at Cr electrode** (anode)

$$Cr^{2+} \longrightarrow Cr^{3+} + e^-$$

Thus, oxidation half-cell is

$$\underset{\ominus}{Cr} \mid Cr^{2+}, Cr^{3+}$$

reduction at Fe electrode (cathode)

$$Fe^{3+} + e^- \longrightarrow Fe^{2+}$$

Thus, reduction half-cell is

$$Fe^{3+}, Fe^{2+} \mid \underset{\oplus}{Fe}$$

Two half-cells are joined by a salt-bridge

$$\underset{\ominus}{Cr} \mid Cr^{2+}, Cr^{3+} \| Fe^{3+}, Fe^{2+} \mid \underset{\oplus}{Fe}$$

17. (a) In a dry-cell battery (also called Leclanche cell)

Anode Outer cover (cylindrical) is made of zinc. Oxidation takes place at anode.

$$Zn \longrightarrow Zn^{2+} + 2e^-$$

Cathode Graphite (carbon) rod in a paste of MnO_2 and carbon black.

Electrolyte Paste of $NH_4Cl/ZnCl_2$ (with starch binder) Reduction takes place.

$$2MnO_2 + 2NH_4^+ + 2e^- \longrightarrow Mn_2O_3 + 2NH_3 + H_2O$$

Thus, I, II and III are correct.

18. (a, b, c) Fe can be protected by making a layer of zinc on it. It is called galvanisation. At the cost of Zn (which is oxidised), Fe is protected.

$$Zn \longrightarrow Zn^{2+} + 2e^-$$
$$Fe \longrightarrow Fe^{2+} + 2e^-$$
$$\downarrow$$
$$\underset{\text{Rust}}{Fe^{3+}}$$
$$Fe^{3+} + 3e^- \longrightarrow \underset{\text{Protected}}{Fe}$$

Thus, Fe which is cathode is protected.

It is called cathodic protection and active metal is thus, sacrified anode.

19. (a, d) If (a) and (d) are combined,

$$E° = +0.89\,V, \Delta G° < 0$$

and $E°_{cell}$ is independent of (OH^-).

20. (b, c) To make cell reaction spontaneous, $E°_{cell} > 0$

Thus,

$$Ni_2O_3(s) + H_2O(l) + 2e^- \longrightarrow 2NiO(s) + 2OH^-(aq); \quad E° = 0.40\,V$$
$$Fe(s) + 2OH^-(aq) \longrightarrow FeO(s) + H_2O(l) + 2e^- = 0.87\,V$$
$$\overline{Ni_2O_3(s) + Fe(s) \longrightarrow FeO(s) + 2NiO(s) \quad E_{cell} = 1.27\,V}$$

Thus, battery emf is independent on $[OH^-]$

$$\Delta G° = -nFE°_{cell}$$
$$= -2 \times 96500 \times 1.27$$
$$= -245100\,J$$
$$= -245.1\,kJ$$

Thus, energy obtained per mole of Ni_2O_3 consumed is -245.1 kJ.

Explanations for Q. Nos. (21-22)

Al metal is used to remove Ag_2S.

$$3Ag_2S + 2Al \longrightarrow 6Ag + 3S^{2-} + 2Al^{3+}$$

$NaHCO_3$ in hot water is used as electrolyte and dissolves displaced S^{2-}.

21. (d) **22.** (c)

Explanations for Q. Nos. (23-24)

In battery H_2SO_4 ionises as

$$H_2SO_4 \rightleftharpoons 2H^+ + SO_4^{2-}$$
$$Pb(s) + PbO_2(s) + \underset{4x}{4H^+(aq)} + \underset{2x}{2SO_4^{2-}(aq)} \longrightarrow PbSO_4(s) + 2H_2O(l)$$

$$\therefore \quad K_{eq} = \frac{1}{[H^+]^4[SO_4^{2-}]^2}$$
$$= \frac{1}{[4x]^4[2x]^2} = \frac{1}{1024x^6}$$

Also, $$E°_{cell} = \frac{2.303RT}{nF}\log K_{eq}$$
$$\log K_{eq} = \frac{E°_{cell}\,nF}{2.303RT}$$
$$= \frac{2.0 \times 2}{0.0591} = 67.68$$
$$\therefore \quad K_{eq} = 4.81 \times 10^{67}$$
$$\frac{1}{1024x^6} = 4.81 \times 10^{67}$$
$$\therefore \quad x = \left(\frac{1}{1024 \times 4.81 \times 10^{67}}\right)^{1/6}$$
$$= 1.65 \times 10^{-12}\,M$$
$$\therefore \quad [SO_4^{2-}] = 2x = 3.30 \times 10^{-12}\,M$$
$$\Delta G° = -nFE°$$
$$= -2 \times 96500 \times 2$$
$$= -38600\,J$$
$$= -386\,kJ$$

23. (c) **24.** (b)

25. (8) $\Delta G° = -nFE° = -2 \times 96500 \times 1.5544\,J$

$$= -3 \times 10^5$$
$$= x \times 10^y$$

Thus, $x + y = 8$

DPP-7 Electrolysis and Faraday's Laws

1. Plan Electronic charge $= \dfrac{\text{Quantity of electricity } (Q)}{\text{Number of electrons } (n)}$

$$Q = It$$

$$n = \frac{\text{Mass}}{\text{Equivalent mass}} \times N_0$$

$$Q = 0.043 \times 6.244 \times 3600 \text{ Coulombs}$$

$$n = \frac{1.27}{254/2} \times 6.02 \times 10^{23}$$

$$\therefore \text{ Electronic charge} = \frac{0.043 \times 6.244 \times 3600\text{C}}{\left(\frac{1.27}{254}\right) \times 6.02 \times 10^{23} \times 2} = 1.606 \times 10^{-19}\text{ C}$$

2. Plan

(i) 1 mole of electrons = N_0 electrons

1 mole of electrons = 1 Faraday = 96500 Coulombs

Thus, charge on one electron $= \dfrac{96500}{N_0} = \dfrac{96500\text{C}}{6.02 \times 10^{23}}$

$= 1.603 \times 10^{-19}\text{C}$

(ii) Charge = current (A) × time (s)

$= 1.0 \times 10 \times 600$ As = 6000 C

Note 1 ampere second = 1C

(iii) **Plan** Energy expended = volt × current × time

$= 230\text{V} \times 2.0\text{A} \times 200\text{s}$

$= 92000$ V As = 92000 J = 92.0 kJ

Note 1 VAs = 1 J

3. (d) $Zn(NO_3)_2(aq) \rightleftharpoons Zn^{2+}(aq) + 2NO_3^-(aq)$

$2NO_3^-(aq) + H_2O \rightleftharpoons 2HNO_3 + O^{2-}(aq)$

At anode $2O^{2-}(aq) \longrightarrow O_2(g) + 4e^-$

In preference to reduction of Zn^{2+} at cathode, $O^{2-}(aq)$ is oxidised to O_2 at anode.

4. (b) $AB \rightleftharpoons \underset{\text{Cation}}{A^+} + \underset{\text{Anion}}{B^-}$

In electrolytic cell,

Cation moves to cathode and anion moves to anode.

At anode $B^- \longrightarrow B + e^-$

At cathode $A^+ + e^- \longrightarrow A$

Electrons move anode to cathode.

5. (b) In electrochemical series (ECS),

$E^\circ_{\text{reduction}}(Ca^{2+} + 2e^- \longrightarrow Ca) < 0$

thus, Ca^{2+} ion is not reduced to Ca instead, H_2O is reduced to H_2.

$$2H_2O + 2e^- \longrightarrow H_2 + 2OH^-$$

Ag, Cu and Cr are below H in ECS ($E^\circ_{\text{reduction}} > 0$), thus they are reduced to metals on electrolysis.

$$Ag^+ + e^- \longrightarrow Ag$$

$$Cu^{2+} + 2e^- \longrightarrow Cu$$

$$Cr^{3+} + 3e^- \longrightarrow Cr$$

Note *In ECS,* Ca, H, Cr, Cu *elements lying above* H *are better reducing agents thus are easily oxidised.*

6. (b) By Faraday's first law

w (amount deposit) $= zQ$

z = Electrochemical equivalent

$= \dfrac{\text{Atomic mass}}{\text{Electrons} \times 96500\text{ C}}$

Q = Coulombs of electricity

$$Ag^+ + e^- \longrightarrow Ag$$

$$w = \frac{108}{1 \times 96500} \times 9650 = 10.8\text{ g}$$

7. (c) When impure metal (M) is anode and pure metal is made cathode, then

At anode (impure)

$M \longrightarrow M^{n+} + ne^-, xE^\circ_0 \qquad E^\circ_{\text{ox}} = xV$

At cathode (pure)

$M^{n+} + ne \longrightarrow M; \qquad E^\circ_{\text{red}} = -xV$

Net $\quad M(\text{anode}) \longrightarrow M(\text{cathode}); \qquad E^\circ_{\text{cell}} = 0$

Thus, impure metal from the anode is deposited at the cathode.

8. (a) $w = zit$

$Ag^+ + e^- \longrightarrow Ag, \; z(Ag) = \dfrac{108}{96500}$

$Mg^{2+} + 2e^- \longrightarrow Mg, \; z(Mg) = \dfrac{24}{2 \times 96500}$

$w(Ag) = z(Ag)\, i_{Ag} t_{Ag}$

$w(Mg) = z(Mg)\, i_{Mg} t_{Mg}$

Amounts and current are equal

$\therefore \quad z(Ag)\, t_{Ag} = z(Mg)\, t_{Mg}$

$\dfrac{108}{96500} \times 1(\text{h}) = \dfrac{1L}{96500} t_{Mg}$

$\therefore \quad t_{Mg} = 9\text{h}$

9. (c) By Faraday's first law of electrolysis

$w = zit$

$Ag^+ + e^- \longrightarrow \underset{(108\text{g})}{Ag}$

$z = \dfrac{108}{96500}$

$w = \dfrac{108}{96500} \times 5\text{A} \times 2 \times 3600\text{ s} = 40.29\text{g}$

But anode is 95% pure.

Hence, actual loss is $= 40.29 \times \dfrac{100}{95} = 42.41\text{g}$

Thus, anode left = 100 − 42.41 = 57.59 g

10. (b) Impure metal is made the anode and oxidation takes place

$\underset{\text{Impure anode}}{Cu(s)} \longrightarrow Cu^{2+} + 2e^-$

Cu^{2+} formed are reduced at cathode and thus, deposited there.

$Cu^{2+} + 2e^- \longrightarrow \underset{\text{Cathode}}{Cu(s)}$

$\underset{\text{Anode}}{Cu(s)} \longrightarrow \underset{\text{Cathode}}{Cu(s)}$

11. (d) Manganate (VII) is MnO_4^- with ON of Mn = + 7

$\underset{\substack{+7 \\ 1\text{ mol}}}{MnO_4^-} + 3e^- \longrightarrow \underset{+4}{MnO_2}$

1 mole required 3F of electricity thus, 1 F will reduce only (1/3) mole MnO_4^- to MnO_2.

Thus, MnO_2 formed = 0.33 mol.

12. (d) When same quantity of electricity is passed, elements/gases are formed in the ratio of their equivalents.

$Al_2O_3 \rightleftharpoons 2Al^{3+} + 3O^{2-}$

At anode $2O^{2-} \longrightarrow O_2 + 4e^-$

Equivalent of $O_2 = \dfrac{M}{4}$, moles $= \dfrac{1}{4}$

At cathode $Al^{3+} + 3e^- \longrightarrow Al$

Equivalent of $Al = \dfrac{M}{3}$, moles $= \dfrac{1}{3}$

Ratio $\quad Al : O_2 = \dfrac{1}{3} : \dfrac{1}{4} = 4 : 3$

13. (d)

	Element formed by 1F
$Ag^+ + e^- \longrightarrow Ag$	1 mole
$Cu^{2+} + 2e^- \longrightarrow Cu$	$\frac{1}{2}$ mole
$Al^{3+} + 3e^- \longrightarrow Al$	$\frac{1}{3}$ mole
$Si^{4+} + 4e^- \longrightarrow Si$	$\frac{1}{4}$ mole

Thus, ratio Ag : Cu : Al : Si

$1 \quad \frac{1}{2} \quad \frac{1}{3} \quad \frac{1}{4}$

$12 \quad 6 \quad 4 \quad 3$

14. (d) H_2SO_4 (acidified water) $\rightleftharpoons 2H^+ + SO_4^{2-}$

At anode

$$SO_4^{2-} + H_2O \rightleftharpoons H_2SO_4 + O^{2-}$$

$$O^{2-} \longrightarrow \frac{1}{2}(O_2) + 2e^-$$

16 g = 0.5 mole by 2F

$$w = zit$$

$$z = \frac{16}{96500 \times \text{electron}}, i = ?, t = 1\text{ s}$$

$$= 8/96500$$

$$w = 0.224 \text{ cc at STP} = \frac{0.224}{22400} \text{ mole at STP}$$

$$= \left(\frac{0.224}{22400} \times 32\right) \text{g at STP}$$

$$\frac{0.224}{22400} \times 32 = \frac{8}{96500} \times i \times 1$$

$\therefore$ $i = 3.86\text{A}$

15. (a) $CuSO_4 \rightleftharpoons Cu^{2+}(aq) + SO_4^{2-}$

Electrons are made of copper. Thus

Oxidation $Cu(s) \longrightarrow Cu^{2+}(aq) + 2e^-$

Reduction $Cu^{2+}(aq) + 2e^- \longrightarrow Cu(s)$

Copper formed = w

$$w = zit = \frac{63.5 \times 10.0 \times 9650}{2 \times 96500} \text{g}$$

$$= \frac{63.5}{2 \times 63.5} = 0.5 \text{ mol}$$

0.5 mole of copper is dissolved from the anode. Thus, its mass decreases. 0.5 mole of copper from the anode is deposited at the cathode. Thus, its mass increases.

Thus, molar concentration of aqueous solution of $CuSO_4$ remains unchanged.

16. (a) $NaCl(aq) \rightleftharpoons Na^+ + Cl^-$

At anode $\frac{1}{2}Cl^- \longrightarrow \frac{1}{2}Cl_2 + e^-$

At cathode $Na^+ + e^- \longrightarrow Na$

$$Na + H_2O \longrightarrow NaOH$$

$$pH = 12.24, pOH = 14 - 12.24 = 1.76$$

$\therefore$ $[OH^-] = 1.74 \times 10^{-2}\text{M}$

Thus, NaCl reacted $= 1.74 \times 10^{-2}\text{M} = 1.74 \times 10^{-2} \text{ mol L}^{-1}$

$= 1.74 \times 10^{-2} \times 0.300$ mol in 300 mL

$= 1.74 \times 10^{-2} \times 300 \times 58.5$ g

$$w = zit$$

$$1.74 \times 10^{-2} \times 0.300 \times 58.5 = \frac{58.5}{96500} \times i \times 6 \times 60 \text{ s}$$

$\therefore$ $i = 1.40\text{A}$

17. (b) LHS RHS

$NH_3 + NH_4^+ | NH_3 + NH_4^+$

RHS OH^- formed combines with NH_4^+

$$NH_4^+ + OH^- \rightleftharpoons NH_4OH \text{ or } NH_3(aq)$$

$$pOH = pK_b + \log\frac{[NH_4^+]}{[NH_3]}$$

LHS H^+ formed combines with NH_3 forming NH_4^+

$$NH_3 + H^+ \rightleftharpoons NH_4^+$$

$$pOH = pK_b + \log\frac{[NH_4^+]}{[NH_3]}$$

In **RHS**, $[NH_3]$ increases hence, pOH decreases

$$pH = 14 - pOH$$

Hence, pH of RHS increases.

In **LHS**, $[NH_4^+]$ increases, hence pOH increases and pH decreases.

18. (d) $NaCl(aq) \rightleftharpoons Na^+(aq) + Cl^-(aq)$

At anode $2Cl^-(aq) \longrightarrow Cl_2(g) + 2e^-$

At cathode $2Na^+(aq) + 2e^- \longrightarrow 2Na(l)$

$$2Na(l) + 2H_2O \longrightarrow 2NaOH(aq) + H_2(g)$$

Cl_2 (g) is formed at anode and NaOH and H_2 are formed at cathode. Since, cathode and anode are separated. Hence, there is no reaction between the products formed at the cathode and anode.

$$Cl_2 + \underset{\text{Cold}}{2NaOH} \longrightarrow NaCl + NaClO + H_2O$$

$$3Cl_2 + \underset{\text{Hot}}{6NaOH} \longrightarrow 5NaCl + NaClO_3 + 3H_2O$$

Note *But these reactions do not occur.*

Note NaCl *is neutral* pH = 7 *initially. After electrolysis,* NaOH *is formed. Hence, solution become, basic.* pH > 7.

19. (a,c,d) Regardless whether a cell is a volatic or an electrolysis cell, anode is the electrode at which oxidation takes place and cathode is the electrode at which reduction takes place.

In electrochemical cell, $Zn \longrightarrow Zn^{2+} + 2e^-$

It is losing electrons thus, anode is negative terminal.

In electrolysis cell,

cation migrates to cathode thus –ve terminal.

anion migrates to anode thus +ve terminal.

20. (a, b, c)

(a) I : $2Cl^-(aq) \longrightarrow Cl_2 + 2e^-$ $\quad E^\circ = 1.36\text{V}$

II : $2H_2O(l) \longrightarrow O_2(g) + 4H^+ + 4e^-$, $\quad E^\circ = 1.23\text{V}$

Thus, E° values of I and II are quite closer – correct.

(b) $E_I = E_I^\circ - \frac{0.0591}{2}\log\frac{1}{[Cl^-]^2} = E_I^\circ + 0.0591\log[Cl^-]$

If $[Cl^-]$ is high, $E_I > E_I^\circ$ thus, favoured.

$$E_{II} = E_{II}^\circ - \frac{0.0591}{4}\log[H^+]^4$$

$$= E_{II}^\circ + 0.0591 \text{ pH}$$

Since, $(-\log[H^+] = pH)$

Thus, if pH is high

$$E_{II} > E_{II}^\circ \text{ thus favoured}$$

Thus, (b) is correct.

(c) Experiment indicates that the applied voltage required is always greater than the voltage calculated from the standard E_{ox}° or E_{red}°. The additional voltage required is called overvoltage. Due to this, Cl_2 and not O_2 is formed.

Thus, (c) is correct.

21. (a,b,c,d)

(a) Oxidation takes place at anode - correct.

$$2H_2O(l) \longrightarrow 4H^+(aq) + O_2(g) + 4e^-$$

(b) Reduction takes place at cathode - correct.

$$2Cu^{2+}(aq) + 4e^- \longrightarrow 2Cu(s)$$

Net $2Cu^{2+}(aq) + 2H_2O \longrightarrow 4H^+(aq) + 2Cu(s) + O_2(g)$;

$$E^\circ_{cell} = -0.89V$$

Since, $E^\circ_{cell} < 0$, in electrolysis.

Cell reaction is spontaneous.

(c) Copper is deposited at the cathode, hence the metal to be electroplated (say Fe), should be made cathode. - correct.

(d) Since, $Cu^{2+}(aq)$ - (blue) is reduced $Cu(s)$, hence blue colour fades – correct.

22. (a, c, d)

(a) $AB \underset{}{\overset{\text{Current}}{\rightleftharpoons}} \underset{\text{Cation}}{A^+} + \underset{\text{Anion}}{B^-}$

Thus, electric current is used to carry out a non-spontaneous reaction.

Thus, (a) is correct.

(b) Since reactions are reverse of electrochemical cell, hence

$$E_{cell} < 0$$
$$\Delta G = +ve$$

Thus, (b) is incorrect.

(c) Cation goes to cathode and anion goes to anode and are discharged. Thus, correct.

At anode $B^- \longrightarrow B + e^-$

At cathode $A^+ + e^- \longrightarrow A$

(d) Some additional voltage called overvoltage is set up and thus other reactions also take place.

Thus, correct.

23. (b,c)

(a) $CuSO_4$ and $CuCl_2$ changes to Cu and solution becomes colourless.

(b) KCl changes to Cl_2 (yellow)

KI changes to I_2 (violet)

Thus, distinguished.

(c) $AgNO_3$ no change

$CuSO_4$ (blue) changes to colourless.

(d) $CuBr_2$ (blue) both change to colourless.

$NiBr_2$ (Blue)

24. (a) By Faraday's second law

$$\frac{\text{Amount of } X}{\text{Amount of } Y} = \frac{\text{Equivalent mass of } X}{\text{Equivalent mass of } Y}$$

$$Zn^{2+} + 2e^- \longrightarrow Zn, \quad E(Zn) = \frac{65}{2}$$

$$Cu^{2+} + 2e^- \longrightarrow Cu, \quad E(Cu) = \frac{63.5}{2}$$

$$Ag^+ + e^- \longrightarrow Ag, \quad E(Ag) = \frac{108}{1}$$

$$2H^+ + 2e^- \longrightarrow H_2, \quad E(H_2) = \frac{2}{2}$$

$$2O^{2-} \longrightarrow O_2 + 4e^-, \quad E(O_2) = \frac{32}{4}$$

$$77 \text{ mL of } O_2 = 0.075 \text{ L } O_2 = \frac{0.077}{22.4} \text{ mol at STP}$$

$$= \frac{0.077}{2 \text{ mol}} \times 32 \text{ g at STP}$$

Reference point is O_2 gas = 0.11g O_2 at STP

Thus,

$$\frac{\text{weight of Zn}}{\text{weight of } O_2} = \frac{\text{equivalent weight of Zn}}{\text{equivalent weight of } O_2}$$

(p) $\therefore$ $w(Zn) = \dfrac{0.11 \times 65}{2} \times \dfrac{1}{8}$

$= 0.45g$ ...(i)

Similarly,

(q) $w(Ag) = \dfrac{\text{equivalent weight of Ag}}{\text{equivalent weight of } O_2} \times \text{weight of } O_2$

$= \dfrac{108}{8} \times 0.11 = 1.485 \text{ g}$ ...(ii)

(r) $w(Cu) = \dfrac{63.5}{2} \times \dfrac{0.11}{8}$

$= 0.44$ g ...(iv)

(s) $w(H_2) = \dfrac{2}{2} \times \dfrac{0.11}{8} = 0.014$ g ...(iii)

Short (Alternate) Method

By electrolysis, amount formed are in equivalent ratio. Thus,

equivalent of Zn formed = equivalent of O_2 formed

O_2 formed = 77 mL O_2 at STP

$= 0.077$ L O_2 at STP

$= \dfrac{0.077}{22.4}$ mol $= \dfrac{0.077}{22.4} \times 4$ equivalents

(p) (Zn formed) $= \dfrac{0.077 \times 4}{22.4}$ equiv

$= \dfrac{0.077 \times 4}{22.4} \times \dfrac{65}{2}$ g $= 0.45$ g

(q) (Ag) $= \dfrac{0.077 \times 4}{22.4}$ equiv $= \dfrac{0.077 \times 4}{22.4} \times 108$ g

$= 1.485$ g

(r) (Cu) = equivalent of O_2

$= \dfrac{0.077 \times 4}{22.4} = \dfrac{0.077 \times 4}{22.4} \times \dfrac{63.5}{2}$ g

$= 0.436$g

(s) (H_2) = equivalent of O_2

$= \dfrac{0.077 \times 4}{22.4} = \dfrac{0.077 \times 4}{22.4} \times 1\text{g } H_2$

$= 0.01375$ g

Explanations for Q. Nos. (25-26)

$$NaCl(aq) \rightleftharpoons Na^+(aq) + Cl^-(aq)$$

At anode $2Cl^-(aq) \longrightarrow Cl_2 + 2e^-$

At cathode $2Na^+(aq) + 2e^- \longrightarrow 2Na(l)$

$$2Na(l) + 2H_2O \longrightarrow 2NaOH + H_2O^+(g)$$

$$2Na^+(aq) + 2H_2O + 2e^- \longrightarrow 2Na^+ + 2OH^- + H_2(g)$$

By Faraday's first law

$$w = zit$$

$$z = \frac{(NaOH)}{\text{electrons} \times 96500}$$

$$= \frac{2 \times 40}{2 \times 96500} = \frac{40}{96500}$$

$$w = \frac{40}{96500} \times 30 \times 60 \times 60$$

$$= 44.77\text{g NaOH} = \frac{44.77}{40}\text{mol}$$

$$Cl_2 \text{ formed} = \frac{\text{NaOH formed}}{2} \text{mol} = \frac{44.77}{40 \times 2} \text{mol}$$

$$= \frac{44.77}{40 \times 2} \times 22.4 \text{ L at STP}$$

$$= 12.54 \text{ L at STP}$$

25. (b) **26.** (a)

Explanations for Q. Nos. (27-28)

H_2 is oxidised at anode and O_2 is reduced at cathode.

H_2 used = 67.2 L at STP = $\frac{67.2}{22.4}$ mol = 3 mol = 6 g

$w = zit$

$z = \frac{2}{2 \times 96500}, t = 15 \text{ min} = 900 \text{ s}, i = ?$

$\therefore \quad 6 = \frac{2 \times i \times 900}{2 \times 96500}$

$i = 643.3 \text{A}$

H_2 consumed = 3 mol = 6 g equivalent

Thus, Cu formed = 6 g equivalent = $\frac{6}{2}$ mol = 3 mol

27. (d) **28.** (c)

29. (2) $M^{n+} + ne^- \longrightarrow M$

By Faraday's first law

$w = zit$

where, z = electrochemical equivalent = $\frac{\text{Atomic mass}}{\text{Electrons} \times 96500}$

$z = \frac{(68 + 69)}{n \times 96500}$

$w = 0.958 \text{g}$

$t = 75 \times 60 \text{ s}$

$i = 0.30 \text{ A}$

$\therefore \quad 0.958 = \frac{137}{n \times 96500} \times 0.3 \times 75 \times 60$

$\therefore \quad n = 2.0$

30. (5) $\underset{+2}{\overset{\uparrow}{Mn(OH)_2}} + H_2O_2 \longrightarrow \underset{+4}{\overset{\uparrow}{MnO_2}} + 2H_2O$

Change in ON = 2

Thus, electrons exchanged = 2

z (electrochemical equivalent of MnO_2) = $\frac{\text{Molar mass}}{n \times 96500}$

$z = \frac{87.0}{2 \times 96500}$

$w = zit, \quad t = 60 \text{s}$

$\therefore \quad i = \frac{w}{zt}$

$i = \frac{0.135 \times 2 \times 96500}{87 \times 60} = 4.99 \approx 5.0 \text{ A}$

31. (9) By Faraday's first law

$w = zit$

$2H^+ + 2e^- \longrightarrow H_2$

$z(H_2) = \frac{2}{2 \times 96500}$

100.8 L at STP = $\frac{100.8}{22.4}$ mol = 4.5 mol = 9 g

$\therefore \quad 9 = \frac{2}{2 \times 96500} \times (i) \times 96500$

$\therefore \quad i = 9 \text{A}$

32. (8) $\frac{\text{Mass (Ag)}}{\text{Volume (plate)}}$ = Density

$\therefore$ Mass of silver = Density × Volume

$= 10.5 \text{ g cm}^{-3} \times 80 \text{ cm}^2 \times \left(\frac{0.010}{10} \text{cm}\right) = 0.84 \text{ g}$

$Ag^+ + e^- \longrightarrow Ag$

$w = zit$

$z = \frac{108}{96500}$

$i = 1.56 \text{A}, t = ?$

$0.84 = \frac{108}{96500} \times 1.56 \times t$

$\therefore \quad t = 481 \text{ s} \approx 8 \text{ min}$

33. (2) Reactions during current is drawn are

$Pb(s) + SO_4^{2-}(aq) \longrightarrow PbSO_4(s) + 2e^-$

$PbO_2(s) + 4H^+(aq) + SO_4^{2-}(aq) + 2e^- \longrightarrow PbSO_4(s) + 2H_2O(l)$

$\therefore$ H_2SO_4 used = 2 moles

Electrons = 2, t = 13.4h

$\therefore \quad w = zit$

$\text{Moles} = \frac{98}{96500 \times 2} \times \frac{6.0}{98} \times 13.4 \times 3600 \text{ mol in 500 mL}$

= 1.50 mol in 500 mL = 3.00 mol in 1 L

Used (H_2SO_4) = 3.0 M

$\therefore \quad (H_2SO_4)_{\text{final}} = 5.0 - 3.0 = 2.0 \text{ M}$

34. (3) $n(H_2) = \frac{pV}{RT} = \frac{2.00 \times 1}{0.082 \times 298} = 0.08175 \text{ mol}$

$w(H_2) = 0.1635 \text{ g}$

Also, $w = zit$

$0.1635 = \frac{1}{96500} \times 1.46 \times t(\text{s})$

$t = 10806 \text{ s} = 3.00 \text{ h}$

Revisal Problems (JEE Main)

1. (b) II. Lead-storage, III. Nickel- Cadmium and IV. Lithium batteries are rechargeable.

2. (a) **With platinum electrodes**

At anode $\quad 2H_2O(l) \longrightarrow O_2(g) + 4H^+ + 4e^-$

At cathode $\quad 2Cu^{2+}(aq) + 4e^- \longrightarrow 2Cu(s)$

With copper electrodes

At anode $\quad Cu(s) \longrightarrow Cu^{2+}(aq) + 2e^-$

At cathode $\quad Cu^{2+}(aq) + 2e^- \longrightarrow Cu(s)$

Net $\quad \underset{\text{Anode}}{Cu(s)} \longrightarrow \underset{\text{Cathode}}{Cu(s)}$

Note *When inert electrodes are used, principle is used in electroplating. Blue colour of $CuSO_4$ gradually fades. With copper electrodes, Cu dissolves from the anode and deposits at cathode. Concentration of copper solution remains unchanged. This makes the basis of purification of metals called refining.*

3. (a) In a galvanic cell, for example

$\underset{\ominus}{Zn} \mid Zn^{2+}(C_1) \mid\mid Cu^{2+}(C_2) \mid \underset{\oplus}{Cu}$

two half-cells are joined by salt-bridge. It does not take part in chemical reaction.

4. (c) Λ_C = molar conductance at given concentration.

Λ_O = molar conductance at infinite dilution (≈ zero concentration)

Thus, $\quad x = \frac{\Lambda_C}{\Lambda_O}$

$\therefore \quad \Lambda_C = x\Lambda_O$

By Ostwald's dilution law

$x = \sqrt{\frac{K_b}{C}} = \sqrt{\frac{1.6 \times 10^{-5}}{0.01}} = 0.04$

$\therefore \quad \Lambda_C = 0.04 \times 2.38 \times 10^{-2} = 9.52 \times 10^{-4} \text{ S m}^2\text{mol}^{-1}$

5. (d) $E_{Mg^{2+}/Mg} = E°_{Mg^{2+}/Mg} - \frac{0.0591}{2} \log \frac{1}{[Mg^{2+}]}$

$$E_{Mg^{2+}/Mg} = E°_{Mg^{2+}/Mg} + \frac{0.0591}{2} \log [Mg^{2+}]$$

This represents a straight line, $y = c + mx$

Thus, y increases as x increases with intercept on Y-axis. When $[Mg^{2+}] = 1\,M$

$$\log [Mg^{2+}] = 0$$

$$E_{Mg^{2+}/Mg} = E°_{Mg^{2+}/Mg} = OA$$

6. (b)(a) $\frac{1}{2}H_2 \longrightarrow H^+ + e^-$

1 bar pH = 1

$[H^+] = 10^{-1}\,M$

$H^+ + e^- \longrightarrow \frac{1}{2}H_2$

pH = 2 1 bar

$[H^+] = 10^{-2}\,M$

$H^+ (10^{-2}M) \longrightarrow H^+ (10^{-1}M)$, $E°_{cell} = 0.00$

Reaction quotient $(Q) = \frac{10^{-1}}{10^{-2}} = 10$

$$E = E° - \frac{0.0591}{1} \log 10 = -0.0591 \text{ V}$$

$(E - E°) = -0.0591$ V

(b) $Cl^- \longrightarrow \frac{1}{2}Cl_2 + e^-$

0.1 M 1 bar

$\frac{1}{2}Cl_2 + e^- \longrightarrow Cl^-$

2 bar 0.01 M

$\frac{1}{2} Cl_2 + Cl^- \longrightarrow \frac{1}{2}Cl_2 + Cl^-$

2 bar 0.1 M 1 bar 0.01 M

$\therefore$ Reaction quotient $(Q) = \left(\frac{(p_{Cl_2})_L}{(p_{Cl_2})_R}\right)^{\frac{1}{2}} \left(\frac{[Cl^-]_R}{[Cl^-]_L}\right)$

$$Q = \left(\frac{1}{2}\right)^{\frac{1}{2}} \left(\frac{0.01}{0.10}\right)$$

$$Q = 0.071$$

$\therefore$ $E = E° - 0.0591 \log Q$

$$E = E° - 0.0591 \log (0.071)$$

$(E - E°) = 0.068$ V

(c) $2Fe^{2+} \longrightarrow 2Fe^{3+} + 2e^-$

$Sn^{4+} + 2e^- \longrightarrow Sn^{2+}$

$2Fe^{2+} + Sn^{4+} \longrightarrow Sn^{2+} + 2Fe^{3+}$

0.1 M 0.2 M 0.1 M 0.2 M

Reaction quotient $(Q) = \frac{[Sn^{2+}][Fe^{3+}]^2}{[Fe^{2+}]^2[Sn^{4+}]} = \frac{(0.1)(0.2)^2}{(0.1)^2(0.2)} = 2$

$$E = E° - \frac{0.0591}{2} \log 2$$

$(E - E°) = -0.009$ V

(d) $Ag + Cl^- \longrightarrow AgCl + e^-$

0.01 M

$AgCl + e^- \longrightarrow Ag + \underset{1M}{Cl^-}$

$Cl^- \longrightarrow Cl^-$

0.01 M 1 M

Reaction quotient $(Q) = \frac{1}{0.01} = 100$

$$E = E° - \frac{0.0591}{1} \log 100$$

$(E - E°) = -0.1182$ V

7. (b) $Ba(OH)_2 = BaCl_2 + 2NaOH - 2NaCl$

$\therefore$ $\Lambda°_{Ba(OH)_2} = \Lambda°(BaCl_2) + 2\Lambda°(NaOH) - 2\Lambda°(NaCl)$

$= [280 + 2 \times 248.0 - 2 \times 126.5] \times 10^{-4}$

$= 523.0 \times 10^{-4}$ S m^{2}mol^{-1}

8. (c) $H^+ + e^- \longrightarrow \frac{1}{2} H_2$

$H^+ (xM) \mid \underset{y\,atm}{Pt(H_2)}$

$$Q \text{ (reaction quotient)} = \frac{(p_{H_2})^{\frac{1}{2}}}{[H^+]}$$

$$E_{SHE} = E°_{SHE} - \frac{0.0591}{1} \log Q = 0 - \frac{0.0591}{1} \log \frac{(p_{H_2})}{[H^+]}$$

$$= -0.0591 \log \frac{(p_{H_2})^{\frac{1}{2}}}{[H^+]}$$

If $\left(\log \frac{(p_{H_2})^{\frac{1}{2}}}{[H^+]}\right)$ is positive then, $E_{SHE} < 0$

(a) $\log \frac{(1)^{\frac{1}{2}}}{(2.0)} = -\log 2 = -0.3010$, $E_{SHE} = +$ ve

(b) $\log \frac{(1)^{\frac{1}{2}}}{(1)} = \log 1 = 0$

(c) $\log \frac{(2)^{\frac{1}{2}}}{(1)} = \log \sqrt{2} = 0.150$, $E_{SHE} < 0$

(d) $\log \frac{(2)^{\frac{1}{2}}}{(2)} = \log \frac{1}{\sqrt{2}} = -\log \sqrt{2} = -0.150$

$E_{SHE} > 0$

9. (d) $Cu^{2+} + 2e^- \longrightarrow Cu$, $E° = 0.34$ V

$Cu + 4NH_3 \longrightarrow [Cu(NH_3)_4^{2+}] + 2e^-$, $E° = 0.12$ V

$Cu^{2+} + 4NH_3 \longrightarrow [Cu(NH_3)_4^{2+}]$, $E° = 0.46$ V

When complex is formed at equilibrium,

$K_{eq} = K$ (stability constant)

$E_{cell} = 0.00$ V

and $E°_{cell} = \frac{0.0591}{x} \log K_{eq}$

$\therefore$ $\log K_{eq} = \frac{0.6 \times x}{0.0591} = \frac{0.46 \times 2}{0.0591} = 15.57$

$K_{eq} = 3.71 \times 10^{15}$

10. (c) This is a quinhydrone electrode in which $[A] = [B]$ where,

A = B =

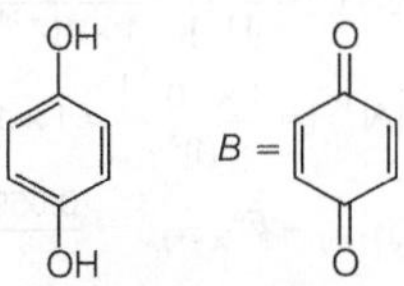

Thus, reaction quotient $= \frac{[H^+]^2 [B]}{[A]} = [H^+]^2$

$\therefore$ $E = E° - \frac{2.303\,RT}{2F} \log [H^+]^2$

$= E° - 0.0591 \log [H^+]$

$= E° + 0.0591\text{pH} = 1.30 + 0.0591 \times 2$

$= 1.30 + 0.12$

$= 1.42$ V

11. (d) $N_2H_4(aq) \longrightarrow N_2(g)$

$\uparrow \qquad \uparrow$

-4 (for two N-atoms) $\quad 0 \quad$ (O.N. of N-atom)

Change in oxidation number (ON) = 4

$$\Delta G° = -nFE°_{cell} = -4 \times 96500 \times 1.56$$
$$= -602160 \text{ J}$$
$$= -602.160 \text{ kJ}$$
$$\Delta G° = 2\Delta G°(H_2O) - \Delta G°(N_2H_4)$$
$$-602.160 = -2 \times 237.2 - \Delta G°(N_2H_4)$$
$$\therefore \quad \Delta G°(N_2H_4) = -474.40 + 602.16$$
$$= 127.76 \text{ kJ mol}^{-1}$$

12. (c) $NaH(aq) \rightleftharpoons Na^+(aq) + H^-(aq)$

At anode

$$2H^-(g) \longrightarrow H_2(g) + 2e^-$$

At cathode

$$2Na^+ + 2e^- \longrightarrow 2Na$$
$$2Na + 2H_2O \longrightarrow 2NaOH + H_2(g)$$

At anode and cathode both, H_2 is evolved.

13. (c) $O_2(g) + 4H^+(aq) + 4e^- \longrightarrow 2H_2O(l)$

Reaction quotient $Q = \dfrac{[H_2O]^2}{p_{O_2}[H^+]^4}$

$$= \frac{1}{1 \times (1 \times 10^{-7})^4} = 1 \times 10^{28}$$
$$E = E° - \frac{2.303RT}{nF}\log Q$$
$$= 1.23 - \frac{0.0591}{4}\log 1 \times 10^{28}$$
$$= 1.23 - \frac{0.0591}{4} \times 28 \log 10$$
$$= 1.23 - 0.0591 \times 7$$
$$= 1.23 - 0.41 = 0.82 \text{ V}$$

14. (a) For the reduction potential

$$Al^{3+} + 3e^- \longrightarrow Al$$
$$Q \text{ (reaction quotient)} = \frac{[Al]}{[Al^{3+}]} = \frac{1}{[Al^{3+}]}$$

Also, $\quad Al(OH)_3(s) \rightleftharpoons Al^{3+}(aq) + 3OH^-(aq)$

$\therefore \quad K_{sp} = [Al^{3+}][OH^-]^3$

$$[Al^{3+}] = \frac{K_{sp}}{[OH^-]^3}$$

When $\quad pH = 14$

$$[H^+] = 10^{-14} \text{ M}$$
$$\therefore \quad [OH^-] = \frac{K_w}{[H^+]} = \frac{1 \times 10^{-14}}{1 \times 10^{-14}} = 1 \text{ M}$$
$$\therefore \quad [Al^{3+}] = \frac{1 \times 10^{-33}}{(1)^3} = 1 \times 10^{-33} \text{ M}$$
$$\therefore \quad E_{Al^{3+}/Al} = E°_{Al^{3+}/Al} - \frac{0.0591}{3}\log(Q)$$
$$E_{Al^{3+}/Al} = -1.66 - \frac{0.0593}{3}\log\frac{1}{(1 \times 10^{-33})}$$
$$= -1.66 - \frac{0.0591}{3}\log 10^{33}$$
$$= -1.66 - \frac{0.0591}{3} \times 33 \log 10$$
$$= -1.66 - \frac{0.0591}{3} \times 33 = -2.31 \text{ V}$$

15. (a) If H_2O_2 is a reducing agent,

$$Fe^{2+} + 2e^- \longrightarrow Fe \qquad E° = -0.44 \text{ V}$$
$$H_2O_2 \longrightarrow O_2 + 2H^+ + 2e^-, \qquad E° = -0.69 \text{ V}$$
$$\overline{Fe^{2+} + H_2O_2 \longrightarrow O_2 + Fe + 2H^+}, \qquad E°_{cell} = -1.13 \text{ V}$$

Since, $E°_{cell} < 0$, hence

H_2O_2 does not reduce Fe^{2+} to Fe.

If H_2O_2 is an oxidising agent,

$$2 \times [Fe^{2+} \longrightarrow Fe^{3+} + e^-], \qquad E° = 0.77 \text{ V}$$
$$H_2O_2 + 2H^+ + 2e^- \longrightarrow 2H_2O, \qquad E° = 1.77 \text{ V}$$
$$\overline{2Fe^{2+} + H_2O_2 + 2H^+ \longrightarrow 2Fe^{3+} + 2H_2O}, \; E°_{cell} = 2.54 \text{ V}$$

Since, $E°_{cell} > 0$, hence Fe^{2+} changes to Fe^{3+}.

16. (b)

$$2 \times [Cu^+ \longrightarrow Cu^{2+} + e^-] \qquad E° = -0.15 \text{ V}$$
$$Cu^{2+} + 2e^- \longrightarrow Cu \qquad E° = +0.34 \text{ V}$$
$$\overline{2Cu^+ \longrightarrow Cu^{2+} + Cu} \qquad E°_{cell} = 0.19 \text{ V}$$

$E°_{cell} > 0$, thus Cu^+ disproportionates to Cu^{2+} and Cu. Thus, copper (II) sulphate exists. Copper (I) sulphate is unstable.

17. (a)

$$Cu^{2+} + e^- \longrightarrow Cu^+, \qquad E° = +0.16 \text{ V}$$
$$\therefore \quad Cu^+ \longrightarrow Cu^{2+} + e^-, \qquad E° = -0.16 \text{ V}$$
$$Cu^+ + e^- \longrightarrow Cu, \qquad E° = +0.52 \text{ V}$$
$$\overline{2Cu^+ \longrightarrow Cu^{2+} + Cu}, \qquad E°_{cell} = +0.36 \text{ V}$$

18. (c) $E°_{D^{2+}/D} = -1.12$ V

Its value being most negative, D is oxidised to D^{2+} most easily and is thus is the best reducing agent. Hence, it is anodic part.

$$D \longrightarrow D^{2+} + 2e^-, \qquad E°_{D/D^{2+}} = 1.12 \text{ V}$$
$$E_{A/A^-} = +0.96 \text{ V}$$

Its value being most positive, A is reduced to A^- by D

$$2A + 2e^- \longrightarrow 2A^- \qquad E°_{A/A^-} = 0.96 \text{ V}$$
$$\overline{D + 2A \longrightarrow D^{2+} + 2A^-} \qquad E°_{cell} = 2.08 \text{ V}$$

19. (c) **Oxidation at anode**

$$Mg(s) \longrightarrow Mg^{2+}(aq) + 2e^-$$

half-cell is $\quad Mg \mid Mg^{2+}(aq)$

Reduction at cathode

$$2H^+(aq) + 2e^- \longrightarrow H_2(g)$$

Half-cell is

$$H^+(aq) \mid Pt(H_2)$$

Complete cell is

$$\underset{\ominus}{Mg} \mid Mg^{2+}(aq) \,||\, H^+(aq) \mid \underset{\oplus}{Pt(H_2)}$$
$$Q \text{ (reaction quotient)} = \frac{[Mg^{2+}]p_{H_2}}{[H^+]^2}$$
$$E_{cell} = E°_{cell} - \frac{2.303RT}{nF}\log Q$$
$$= E°_{cell} - \frac{0.0591}{2}\log\frac{[Mg^{2+}]p_{H_2}}{[H^+]^2}$$

On adding H^+, $[H^+]$ increases, Q decreases.

Equilibrium is displaced towards right and E_{cell} increased.

20. (d) For spontaneous reaction, $E^\circ_{cell} > 0$.

(a) $X = Ni$, $Y = Fe$

$$Ni + Fe^{2+} \longrightarrow Fe + Ni^{2+}$$

$$E^\circ_{cell} = E^\circ_{Ni/Ni^{2+}} + E^\circ_{Fe^{2+}/Fe}$$
$$= -E^\circ_{Ni^{2+}/Ni} + E^\circ_{Fe^{2+}/Fe}$$
$$= -(-0.23) - 0.44$$
$$= -0.21\ V$$

$E^\circ_{cell} < 0$ thus, non-spontaneous.

(b) $X = Ni$, $Y = Zn$

$$Ni + Zn^{2+} \longrightarrow Zn + Ni^{2+}$$

$$E^\circ_{cell} = E^\circ_{Ni/Ni^{2+}} + E^\circ_{Zn^{2+}/Zn}$$
$$= -E^\circ_{Ni^{2+}/Ni} + E^\circ_{Zn^{2+}/Zn}$$
$$= -(-0.23) - 0.76$$
$$= +0.23 - 0.76 = -0.53\ V$$

$E^\circ_{cell} < 0$, thus non-spontaneous.

(c) $X = Fe$, $Y = Zn$

$$Fe + Zn^{2+} \longrightarrow Fe^{2+} + Zn$$

$$E^\circ_{cell} = E^\circ_{Fe/Fe^{2+}} + E^\circ_{Zn^{2+}/Zn}$$
$$= -E^\circ_{Fe^{2+}/Fe} + E^\circ_{Zn^{2+}/Zn}$$
$$= -(-0.44) - 0.76$$
$$= +0.44 - 0.76 = -0.32\ V$$

$E^\circ_{cell} < 0$, thus, non-spontaneous

(d) $X = Zn$, $Y = Ni$

$$Zn + Ni^{2+} \longrightarrow Zn^{2+} + Ni$$

$$E^\circ_{cell} = E^\circ_{Zn/Zn^{2+}} + E^\circ_{Ni^{2+}/Ni}$$
$$= -E^\circ_{Zn^{2+}/Zn} + E^\circ_{Ni^{2+}/Ni}$$
$$= -(-0.76) + (-0.23)$$
$$= 0.76 - 0.23 = 0.43\ V$$

$E^\circ_{cell} > 0$, thus spontaneous.

21. (c) $Zn(s) + Cu^{2+}(aq) \rightleftharpoons Cu(s) + Zn^{2+}(aq)$

$$\text{Reaction quotient } (Q) = \frac{[Zn^{2+}]}{[Cu^{2+}]}$$

During discharging, $[Zn^{2+}]$ increases, $[Cu^{2+}]$ decreases. Thus, $[Zn^{2+}]/[Cu^{2+}]$ increases. E_{cell} decreases.

$$E_{cell} = E^\circ_{cell} - \frac{0.0591}{2}\log\frac{[Zn^{2+}]}{[Cu^{2+}]}$$

$$0.8634 = 1.10 - \frac{0.0591}{2}\log\frac{[Zn^{2+}]}{[Cu^{2+}]}$$

$$\log\frac{[Zn^{2+}]}{[Cu^{2+}]} = 8.00$$

22. (d) When AgI(s), a sparingly soluble salt is dissolved, following equilibrium is set up

$$AgI(s) \rightleftharpoons Ag^+(aq) + I^-(aq)$$
$$K_{sp} = [Ag^+][I^-]$$

Given,

$$AgI(s) + e^- \longrightarrow Ag + I^-, \quad E^\circ = -0.150\ V$$
$$Ag \longrightarrow Ag^+ + e^-, \quad E^\circ = -0.800\ V$$

Required equation

$$AgI(s) \longrightarrow Ag^+ + e^-, \quad E^\circ_{cell} = -0.950\ V$$

At equilibrium, $E_{cell} = 0$

Then, $$E_{cell} = E^\circ_{cell} - \frac{2.303\,RT}{F}\log K$$

$$0 = -0.95 - 0.0592\log K_{sp}$$

$\therefore$ $$\log K_{sp} = -\frac{0.95}{0.0592} = -16.0$$

$\therefore$ $$K_{sp} = \text{antilog of } (-16) = 10^{-16}$$

23. (d) $\Lambda_{(Na_2SO_4)} = 2\Lambda_{(Na^+)} + \Lambda_{(SO_4^{2-})}$

$$\Lambda_{(Na_2SO_4)} = \frac{\text{Conductivity } (S\ m^{-1})}{\text{Concentration } (mol\ m^{-3})}$$

$$0.001\ mol\ dm^{-3} = 0.001 \times 1000\ mol\ m^{-3}$$
$$= 1\ mol\ m^{-3}$$

$$\Lambda_{(Na_2SO_4)} = \frac{2.6 \times 10^{-2}\ S\ m^{-1}}{1\ mol\ m^{-3}}$$
$$= 2.6 \times 10^{-2}\ S\ m^2\ mol^{-1}$$
$$= 26 \times 10^{-3}\ S\ m^2 mol^{-1}$$

$$\Lambda_{(Na_2SO_4)} = 2\Lambda_{(Na^+)} + \Lambda_{(SO_4^{2-})}$$
$$26 \times 10^{-3} = 2 \times 5 \times 10^{-3} + \Lambda_{(SO_4^{2-})}$$

$\therefore$ $$\Lambda_{(SO_4^{2-})} = 16 \times 10^{-3} S\ m^2 mol^{-1}$$

24. (a) In saturated solution of sparingly soluble salt,

$$\underset{C \longrightarrow 0}{\text{Limit}}\ \Lambda^\circ = \Lambda_O (= \Lambda_{equiv}) = \frac{1000 \times \text{Specific conductance}}{\text{Normality}}$$

$$BaSO_4(\text{saturated}) \rightleftharpoons \underset{s}{Ba^{2+}} + \underset{s}{SO_4^{2-}}$$

$$K_{sp} = [Ba^{2+}][SO_4^{2-}]$$
$$= s^2, \text{ where } s = \text{solubility in mol } L^{-1}$$

$$\text{Molarity} \times 2 = \text{Normality}$$

$\therefore$ $$\text{Molarity} = s = \left(\frac{\text{Normality}}{2}\right)$$

From given data

$$\Lambda^\circ(BaSO_4) = \Lambda^\circ(BaCl_2) + \Lambda^\circ(H_2SO_4) - 2\Lambda^\circ(HCl)$$
$$= 100 + 250 - 300$$
$$= 50\ \Omega^{-1}cm^2 equiv^{-1}$$

$$50 = \Lambda^\circ(BaSO_4) = \frac{1000 \times 500}{\text{Normality}} 1 \times 10^{-6}$$

$\therefore$ $$\text{Normality} = \frac{1000 \times 1 \times 10^{-6}}{50} = 2 \times 10^{-5}\text{ g equiv } L^{-1}$$

$\therefore$ $$\text{Molarity} = 1 \times 10^{-5}\ mol\ L^{-1} = \text{Solubility}$$

$\therefore$ $$K_{sp} = s^2 = 1 \times 10^{-10}\ M^2$$

25. (b) Resistance $(R) \propto \rho$ (Distance between electrodes)

$$\propto \frac{1}{a} \text{ (Area between electrodes)}$$

$$R \propto \frac{\rho}{a}$$

$$R = k\left(\frac{\rho}{a}\right) = kC$$

$\left(\frac{\rho}{a}\right)$ is called cell constant (C) and k is called specific resistance.

$\log(R) = \log k + \log(C)$ – a straight line (y = constant + mx)

Thus, graph between $\log R$ and $\log C$ is of the type

Slope = 1 = tan 45°

$OA = \log k$

when $C = 1 = \frac{\rho}{a}$ $(\rho = a)$

then $\log R = \log k$

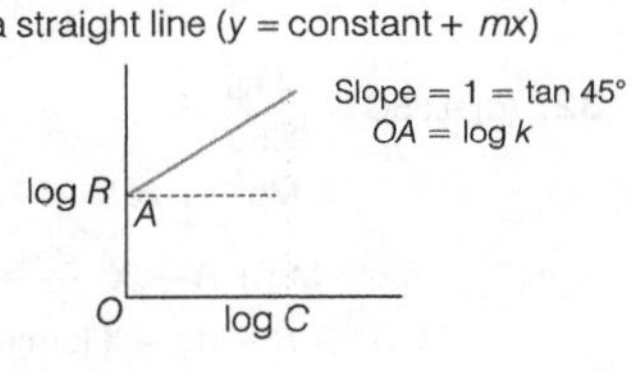

26. (a) By Kohlrausch's law at infinite dilution,

$$\Lambda^\circ_{NaBr} = \Lambda^\circ_{Na^+} + \Lambda^\circ_{Br^-}$$
$$= \Lambda^\circ_{NaCl} + \Lambda^\circ_{KBr} - \Lambda^\circ_{KCl}$$
$$= 126 + 152 - 150$$
$$= 278 - 150$$
$$= 128\ S\ cm^2 mol^{-1}$$

27. (d) Λ_m (molar conductance) $= \dfrac{\text{Specific conductance}}{\text{Molarity}}$

Molarity is same hence, equal molar conductance.

Λ_{equiv} (equivalent conductance) $= \dfrac{\text{Specific conductance}}{\text{Normality}}$

$$\text{Normality} = \text{Basicity} \times \text{Normality}$$

$$0.01\,M\,HCl = 0.01\,N\,HCl$$

$$0.01\,M\,H_2SO_4 = 0.02\,N\,H_2SO_4$$

$$0.01\,M\,H_3PO_4 = 0.03\,N\,H_3PO_4$$

Thus, Λ_{equiv} of HCl is maximum.

28. (a) Radius $= \dfrac{2.0}{2} = 1.0\text{ mm} = 1 \times 10^{-3}\text{ m}$

Area of cross-section and resistance (R) is related to the resistivity as,

(Specific resistance) , the length and area by

$$R = \text{Resistivity} \times \frac{\text{Length}}{\text{Area}}$$

$$= \frac{2.824 \times 10^{-8}\,\Omega\text{m} \times 1.00\text{ m}}{\pi(1.0 \times 10^{-3}\text{m})^2}$$

$$= 9.0 \times 10^{-3}\,\Omega$$

$$\text{Conductance} = \frac{1}{\kappa} = \frac{1}{9.0 \times 10^{-3}} = 111\,\Omega^{-1}\text{(S)}$$

29. (c) By Faraday's first law,

$$w\text{(amount)} = zit = zQ$$

where, $it = Q$ = quantity of electricity

z = electrochemical equivalent

$$= \frac{\text{Atomic mass}}{\text{Number of electrons} \times 96500\text{ C}}$$

$$\therefore\ 5.12 \times 10^3\text{g} = \frac{27}{3 \times 96500} \times Q$$

$$\therefore\ Q = \frac{5.12 \times 10^3 \times 96500}{27} = 1.83 \times 10^7\text{C}$$

30. (a) In the electrosynthesis (as electrolysis), ions are discharged at the electrodes thus, reverse reaction takes place.

$$M \longrightarrow M^{n+} + ne^- \qquad E > 0$$

Thus, $M^{n+} + ne \longrightarrow M \qquad E < 0$

Thus, $E_{cell} = -\text{ve}$

$\therefore\ \Delta G = -nFE_{cell} = +\text{ve}$

31. (b) Silver formed $= 108\text{ g} = \dfrac{108\text{ g}}{108\text{ g mol}^{-1}} = 1.0\text{ mol}$

Number of Ag atoms in 1 mole Ag = N_0 atoms

Total charge carried by N_0 atoms $= N_0 \times 1.6 \times 10^{-19}$C

$\therefore N_0 \times 1.6 \times 10^{-19} = 96500$ C

$$\therefore\ N_0 = \frac{96500}{1.0 \times 10^{-19}} = 6.03 \times 10^{23}$$

32. (d) $9.65\text{C} = \dfrac{9.65}{96500}F$

$$Mg^{2+} + 2e^- \longrightarrow Mg$$

$$Mg + R\text{—}X \xrightarrow{\text{Ether}} R\text{—}Mg\text{—}X$$

$R - Mg - X$ formed = Mg formed

Mg formed $= \left(\dfrac{1}{2}\right)$ of Faraday passed

$$= \frac{9.65}{96500 \times 2} = 5 \times 10^{-5}\text{ mol}$$

33. (a)

$$2H_2O + 2e^- \longrightarrow H_2 + 2OH^- \qquad \ldots\text{(i)}$$

$$2H_2O \longrightarrow O_2 + 4H^+ + 4e^- \qquad \ldots\text{(ii)}$$

From Eq. (i), 1 mole H_2 = 22.4 L H_2 is obtained by 2 units

$XLH_2 = \dfrac{2X}{22.4}$ units costing ₹ = x

$\therefore$ 1 unit cost $= \dfrac{22.4}{2} = ₹11.2$

From Eq. (ii), 22.4 L O_2 at STP = 4 units

XLO_2 at STP $= \dfrac{4}{22.4}X$ units $= \dfrac{4X}{22.4} \cdot ₹11.2 = ₹2X$

34. (a)

$$\underset{+2}{Mn^{2+}} \longrightarrow \underset{0}{Mn} + \underset{+3}{Mn^{3+}}$$

ON: Mn^{2+} (+2) → Mn (0): Reduction; Mn^{2+} (+2) → Mn^{3+} (+3): Oxidation

Mn^{2+} is oxidised as well as reduced. Thus, it is a disproportionation reaction.

$$2 \times [Mn^{2+} \longrightarrow Mn^{3+} + e^-] \qquad E° = -1.15\text{V}$$

$$Mn^{2+} + 2e^- \longrightarrow Mn \qquad E° = -1.18\text{ V}$$

$$\therefore\ 3Mn^{2+} \longrightarrow 2Mn^{3+} + Mn \qquad E° = -2.69\text{ V}$$

Since, $E° < 0$, hence reaction will not occur.

35. (d) Fe is active metal and is oxidised to Fe^{2+}; Fe^{2+} is dissolved in O_2 in acidic medium (or alkaline medium) forming $Fe_2O_3(s)$ - rust.

36. (a) **Anode** $\quad Zn \longrightarrow \underset{0.01\text{ M}}{Zn^{2+}} + 2e^-$

Cathode $\quad \underset{0.1\text{ M}}{Zn^{2+}} + 2e^- \longrightarrow Zn$

$$Zn^{2+}\ (0.1\text{ M}) \longrightarrow Zn^{2+}\ (0.01\text{ M})$$

$$Q = \frac{0.01}{0.10} = \frac{1}{10}$$

$$E°_{cell} = 0$$

$$E_{cell} = 0 - \frac{0.0591}{2}\log\frac{1}{10} = \frac{0.0591}{2}\log 10 = \frac{0.0591}{2}\text{ V}$$

$E_{cell} > 0$ and spontaneous. Thus, Statement I is correct.

Zn^{2+} ion flows from 0.1 M (concentrated) to 0.01 M (dilute). Thus, concentrated solution becomes more dilute and dilute becomes more concentrated. Thus,Statements I and II both are correct and Statement II is the correct explanation of Statement I.

37. (c) Iron materials as electricity poles can be protected from rusting by connecting it to active metals as zinc. This is a type of cathodic protection and zinc which is oxidised is the sacrified anode. It is not the galvanisation. Thus, Statement I is correct but Statement II is incorrect.

38. (c) Salt-bridge connects two half-cells and it allows cations and anions to move between two half-cells.

Electrolyte is generally $NaNO_3$, NH_4NO_3, KNO_3. NaCl, KCl cannot be used with silver electrodes as it react with silver forming AgCl. Thus, Statement I is correct and Statement II is incorrect.

39. (d)

$$E°_{Cu^{2+}/Cu} = 0.34\text{V}$$

$$E°_{Zn^{2+}/Zn} = -0.76\text{V}$$

$$Cu^{2+} + 2e^- \longrightarrow Cu \qquad E° = 0.34$$

$$Zn \longrightarrow Zn^{2+} + 2e^- \qquad E° = 0.76$$

$$Cu^{2+} + Zn \longrightarrow Zn^{2+} + Cu \qquad E°_{cell} = 1.10\text{V}$$

Since, $E°_{cell} > 0$, thus $CuSO_4$ is reduced to Cu.

Thus, $CuSO_4$ cannot be stored in a vessel made of zinc. Thus, Statement I is incorrect and Statement II is correct (Standard $E°_{red}$ values)

40. (d) If different number of electrons are involved,

$$E°_3 \neq |E°_1| + |E°_2|$$

but $\quad \Delta G°_3 = |\Delta G°_1| + |\Delta G°_2|$

Hence, Statement I is incorrect and Statement II is correct.

Revisal Problems (JEE Advanced)

1. (d) From potentiometric curve, EMF at the equivalent point = 2.0 V.

$\therefore \quad E^\circ_{cell} = 2.00\text{ V}$

$$E^\circ_{cell} = \frac{0.0591}{n}\log K_{eq}$$

$n = 10$ = electrons exchanged by 2 moles of MnO_4^-

$$2.0 = \frac{0.0591}{10}\log K_{eq}$$

$$\therefore \quad \log K_{eq} = \frac{10 \times 2}{0.0591} = \frac{20}{0.0591}$$

2. (c) $NiO_2(s) + 2H_2O + 2e^- \longrightarrow Ni(OH)_2 + 2OH^-, E^\circ = -0.490\text{ V}$

$Ni^{2+} + 2H_2O \longrightarrow NiO_2(s) + 4H^+ + 2e^-, \quad E^\circ = -1.678\text{ V}$

Adding

$Ni^{2+} + 4H_2O \longrightarrow Ni(OH)_2 + 2OH^- + 4H^+, \quad E^\circ = -2.168\text{ V}$

As $2H^+ + 2OH^- \longrightarrow 2H_2O$

Thus,

$Ni^{2+} + 4H_2O \longrightarrow Ni(OH)_2 + 2H_2O + 2H^+$

$\therefore \quad Ni^{2+} + 2H_2O \longrightarrow Ni(OH)_2 + 2H^+, \qquad E^\circ = -2.168\text{ V}$

$$\Delta G^\circ = -nFE^\circ$$
$$= -2 \times 96500 \times (-2.168)$$
$$= 418424\text{ J}$$
$$= 418.424\text{ kJ}$$

($\Delta G^\circ > 0$, reaction is non-spontaneous)

3. (a) With Ag electrodes

Anode

$Ag + NO_3^- \longrightarrow Ag^+ + NO_3^- + e^-$

Cathode $\quad Ag^+ + e^- \longrightarrow Ag$

Ag electrode dissolves at anode and equivalent amount is deposited at the cathode. Thus, there is no change in concentration of $AgNO_3$.

With Pt electrodes

Anode

$NO_3^- + H_2O \longrightarrow 2H^+ + NO_3^- + \frac{1}{2}O_2 + 2e^-$

Cathode $\quad Ag^+ + e^- \longrightarrow Ag$

Ag is deposited at the cathode and thus, concentration of $AgNO_3$ decreases.

z (electrochemical equivalent) $= \dfrac{108}{96500 \times \text{electrons}}$

Ag deposited $= izt$

$$= 2 \times \frac{108}{96500} \times 2 \times 3600$$
$$= 16.12\text{ g} = 0.15\text{ mol}$$

$AgNO_3$ left $= 1.0 - 0.15 = 0.85$ mol in 1 kg water

Molality $= 0.85\text{ mol kg}^{-1}$

$= 0.85$ molal

4. (d) **In I,** $Ag^+ + e \longrightarrow Ag$

Ag deposited by 1F = 1mol

Ag deposited by 0.5F = 0.5 mol

Thus, $[Ag^+]_I = 1.0 - 0.5 = 0.5\text{ mol L}^{-1}$

In II, Ag deposited by 1F = 1mol

Thus, $\quad [Ag^+]_{II} = 2.0 - 1.0 = 1.0\text{ mol L}^{-1}$

Initially cell is

$$\underset{-}{Ag} \mid Ag^+\,(1M) \parallel Ag^+\,(2M) \mid \underset{+}{Ag}$$

Anode $\quad Ag \longrightarrow (Ag^+)_L + e^-$

Cathode $\quad (Ag^+)_R + e^- \longrightarrow Ag$

$$\underset{2M}{(Ag^+)_R} \longrightarrow \underset{1M}{(Ag^+)_L}$$

$$Q = \frac{[Ag^+]_L}{[Ag^+]_R} = \left(\frac{1}{2}\right)$$

$$E = E^\circ - \frac{0.0591}{1}\log\left(\frac{1}{2}\right)$$

$$E = 0 + 0.0591\log 2$$

After current is passed cell is

$$\underset{-}{Ag} \mid Ag^+\,(0.5M) \parallel Ag^+\,(1M) \mid \underset{+}{Ag}$$

$$\therefore \quad Q = \frac{[Ag]_L}{[Ag^+]_R} = \frac{0.5}{1.0} = \left(\frac{1}{2}\right)$$

$$\therefore \quad E' = 0 + 0.0591\log 2$$

Thus, no change in (EMF)

5. (b) Initially

$$Cu + 2Ag^+(aq) \longrightarrow 2Ag + Cu^{2+}(aq)$$

$$E^\circ_{cell} = E^\circ_{Cu/Cu^{2+}} + E^\circ_{Ag^+/Ag}$$
$$= -E^\circ_{Cu^{2+}/Cu} + E^\circ_{Ag^+/Ag}$$
$$= -0.34 + 0.80 = 0.46\text{ V}$$

$$E^\circ_1 = E^\circ_{cell}$$
$$E^\circ_2 = E^\circ_{cell}\text{ (at 10° higher)} = 0.48\text{ V}$$

$$\therefore \quad dE^\circ_{cell} = 0.48 - 0.46 = 0.02\text{ V}$$
$$dT = 10^\circ$$
$$\left(\frac{dE^\circ_{cell}}{dT}\right) = \frac{0.02}{10}\text{ VK}^{-1}$$

$$\Delta G^\circ = \Delta H^\circ - T\Delta S^\circ$$

Also, $\quad \Delta G^\circ = \Delta H^\circ + T\left[\dfrac{d(\Delta G)}{dT}\right]_p$

$$\therefore \quad \Delta S^\circ = -\left[\frac{d(\Delta G)}{dT}\right]_p = -\left[\frac{d(-nFE^\circ_{cell})}{dT}\right]_p$$
$$= nF\left[\frac{dE_{cell}}{dT}\right]_p$$

$$n = 2$$
$$F = 96500\text{ C mol}^{-1}$$

$$\therefore \quad \Delta S^\circ = 2 \times 96500 \times \frac{0.02}{10}\text{ VK}^{-1}\text{C mol}^{-1}$$
$$= 386\text{ VK}^{-1}\text{C mol}^{-1} = 386\text{ JK}^{-1}$$

6. (d),

(a) $\times 4[Fe^{2+} \longrightarrow Fe^{3+} + e^-, \quad E^\circ = -0.77\text{ V}$

$O_2 + 4H^+ + 4e^- \longrightarrow 2H_2O, \quad E^\circ = +1.23\text{ V}$

$4Fe^{2+} + O_2 + 4H^+ \longrightarrow 4Fe^{3+} + 2H_2O \quad E^\circ_{cell} = 0.46\text{ V}$

$E^\circ_{cell} > 0$, hence Fe^{2+} is oxidised to Fe^{3+} by air.

Thus, (a) is correct

(b) $\times 2\ [Fe^{3+} + e^- \longrightarrow Fe^{2+}, \qquad E^\circ = +0.77\text{ V}$

$Fe \longrightarrow Fe^{2+} + 2e^- \qquad E^\circ = +0.44\text{ V}$

$2Fe^{3+} + Fe \longrightarrow 3Fe^{2+} \qquad E^\circ_{cell} = 1.21\text{ V}$

$E^\circ_{cell} > 0$, thus spontaneous

Thus, Fe^{3+} is reduced to Fe^{2+} by iron nails.

Then (b) is correct.

(c) $O_2 + 4H^+ + 2Fe \longrightarrow 2Fe^{2+} + 2H_2O, E^\circ_{cell} = 1.67\text{ V}$

$E^\circ_{cell} > 0$, thus spontaneous.

Hence, corrosion of iron takes place in air.

7. (d)

Anode

$$Pb(s) \rightleftharpoons Pb^{2+}(aq) + 2e^- \qquad E°_{OX} = x$$

Cathode

$$\underset{\text{(Mercurous)}}{Hg_2^{2+}(aq)} + 2e^- \rightleftharpoons 2Hg(l) \qquad E°_{red} = y$$

$$Pb(s) + Hg_2^{2+}(aq) \rightleftharpoons Pb^{2+}(aq) + 2Hg(l)$$

$$E°_{cell} = x + y = 0.92\text{ V}$$

$$\text{Reaction quotient, } Q = \frac{[Pb^{2+}]}{[Hg_2^{2+}]}$$

Pb^{2+} is from $PbSO_4 \rightleftharpoons Pb^{2+} + SO_4^{2-}$

$$[Pb^{2+}] = \sqrt{K_{sp}(PbSO_4)}$$

$$[Pb^{2+}] = \sqrt{2 \times 10^{-8}}\text{ M}$$

Hg_2^{2+} is from

$$Hg_2SO_4 \rightleftharpoons Hg_2^{2+} + SO_4^{2-}$$

$$[Hg_2^{2+}] = \sqrt{K_{sp}(Hg_2SO_4)} = \sqrt{1 \times 10^{-6}}$$

$$\therefore \quad Q = \frac{\sqrt{2 \times 10^{-8}}}{\sqrt{1 \times 10^{-6}}}$$

$$= \sqrt{2 \times 10^{-2}} = 0.141$$

$$E_{cell} = E°_{cell} - \frac{2.303RT}{nF}\log Q$$

$$= 0.92 - \frac{0.0591}{2}\log 0.141$$

$$= 0.92 - 0.0295\log 0.141$$

$$= 0.92 - 0.0295(-0.85)$$

$$= 0.92 + 0.025 = 0.95\text{ V}$$

8. (c) $BaSO_4 = BaCl_2 + H_2SO_4 - 2HCl$

$$\therefore \Lambda°_{equiv}(BaSO_4) = \Lambda°_{BaCl_2} + \Lambda°_{H_2SO_4} - 2\Lambda°_{HCl}$$

$$= (x_1 + x_2 - 2x_3)$$

$\Lambda°_{equiv} = \Lambda_{equiv}$ for sparingly soluble salt and solubility = molarity as $BaSO_4$

$$\therefore \quad (x_1 + x_2 - 2x_3) = \frac{1000 \times \text{specific conductance}}{\text{Normality}}$$

$$\therefore \quad \text{Normality} = \frac{1000 \times y}{(x_1 + x_2 - 2x_3)}$$

$$\because \quad \text{Molarity} = \frac{1000 \times y}{2(x_1 + x_2 - 2x_3)} = \text{solubility}$$

$$\therefore \quad K_{sp} = [Ba^{2+}][SO_4^{2-}] = s^2$$

$$= \left[\frac{1000\,y}{2(x_1 + x_2 - 2x_3)}\right]^2$$

$$= \frac{10^6 y^2}{4(x_1 + x_2 - 2x_3)^2}$$

9. (a) Let solubility of AgBr in presence of $AgNO_3 = x$

$$AgBr(s) \rightleftharpoons \underset{x}{Ag^+} + \underset{x}{Br^-}$$

$$AgNO_3 \rightleftharpoons \underset{1 \times 10^{-7}}{Ag^+} + \underset{1 \times 10^{7}}{NO_3^-}$$

$$\therefore \quad [Ag^+] = (x + 1 \times 10^{-7})$$

$$[Br^-] = x$$

$$\therefore \quad [Ar^+][Br^-] = K_{sp}$$

$$(x + 1 \times 10^{-7})x = 12 \times 10^{-14}$$

$$x^2 + x \times 10^{-7} - 12 \times 10^{-14} = 0$$

$$x = \frac{-1 \times 10^{-7} \pm \sqrt{1 \times 10^{-14} + 48 \times 10^{-14}}}{2}$$

$$= \frac{-1 \times 10^{-7} + 7 \times 10^{-7}}{2}$$

$$x = 3 \times 10^{-7}\text{ M}$$

$$\therefore \quad [Ag^+] = 3 \times 10^{-7}\text{M} + 1 \times 10^{-7}$$

$$= 4 \times 10^{-7}\text{M}$$

$$[Br^-] = 3 \times 10^{-7}\text{M}$$

$$[NO_3^-] = 1 \times 10^{-7}\text{M}$$

Total specific conductance

$$= \lambda°_{Br^-} \times [Br^-] + \lambda°_{Ag^+} \times [Ag^+] + \lambda°_{NO_3^-} \times [NO_3^-]$$

$$= 8 \times 10^{-3} \times 3 \times 10^{-7} + 6 \times 10^{-3} \times 4 \times 10^{-7} + 7 \times 10^{-3} \times 1 \times 10^{-7}$$

$$= 55 \times 10^{-10}\text{ S m}^2\text{mol}^{-1}\text{mol L}^{-1}$$

$$= 55 \times 10^{-10}\text{ S m}^2\text{L}^{-1}$$

$$= 55 \times 10^{-10}\text{S m}^2 \times 1000\text{ m}^3$$

$$= 55 \times 10^{-7}\text{ S m}^{-1}$$

$$= 5.5 \times 10^{-6}\text{Sm}^{-1}$$

Note $1L = 1\text{dm}^3 = \frac{1}{1000}\text{m}^3$

10. (c)

(a) $\Delta G°_I = -nFE° = -2 \times 96500(-2.71)$ J

$= 523030$ J $= 5.23 \times 10^5$ kJ

$\Delta G°_{II} = -2 \times 96500 \times 0.83$ J

$= 160190$ J $= 1.602 \times 10^5$kJ

$\Delta G°_{II}$ is less positive than $\Delta G°_I$

Hence, II is preferred-true.

(b) $Na^+ + e^- \longrightarrow Na$

$$Hg + Na \longrightarrow \underset{\text{Amalgam}}{Hg(Na)}$$

If amalgam is formed, I is preferred thus (a), (b) both are correct.

11. (c)

$$E_1 = E°_1 - \frac{0.0591}{p}\log\frac{[RED_1]}{[OX_1]}$$

$$E_2 = E_2^\circ - \frac{0.0591}{q}\log\frac{[RED_2]}{[OX_2]}$$

At equivalence point, $E_1 = E_2 = E_{equiv}$

and $\frac{[RED_1][RED_2]}{[OX_1][OX_2]} = 1$

$$E_1 p = E°_1 p - 0.0591\log\frac{[RED_1]}{[OX_1]}$$

$$E_2 q = E°_2 q - 0.0591\log\frac{[RED_2]}{[OX_2]}$$

$$\therefore \quad E_1 p + E_2 q = (E_{equiv}p + E_{equiv}q)$$

$$E_{equiv}(p + q) = E_1^\circ p + E_2^\circ q - 0.0591\log\frac{[RED_1][RED_2]}{[OX_1][OX_2]}$$

$$= E_1^\circ p + E_2^\circ q - 0.0591\log 1$$

$$\therefore \quad E_{equiv} = \left(\frac{E°_1 p + E°_2 q}{p + q}\right)$$

12. (a)

$$\frac{dE}{dT} = -7 \times 10^{-14}$$

$$E = -\int_{T_1}^{T_2}\left(\frac{dE}{dT}\right)_{dT} - \int_{T_1}^{T_2}(-7 \times 10^{-4})dT$$

$$= +7 \times 10^{-4}(1050 - 850) = 0.14\text{ V}$$

13. (a) $CH_3COOH + NaOH$ (being added) $\rightleftharpoons CH_3COONa + H_2O$

As CH_3COONa is formed, it is hydrolysed at the same time forming CH_3COOH (weak acid) and NaOH (strong base). Due to formation of strong base, conductance increases at every addition of NaOH. After CH_3COOH is completely neutralised further addition of NaOH causes rapid increase in conductance (no CH_3COOH at this stage)

$$CH_3COONa + H_2O \xrightleftharpoons{\text{Hydrolysis}} CH_3COOH + NaOH$$

Thus, both Statements I and II are correct and Statement II is the correct explanation of Statement I.

14. (a) Oxidation

$$2Cl^- \longrightarrow Cl_2 + 2e^-, \quad E° = 1.36\text{ V}$$

$$2H_2O \longrightarrow O_2 + 4H^+ + 4e^-, \quad E° = 1.23\text{ V}$$

In concentrated NaCl, Cl_2 is favoured and in dilute NaCl, O_2 is favoured. However, product is Cl_2. It is due to over-voltage. Experiments indicate that the applied voltage required is always greater than the voltage calculated from the standard potentials. The additional voltage required is called over-voltage. Thus, Statements I and II are correct and Statement II is the correct explanation of Statement I.

15. (b,c,d)

(a) From (I) and (IV)

$$2MnO_4^- + 16H^+ + 10Cl^- \longrightarrow 2Mn^{2+} + 5Cl_2 + 8H_2O$$

$$E°_{cell} = 1.51 - 1.40 = 0.11\text{ V}$$

This cell reaction is spontaneous since, $E°_{cell} > 0$.

Thus, MnO_4^- cannot be used in aqueous HCl.

(b) Combining (II) and (IV)

$$Cr_2O_7^{2-} + 14H^+ + 6Cl^- \longrightarrow 2Cr^{3+} + 3Cl_2 + 7H_2O$$

$$E°_{cell} = 1.38 - 1.40 = -0.02\text{ V}$$

$$E°_{cell} < 0$$

Thus, $Cr_2O_7^{2-}$ is not affected in HCl medium. Thus, HCl can be used in $Cr_2O_7^{2-}$ solution.

(c) In I, source of H^+ is H_2SO_4, SO_4^{2-} is not further oxidised by MnO_4^- hence, H_2SO_4 can be used.

(d) same as (c).

16. (a,b,c) κ (specific conductance of water) $= 5.7 \times 10^{-6}\text{ Sm}^{-1}$

$$\kappa = L\text{ (conductance)} \times \text{cell constant}$$

For H_2O, in $L = \dfrac{\kappa}{\text{cell constant}}$

For solution conductance $= 10L = \dfrac{10\kappa}{\text{cell constant}}$

$\therefore$ Conductance of pure KCl solution $= L$ (solution) $- L$ (water)

$$= \frac{9\kappa}{\text{cell constant}}$$

$\therefore$ Specific conductivity of solution $= L \times$ cell constant

$$= \frac{9\kappa}{\text{cell constant}} \times \text{cell constant} = 9\,\kappa = 9 \times 5.7 \times 10^{-6}\text{ Sm}^{-1}$$

$$\Lambda_m\text{ (molar conductance)} = \frac{\text{specific conductance}}{\text{concentration (mol m}^{-3})}$$

$$\text{Concentration} = \frac{9 \times 5.7 \times 10^{-6}\text{ Sm}^{-1}}{0.0147\text{ Sm}^2\text{mol}^{-1}}$$

$$= 3.5 \times 10^{-3}\text{mol m}^{-3}$$

$$= \frac{3.5 \times 10^{-3}\text{ mol}}{\text{m}^3} = \frac{3.5 \times 10^{-3}\text{mol}}{1000\text{ dm}^3}$$

$$= 3.5 \times 10^{-6}\text{ mol dm}^{-3} = 3.5 \times 10^{-6}\text{ mol L}^{-1}$$

$$= 3.5 \times 10^{-9}\text{mol cm}^{-3}$$

17. (c)

(a) Due to dilution, ionisation increases hence, conduction increases.

(b) Increase in temperature, increases ionisation thus, conduction increases.

(c) In higher concentration range, ions are closer to one another thus interionic attraction increases. Hence, conduction decreases.

18. (a,b,d) $(N)_{red}$ is the reduction product of NO_3^-.

$$V \longrightarrow V^{2+} + 2e^-, \quad E° = +1.19\text{ V}$$

$$NO_3^- + e^- \longrightarrow (N)_{red}, \quad E° = +0.96\text{ V}$$

$$E°_{cell} = 2.15\text{ V}, \quad E°_{cell} > 0$$

Hence, V is oxidised to V^{2+} by NO_3^- ion

$$Hg \longrightarrow Hg^{2+} + 2e^-, \quad E° = -0.80\text{ V}$$

$$NO_3^- + e^- \longrightarrow (N)_{red}, \quad E° = +0.96\text{ V}$$

$$E°_{cell} = 0.16\text{ V}, \quad E°_{cell} > 0$$

Thus, Hg is also oxidised to Hg^{2+} by NO_3^- ion.

$$Fe \longrightarrow Fe^{3+} + 3e^-, \quad E° = +0.04\text{V}$$

$$NO_3^- + e^- \longrightarrow (N)_{red}, \quad E° = 0.96$$

$$E°_{cell} = 1.00\text{V}, E°_{cell} > 0$$

Hence, Fe is oxidised to Fe^{3+}

$$Au \longrightarrow Au^{3+} + 3e^-, \quad E° = -1.40\text{ V}$$

$$NO_3^- + e^- \longrightarrow (N)_{red}, \quad E° = 0.96\text{ V}$$

$$E°_{cell} = -0.44\text{ V}$$

Since, $E°_{cell} < 0$, hence Au is not oxidised to Au^{3+} by NO_3^- ion.

Thus, (a) V and Hg, (b) Hg and Fe, (d) Fe and V pairs are oxidised by NO_3^- in aqueous medium.

19. (a,b,c)

(a) Fe^{3+} is reduced to Fe^{2+} by Fe which is oxidised thus, reaction can take place even in the absence of second half-cell or a current-correct.

$$2FeCl_3 + Fe \longrightarrow 3FeCl_2$$

(b) $Fe \longrightarrow Fe^{2+}$, thus Fe is anode

$$Fe \mid Fe^{2+}$$

$$Fe^{3+} + e^- \longrightarrow Fe^{2+}$$

Since, oxidised species and reduced species are ions, hence Pt cathode is used, $Fe^{3+}, Fe^{2+} \mid$ Pt-correct.

(c) Fe as cathode would directly react with Fe^{3+} thus, short-circuiting the cell-correct.

(d) Cell set up is

$$Fe \mid Fe^{2+} \parallel Fe^{3+}, Fe^{2+} \mid Pt$$

Thus, given cell is incorrect.

20. (b)

(i) Before current is passed, mixture of $H_2PO_4^-$ and HPO_4^{2-} is a buffer of equimolar value.

$$H_2PO_4^- \rightleftharpoons H^+ + HPO_4^{2-}$$

Thus, by Hasselbalch Henderson equation

$$pH = pK_a + \log\frac{[HPO_4^{2-}]}{[H_2PO_4^-]}$$

$$= pK_a + \log\left(\frac{1}{1}\right)$$

$$= pK_a = 2.15$$

Thus, **(i)** $\rightarrow$ **(s)**

(ii) In anode side

$$HPO_4^{2-} + H^+ \rightleftharpoons H_2PO_4^-$$

	HPO_4^{2-}	H^+	$H_2PO_4^-$
Initial	1.0 M	0	1.0 M
$[H^+]$ added		0.156 M	
After	(1.0 − 0.156M)		(1.0 + 0.156M)
electrolysis	0.844 M	0	1.156 M

∴ pH of the buffer in anodic side

$$pH = pK_a + \log\frac{[HPO_4^{2-}]}{[H_2PO_4^-]} = 2.15 + \log\frac{0.844}{1.156} = 2.01$$

Thus, **(ii) → (r)**

(iii) At the cathodic side $[OH^-]$ is formed

$$H_2PO_4^- + OH^- \rightleftharpoons HPO_4^{2-} + H_2O$$

	$H_2PO_4^-$	OH^-	HPO_4^{2-}
Initial	1.0 M	0	1.0 M
$[OH^-]$ added		0.156 M	
After electrolysis	(1.0 − 0.156)	0	(1.0 + 0.156)
	0.844 M		1.156 M

$$\therefore\ pH = pK_a + \log\frac{[HPO_4^{2-}]}{[H_2PO_4^-]} = 2.15 + \log\frac{1.156}{0.844} = 2.29$$

Thus, **(iii) → (p)**

(iv) On electrolysis,

Anodic reaction

$$2H_2O(l) \xrightarrow{\text{Current}} 4H^+ + O_2 + 4e^-$$

Cathodic reaction

$$4H_2O(l) + 4e^- \longrightarrow 4OH^- + 2H_2$$

$$H^+ \text{ formed} = izt = 1.25 \times \frac{1 \times 201 \times 60}{96500}$$

$$= 0.156\ g\ L^{-1} = 0.156\ M \text{ at anode}$$

Thus, OH^- formed = 0.156 M at cathode.

Thus **(iv) → (q)**

21. (c) $\Lambda^\circ(H_2O) = \lambda^\circ(H^+) + \lambda^\circ(OH^-)$

$$= (3.5 \times 10^{-2} + 2.0 \times 10^{-2})\ S\ m^2 mol^{-1}$$

$$= 5.5 \times 10^{-2} S\ m^2 mol^{-1}$$

Density of water = 1.0 g cm^{-3} = 1000 g dm^{-3}

$$\therefore\quad \text{Molarity } (H_2O) = \frac{1000}{18} = 55.55\ mol\ dm^{-3}$$

$$= 55.55 \times 10^3\ mol\ m^{-3}$$

$$\Lambda_c (\text{molal conductance}) = \frac{\kappa}{\text{Molarity } (mol\ m^{-3})}$$

$$= \frac{5.7 \times 10^{-6} S\ m^{-1}}{55.5 \times 10^3 mol\ m^{-3}}$$

$$= 1.027 \times 10^{-10} S\ m^2 mol^{-1}$$

$$\therefore\quad p_{\Lambda_c} = 9.988 \approx 10.0$$

Thus, **(ii) → (p)**

$$x \text{ (degree of ionisation)} = \frac{\Lambda_c}{\Lambda^\circ} = \frac{1.027 \times 10^{-10}}{5.55 \times 10^{-2}} = 1.867 \times 10^{-9}$$

$$p_x = 8.73$$

Thus, **(i) → (q)**

In H_2O at 298 K

$$[H^+] = [OH^-] = Cx = 55.55\ M \times 1.867 \times 10^{-9}$$

$$= 1.03 \times 10^{-7}\ M$$

$$\therefore\quad pH = -\log[H^+] = 6.98 \approx 7.0$$

Thus, **(iii) → (s)**

$$K_{av} = [H^+][OH^-] = (1.03 \times 10^{-7})^2 = 1.061 \times 10^{-14}$$

$$\therefore\quad pK_w = 14.0$$

Thus, **(iv) → (r)**

22. (a) $x = \frac{\Lambda_m}{\Lambda^\circ_m}$

$$\therefore\quad \Lambda_m = x\Lambda^\circ_m$$

By Ostwald's dilution law $x = \sqrt{\frac{K_a}{C}}$

(i) $\Lambda_m = \sqrt{\frac{K_a}{C}}\,\Lambda^\circ_m = \sqrt{\frac{1.8 \times 10^{-5}}{0.04}}\ 390.5$

$$= 8.28\ S\ cm^2 mol^{-1} = 8.28 \times 10^{-4} S\ m^2 mol^{-1}$$

Thus, **(i) → (p)**

(ii) $\frac{\kappa(\text{conductivity}) \times 1000}{\text{Molarity}} = \Lambda_m$

$$\therefore\quad \kappa(\text{conductivity}) = \frac{\Lambda_m \times \text{molarity}}{1000}$$

$$= \frac{8.28 \times 0.04}{1000}$$

$$= 3.312 \times 10^{-4}\ S\ cm^{-1}$$

$$= 3.312 \times 10^{-4} \times 100\ Sm^{-1}$$

$$= 3.312 \times 10^{-2} S\ m^{-1}$$

Thus, **(ii) → (r)**

(iii) Also $\kappa(\text{conductivity}) = \frac{\text{cell constant}}{\text{resistance}}$

$$\therefore\quad \text{Resistance} = \frac{\text{cell constant } (cm^{-1})}{\kappa \text{ (conductivity) } Scm^{-1}}$$

$$= \frac{0.206\ cm^{-1}}{3.312 \times 10^{-4}\ S\ cm^{-1}}$$

$$= 622\ \Omega = 0.622\ k\ ohm$$

Thus, **(iii) → (s)**

(iv) $\text{Conductance} = \frac{1}{\text{resistance}} = \frac{1}{622} = 1.61 \times 10^{-3} ohm^{-1}$

$$= 1.61 \times 10^{-3} S$$

Thus, **(iv) → (q)**

Explanations for Q. Nos. (23-24)

During charging

$$Pb + SO_4^{2-} \longrightarrow PbSO_4 + 2e^-$$

During discharging

$$PbO_2 + 4H^+ + SO_4^{2-} + 2e^- \longrightarrow PbSO_4 + 2H_2O$$

H_2SO_4	Density	% by weight	H_2SO_4 per litre
Initial	1.294g mL^{-1}	39%	$1294 \times \frac{39}{100} = 505$g
Final	1.139g mL^{-1}	20%	$1139 \times \frac{20}{100} = 228$ g

Thus, H_2SO_4 consumed per litre = (505 − 228) g = 277 g

Total acid in 3.5 L = 969.5 g

H_2SO_4 consumed = 9.89 mol

Electrons consumed = 2 mol for 2 mol H_2SO_4

= 1 mol for 1 mol H_2SO_4

Thus, electrons = 9.89 mol

$$\therefore\quad \text{Ampere hour} = \frac{9.89 \times 96500}{3600}$$

$$= 265.1\ Ah$$

23. (a) **24.** (b)

Explanations for Q. Nos. (25-26)

(a) $\Lambda_m = \Lambda^\circ_m - (A + B\Lambda^\circ_m)\sqrt{C}$

Thus, $\Lambda^\circ_m > \Lambda_m$ correct

(b) As $C \longrightarrow 0$, $\Lambda_m = \Lambda^\circ_m$

Thus, correct.

(c) $A = \dfrac{82.4}{(DT)^{1/2}\eta}$, $B = \dfrac{8.2 \times 10^5}{(DT)^{3/2}}$

where, D = Dielectric constant

T = Temperature

η = Coefficient of viscosity

Thus, correct.

I. As solution is diluted, C decreases and *vice-versa*. Increase in concentration, increases interionic attraction between ions, hence Λ_m decreases as shown for strong electrolytes.

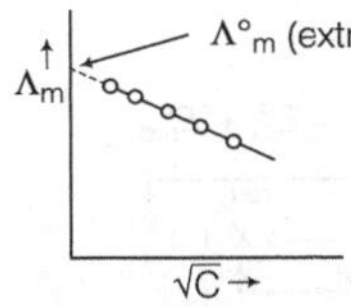

II. By Oswald's dilution law

$$x \text{ (degree of ionisation)} \propto \frac{1}{\sqrt{C}}$$

Thus, in a very dilute solution of weak salt, ionisation is very rapid (as compared to strong salt.) In such case $\Lambda°_m$ cannot be obtained by extrapolation to zero concentration.

III. For strong electrolytes, Λ_m *vs* $\sqrt{C}$ is uniform at all concentration range, hence $\Lambda°_m$ can be obtained by extrapolation of graph between Λ_m and $\sqrt{C}$ to zero concentration.

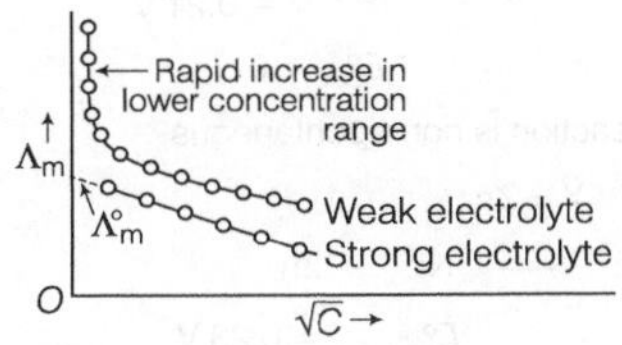

Thus, I, II and III are correct.

25. (d) **26.** (a)

27. (1) At Ag | Ag^+ (anode)

Oxidation takes place

$$Ag \longrightarrow Ag^+ + e^-$$

At Cu^{2+}|Cu (cathode)

Reduction takes place

$$Cu^{2+} + 2e^- \longrightarrow Cu$$

Net $\quad Cu^{2+} + 2Ag \longrightarrow 2Ag^+ + Cu$

Thus, $[Cu^{2+}]$ decreases and that of $[Ag^+]$ increases.

Reaction quotient

$$Q = \frac{[Ag^+]^2}{[Cu^{2+}]} = \frac{(1)^2}{(1)} = 1.0$$

$$\therefore \quad E_{cell} = E°_{cell} - \frac{0.0591}{2}\log 1$$

$$= E°_{cell} - 0 = E°_{cell}$$

Current = 9.65 A

Time = 3600 s (=1 h)

Quantity of current passed = It

$$= 9.05 \times 3600 \text{ C}$$

$$= \frac{9.65 \times 3600}{96500} \text{ F} = 0.36 \text{ F}$$

Thus, $[Ag^+]$ formed = 0.36 equivalent = 0.36 mol

$\therefore$ $[Ag^+]$ after current is passed = 1.36 M

$[Cu^{2+}]$ reduced = 0.36 equivalent = $\dfrac{0.36}{2}$ = 0.18 mol

$\therefore$ $[Cu^{2+}]$ after current is passed = 1.0 − 0.18 = 0.82 M

$\therefore$ Reaction quotient, is thus

$$Q = \frac{[Ag^+]^2}{[Cu^{2+}]} = \frac{(1.36)^2}{0.82} = 2.26$$

$$\therefore \quad E'_{cell} = E°_{cell} - \frac{0.0591}{2}\log Q$$

$$= E°_{cell} - \frac{0.0591}{2}\log 2.26 = E°_{cell} - 0.01$$

Also, $\quad E_{cell} = E°_{cell}$

$\therefore \quad E_{cell} - E'_{cell} = 0.01 \text{ V} = 0.01 \times 100 = 1$ centivolt

28. (7) In electrolysis, cation goes to cathode and anion to anode

$$Cu^{2+} + 2e^- \longrightarrow Cu$$

equivalent weight of Cu = $\left(\dfrac{63.5}{2}\right)$

$$SO_4^{2-} + H_2O \rightleftharpoons H_2SO_4 + O^{2-}$$

$$O^{2-} \longrightarrow \frac{1}{2}O_2 + 2e^-$$

equivalent weight of $O_2 = \dfrac{32}{4} = 8$

Cu deposited = 0.4 g = $\dfrac{0.4}{31.75}$ g equivalent

O_2 formed = Cu deposited

$$= \frac{0.4}{31.75} \text{ g equivalent} = \frac{0.4}{31.75} \times 8 \text{ g}$$

$$= \frac{0.4 \times 8}{31.75 \times 32} \text{ mol} = \frac{0.4 \times 8 \times 22.4}{31.75 \times 32} \text{ L at STP}$$

$$= 0.07 \text{ L} = 7 \text{ centilitre}$$

29. (8)

At anode, $\quad \dfrac{1}{2}H_2 \longrightarrow H^+ + e^-, \quad E°_{SHE} = 0.00$ V

At cathode, $\dfrac{1}{2}Hg_2Cl_2 + e^- \longrightarrow Hg + Cl^- \quad E°_{Cl^-/Hg_2Cl_2,\,Hg}$

$$\frac{1}{2}H_2 + \frac{1}{2}Hg_2Cl_2 \longrightarrow H^+ + Cl^- + Hg$$

$$, -E°_{Hg,\,Hg_2Cl_2/Cl^-} = +0.2802 \text{ V}$$

Reaction quotient (Q) = $[H^+][Cl^-]$

$\therefore \quad = [H^+] \; E° = 0.2802$

$$E_{cell} = E°_{cell} - \frac{0.0591}{1}\log[H^+]$$

$$0.7530 = 0.2802 + 0.0591\text{pH}$$

$$\frac{0.7630 - 0.2802}{0.0591} = \text{pH}$$

$$\text{pH} = 8$$

30. (3) $Ag^+ + e^- \longrightarrow Ag$

$$z \text{(electrochemical equivalent)} = \frac{\text{Atomic mass}}{\text{Electrons} \times 96500} = \frac{108}{1 \times 96500}$$

w (silver deposited) = zit

$$= \frac{108}{96500} \times 0.20 \times 3 \times 3600 = 2.417 \text{ g}$$

$$\frac{\text{Mass}}{\text{Volume}} = \text{Density}$$

$$\therefore \quad \text{Volume} = \frac{\text{Mass}}{\text{Density}} = \frac{2.417 \text{ g}}{10.20 \text{ g cm}^{-3}}$$

$$\text{Volume} = 0.2370 \text{ cm}^3$$

Area × thickness = Volume

800 × thickness = 0.2370

$$\therefore \quad \text{thickness} = \frac{0.2370}{800} \text{ cm}$$

$$= 2.96 \times 10^{-4} \text{ cm} \approx 3.0 \times 10^{-4} \text{ cm}$$

JEE Main & AIEEE Archive

1. (c) $Cu^{2+} + 2e^- \longrightarrow Cu$

1 mol 2 F 1 mol

Cu formed by two Faraday = 1 mole

= 63.5 g

2. (a) For 0.2 M solution, $R = 50\,\Omega$

$$\kappa = 1.4\ \text{S m}^{-1} = 1.4 \times 10^{-2}\ \text{S cm}^{-1}$$

The resistivity of the solution is related to specific conductance by

$$\rho = \frac{1}{\kappa} = \frac{1}{1.4 \times 10^{-2}}\ \Omega\ \text{cm}$$

It is known that, $R = \rho \frac{l}{a}$

$$\Rightarrow \quad \frac{R}{\rho} = \frac{l}{a}$$

$$\Rightarrow \quad \frac{l}{a} = R \times \frac{1}{\rho} = 50 \times 1.4 \times 10^{-2}\ \text{cm}$$

For 0.5 M solution,

$$R = 280\,\Omega$$

$$\kappa = ?$$

$$\frac{l}{a} = 50 \times 1.4 \times 10^{-2}\text{cm}$$

The specific conductance of the solution is given by

$$\kappa = \frac{1}{\rho} = \frac{1}{R} \times \frac{l}{a}$$

$$\kappa = \frac{1}{280} \times 50 \times 1.4 \times 10^{-2}$$

$$= 2.5 \times 10^{-3}\ \text{S cm}^{-1}$$

Now, the molar conductivity of the solution is given by

$$\Lambda_m = \frac{\kappa \times 1000}{M} = \frac{2.5 \times 10^{-3} \times 1000}{0.5}$$

$$= 5.5\ \text{cm}^2\text{mol}^{-1} = 5 \times 10^{-4}\ \text{S m}^2\ \text{mol}^{-1}$$

Hence, the molar conductivity of 0.5 M solution of the electrolyte is $5 \times 10^{-4}\ \text{S m}^2\ \text{mol}^{-1}$.

3. (c) Debye-Huckel Onsager equation can be written as

$$\lambda_c = \lambda_\infty - (B)\sqrt{C}$$

Where,

λ_c = Molar conductivity of the solution at certain concentration

λ_∞ = Limiting molar conductivity

C = Concentration

B = Constant that depends on temperature, charges on the ions and dielectric constant as well as viscosity of the solution

4. (a)

$$Mn^{2+} + 2e^- \longrightarrow Mn, \quad E^\circ = -1.18\ \text{V} \quad \ldots\text{(i)}$$

$$2\,(Mn^{3+} + e^- \longrightarrow Mn^{2+}), \quad E^\circ = +1.51\ \text{V} \quad \ldots\text{(ii)}$$

Subtracting Eq. (ii) from Eq. (i), we get

$$3\,Mn^{2+} \longrightarrow Mn + 2Mn^{3+}$$

$$E^\circ = -1.18 - (+1.51)$$

$$= -2.69\ \text{V}$$

Since, the value of E° is –ve, therefore the reaction is non-spontaneous.

5. (d) Higher the SRP (Standard Reduction Potential), stronger is the oxidising agent. Among the given, electrode potentials, $E^\circ_{MnO_4^-/Mn^{2+}}$ is highest. Hence, MnO_4^- is the strongest oxidising agent.

6. (d) A cell reaction is spontaneous, if $\Delta G^\circ < 0$.

Since, $\Delta G^\circ = -nFE^\circ_{cell}$

Thus, $E^\circ_{cell} > 0$

$$E^\circ_{cell} = E^\circ_{ox} + E^\circ_{red}$$

$$X^{2+} + Y \longrightarrow X + Y^{2+}$$

(red: $X^{2+} \to X$; Oxi: $Y \to Y^{2+}$)

(a) X = Ni, Y = Fe

$$Ni + Fe^{2+} \longrightarrow Ni^{2+} + Fe$$

$$E^\circ_{Ni/Ni^{2+}} = +0.23\ \text{V}$$

$$E^\circ_{Fe^{2+}/Fe} = -0.44\ \text{V}$$

Thus, $E^\circ_{cell} = E^\circ_{Ni/Ni^{2+}} + E^\circ_{Fe^{2+}/Fe}$

$$= -0.21\ \text{V}$$

$$E^\circ_{cell} < 0$$

⇒ Reaction is non-spontaneous.

(b) X = Ni, Y = Zn

$$Ni + Zn^{2+} \longrightarrow Ni^{2+} + Zn$$

$$E^\circ_{Ni/Ni^{2+}} = 0.23\ \text{V}$$

$$E^\circ_{Zn^{2+}/Zn} = -0.76\ \text{V}$$

∴ $E^\circ_{cell} = 0.23 + (-0.76)$

$$= -0.53\ \text{V}$$

$$E^\circ_{cell} < 0$$

⇒ Non-spontaneous reaction.

(c) X = Fe, Y = Zn

$$Fe + Zn^{2+} \longrightarrow Fe^{2+} + Zn$$

$$E^\circ_{cell} = E^\circ_{Fe/Fe^{2+}} + E^\circ_{Zn^{2+}/Zn}$$

$$= 0.44 - 0.76$$

$$= -0.32\ \text{V}$$

$$E^\circ_{cell} < 0$$

⇒ Non-spontaneous reaction.

(d) X = Zn, Y = Ni

$$Zn + Ni^{2+} \longrightarrow Zn^{2+} + Ni$$

$$E^\circ_{cell} = E^\circ_{Zn/Zn^{2+}} + E^\circ_{Ni^{2+}/Ni}$$

$$= 0.76 - 0.23 = 0.53\ \text{V}$$

$$E^\circ_{cell} > 0$$

⇒ Spontaneous reaction.

7. (c) For hydrogen electrode,

$$H^+(aq) + e^- \longrightarrow \frac{1}{2}H_2(g)$$

$$E = E° - \frac{0.0591}{n}\log\frac{(p_{H_2})^{1/2}}{[H^+]}$$

Here, n = number of electrons involved in the reaction

$$E = E° - \frac{0.0591}{1}\log\frac{(pH_2)^{1/2}}{[H^+]}$$

$$E = -0.0591\log\frac{(p_{H_2})^{1/2}}{[H^+]}$$

[$\because E° = 0$ for H – electrode]

$\Rightarrow$ E will be –ve when $\frac{(p_{H_2})^{1/2}}{[H^+]}$ is +ve.

i.e. $(p_{H_2})^{1/2} > [H^+]$

	p_{H_2}	$[H^+]$	$\frac{(p_{H_2})^{1/2}}{[H^+]}$	E_{red}
(a)	1 atm	2.0 M	0.25	+ve
(b)	1 atm	1.0 M	1.0	0
(c)	2 atm	1.0 M	2.0	–ve
(d)	2 atm	2.0 M	0.50	+ve

Hence, when $p(H_2)$ is 2 atm and $[H^+]$ is 1.0 M, E_{red} is negative.

8. (b) For 0.2 M solution,

Specific conductance (κ)

$$= \text{Conductance}\left(\frac{1}{R}\right) \times \text{Cell constant } (l/a)$$

$$\Rightarrow \quad \kappa = \frac{1}{50} \times \frac{l}{a} = 1.3 \text{ S m}^{-1}$$

$$\Rightarrow \quad \frac{l}{a} = 1.3 \times 50 \text{ m}^{-1}$$

$$\frac{l}{a} = 1.3 \times 50 \times 10^{-2} \text{ cm}^{-1}$$

For 0.4 M solution, $R = 260\ \Omega$

$$R = \rho\frac{l}{a}$$

$$\Rightarrow \quad \frac{1}{\rho} = \kappa = \frac{1}{R} \times \frac{l}{a}$$

$$\kappa = \frac{1}{260} \times 1.3 \times 50 \times 10^{-2} \text{ S cm}^{-1}$$

Molar conductivity of the solution is given by

$$\Lambda_m = \frac{\kappa \times 1000}{M}$$

$$= \frac{1}{260} \times \frac{1.3 \times 50 \times 10^{-2} \times 1000}{4}$$

$$= 6.25 \text{ S cm}^2 \text{ mol}^{-1}$$

$$= 6.25 \times 10^{-4} \text{ S m}^2 \text{ mol}^{-1}$$

9. (c) $\frac{2}{3}Al_2O_3 \longrightarrow \frac{4}{3}Al + O_2$

Thus, $\frac{2}{3} \times 3\ (O^{2-})$

i.e. $2O^{2-} \longrightarrow O_2 + 4e^{-1}$ [$\because n = 4$]

$$\Delta G = +966 \text{ kJ mol}^{-1}$$

$$= 966 \times 10^3 \text{ J mol}^{-1}$$

$$\Delta G = -nFE_{cell}$$

$$966 \times 10^3 = -4 \times 96500 \times E_{cell}$$

$$E_{cell} = -2.5 \text{ V}$$

10. (c) It is given that,

$Fe^{3+} + 3e^- \longrightarrow Fe,\ E_1° = -0.036$ V, $n_1 = 3$ ...(i)

$Fe^{2+} + 2e^- \longrightarrow Fe,\ E_2° = -0.439$ V, $n_2 = 2$ ...(ii)

We need to calculate

$Fe^{3+} + e^- \longrightarrow Fe^{2+},\ E_3° = ?$ $n_3 = 1$ ...(iii)

Equation (i)–(ii) = (iii)

$$\therefore \quad \Delta G°_3 = \Delta G°_1 - \Delta G°_2$$

$$-n_3FE°_3 = -n_1FE°_1 + n_2FE°_2$$

$$\therefore \quad E°_3 = \frac{n_1E°_1 - n_2E°_2}{n_3}$$

$$= \frac{3 \times (-0.036) - 2(-0.439)}{1}$$

$$= 0.77 \text{ V}$$

11. (a) $Cr \mid Cr^{3+}$ (0.1 M) $\mid\mid Fe^{2+}$ (0.01 M) $\mid Fe$

Oxidation half-cell

$$Cr \longrightarrow Cr^{3+} + 3e^-] \times 2$$

Reduction half-cell

$$Fe^{2+} + 2e^- \longrightarrow Fe] \times 3$$

Net cell reaction

$$2Cr + 3Fe^{2+} \longrightarrow 2Cr^{3+} + 3Fe$$

$$n = 6$$

$$E°_{cell} = E°_{oxi} + E°_{red} = 0.72 - 0.42 = 0.30\text{V}$$

$$E_{cell} = E°_{cell} - \frac{0.0591}{n}\log\frac{[Cr^{3+}]^2}{[Fe^{2+}]^3}$$

$$= 0.30 - \frac{0.0591}{6}\log\frac{(0.1)^2}{(0.01)^3}$$

$$= 0.30 - \frac{0.0591}{6}\log\frac{10^{-2}}{10^{-6}}$$

$$= 0.30 - \frac{0.0591}{6}\log 10^4$$

$$E_{cell} = 0.2606 \text{ V}$$

12. (a) Molar conductivity of aqueous CH_3COOH solution can be calculated from Kohlraush's law as

$$\lambda°_{CH_3COOH} = \lambda°_{CH_3COONa} + \lambda°_{HCl} - \lambda°_{NaCl}$$

Hence, for determining the molar conductivity of CH_3COOH solution, $\lambda°_{NaCl}$ is required.

13. (c) Cell is completely discharged. It means equilibrium gets established, $E_{cell} = 0$

$Zn \mid Zn^{2+}$ (1 M) $\mid\mid Cu^{2+}$ (1 M) $\mid Cu$

Cell reaction : $Zn + Cu^{2+} \rightleftharpoons Zn^{2+} + Cu$

$$K_{eq} = \frac{[Zn^{2+}]}{[Cu^{2+}]}$$

We know, $E_{cell} = E°_{cell} - \frac{0.0591}{n}\log K_{eq}$

$$E°_{cell} = \frac{0.0591}{n}\log K_{eq}$$

or $1.10 = \frac{0.0591}{2}\log K_{eq}$

$$\log K_{eq} = 37.23$$

$$K_{eq} = \frac{[Zn^{2+}]}{[Cu^{2+}]}$$

$$= \text{antilog}\frac{2.20}{0.0591}$$

$$= \text{antilog } 37.3$$

14. (d) **For 0.1 M solution,**

$$R = 100\,\Omega$$

$$\kappa = \frac{1}{R}\left(\frac{l}{a}\right)$$

$\frac{l}{a}$ (cell constant) $= 1.29 \times 100\ m^{-1}$

For 0.2 M solution

Given, $R = 520\,\Omega$, $C = 0.2$ M,

μ (molar conductivity) = ?

Molar conductivity, $\Lambda_m = \frac{\kappa \times 1000}{M}$

κ can be calculated as $\kappa = \frac{1}{R}\left(\frac{l}{a}\right)$, now cell constant is known]

Hence, $\Lambda_m = \frac{1}{520} \times 129 \times \frac{1000}{0.2} \times 10^{-6}\ m^3$

$= 12.4 \times 10^{-4}\ S\ m^2\ mol^{-1}$

15. (d) According to Kohlrausch's law

$$\lambda^\circ_{CH_3COOH} = \lambda^\circ_{CH_3COO^-} + \lambda^\circ_{H^+} \quad \text{...(i)}$$

$$\lambda^\circ_{HCl} = \lambda^\circ_{H^+} + \lambda^\circ_{Cl^-} \quad \text{...(ii)}$$

$$\lambda^\circ_{CH_3COONa} = \lambda^\circ_{CH_3COO^-} + \lambda^\circ_{Na^+} \quad \text{....(iii)}$$

Thus, on adding Eqs. (ii) and (iii), if $\lambda^\circ_{Na^+}$ and $\lambda^\circ_{Cl^-}$ are subtracted, we can obtained the value of λ°_{HOAc}. Thus, additional value required is λ°_{NaCl}.

16. (d) $AgI(s) + e^- \rightleftharpoons Ag(s) + I^-;\ E_1^\circ = -0.152$ V

$Ag(s) \longrightarrow Ag^+ + e^-\ ;\ E_2^\circ = -0.8$ V

$AgI(s) \longrightarrow Ag^+ + I^-\ ;\ E^\circ = -0.952$

$$E^\circ_{cell} = \frac{0.059}{n}\log K_{sp}$$

$$-0.952 = \frac{0.059}{1}\log K_{sp}$$

$$\log K_{sp} = \frac{-0.952}{0.059} = -16.135$$

17. (b) $\Lambda^\infty_{AcOH} = \Lambda^\infty_{AcONa} + \Lambda^\infty_{HCl} - \Lambda^\infty_{NaCl}$

$= 91.0 + 426.2 - 126.5 = 390.7$

18. (d) $Al^{3+} + 3e^- \longrightarrow Al$

$w = zQ$

where, w = amount of metal = 5.12 kg = 5.12×10^3 g

z = electrochemical equivalent

$= \frac{\text{equivalent weight}}{96500} = \frac{\text{atomic mass}}{\text{electrons} \times 96500} = \frac{27}{3 \times 96500}$

$$5.12 \times 10^3 = \frac{27}{3 \times 96500} \times Q$$

$$Q = \frac{5.12 \times 10^3 \times 3 \times 96500}{27}\ C = 5.49 \times 10^7\ C$$

19. (a) By Kohlrausch's law.

$$\Lambda^\circ_{NaBr} = \Lambda^\circ_{NaCl} + \Lambda^\circ_{KBr} - \Lambda^\circ_{KCl}$$

$= 126 + 152 - 150$

$= 128\ S\ cm^2\ mol^{-1}$

20. (a) $E^\circ_{Cr^{3+}/Cr^{2+}} = -0.41$ V

$E^\circ_{Mn^{3+}/Mn^{2+}} = +1.57$ V

$E^\circ_{Fe^{3+}/Fe^{2+}} = +0.77$ V

$E^\circ_{Co^{3+}/Co^{2+}} = +1.97$ V

More negative value of E°_{red} indicates better reducing agent. Thus, easily oxidised. Hence, oxidation of Cr^{2+} to Cr^{3+} is the easiest.

21. (c) $Sn(s) + 2Fe^{3+}(aq) \longrightarrow 2Fe^{2+}(aq) + Sn^{2+}(aq)$

$$E^\circ_{cell} = E^\circ_{oxi} + E^\circ_{red}$$

$$= E^\circ_{Sn/Sn^{2+}} + E^\circ_{Fe^{3+}/Fe^{2+}}$$

Given, $E^\circ_{Sn^{2+}/Sn} = -0.14$ V

$\therefore$ $E^\circ_{Sn/Sn^{2+}} = +0.14$ V

$E^\circ_{Fe^{3+}/Fe^{2+}} = 0.77$ V

$\therefore$ $E^\circ_{cell} = 0.14 + 0.77$

$= 0.91$ V

22. (c) $Zn(s) + 2H^+ \longrightarrow Zn^{2+}(aq) + H_2(g)$

Reaction quotient, $Q = \frac{[Zn^{2+}]}{[H^+]^2}$

Corresponding cell is

$Zn \mid Zn^{2+}(C_1) \mid\mid H^+(aq) \mid Pt(H_2)$

Anode (−) Cathode (+)

and $E_{cell} = E^\circ_{cell} - \frac{0.0591}{2}\log K$

$$= E^\circ_{cell} - \frac{0.0591}{2}\log\frac{[Zn^{2+}]}{[H^+]^2}$$

$$= E^\circ_{cell} + \frac{0.0591}{2}\log\frac{[H^+]^2}{[Zn^{2+}]}$$

If H_2SO_4 is added to cathodic compartment, (towards reactant side), then Q decrease (due to increase in H^+).

Hence, equilibrium is displaced towards right and E_{cell} increases.

23. (c) Relation between K_{eq} and E°_{cell} is

$$E^\circ_{cell} = \frac{2.303\,RT}{nF}\log K_{eq}$$

$$E^\circ_{cell} = \frac{0.0591}{n}\log K_{eq}\ \text{(at 298 K)}$$

$$0.591 = \frac{0.0591}{1}\log K_{eq}$$

$\therefore$ $\log K_{eq} = 10$

$\therefore$ $K_{eq} = 1 \times 10^{10}$

24. (b) $Ag^+ + e^- \longrightarrow Ag$

Number of Faradays $= \frac{9650}{96500} = \frac{1}{10}$

$Ag^+ + e^- \longrightarrow Ag$

$\therefore$ Mass of Ag produced $= \frac{1}{10} \times 108$

$= 10.8$ g

25. (a) More negative value of E° means higher reducing power.

26. (d) The standard emf of the cell,

$$E^\circ_{cell} = \frac{2.303RT}{nF}\log K_{eq} \Rightarrow 0.295 = \frac{0.0591}{2}\log K_{eq}$$

$\therefore \log K_{eq} = 10$

$\therefore \quad K_{eq} = 1\times 10^{10}$

27. (c) $S \propto$ area $(m^2) \propto$ conc $(mol/m^3) \propto \frac{1}{length}$ (m^{-1})

$$S = \kappa m^2 \frac{mol}{m^3} m^{-1}$$

$$\kappa = S\ mol^{-1}\ m^2$$

28. (c) At LHS (oxidation)

$$2\times (Ag \longrightarrow Ag^+ + e^-), \qquad E^\circ_{oxi} = -x$$

At RHS (reduction)

$$Cu^{2+} + 2e^- \longrightarrow Cu, \qquad E^\circ_{red} = +y$$

$$2Ag + Cu^{2+} \longrightarrow Cu + 2Ag^+, E^\circ_{cell} = (y-x)$$

Note *E° values remain constant when half-cell equation is multiplied/divided.*

29. (b) LHS half-cell,

$$\underset{p_1}{H_2(g)} \longrightarrow 2H^+\ (1\ M) + 2e^-$$

RHS half cell,

$$2H^+\ (1\ M) + 2e^- \longrightarrow \underset{p_2}{H_2(g)}$$

$$\underset{p_1}{H_2(g)} \longrightarrow \underset{p_2}{H_2(g)}$$

$$E^\circ_{cell} = 0.00\ V,$$

$$K = \frac{p_2}{p_1}$$

$$n = 2$$

$$E_{cell} = E^\circ_{cell} - \frac{RT}{nF}\log_e K$$

$$= 0 - \frac{RT}{2F}\log_e \frac{p_2}{p_1}$$

$$E_{cell} = \frac{RT}{2F}\log_e \frac{p_1}{p_2}$$

JEE Advanced & IIT JEE Archive

1. (a) This problem is based on the characteristics of salt-bridge.

Functions of salt-bridge

(i) It connects the two half-cells and completes the cell circuit.

(ii) It keeps the solutions of two half-cells and complete the cell circuit but does not participate chemically in the cell reaction.

(iii) It maintains the diffusion of ions from one electrode to another electrode.

(iv) A cell reaction may also occur in the absence of salt-bridge. Sometimes, both the electrodes dip in the same electrolyte solution and in such cases we do not require a salt-bridge." So, option (c) is incorrect.

(v) This prevent mixing of two electrolytic solutions hence, option (d) is incorrect.

2. (a)

(i) X and Y both are weak electrolytes, hence have minimum conductance. When X $[(C_2H_5)_3N]$ is added to Y (CH_3COOH), ions are formed hence conductance increases.

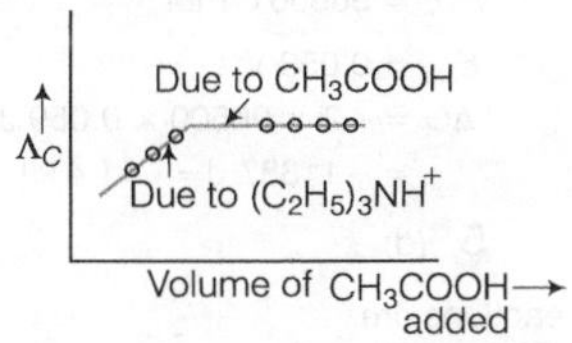

$$\underset{(Y)}{CH_3COOH} + \underset{(X)}{(C_2H_5)_3N} \longrightarrow CH_3COO^- + (C_2H_5)_3NH^+$$

After complete reaction further addition of Y, makes no change in conductance.

Thus, **(i) → (r)**

(ii) Net reaction is

$$\underset{(X)}{I^-} + \underset{(Y)}{Ag^+} \longrightarrow \underset{\text{Insoluble}}{AgI(s)}$$

where $X(I^-)$ is added to $Y(Ag^+)$ insoluble AgI is formed. Hence, there is no change in conductance. After complete reaction of Ag^+, further addition of I^- causes increase in conductance.

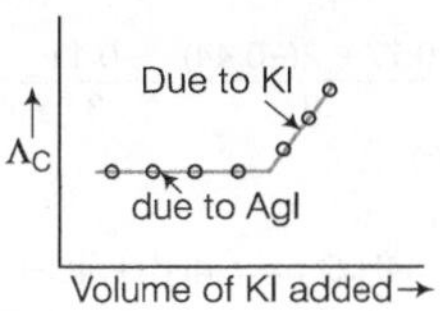

Thus, **(ii) → (s)**

(iii) KOH is a strong electrolyte. As $CH_3COOH(X)$ is added, OH^- is neutralised by CH_3COOH.

$$\underset{(Y)}{OH^-} + \underset{(X)}{CH_3COOH} \longrightarrow CH_3COO^- + H_2O$$

CH_3COO^- is hydrolysed at the same time.

$$CH_3COO^- + H_2O \longrightarrow CH_3COOH + OH^-$$

But being in small extent, conductance decreases. After complete reaction of KOH, further addition of CH_3COOH, (weak) there is no change in conductance.

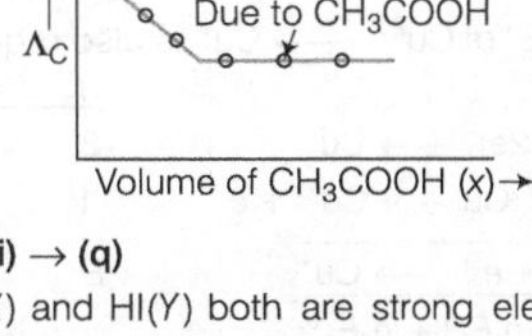

Thus, **(iii) → (q)**

(iv) NaOH(X) and HI(Y) both are strong electrolytes. Thus, net reaction is

$$\underset{(X)}{OH^-} + \underset{(Y)}{H^+} \longrightarrow H_2O$$

When (X) is added to (Y), H_2O is formed hence conductance decreases. After complete reaction of OH^-, further addition of NaOH (X), causes increases in conductance.

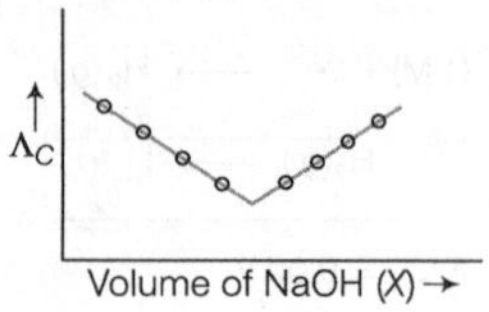

Thus, **(iv) → (p)**

Hence, (i) → (r), (ii) → (s), (iii) → (q), (iv) → (p)

3. (d) **Plan** When different number of electrons are involved in a redox reaction

$$\Delta G^\circ_{net} = \Delta G^\circ_1 + \Delta G^\circ_2$$

$$-n_3FE^\circ_3 = -n_1FE^\circ_1 - n_2FE^\circ_2$$

$$\therefore \quad E^\circ_3 = \frac{n_1E^\circ_1 + n_2E^\circ_2}{n_3}$$

(i) Fe^{3+} / Fe

Net reaction $Fe^{3+} \longrightarrow Fe$

is obtained from

	n	E°
$Fe^{3+} + e^- \longrightarrow Fe^{2+}$	$n_1 = 1$	$E_1^{od} = 0.77$ V
$Fe^{2+} + 2e^- \longrightarrow Fe$	$n_2 = 2$	$E_2^\circ = -0.44$ V
$\because Fe^{3+} + 3e^- \longrightarrow Fe$	$n_3 = 3$	$E_3^\circ = ?$

$$E^\circ_3 = \frac{n_1E^\circ_1 + n_2E^\circ_2}{n_3}$$

$$= \frac{0.77 + 2(-0.44)}{3} = \frac{-0.11}{3} = -0.04 \text{ V}$$

Thus, **(i) → (r)**

(ii) **Net reaction**

$$4H_2O \rightleftharpoons 4H^+ + 4OH^-$$

is obtained from

	n	E°
$2H_2O \longrightarrow O_2 + 4H^+ + 4e^-$	$n_1 = 4$	-1.23 V
$2H_2O + O_2 + 4e^- \longrightarrow 4OH^-$	$n_2 = 4$	$+0.40$ V
$4H_2O \longrightarrow 4H^+ + 4e^-$	$n_3 = 4$	?

$$E^\circ_3 = \frac{n_1E^\circ_1 + n_2E^\circ_2}{n_3}$$

$$= E^\circ_1 + E^\circ_2 = -1.23 + 0.40 = -0.83 \text{ V}$$

Thus, **(ii) → (s)**

(iii)

Reduction

$$Cu^{2+} + Cu \longrightarrow 2Cu^+$$

Oxidation

For thus E° of $Cu^{2+} \longrightarrow Cu^+$ is also required.

	n	E°
$Cu^{2+} + 2e^- \longrightarrow Cu$	2	0.34 V
$Cu \longrightarrow Cu^+ + e^-$	1	-0.52 V
$Cu^{2+} + e^- \longrightarrow Cu^+$	E°_3	?

$$E_3^\circ = \frac{n_1E_1^\circ + n_2E_2^\circ}{n_3}$$

$$= \frac{2 \times 0.34 + 1 \times (-0.52)}{1} = 0.16 \text{ V}$$

Also,

	n	E°
$Cu \longrightarrow Cu^+ + e^-$	$n_1 = 1,$	-0.52 V
$Cu^{2+} + e^- \longrightarrow Cu^+$	$n_2 = 1$	0.10 V
$Cu^{2+} + Cu \longrightarrow 2Cu^+$		

$$E^\circ = -0.52 + 0.16 = -0.36 \text{ V}$$

Thus, **(iii) → (p)**

(iv) $Cr^{3+} \longrightarrow Cr^{2+}$

is obtained from

	n	E°
$Cr^{3+} + 3e^- \longrightarrow Cr$	3	-0.74 V
$Cr \longrightarrow Cr^{2+} + 2e^-$	2	$+0.91$ V
$Cr^{3+} + e^- \longrightarrow Cr^{2+}$	1	?

$$E_3^\circ = \frac{-0.74 \times 3 + 2 \times 0.91}{1}$$

$$= -0.4 \text{ V}$$

Thus, **(iv) → (q)**

Explanations for Q. Nos. (4-5)

$$M\text{ (LHS)} \rightleftharpoons \underset{(\text{in } MX_2)}{M^{2+}} + 2e^-, \qquad E^\circ_{ox} = xV$$

$$\underset{(0.001\text{ M})}{M^{2+}\text{ (RHS)}} + 2e^- \longrightarrow M, \qquad E^\circ_{red} = -XV$$

$$M^{2+}\text{ (0.001M)} \longrightarrow M^{2+}\text{ (in } MX_2), \qquad E^\circ_{cell} = 0.00 \text{ V}$$

$$\therefore \quad Q = \frac{[M^{2+}](\text{in } MX_2)}{[M^{2+}](0.001\text{ M})}$$

By Nernst equation, $E_{cell} = E^\circ_{cell} - \frac{2.303RT}{nF}\log Q$

$$0.059 = 0 - \frac{0.059}{2}\log\frac{[M^{2+}]}{(0.001)}$$

$$\log\frac{[M^{2+}]}{(0.001)} = -2$$

$$\frac{[M^{2+}]}{(0.001)} = 10^{-2}$$

$$\therefore \quad [M^{2+}] = 10^{-2} \times 10^{-3} = 10^{-5} \text{ mol dm}^{-3}$$

$$MX_2(s) \rightleftharpoons M^{2+} + 2X^-$$

$$\therefore \quad [X^-] = 2 \times 10^{-5} \text{ mol dm}^{-9}$$

$$K_{sp} = [M^{2+}][X^-]^2 = (10^{-5})(2 \times 10^{-5})^2$$

$$K_{sp} = 4 \times 10^{-15} \text{ mol}^3\text{dm}^{-9}$$

Also, $\Delta G = -nFE_{cell}$

$n = 2$

$F = 96500 \text{ C mol}^{-1}$

$E_{cell} = 0.059$ V

$$\therefore \quad \Delta G = -2 \times 96500 \times 0.059 \text{ J}$$

$$= -11387 \text{ J} = -11.4 \text{ kJ}$$

4. (b) **5.** (d)

6. (d) The half reactions are

$$Fe(s) \longrightarrow Fe^{2+}(aq) + 2e^- \times 2$$

$$O_2(g) + 4H^+ + 4e^- \longrightarrow 2H_2O$$

$$2Fe(s) + O_2(g) + 4H^+ \longrightarrow 2Fe^{2+}(aq) + 2H_2O(l);$$

$$E = E^\circ - \frac{0.059}{4}\log\frac{(10^{-3})^2}{(10^{-3})^4(0.1)}$$

$$= 1.57 \text{ V}$$

7. (d) As $AgNO_3$ is added to solution, KCl will be displaced according to the following reaction

$$AgNO_3(aq) + KCl(aq) \longrightarrow AgCl(s) + KNO_3(aq)$$

For every mole of KCl displaced from solution, one mole of KNO_3 comes in solution resulting in almost constant conductivity. As the end point is reached, added $AgNO_3$ remain in solution increasing ionic concentration, hence conductivity increases.

8. (b) $M^+ (aq, 1\,M) \longrightarrow M^+ (aq, 0.05\,M)$

$$E_{cell} = 0 - \frac{2.303\,RT}{F} \log \frac{0.05}{1} > 0$$

Hence, $|E_{cell}| = E_{cell} = 0.70$ V and $\Delta G < 0$ for spontaneity of reaction.

$$2.303\,RT / F = \frac{0.70}{\log 0.05} = 0.0538$$

9. (c) $E_{cell} = E° - \frac{0.0538}{1} \log 0.0025$

$$= 0.139 \text{ V} = 139 \text{ V}$$

10. (a,b,d) Metals with $E°$ value less than 0.96 V will be able to reduce NO_3^- in aqueous solution.Therefore, metals V ($E°$ = – 1.19 V), Fe ($E°$ = – 0.04 V), Hg ($E°$ = 0.86 V) will all reduce NO_3^- but Au ($E°$ = 1.40 V) cannot reduce NO_3^- in aqueous solution.

11. (b) 0.01 mol H_2 = 0.02 g equivalent

$\Rightarrow$ Coulombs required = 0.02 × 96500 = 1930 C

$$\Rightarrow \quad Q = It = 1930 \text{ C}$$

$$\Rightarrow \quad t = \frac{1930}{10 \times 10^{-3}} = 19.3 \times 10^4 \text{ s}$$

12. (c) For spontaneous redox reaction : $E°_{cell} > 0$

For $\quad 2I^- + Cl_2 \longrightarrow 2Cl^- + I_2$

$$E° = 1.36 - 0.54 = 0.82 \text{ V} > 0$$

i.e. Cl_2 will spontaneously oxidise I^-.

In other cases $E°_{cell} < 0$, they are non-spontaneous.

13. (d) For the reaction,

(i) $4Fe^{3+} + 2H_2O \longrightarrow 4Fe^{2+} + 4H^+ + O_2;\ E° = -0.46$ V

(ii) $4Mn^{3+} + 2H_2O \longrightarrow 4Mn^{2+} + 4H^+ + O_2;\ E° = +0.27$ V

As evidenced above, reaction (i) is non-spontaneous, therefore, Fe^{3+} is stable in acid solution. However, reaction (ii) is spontaneous Mn^{3+} oxidises H_2O to O_2 and itself reduced to Mn^{2+} in acidic medium.

14. (b) $E°$ for $2Ag^+ + C_6H_{12}O_6 + H_2O \rightleftharpoons 2Ag\,(s) + C_6H_{12}O_7 + 2H^+$ is 0.75 V

Also, $$E° = \frac{0.0592}{2} \log K$$

$$\Rightarrow \quad \log K = \frac{2E°}{0.0592} = 25.33$$

$$\Rightarrow \quad \ln K = 2.303 \log K = 58.35$$

15. (c) On increasing concentration of NH_3, the concentration of H^+ ion decreases, therefore,

$$E_{red} = E°_{red} - \frac{0.0592}{2} \log [H^+]^2$$

$$= 0 - \frac{0.0592}{2} \times 2 \log 10^{-11} = 0.65 \text{ V}$$

i.e. E_{red} increases by 0.65 V.

16. (d) NH_3 has no effect on the $E°$ of glucose/gluconic acid electrode.

17. The solubility of AgBr in 10^{-7} M $AgNO_3$ solution is determined as

$$\underset{}{AgBr} \rightleftharpoons \underset{S + 10^{-7}\,M}{Ag^+} + \underset{S\,M}{Br^-}$$

$$AgNO_3 \longrightarrow \underset{10^{-7}\,M}{Ag^+} + \underset{10^{-7}\,M}{NO_3^-}$$

$$K_{sp} = 14 \times 10^{-14} = S\,(S + 10^{-7})$$

Solving for S gives

$$S = 3 \times 10^{-7} \text{ M}$$

$$\Rightarrow \quad [Br^-] = 3 \times 10^{-7} \text{ M}$$

$$[Ag^+] = 4 \times 10^{-7} \text{ M},$$

$$[NO_3^-] = 10^{-7} \text{ M}$$

$\Rightarrow \kappa$ (specific conductance) $= \kappa_{Br^-} + \kappa_{Ag^+} + \kappa_{NO_3^-}$

$= [8 \times 10^{-3} \times 3 \times 10^{-7} + 6 \times 10^{-3} \times 4 \times 10^{-7} + 7 \times 10^{-3} \times 10^{-7}]\,1000$

$= 24 \times 10^{-7} + 24 \times 10^{-7} + 7 \times 10^{-7}$

$= 55 \times 10^{-7}$ S m^{-1} = 55 (in terms of 10^{-7} S m^{-1})

18. (a) $\quad \Delta G° = \Sigma \Delta G_f°$ (products) $- \Sigma \Delta G_f°$ (reactants)

$= -109 - (-129 + 77)$ kJ

$= -57$ kJ

Cell : Ag | AgCl, Cl^- || Ag^+| Ag

For K_{sp}, reaction is

$$AgCl\,(s) \rightleftharpoons Ag^+ + Cl^-$$

$$\Delta G° = +57 \text{ kJ}$$

$$\Rightarrow \quad \Delta G° = -RT \ln K_{sp}$$

$$\Rightarrow \quad \log K_{sp} = -\frac{\Delta G°}{2.3\,RT}$$

$$= -\frac{57 \times 1000}{2.3 \times 8.314 \times 298} = -10$$

Now, $\quad E°$ of $Ag^+ + Cl^- \rightleftharpoons AgCl$

$$E° = -\frac{\Delta G°}{nF} = \frac{57000}{96500} = 0.59 \text{ V}$$

(b) The cell reaction is :

$$Zn + 2Ag^+ \rightleftharpoons Zn^{2+} + 2Ag \quad E° = 1.56 \text{ V}$$

$$\Rightarrow \quad 0 = E° - \frac{0.059}{2} \log \frac{[Zn^{2+}]}{[Ag^+]^2}$$

$$\Rightarrow \quad \log \frac{[Zn^{2+}]}{[Ag^+]^2} = \frac{2E°}{0.059} = \frac{2 \times 1.56}{0.059} = 52.88$$

Moles of Zn added $= \frac{6.539 \times 10^{-2}}{65.39} = 10^{-3}$

$\Rightarrow$ Moles of Ag formed = 2 × 10^{-3}.

19. (b) The net reaction is

$$2H^+ + \frac{1}{2}O_2 + Fe \longrightarrow H_2O + Fe^{2+}; \quad E° = 1.67 \text{ V}$$

$$\Delta G° = -nE°F$$

$$= -\frac{2 \times 1.67 \times 96500}{1000} \text{ kJ} = -322.31 \text{ kJ}$$

4. Chemical Kinetics

DPP-1 Rate Law, Molecularity and Order of Reaction

1. Plan For a reaction,

$$aA + bB \longrightarrow cC + dD$$

$$\frac{dx}{dt} = -\frac{1}{a}\frac{d[A]}{dt} = -\frac{1}{b}\frac{d[B]}{dt} = \frac{1}{c}\frac{d[C]}{dt} = \frac{1}{d}\frac{d[D]}{dt}$$

Thus, for the reaction

$$A + 2B \longrightarrow 3C + D$$

$$\frac{dx}{dt} = 1.0\ \text{Ms}^{-1} = -\frac{d[A]}{dt} = -\frac{1}{2}\frac{d[B]}{dt} = \frac{1}{3}\frac{d[C]}{dt} = \frac{d[D]}{dt}$$

$\therefore$

$$-\frac{d[A]}{dt} = 1.0\ \text{M s}^{-1}$$

$$-\frac{d[B]}{dt} = 2.0\ \text{M s}^{-1}$$

$$+\frac{d[C]}{dt} = 3.0\ \text{M s}^{-1}$$

$$\frac{d[D]}{dt} = 1.0\ \text{M s}^{-1}$$

2. Plan Molar concentration = mol L^{-1}

Thus, if volume is changed, $\left(\frac{dx}{dt}\right)$ is also affected.

$$\left(\frac{dx}{dt}\right)_1 = k[A][B] = k\left(\frac{n_A}{V}\right)\left(\frac{n_B}{V}\right)$$

$$\left(\frac{dx}{dt}\right)_2 = k\left(\frac{n_A}{V/4}\cdot\frac{n_B}{V/4}\right) = 16k\left(\frac{n_A}{V}\right)\left(\frac{n_B}{V}\right) = 16\left(\frac{dx}{dt}\right)_1$$

Thus, rate becomes 16 times.

3. Plan If concentration of a reactant is made m-times, new rate of the reaction of the nth order becomes m^n-times.

i. $$\left(\frac{dx}{dt}\right)_1 = k[NO_2^-][H^+]^2[I^-]$$

$$\left(\frac{dx}{dt}\right)_2 = k[2NO_2^-][H^+]^2[I^-] = 2k[NO_2^-][H^+]^2[I^-] = 2\left(\frac{dx}{dt}\right)$$

Rate is doubled.

ii. $$\left(\frac{dx}{dt}\right)_3 = k[NO_2^-][2H^+]^2[I^-]$$

$\therefore$ $$\left(\frac{dx}{dt}\right)_3 = 4k[NO_2^-][H^+]^2[I^-] = 4\left(\frac{dx}{dt}\right)_1$$

The rate becomes four times.

iii. $[H^+] = 10^{-pH}$

$$[H^+]_1 = 10^{-pH}$$

$$[H^+]_2 = 10^{-pH-1}$$

$$\frac{[H^+]_2}{[H^+]_1} = 10^{-pH-1+pH} = 10^{-1} = \frac{1}{10}$$

$\therefore$ $$[H^+]_2 = \frac{1}{10}[H^+]_1$$

$\therefore$ $$\left(\frac{dx}{dt}\right)_4 = k[NO_2^-]\left[\frac{1}{10}[H^+]\right]^2[I^-]$$

$$= \frac{1}{100}k[NO_2^-][H^+]^2[I^-] = \frac{1}{100}\left(\frac{dx}{dt}\right)_1$$

Rate becomes $\frac{1}{100}$ times.

iv. $$\left(\frac{dx}{dt}\right)_5 = k[2NO_2^-][2[H^+]^2]^2[I^-]^2$$

$$= 16k[NO_2^-][H^+]^2[I^-] = 16\left(\frac{dx}{dt}\right)_1$$

Rate becomes sixteen times.

4. (c) By stoichiometry of the reaction

$$-\frac{1}{5}\frac{d[O_2]}{dt} = -\frac{1}{4}\frac{d[NH_3]}{dt}$$

$\therefore$ $$-\frac{d[O_2]}{dt} = -\frac{5}{4}\frac{d[NH_3]}{dt} = -1.25\frac{d[NH_3]}{dt}$$

Similarly, $$\frac{d[NO]}{dt} = -\frac{d[NH_3]}{dt}$$

$$\frac{d[H_2O]}{dt} = -\frac{6}{4}\frac{d[NH_3]}{dt} = -1.50\frac{d[NH_3]}{dt}$$

5. (b) $$\frac{dx}{dt} = \frac{1}{2}\frac{d[NO]}{dt} = \frac{1}{2}\times 1.6\times 10^{-4}\ \text{M s}^{-1}$$

$$= 8.0\times 10^{-5}\ \text{M s}^{-1}$$

6. (b) $2SO_2 + O_2 \longrightarrow 2SO_3$

In terms of stoichiometric coefficients

$$-\frac{1}{2}\frac{d[SO_2]}{dt} = -\frac{d[O_2]}{dt} = +\frac{1}{2}\frac{d[SO_3]}{dt}$$

Thus, $$-\frac{1}{2}\frac{d[SO_2]}{dt} = +\frac{1}{2}\frac{d[SO_3]}{dt}$$

$$-\frac{d[SO_2]}{dt} = \frac{d[SO_3]}{dt}$$

Rate of disappearance of SO_2 = Rate of formation of SO_3

But quantity is to be expressed in terms of moles.

Rate of formation of SO_3 = 100 kg SO_3 min^{-1}

$$= \frac{100\times 10^3}{80}\ \text{mol } SO_3\ \text{min}^{-1}$$

Thus, rate of disappearance of SO_2

$$= \frac{100\times 10^3}{80}\ \text{mol } SO_2\ \text{min}^{-1}$$

$$= \frac{100\times 10^3}{80}\times 64\ \text{g } SO_2\ \text{min}^{-1}$$

$$= \frac{100}{80}\times 64\ \text{kg } SO_2\ \text{min}^{-1}$$

$$= 80\ \text{kg } SO_2\ \text{min}^{-1}$$

7. (b) $2A + -1B \underset{k_2}{\overset{k_1}{\rightleftharpoons}} C$

For a forward reaction, $$\left(\frac{dx}{dt}\right)_f = k_1[A]^2[B]^{-1}$$

and for a backward reaction, $$\left(\frac{dx}{dt}\right)_b = k_2[C]$$

and thus net rate is

$$\left(\frac{dx}{dt}\right)_{net} = \left(\frac{dx}{dt}\right)_f - \left(\frac{dx}{dt}\right)_b = k_1[A]^2[B]^{-1} - k_2[C]$$

8. (b) For graph, rate of forward reaction is independent of concentration of $\left(\frac{dx}{dt}\right)_f = k_1[A]^x = k_1[A]^0 = k_1$.

For backward reaction, $$\left(\frac{dx}{dt}\right)_b = k_2[B]^n$$

$$2\times 10^{-2} = (0.01)^n$$

$$4\times 10^{-2} = (0.02)^n$$

$(2) = (2)^n$ thus, $n = 1$

$\therefore$ $$\left(\frac{dx}{dt}\right)_b = k_2[B]$$

$\therefore$ Net rate = $k_1 - k_2[B]$

9. (c) $-\frac{1}{x}\frac{d[A]}{dt} = \frac{1}{y}\frac{d[B]}{dt}$

$\therefore \quad -\frac{d[A]}{dt} = \frac{x}{y}\frac{d[B]}{dt} = \frac{1}{4}\frac{d[B]}{dt}$

	x	y	x/y
(a)	1	2	$\frac{1}{2}$
(b)	2	4	$\frac{1}{2}$
(c)	$\frac{1}{2}$	2	$\frac{1}{4}$
(d)	4	1	$\frac{4}{1}$

10. (a) $\left(\frac{dx}{dt}\right)_f = k_f\,[H_2]\,[A_2]$ order = 2

$\left(\frac{dx}{dt}\right)_b = k_b[H]^2[A_2]$ order = 3

At equilibrium

$$\left(\frac{dx}{dt}\right)_f = \left(\frac{dx}{dt}\right)_b$$

$\therefore \quad k_f[H_2][Ar] = k_b[H]^2[Ar]$

$$K_c = \frac{k_f}{k_b} = \frac{[H]^2[A_2]}{[H_2][A_2]}$$

$\therefore \quad k_b = \frac{k_f}{K_c} = \frac{2.2 \times 10^4\,\text{L mol}^{-2}\text{s}^{-1}}{1.00 \times 10^{-4}}$

$$= 2.2 \times 10^8\ \text{L}^2\text{mol}^{-2}\text{s}^{-1}$$

11. (c) For the reaction, $xA + yB \longrightarrow zC$

$$-\frac{1}{x}\frac{d[A]}{dt} = -\frac{1}{y}\frac{d[B]}{dt} = \frac{1}{z}\frac{d[C]}{dt}$$

$$\frac{1}{x} = 1, \frac{1}{y} = 1, \frac{1}{z} = 1.5 = \frac{3}{2}$$

$x : y : z = 3 : 2 : 2$

12. (b) For the reaction, $2N_2O_5 \longrightarrow 4NO_2 + O_2$

$$-\frac{1}{2}\frac{[N_2O_5]}{\Delta t} = \frac{1}{4}\frac{\Delta[NO_2]}{\Delta t} = \frac{\Delta[O_2]}{\Delta t}$$

$$\frac{k_1}{2}[N_2O_5] = \frac{k_2}{4}[N_2O_5] = k_3[N_2O_5]$$

$\therefore \quad \frac{k_1}{2} = \frac{k_2}{4} = k_3$

$$2k_1 = k_2 = 4k_3$$

13. (c) $pV = nRT$

$$p = \frac{n}{V}RT = \frac{\text{mol}}{\text{L}}RT = [CH_3CHO]RT$$

$$-\frac{d[CH_3CHO]}{dt} = -\left(\frac{dp}{dt}\right)\frac{1}{RT}$$

14. (d) A is present in excess, hence its concentration remains unchanged. Thus, order w.r.t. $A = 0$.

order w.r.t. $B = b$, Rate $k[B]^b$

Rate = R, when $[B] = 0.01$ M

Rate = $2R$, when $[B] = 0.04$ M

$\therefore \quad \frac{2R}{R} = \frac{k(0.04)^b}{k(0.01)^b} = (4)^b$

$$2 = (4)^b$$

$$(4)^{1/2} = (4)^b$$

$\therefore \quad b = \left(\frac{1}{2}\right)$

$\therefore \quad \left(\frac{dx}{dt}\right) = k[A][B]^{1/2}$

15. (a) Rate law for the reaction,

$$A \longrightarrow B$$

$$-\frac{d[A]}{dt} = 2\frac{d[B]}{dt}$$

or $\quad -\frac{1}{2}\frac{d[A]}{dt} = \frac{d[B]}{dt}$

Thus, actual stoichiometry of the reactants

$$2A \longrightarrow B$$

$\therefore \quad -\frac{d[A]}{dt} = k[A]^2$

16. (a) For a gaseous state, $pV = nRT$

$\therefore \quad = \frac{n}{V}RT$

$\therefore \quad \left(\frac{dp}{dt}\right) = \frac{RT}{V}\left(\frac{dn}{dt}\right)$

$\therefore \quad \frac{1}{RT}\left(\frac{dp}{dt}\right) = \frac{1}{V}\left(\frac{dn}{dt}\right)$

$$p = \frac{n}{V}RT = \frac{\text{mol}}{\text{L}}RT = CRT$$

where, C is the molal concentration.

$\therefore \quad \frac{dp}{dt} = RT\left(\frac{dC}{dt}\right)$

$\therefore \quad \frac{1}{RT}\left(\frac{dp}{dt}\right) = \left(\frac{dC}{dt}\right)$

Thus, $\quad \frac{1}{RT}\left(\frac{dp}{dt}\right) = \frac{1}{V}\left(\frac{dn}{dt}\right) = \left(\frac{dC}{dt}\right)$

17. (c) Let order w.r.t. $A = a$, order w.r.t. $B = b$

$$-\frac{d[B]}{dt} = k[A]^a[B]^b$$

$$0.25 = k(0.1)^a(0.1)^b \quad \ldots\text{(i)}$$

$$0.50 = k(0.2)^a(0.1)^b \quad \ldots\text{(ii)}$$

$$0.25 = k(0.1)(0.2)^b \quad \ldots\text{(iii)}$$

From Eqs. (i) and (ii), we get

$$2 = (2)^a$$

$\therefore \quad a = 1$

From Eqs. (i) and (iii), we get

$$1 = (2)^b$$

$$(2)^0 = (2)^b$$

$\therefore \quad b = 0$

Thus, rate law $= k[A][B]^0 = k[A]$

18. (d) If concentration of the nth order reaction is made m time, rate becomes m^n times.

Here, $\quad m = 2$

$$m^n = \frac{1}{2}$$

$\Rightarrow \quad 2^n = \frac{1}{2} = 2^{-1}$

$\therefore \quad n = -1$

Explanations for Q. Nos. (19-20)

If rate $= k[NO_2]$ at $[NO_2] = 0.01$ mol L^{-1}s^{-1}

5.4×10^{-5} mol L^{-1}s^{-1} $= k(0.01)$ molL^{-1}s^{-1}

then $\quad k = 5.4 \times 10^{-3}\text{s}^{-1}$

and $\quad$ rate $= 5.4 \times 10^{-3}\text{s}^{-1}$ (0.005) mol L^{-1}

$= 5.4 \times 10^{-5} \times 0.005 = 2.7 \times 10^{-7}$ mol L^{-1}s^{-1}

If rate $= k[NO_2]^2$ at 0.01 molL^{-1}

then $\quad 5.4 \times 10^{-5}$ mol L^{-1}s^{-1} $= k(0.01)^2$

$= k \times 1 \times 10^{-4}$ mol^{2}L^{-2}

$\therefore \quad k = 5.4 \times 10^{-1}\ \text{L mol}^{-1}\ \text{s}^{-1}$

and at $\quad [NO_2] = 0.005\ \text{mol L}^{-1}$

$$\text{Rate} = 5.4 \times 10^{-1}\ \text{L mol}^{-1}\ \text{s}^{-1}\ (0.005)^2\ \text{mol}^2\text{L}^{-2}$$
$$= 1.35 \times 10^{-5}\ \text{mol L}^{-1}\text{s}^{-1}$$

Which is same as given.

Thus, $\quad$ rate law $= k[NO_2]^2$

and $\quad k = 0.54\ \text{L mol}^{-1}\ \text{s}^{-1}$

19. (b) $\qquad$ **20.** (a)

21. (2) $xA \longrightarrow yB$

$$-\frac{1}{x}\frac{d[A]}{dt} = \frac{1}{y}\frac{d[B]}{dt}$$

$$\therefore \quad -\frac{d[A]}{dt} = \frac{d[B]}{dt}\left(\frac{x}{y}\right)$$

$$\log\left[-\frac{d[A]}{dt}\right] = \log\left[\frac{d[B]}{dt}\right] + \log\left(\frac{x}{y}\right) = \log\left[\frac{d[B]}{dt}\right] + 0.3$$

$$\therefore \quad \log\left(\frac{x}{y}\right) = 0.3 = \log(2)$$

$$\therefore \quad \left(\frac{x}{y}\right) = 2$$

22. (2) By stoichiometry of the reaction,

$$-\frac{1}{3}\frac{d[H_2]}{dt} = \frac{1}{2}\frac{d[NH_3]}{dt}$$

$$\therefore \quad \frac{d[NH_3]}{dt} = -\frac{2}{3}\frac{d[H_2]}{dt}$$

$$-\frac{d[H_2]}{dt} = 6\ \text{g min}^{-1}$$
$$= 3\ \text{mol min}^{-1}$$

$$\therefore \quad \frac{d[NH_3]}{dt} = \frac{2}{3} \times 3 = 2\ \text{mol min}^{-1}$$

23. (5) $\quad \Delta(O_3) = 7.2 \times 10^{-15}\text{M}$

$$\Delta t = 1\ \text{h} = 3600\ \text{s}$$

$$-\frac{\Delta(O_3)}{\Delta t} = k[O_3]^2$$

$$\frac{7.2 \times 10^{-15}\text{M}}{60 \times 60\ \text{s}} = k[2.0 \times 10^{-6}]^2\ \text{M}^2$$

$$\therefore \quad k = 5.0 \times 10^{-7}\text{M}^{-1}\text{s}^{-1}$$
$$= x \times 10^{-7}\text{M}^{-1}\text{s}^{-1}$$

Thus, $\quad x = 5$

DPP-2 Zeroth-Order and First-Order Reactions

1. (d) For zeroth order reaction

$$k = \frac{x_2 - x_1}{t_2 - t_1} = \frac{\Delta x}{\Delta t} = \frac{0.01 - 0.008}{10 - 0}$$
$$= \frac{0.002}{10} = 2 \times 10^{-4}\ \text{M min}^{-1}$$

When reaction is 50% completed.

$$[A]_t = [A_0]/2 = \frac{0.01}{2}$$

$$t_{50} = \frac{[A_0]}{2k} = \frac{0.01\text{M}}{2 \times 2 \times 10^{-4}\ \text{M min}^{-1}} = 25\ \text{min}$$

2. (c) For zeroth order reaction

$$\frac{dx}{dt} = k[A]^n = k[A]^0 = k$$

$$\therefore \quad dx = k \cdot dt$$

$$\therefore \quad x = kt$$

x (product) formed after 15 s = $0.025\ \text{Ms}^{-1} \times 15\text{s} = 0.375$ M

Then, reactant left = 0.500 − 0.375 = 0.125 M

3. (a) For zeroth order reaction $k = \dfrac{x}{t}$

Slope $= k = \tan 60° = \sqrt{3}$

$$\text{Half-life} = \frac{a}{2k} = \frac{17.32}{2\sqrt{3}}\ \text{min} = 5\ \text{min}$$

4. (b) $\log C_t = -kt + \log C_0$

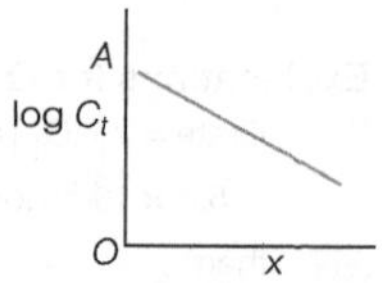

$$y = -mx + C$$

This represents a straight line.

$$y = \log C_t$$
$$-m = -k$$
$$C = \log C_0$$
$$OA = \log C_0 = C$$

5. (a) For first order reaction,

time = 15 min to change its concentration from 0.8 m to 0.4 m

Thus, 50% reaction.

Thus, $\quad t_{50} = 15$ min

Concentration changes from 0.1 M to 0.025 M

$$\text{Amount reacted} = \frac{0.1 - 0.025}{0.1} = 75\%$$

$$t_{75} = 2 \times t_{50}$$
$$= 2 \times 15 = 30\ \text{min}$$

Alternate Method

$$k = \frac{2.303}{t}\log\frac{0.8}{0.4}$$

$$\therefore \quad = \frac{2.303}{15}\log 2$$

$$k = \frac{2.303}{t}\log\frac{0.10}{0.025}$$
$$= \frac{2.303}{t}\log 4 = \frac{2.303}{t}\log 2^2$$

$$t = \left(\frac{2.303}{t}\log 2\right) \times 2$$
$$= 15 \times 2 = 30\ \text{min}$$

6. (d) For first order reaction,

$$k = \frac{1}{t}\log_e\frac{C_0}{C}$$

$$-kt = \log_e\frac{C}{C_0}$$

$$\therefore \quad \frac{C}{C_0} = e^{-kt}$$

$$C = C_0 e^{-kt}$$

$$t_{av} = \frac{1}{k}$$

$$\therefore \quad \text{time},\ t = 2t_{av} = \frac{2}{k}$$

$$\therefore \quad C = C_0 e^{-\frac{k2}{k}}$$
$$= C_0 e^{-2} = \frac{C_0}{e^2}$$

If $C_0 = 100$, then $\quad C = 100/e^2$

7. (c) $k = \frac{2.303}{t} \log \frac{a}{(a-x)}$

when $a = 100$,

$$(a - x) = \frac{a}{2} = 50$$

$$t = 6.93 \text{ min } (= t_{50})$$

$$\therefore \quad k = \frac{2.303}{6.93 \text{ min}} \times \log \frac{100}{50}$$

$$= \frac{2.303 \times 0.3010}{6.93} = \frac{0.693}{6.93} = 0.1 \text{ min}^{-1}$$

when, $a = 100$,

$$(a - x) = 100 - 99 = 1, t = t_{99}$$

$$\therefore \quad t_{99} = \frac{2.303}{0.1} \log \frac{100}{1} = \frac{2.303}{0.1} \times 2 = 46.06 \text{ min}$$

8. (b) For the first order kinetic,

$$-\frac{d[A]}{dt} = k[A]$$

$$k = \frac{2.303}{t} \log \left(\frac{a}{a-x}\right)$$

$$= \frac{2.303}{40} \log \frac{0.1}{0.025}$$

$$= 0.0347 \text{ min}^{-1}$$

$$\therefore \quad -\frac{d[A]}{dt} = 0.0347 \text{ min}^{-1} \times 0.01 \text{ M}$$

$$= 3.47 \times 10^{-4} \text{ Mmin}^{-1}$$

9. (d) $-\frac{d[A]}{dt} = k = \text{constant}$

Then, $\frac{\Delta x}{\Delta t} = \frac{2}{5}$ (in each case)

At 20 min, $\Delta x = 20$ mol

and $[A] = 12$ mol

Taking initial and final change

$$\Delta x = 20 - 12 = 8 \text{ M}$$

$$\Delta t = 20 - 0 = 20 \text{ min}$$

$$\therefore \quad \frac{\Delta x}{\Delta t} = \frac{8}{20} = 0.4 \text{ M min}^{-1}$$

10. (a) $k = \frac{2.303}{t} \log \frac{C_0}{C}$

$$k = \frac{1}{t} \log_e \left(\frac{C_0}{C}\right)$$

$$kt = \log_e \left(\frac{C_0}{C}\right)$$

$$\therefore \quad \frac{C_0}{C} = e^{kt}$$

$$\frac{C}{C_0} = e^{-kt}$$

$$1 - \frac{C}{C_0} = 1 - e^{-kt}$$

$$\frac{(C_0 - C)}{C_0} = (1 - e^{-kt})$$

$$(C_0 - C) = C_0 [1 - e^{-kt}]$$

11. (c) Disintegration follows first-order kinetics

t = total time = 30 min

t_{50} = half-life = 10 min

y = number of half-life $= \frac{30}{10} = 3$

$\therefore$ Activity after y half-life

$$C = C_0 \left(\frac{1}{2}\right)^y = 40 \left(\frac{1}{2}\right)^3 = \frac{40}{8} = 5 \mu\text{Ci}$$

12. (a) $2A(g) \longrightarrow B(g) + C(s)$

	$2A(g)$	$B(g)$	
Initial	p_i	0	(No pressure being solid)
At time t,	$(p_i - 2x)$	x	
At complete reaction	$(p_i - p_i)$	$\frac{p_i}{2}$	

$\therefore$ Pressure at time $t = p_t = p_i = -2x + x$

$$= (p_i - x) = 300 \text{ pascal}$$

Pressure at the end $= p_\infty = \frac{p_i}{2} = 200$ pascal

$$\therefore \quad p_i = 400 \text{ pascal} = a$$

$$x = 100 \text{ pascal}$$

$$\therefore \quad (p_i - 2x) = (400 - 200) = 200 \text{ pascal}$$

$$\therefore \quad k = \frac{2.303}{t} \log \frac{p_i}{(p_i - 2x)}$$

$$= \frac{2.303}{10 \text{ min}} \log \frac{400}{200} = \frac{2.303 \log 2}{10 \text{ min}}$$

$$= \frac{2.303 \times 0.3010}{10 \text{ min}} = \frac{0.693}{10 \text{ min}} = 0.0693 \text{ min}^{-1}$$

13. (a) When there is no radioactive substance in G.M. Counter, it records = 3 dps. Thus, it is the **zero-error** in the counter and in each it is added. Thus, initial counts = 23 − 3 = 20 dps.

At time t (after 10 min) = 13 − 3 = 10 dps

Thus, half-life is 10 min = y

Hence, after next 10 min, dps = 5

Hence, reading in G.M. Counter = 5 + 3 = 8 dps = x

14. (a) r_0 = rotation at the start, r_t = rotation at time t

r_∞ = rotation at the end

then $(r_\infty - r_0) = a$ and $(r_\infty - r_t) = (a - x)$

When 50% inversion has taken place

$$a - x = \frac{a}{2}$$

$$\therefore \quad 2(a - x) = a$$

$$2(r_\infty - r_t) = r_\infty - r_0$$

$$2r_\infty - 2r_t = r_\infty - r_0$$

$$\therefore \quad r_0 = (2r_t - r_\infty)$$

15. (c) When B is formed from A, two α-particles (He) are formed.

A taken initially = 1 mol

A decomposed in two half-lifes = 75% $= \frac{3}{4}$ mol

Thus, helium formed $= 2 \times \frac{3}{4}$ mol $= \frac{3}{2}$ mol

This helium remains in closed vessel.

Thus, pressure set up is due to helium.

$$p = \frac{n}{V} RT$$

$$= \frac{3}{2} \times \frac{\text{mol} \times 0.0821 \text{ L atm mol}^{-1} \text{ K}^{-1} \times 300 \text{ K}}{1 \text{L}}$$

$$= 36.945 \text{ atm} \approx 37.0 \text{ atm}$$

16. (c) $T_{av} = \frac{1}{k}$, $\quad T_{50} = \frac{0.693}{k} = 0.693\, T_{av}$

$$T_{75} = 2T_{50}$$

$$T_{50} = \left(\frac{T_{75}}{2}\right)$$

$$T_{av} : T_{50} : T_{75}$$

$$\frac{T_{50}}{0.693} : T_{50} : 2\,T_{50}$$

$$\frac{1}{0.693} : 1 : 2$$

$$1.44 : 1 : 2$$

17. (d) For the first order reaction,

$$k = \frac{2.303}{t}\log\left(\frac{a}{a-x}\right) \quad \text{(Equation III)}$$

$$kt = \log_e\left(\frac{a}{a-x}\right)$$

$$-kt = \log_e\left(\frac{a-x}{a}\right)$$

$$\therefore \quad \left(\frac{a-x}{a}\right) = e^{-kt}$$

$$(a-x) = ae^{-kt}$$

$$x = a - ae^{-kt} = a(1-e^{-kt}) \quad \text{(Equation I)}$$

when $t = t_{50}$,

$$(a-x) = \left(\frac{a}{2}\right)$$

$$\therefore \quad k = \frac{2.303}{t_{50}}\log\frac{a}{(a/2)} = \frac{2.303}{t_{50}}\log 2$$

$$t_{50} = \frac{2.303}{k}\log(2)$$

$$\therefore \quad \frac{2.303\log 2}{t_{50}} = \frac{2.303}{t}\log\left(\frac{a}{a-x}\right)$$

$$\frac{t}{t_{50}}\log\left(\frac{1}{2}\right) = \log\left(\frac{a-x}{a}\right)$$

$$\log\left(\frac{1}{2}\right)^y = \log\left(\frac{a-x}{a}\right) \quad \text{when} \quad y = \frac{t}{t_{50}}$$

$$\therefore \quad (a-x) = a\left(\frac{1}{2}\right)^y \quad \text{(Equation II)}$$

18. (b) $C_0 = 1\,\text{M}$

$$k = 0.0693\ \text{min}^{-1}$$

$$\therefore \quad t_{50} = \frac{0.693}{0.0693} = 10\ \text{min}$$

Time = 30 min

$$y = \text{number of half-life } \frac{30}{10} = 3$$

Concentration after time $t = C$

$$\therefore \quad C = C_0\left(\frac{1}{2}\right)^y = 1\left(\frac{1}{2}\right)^3 = \frac{1}{8}\text{M}$$

$$\therefore \quad \frac{dx}{dt} = kC = \frac{0.0693}{8} = 8.66 \times 10^{-3}\ \text{M min}^{-1}$$

19. (a) $C_6H_5N_2Cl + H_2O \xrightarrow{\Delta} C_6H_5OH + N_2 + HCl$

Initial	a	0
At time t	$(a-x)$	x

$$50\ \text{mL of 1M } A \equiv 50 \times 1\ \text{millimol}$$

$$= \frac{50 \times 1}{1000}\ \text{mol} = 0.05\ \text{mol}$$

$$\therefore \quad a = 0.05\ \text{mol}$$

At time t, $\quad N_2$ formed $= x = \dfrac{pV}{RT} = \dfrac{1 \times 1.226}{0.0821 \times 300}$

$$= 0.0498\ \text{mol}$$

$$\therefore \quad (a-x) = 0.050 - 0.0498$$

$$= 2 \times 10^{-4}\ \text{mol}$$

$$k = \frac{2.303}{t}\log\left(\frac{a}{a-x}\right)$$

$$= \frac{2.303}{60\ \text{min}}\log\frac{0.05}{2 \times 10^{-4}} = \frac{2.303}{60}\log 250$$

$$= \frac{2.303}{60} \times 2.40$$

$$\text{Half-life } (t_{50}) = \frac{2.303\log 2}{k} = \frac{2.303 \times 0.3010}{2.303 \times 2.4} \times 60 = 7.53\ \text{min}$$

20. (d) $k = \dfrac{2.303}{t}\log\left(\dfrac{a}{a-x}\right)$

If $a = 100$, $\quad x = 0.1$ in 3 h

$$\therefore \quad k = \frac{2.303}{3}\log\left(\frac{100}{100-0.1}\right)$$

$$= 3.336 \times 10^{-4}\ \text{h}^{-1}$$

$$t_{50} = \frac{0.693}{k} = \frac{0.693}{3.336 \times 10^{-4}\ \text{h}^{-1}}$$

$$= 2077\ \text{h} = \frac{2077}{24}\ \text{day} = 86.54\ \text{day}$$

21. (c) Rate constant, $k = \dfrac{0.693}{t_{50}} = \dfrac{0.693}{1.28 \times 10^9\ \text{yr}}$

$$= 5.414 \times 10^{-10}\ \text{yr}^{-1}$$

Number of 40 K atoms in 1 g KCl

$$= \frac{1}{74.5}\ \text{mol}$$

$$= \frac{1}{74.5} \times 6.02 \times 10^{23}\ \text{molecules}$$

$$= \frac{1}{74.5} \times 6.02 \times 10^{23} \times \frac{0.01}{100}\ {}^{40}\text{K atoms}$$

$$= 8.08 \times 10^{17}\ \text{atoms}$$

$$\therefore \quad \text{Activity} = k\lambda = 5.414 \times 10^{-10} \times 8.08 \times 10^{17}$$

$$= 4.37 \times 10^8\ \text{yr}^{-1}$$

Explanations for Q. Nos. (22-23)

$$k(^{14}\text{O}) = \frac{0.693}{t_{50}} = \frac{0.693}{71.0\ \text{s}}$$

$$k(^{15}\text{O}) = \frac{0.693}{t_{50}} = \frac{0.691}{124\ \text{s}}$$

$$\therefore \quad \frac{k(^{14}\text{O})}{k(^{15}\text{O})} = 1.75$$

$$N(^{14}\text{O}) = N_0\,e^{-kt} \quad (t = 1\text{h} = 3600\ \text{s})$$

$$= N_0\,e^{-0.693 \times 3600/71}$$

$$N(^{15}\text{O}) = N_0\,e^{-0.693 \times 3600/124}$$

$$\therefore \quad \frac{N(^{14}\text{O})}{N(^{15}\text{O})} = e^{\left(-0.693 \times 3640/71 + \frac{0.693 \times 3600}{124}\right)}$$

$$= e^{-15.02} = 3.0 \times 10^{-7}$$

22. (a) **23.** (b)

Explanations for Q. Nos. (24-25)

$$CH_3COOC_2H_5 + H_2O \xrightarrow{H^+} CH_3COOH + C_2H_5OH$$

at $t = 0$	a	0	0
at time t,	$(a-x)$	x	x
at ∞ time	$(a-a)$	a	a

At time $t = 0$, volume of NaOH is due to H^+ (catalyst) $= V_0$ and remains constant

at time t, volume of NaOH is due to H^+ and $CH_3COOH = V_t$

$$\therefore \quad V_t = x + H^+ = x + V_0$$

$$x = (V_t - V_0)$$

At infinite time (complete reaction), volume of NaOH is due to total CH_3COOH (or ester) and $H^+ = a + V_0$

$$\therefore \quad V_\infty = a + V_0$$

$$\therefore \quad a = (V_\infty - V_0)$$

$$(a-x) = (V_\infty - V_0) - (V_t - V_0) = (V_\infty - V_t)$$

Thus, $\quad k = \dfrac{2.303}{t}\log\left(\dfrac{a}{a-x}\right)$

$$k = \frac{2.303}{t}\log\frac{(V_\infty - V_0)}{(V_\infty - V_t)}$$

If ester is 50% hydrolysed,

$$x = \frac{a}{2}$$

$$\therefore \quad V_t = x + V_0 = \frac{a}{2} + V_0$$

$$\frac{a}{2} = V_t - V_0$$

$$a = 2(V_t - V_0)$$

$$(V_\infty - V_0) = 2V_t - 2V_0$$

$$V_\infty = (2V_t - V_0)$$

24. (b) **25.** (c)

Explanations for Q. Nos. (26-27)

$$\left(-\frac{d[A]}{dt}\right)_{I} = k(6)$$

$$\left(-\frac{d[A]}{dt}\right)_{II} = k(12)$$

$$\left(-\frac{d[A]}{dt}\right)_{III} = k(a)$$

Thus, ratio fo rates = 6 : 12 : 9 = 1 : 2 : 15

For first order reaction T_{50} is independent of initial concentration. Thus, T_{50} will be equal to 1 : 1 : 1.

26. (a) **27.** (a)

28. (2)

Permissible element = 10 times the initial value

Activity at time t = $[A] = \frac{[A]_0}{10}$

$$\therefore \quad [A]_0 \text{ (initial)} = 10[A]$$

Also, $$k = \frac{2.303 \log 2}{30}$$

Also, $$k = \frac{2.303}{t} \log \frac{[A_0]}{[A]}$$

$$\frac{2.303 \log 2}{30} = \frac{2.303}{t} \log \frac{[A_0]}{[A_0]/10}$$

$$t = \frac{30}{\log 2} \log(10)$$

$$= \frac{30}{0.3010} = 99.67 \text{ days}$$

$$= 100 \text{ days} = 10^2 \text{ days} = 10^n \text{ days}$$

Thus, $n = 2$

29. (2)

$$k = \frac{2.303}{t} \log\left(\frac{a}{a-x}\right)$$

$$\frac{kt}{2.303} = \log a - \log(a - x)$$

$$\therefore \quad \log(a - x) = -\frac{kt}{2.303} + \log a$$

This is a straight line with slope = $\frac{-k}{2.303} = \tan\theta = \tan 45°$

$$\therefore \quad k = 2.303 \tan 45° = 2.303 \text{ min}^{-1}$$

$$\text{Half-life} = \frac{2.303 \log 2}{k} = \frac{2.303 \log 2}{2.303}$$

$$= \log 2 = \log x$$

Thus, $x = 2$ min

30. (5) At time $t = 0$, $x = 0$

$$\log(a - x) = \log a = OA$$

$$\log a = 1.3010$$

$$\therefore \quad a = 20.0 \text{ mol dm}^{-3}$$

Half-life = log2 = 0.3010 min

After two half-lifes (0.6021 min)

Reactant left = $\frac{a}{4} = \frac{20}{4} = 5 \text{ mol dm}^{-3}$

DPP-3 Determination of Order of Reaction

1. (d) Let rate law be $\frac{-d[A]}{dt} = k[A]^n$

$$1 \times 10^{-4} = k(0.01)^n$$

$$1.41 \times 10^{-4} = k(0.02)^n$$

$$\therefore \quad 1.41 = (2)^n$$

$$(2)^{1/2} = (2)^n \qquad (\text{Since, } \sqrt{2} = 1.41)$$

$$\therefore \quad n = \frac{1}{2}$$

Thus, rate law is $\frac{-d[A]}{dt} = k[A]^{1/2}$

2. (d) $\left(\frac{dx}{dt}\right)_1$ = rate = $k[A]^n[B]^m$

$$\left(\frac{dx}{dt}\right)_2 = k[2A]^n \left[\frac{B}{2}\right]^m = k[2A]^n [2B]^{-m}$$

$$\frac{\left(\frac{dx}{dt}\right)_2}{\left(\frac{dx}{dt}\right)_1} = (2^{n-m})$$

3. (c) Rate = $k[A][B]$

The given reaction is first order in A and first order is B.

Thus, total order = 2

(a) Unit of k = conc^{1-n} time^{-1} = conc^{-1} time^{-1}

Thus, (a) is false.

(b) $t_{1/2} = \frac{1}{k_a}$ of second-order reaction, thus (b) is false.

(c) $\frac{d[C]}{dt} = -\frac{1}{2}\frac{d[A]}{dt}$

Thus, (c) is correct.

Unit of k = conc$^{(1-n)}$ time^{-1} $(n = 2)]$ = conc^{-1} time^{-1}

Thus, value of k is dependent on the concentration of A and B. Thus, (d) is false.

4. (a) $t_{50} \propto \left(\frac{1}{a}\right)^{n-1}$

t_{50} is independent of concentration of B then w.r.t. B, $(n - 1) = 0$. Thus, $n = 1$

Thus, order w.r.t. B = 1

Also, if concentration is made m-times, rate becomes, m^n-times.

Hence, w.r.t. A

$$m = 2, m^n = 2$$

$$\therefore \quad 2^n = 2 = 2^1$$

Thus, $n = 1$

Thus, order w.r.t. A = 1

Thus, total order = 2 = n

Unit of nth order reaction = conc$^{(1-n)}$ time^{-1}

conc in molL^{-1} and time in seconds.

Then, unit = $(\text{mol L}^{-1})^{1-2}\text{s}^{-1} = (\text{mol L}^{-1})^{-1}\text{s}^{-1}$

$$= \text{L mol}^{-1}\text{s}^{-1}$$

5. (a) I, II, V — represent first order equation.

III, VI — zeroth order equation

IV—second order equation

6. (d) For nth order reaction ($n > 1$)

$$k = \frac{1}{(n-1)t}\left[\frac{1}{(a-x)^{n-1}} - \frac{1}{a^{n-1}}\right]$$

when, $t = t_{50}$, $x = \frac{a}{2}$, $(a - x) = \left(\frac{a}{2}\right)$

$$\therefore \quad k = \frac{1}{(n-1)t_{50}}\left[\frac{1}{\left(\frac{a}{2}\right)^{n-1}} - \frac{1}{a^{n-1}}\right]$$

$$\therefore \quad t_{50} = \frac{1}{(n-1)k}\left[\frac{2^{n-1}-1}{a^{n-1}}\right]$$

when $t = t_{75}$, $x = \frac{3a}{4}$,

$$(a - x) = \frac{a}{4}$$

$$\therefore \quad t_{75} = \frac{1}{(n-1)k}\left[\frac{4^{n-1}-1}{a^{n-1}}\right]$$

Given, $t_{75} = 3\,t_{50}$

$$\therefore \quad \frac{1}{(n-1)k}\left[\frac{4^{n-1}-1}{a^{n-1}}\right] = \frac{3}{(n-1)k}\left[\frac{2^{n-1}-1}{a^{n-1}}\right]$$

$$\therefore \quad 4^{n-1} - 1 = 3\,(2^{n-1} - 1)$$

This equation is true, if $n = 2$

$$(4^{2-1} - 1) = 3\,(2^{2-1} - 1)$$

$$(4 - 1) = 3\,(2 - 1)$$

$$3 = 3$$

7. (b) $\log\left(\frac{dx}{dt}\right) = \log k + n/\,y[A]$

n = order of the reaction = slope of the line

Larger the value of the slope of the line, larger the order.

Thus, $P = 3, Q = 2, R = 1, S = 0$.

8. (d) $\left(\frac{dx}{dt}\right) = k[A]^x$

$$\left(\frac{dx}{dt}\right) = k[A]^0 = k$$

rate = rate constant

Thus, order = zero

Rate is independent of concentration of A.

Thus, rate remains constant.

9. (c) For nth order reaction, half-life t_{50} is $t_{50} \propto \left(\frac{1}{a}\right)^{n-1}$

(T_{50}) is t_1 at $[A] = a_1$ and is t_2 at $[A] = a_2$

$$\therefore \quad \frac{t_1}{t_2} = \left(\frac{a_2}{a_1}\right)^{n-1} \Rightarrow \left(\frac{52}{26}\right) = \left(\frac{1.004}{0.502}\right)^{n-1}$$

$$(2) = (2)^{n-1}$$

$$\therefore \quad n - 1 = 1, n = 2$$

10. (c) T_{50} is independent of concentration of $[A]$.

Thus, it follows first order kinetics.

$$T_{50} = \frac{2.303 \log 2}{k}$$

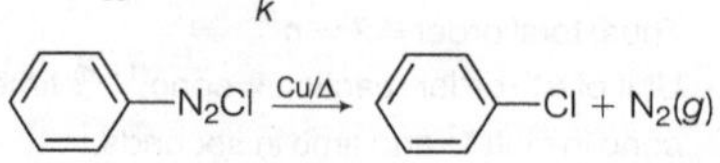

at time $t = 0$	a	0	0
at time t = 10 min	$(a - x)$	x	x
at $t = \infty$ (complete)	$(a - a)$	a	a

Thus, $a = 50$ L, $x = 10$ L

$$(a - x) = 50 - 10 = 40 \text{ L}$$

For first order reaction

$$k = \frac{2.303}{t}\log\left(\frac{a}{a-x}\right) = \frac{2.303}{10}\log\frac{50}{40}$$

$$= \frac{2.303}{10}\log 1.25 = \frac{2.303}{10} \times 0.0969$$

$$T_{50} = \frac{2.303 \log 2}{k} = \frac{2.303 \times 0.3010 \times 10}{2.303 \times 0.0969} = 31 \text{ min}$$

11. (b) Cl-atom attached to N-atom, is an oxidising agent and can oxidise KI to I_2.

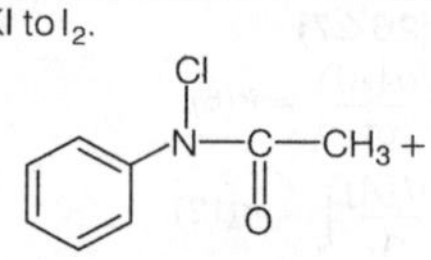 $+ 2I^- \rightarrow I_2 + 2Cl^-$

I_2 can be titrated using hypo in iodometric titration.

$$I_2 + \underset{\text{Hypo}}{2S_2O_3^{2-}} \longrightarrow S_4O_6^{2-} + 2I^-$$

Cl-atom attached to benzene nucleus is not an oxidising agent.

12. (b) $A \longrightarrow B$

$$\frac{-d[A]}{dt} = \frac{d[B]}{dt}$$

$[B]$ increases uniformly with time t. Order is zero. Thus, at every point, $\frac{d[B]}{dt}$ = constant

Hence, $\frac{-d[A]}{dt}$ = constant with time

13. (c)

	Reaction	Time
0.12M ⟶ 0.06M	50%	10 h
0.12 M ⟶ 0.03 M	75 %	20 h

We find that, $t_{75} = 2 \times t_{50}$

Thus, given reaction is of first-order.

14. (c)

(i) Rate remains constant

$$\left(\frac{dx}{dt}\right) = k[A]^n = k[A]^0 = k$$

Thus, order = 0

Thus, **(i) → (s)**

(ii) Half-life period is independent of concentration

$$T_{50} \propto \left(\frac{1}{a}\right)^{n-1}$$

$$\therefore \quad (n - 1) = 0$$

$$\therefore \quad n = 1$$

Thus, **(ii) → (r)**

(iii) For $(n > 1)$, $k = \frac{1}{(n-1)t}\left[\frac{1}{(a-x)^{n-1}} - \frac{1}{a^{n-1}}\right]$

if $n = 2$, $k = \frac{1}{t}\left[\frac{1}{(a-x)} - \frac{1}{a}\right]$

Thus, $\frac{1}{(a-x)} = kt + \frac{1}{a}$

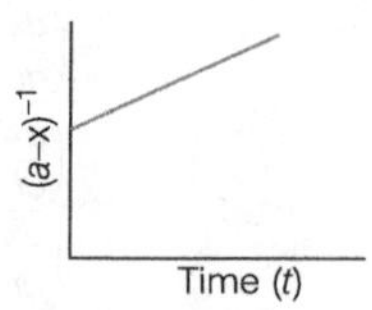

Thus, graph is for second-order reactions

Thus, **(iii) → (q)**

Explanations for Q. Nos. (15-16)

Time : [PAN]

0 min	5.0×10^{14}	molecules L^{-1}
30 min	2.5×10^{14}	molecules L^{-1}
60 min	1.25×10^{14}	molecules L^{-1}

At 30 min, 50% reaction takes place.

At 60 min, 75% reaction takes place.

Thus, $T_{75} = 2 \times T_{50}$

It is valid for first order reaction. Thus, order = 1

For first order reactions, $T_{50} = \dfrac{0.693}{k}$

$$\therefore \quad k = \frac{0.693}{T_{50}} = \frac{0.693}{30} = 0.0231 \text{ min}^{-1}$$

15. (b) **16.** (a)

Explanations for Q. Nos. (17-18)

$$\frac{-dp}{dt} = k(p_{NO})^a (p_{H_2})^b$$

$$0.0012 = k(0.25)^a (0.2)^b \quad \ldots\text{(i)}$$

$$0.0024 = k(0.50)^a (0.1)^b \quad \ldots\text{(ii)}$$

$$0.0048 = k(0.50)^a (0.2)^b \quad \ldots\text{(iii)}$$

From Eqs. (i) and (iii), we get

$$4 = (2)^a$$

$$\therefore \quad (2)^2 = (2)^a$$

$$\therefore \quad a = 2$$

Thus w.r.t. NO, order = 2

From Eqs. (ii) and (iii), we get

$$2 = (2)^b$$

$$(2)^1 = (2)^b$$

$$\therefore \quad b = 1$$

Thus, w.r.t. H_2, order = 1

Total order = 3

From Eq. (i),

$$0.0012 = k(0.25)^2(0.2)^1 = k(0.25)^2 (0.2)$$

$$\therefore \quad k = 0.0012/(0.25)^2 (0.2) = 0.096 \text{ atm}^{-2}\text{min}^{-1}$$

17. (c) **18.** (c)

Explanations for Q. Nos. (19-20)

Rate law $-\dfrac{\Delta[A]}{\Delta t} = k[A]^a[B]^b$

where, a = order w.r.t. A, b = order w.r.t. B

Set	$[A]_0$	$[A]_t$ M	$[B]_0$	$\Delta[A]$	Δt h	$-\dfrac{\Delta[A]}{\Delta t}$ M h^{-1}
1	0.10	0.0975	1.0	0.0025	0.50	5.0×10^{-3}
2	0.10	0.0900	2.0	0.0100	0.50	2.0×10^{-2}
3	0.05	0.04500	1.0	0.0050	2.00	2.5×10^{-3}

(i) $5.0 \times 10^{-3} = k[0.1]^a [1.0]^b$, (ii) $2.0 \times 10^{-2} = k[0.1]^a [2.0]^b$

(iii) $2.5 \times 10^{-3} = k[0.05]^a [1.0]^b$

From Eqs. (i) and (ii), we get

$$\frac{2.0 \times 10^{-2}}{5.0 \times 10^{-3}} = (2)^b \quad \Rightarrow \quad (4) = (2)^b$$

$$(2)^2 = (2)^b$$

$$\therefore \quad b = 2$$

Order w.r.t. B = 2

From Eqs. (i) and (iii), we get

$$\frac{5.0 \times 10^{-3}}{2.5 \times 10^{-3}} = \left(\frac{0.10}{0.05}\right)^a$$

$$(2) = (2)^a$$

$$a = 1$$

Order w.r.t. A = 1

From Eq. (ii), $5.0 \times 10^{-3} = k(0.1)(1.0)^2$

$$\therefore \quad k = 5.0 \times 10^{-2} \text{ M}^{-2}\text{h}^{-1}$$

19. (b) **20.** (b)

21. (2) For nth order reactions,

$$T_{50} \propto \left(\frac{1}{a}\right)^{n-1}$$

$$T_{50} = k\left(\frac{1}{a}\right)^{n-1}$$

$$\therefore \quad \log T_{50} = (n-1)\log\left(\frac{1}{a}\right) + \log k$$

For a straight line, $y = mx + c$

Slope $= m = \tan\theta = \tan 45^\circ = (n-1)$

$$1 = n - 1$$

$$\therefore \quad n = 2$$

22. (3) Half-life $T_{50} \propto \left(\dfrac{1}{\text{conc}}\right)^{a-1}$

$$T_{50} \propto \left(\frac{1}{[A]}\right)^{a-1}$$

T_{50} is independent of $[A]$, then $(a-1) = 0$

$$a = 1$$

$$T_{50} \propto \left(\frac{1}{H^+}\right)^{b-1} \quad \Rightarrow \quad \frac{T_1}{T_2} = \left(\frac{(H^+)_2}{(H^+)_1}\right)^{b-1}$$

$$T_1 = 100 \text{ min}, [H^+]_1 = 10^{-pH} = 10^{-5} \text{ M}$$

$$T_2 = 10 \text{ min}, [H^+]_2 = 10^{-pH} = 10^{-4} \text{ M}$$

$$\therefore \quad \frac{100}{10} = \left(\frac{10^{-4}}{10^{-5}}\right)^{b-1}$$

$$\Rightarrow \quad (10) = (10)^{b-1}$$

$$b - 1 = 1$$

$$b = 2$$

$$\therefore \quad a + b = 3$$

23. (2)

$$T_{50} \propto \left(\frac{1}{a}\right)^{n-1}$$

$$\therefore \quad \frac{T_1}{T_2} = \left(\frac{a_2}{a_1}\right)^{n-1}$$

$$\left(\frac{200}{50}\right) = \left(\frac{0.4}{0.1}\right)^{n-1}$$

$$(4) = (4)^{n-1}$$

$$n - 1 = 1$$

$$n = 2$$

24. (7)

$$\frac{-d[A]}{dt} = k[A]^a [B]^b$$

(i) $1 \times 10^{-4} = k[a]^a [0.01]^b$

(ii) $2 \times 10^{-4} = k[2]^a [0.01]^b$

(iii) $8 \times 10^{-4} k[2]^a [0.02]^b$

(iv) $8 \times 10^{-y} = k[0.2]^a [0.002]^b$

From (i) and (ii),

$$(2) = (2)^a \Rightarrow a = 1$$

From (ii) and (iii),

$$4 = (2)^b \Rightarrow b = 2$$

$$\therefore \quad \frac{-d[A]}{dt} = k[A][B]^2 \text{ from (i)}$$

$$k = 1 \, M^{-2}s^{-1} \text{ from (iv)}$$

$$\frac{-d[A]}{dt} = 1\,(0.2)\,(0.002)^2$$

$$= 8 \times 10^{-7} = 8 \times 10^{-y}$$

Thus, $y = 7$

DPP-4 Energy of Activation

1. Plan (i) By Arrhenius equation

$$\log k = \log A - \frac{E_a}{2.303RT}$$

$$E_a = \frac{2.303RT_1T_2}{(T_2 - T_1)} \log \frac{k_2}{k_1}$$

at temperatures T_1 and T_2

and $$E_a = \frac{2.303\,RT_1T_3}{(T_3 - T_1)} \log \frac{k_3}{k_1}$$

at temperatures T_1 and T_3

$$\therefore \quad \frac{2.303RT_1T_2}{(T_2 - T_1)} \log \frac{k_2}{k_1} = \frac{2.303\,RT_1T_3}{(T_3 - T_1)} \log \frac{k_3}{k_1}$$

$$\frac{T_2}{(T_2 - T_1)} \log \frac{k_2}{k_1} = \frac{T_3}{(T_3 - T_1)} \log \frac{k_3}{k_1}$$

T_1, T_2, T_3 and k_1, k_2 are known. Thus, value of k_3 is derived.

$$\frac{310}{10} \log \frac{4 \times 10^{-5}}{2 \times 10^{-5}} = \frac{320}{20} \log \frac{k_3}{(2 \times 10^{-5})}$$

$$\log\left(\frac{k_3}{2 \times 10^{-5}}\right) = \frac{31}{16} \log 2$$

$$\log\left(\frac{k_3}{2 \times 10^{-5}}\right) = 0.5832$$

$$\therefore \quad k_3 = 7.66 \times 10^{-5} \text{s}^{-1}$$

(ii) $$E_a = \frac{2.303\,RT_1T_2}{(T_2 - T_1)} \log \frac{k_2}{k_1}$$

$$= \frac{2.303 \times 8.314 \times 10^{-3}\,\text{kJ mol}^{-1}\text{K}^{-1} \times 300\,\text{K} \times 310\,\text{K}}{10\text{K}} \log 2$$

$$= 53.60\ \text{kJ mol}^{-1}$$

2. Plan By Arrhenius equation

$$\log k = \log A - \frac{E_a}{2.303RT}$$

If E_a is known, log A hence value of A (Arrhenius pre-exponential factor) can be calculated.

(i) $$E_a = \frac{2.303\,RT_1T_2}{(T_2 - T_1)} \log \frac{k_2}{k_1}$$

$$= \frac{2.303 \times 8.314 \times 10^{-3}\,\text{kJ mol}^{-1}\text{K}^{-1} \times 600\,\text{K} \times 700\text{K}}{(700 - 600)\,\text{K}}$$

$$\log \frac{1.52 \times 10^3}{3.8 \times 10^2} = 48.416\ \text{kJ mol}^{-1}$$

(ii) $$\log k_1 = \log A - \frac{E_a}{2.303RT_1}$$

$$\log(3.8 \times 10^2) = \log A - \frac{48.416\,\text{kJ mol}^{-1}}{2.303 \times 8.314 \times 10^{-3}\,\text{kJ mol}^{-1}\text{K}^{-1} \times 600\,\text{K}}$$

$$\log A = 6.7942$$

$$A = 6.225 \times 10^6\ \text{dm}^3\ \text{mol}^{-1}\ \text{s}^{-1}$$

3. (a) $$E_a = \frac{2.303RT_1T_2}{(T_2 - T_1)} \log\left(\frac{k_2}{k_1}\right)$$

$$\frac{k_2}{k_1} = 2$$

$$\therefore\ E_a = \frac{2.303 \times 8.314 \times 10^{-3}\,\text{kJ mol}^{-1}\,\text{K}^{-1} \times 300\text{K} \times 310\text{K}}{(310 - 300)\text{K}} \log 2$$

$$= 53.6\ \text{kJ mol}^{-1}$$

4. (c) By Arrhenius equation

$$k = Ae^{-E_a/RT}$$

$$k = (5.0 \times 10^{11}\ \text{L mol}^{-1}\text{s}^{-1})\,e^{-29000/T}$$

Thus, $$\frac{E_a}{R} = 29000\text{K}$$

$$E_a = (29000 \times R)\text{K}$$

$$= 29000 \times 2 \times 10^{-3}\ \text{kcal mol}^{-1}\ \text{K}^{-1}\text{K}$$

$$= 58.0\text{kcal mol}^{-1}$$

5. (b)

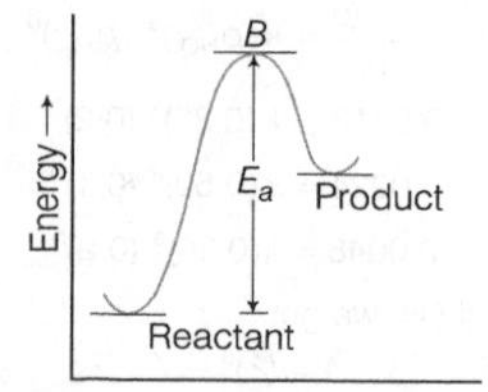

E_a (activation energy) is the energy required to activate the reactant so that they may collide and give the final product. Below point B, the reactants will not react at all.

6. (c)

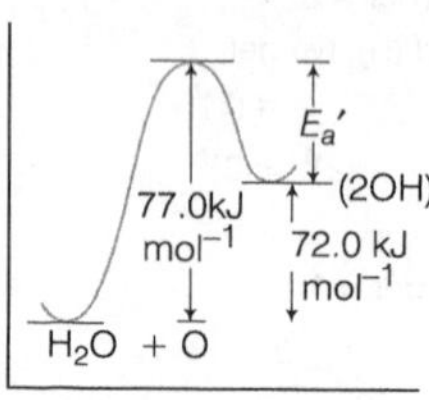

$$E_a'\ (2\text{OH} \longrightarrow H_2O + O)$$

$$= 77.0 - 72.0$$

$$= 5.0\ \text{kJ mol}^{-1}$$

7. (c) k_1 (no catalyst) $= Ae^{-E_a/RT}$

$$k_2 \text{ (with catalyst)} = Ae^{-(E_a - 2)/RT}$$

$$\therefore \quad \frac{k_2}{k_1} = e^{2/RT} = e^{2/0.002 \times 300} = 28.03$$

$$\therefore \quad k_2 = k_1 \times 28.03$$

Thus, rate is affected (28.03) times.

8. (d) On increasing temperature, number of collisions hence kinetic energy increases.

$$\text{KE} = \frac{1}{2}mu^2 = \frac{3}{2}RT$$

Thus, $$u^2 \propto T$$

$$u \propto \sqrt{T}$$

Thus rate of the reaction or rate constant increases with increase in temperature.

Half-life of any reaction $\propto \frac{1}{k}$

Thus, half-life decreases with increase in temperature, reaction may be exothermic or endothermic.

Thus, $T_2 > T_1$

9. (d) k varies with temperature as given by Arrhenius equation

$$k = Ae^{-E_a/RT}$$

$$= Ae^{RT/E_a}$$

k has least value at low T and high E_a.

10. (b) $k_1 = e^{-E_{a_1}/RT}$

$$k_2 = e^{-E_{a_2}/RT} \qquad (\text{let } E_{a_2} > E_{a_1})$$

$$k_1 = k_2 \text{ (rates equal at room temperature } T)$$

$$k_1' = e^{-E_{a_1}/RT'} = e^{RT'/E_{a_1}}$$

$$k_2' = e^{-E_{a_2}/RT'} e^{RT'/E_{a_2}}$$

If temperature is increased to T',

$$\frac{k_1'}{k_1} = e^{R(T'-T)/E_{a1}}$$

$$\frac{k_2'}{k_2} = e^{R(T'-T)/E_{a2}}$$

If $\quad E_{a_2} > E_{a_1}$, then $\left(\frac{k_2'}{k_2}\right) < \left(\frac{k_1'}{k_1}\right)$

Thus, reaction with smaller activation energy will be faster on increasing temperature.

11. (b) $k = Ae^{-E_a/RT}$

$$\log k = \log A - \frac{E_a}{RT}$$

$$\log_{10} k = \log_{10} A - \frac{E_a}{2.303RT}$$

$$= \log_{10} A - \frac{2.303RT}{2.303RT}$$

$$\log_{10} k = \log_{10} A - 1$$

$$\log_{10} k - \log_{10} A = 1$$

$$\log_{10}\left(\frac{k}{A}\right) = -1$$

$$\left(\frac{k}{A}\right) = \text{antilog of } (-1) = 0.1$$

12. (a) Energy required to take the reactant to the activated state is $(E_1 + E_2)$. Since, reactions is endothermic, product is less stable.

13. (c) Number of ten degrees in $50° = \frac{50}{10} = 5$

Increase in rate due to 10° rise in temperature = 2

Increase in rate due to 50 ° rise in temperature = 2^5 = 32 times

14. (b) $A \xrightarrow{k_1} B$ $k_1 = A'e^{-E_{a_1}/RT}$ (A' = Arrhenius constant)

$$A \xrightarrow{k_2} B$$

$$k_2 = A'e^{-E_{a_2}/RT} = A'e^{-2E_{a_1}/RT} \text{ (Since, } E_{a_2} = 2E_{a_1})$$

$$\frac{k_1}{k_2} = e^{(-E_{a_1} + 2E_{a_1})/RT} = e^{E_{a_1}/RT}$$

$$\therefore \quad k_1 = k_2 e^{E_{a_1}/RT}$$

15. (c, d) Given reaction is an endothermic reaction $\Delta H > 0$.

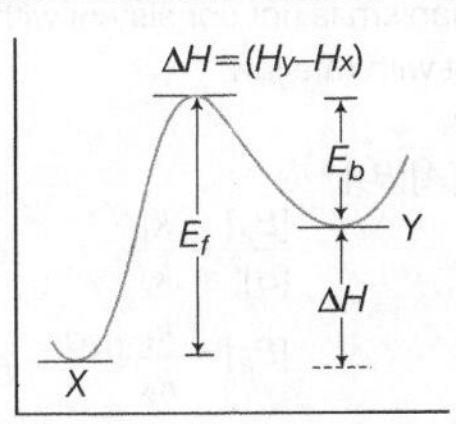

$$E_f > E_b$$

Thus, (d) is correct.

$$E_b + \Delta H = E_f$$

Thus, (c) is correct.

16. (a, b, c, d) All the given facts (a), (b), (c) and (d) are true for **Collision Theory.**

17. (c) By Arrhenius equation, $\log_{10} k = \log_{10} A - \frac{E_a}{2.303RT}$

$$\log_{10} A = 5$$

$$\therefore \quad A = 10^5 \text{ min}^{-1}$$

Thus, **(i) → (r)**

Unit of k is in min^{-1}.

Thus, **(ii) → (p)**

At 500 K, $\quad \log_{10} k = 5 - \frac{2 \times 10^3}{500} = 1$

$$k = 10\text{min}^{-1}$$

$$\therefore \quad \frac{-d[A]}{dt} = k[A]_0^n = k[A]_0^1$$

$$= 10 \text{ min}^{-1} (0.2\text{M}) = 2 \text{ M min}^{-1}$$

Thus, **(iii) → (q)**

$$\frac{E_a}{2.303R} = 2 \times 10^3\text{K}$$

$$E_a = 2.303 \times R \times 2 \times 10^3\text{K}$$

$$= 2.303 \times 0.002 \text{ kcal mol}^{-1}\text{K} \times 2 \times 10^3\text{K}$$

$$= 9.2 \text{ kcal mol}^{-1}$$

Thus, **(iv) → (s)**

Explanations for Q. Nos. (18-19)

$$\log k(\text{min}^{-1}) = -\frac{11067\text{K}}{T} + 31.330$$

At 298 K, $\quad \log k(\text{min}^{-1}) = -\frac{11067}{298} + 31.330 = -5.8076 \text{ min}^{-1}$

$$k = 1.5574 \times 10^{-6} \text{ min}^{-1}$$

$$k = 2.596 \times 10^{-8}\text{s}^{-1}$$

$$\log k = \log A - \frac{E_a}{2.303RT}$$

On comparison with given equation

$$\frac{E_a}{2.303R} = 11067\text{K}$$

$$\therefore \quad E_a = 2.303 R \times 11067\text{K}$$

$$= 2.303 \times 8.314 \times 10^{-3} \text{ kJ mol}^{-1}\text{K}^{-1} \times 11067 \text{ K}$$

$$= 211.90 \text{ kJ mol}^{-1}$$

18. (b) **19.** (a)

Explanations for Q. Nos. (20-21)

$$A + B \underset{k_2}{\overset{k_1}{\rightleftharpoons}} C + D$$

$$K_c = \frac{k_1}{k_2} = \frac{[C][D]}{[A][B]} = \frac{5 \times 4}{1 \times 2} = 10$$

$$k_1 = 10k_2$$

Also, $\quad k_1 = Ae^{-Ea/RT} = 1.0 \times 10^{10}\text{M}^{-1}\text{s}^{-1} e^{-100/8.314 \times 10^{-3} \times 700}$

$$= 345 \text{ M}^{-1} \text{ s}^{-1}$$

$$\therefore \quad k_2 = \frac{k_1}{10} = 34.5 \text{ M}^{-1}\text{s}^{-1}$$

Also, $\quad E_a = \frac{2.303R\, T_1 T_2}{(T_2 - T_1)} \log \frac{k_1' \text{ (at 1000 K)}}{k_1 \text{ (at 700 K)}}$

$$100 = \frac{2.303 \times 8.314 \times 10^{-3} \times 700 \times 1000}{(1000 - 700)} \log \frac{k_1'}{345 \text{ M}^{-1} \text{ s}^{-1}}$$

$$\log \frac{k_1'}{345 \text{ M}^{-1}\text{s}^{-1}} = 2.2383$$

$$\frac{k_1'}{345} = 173.103$$

$$\therefore \quad k_1 = 5.97 \times 10^4\text{M}^{-1}\text{s}^{-1}$$

20. (b) **21.** (d)

22. (8) Rate of souring of milk at 27°C (300 K) is $\left(\frac{1}{x}\right)$ times faster than at –3°C (270K).

$$\therefore \quad \frac{k_{300}}{k_{270}} = x$$

$$\therefore \quad E_a = \frac{2.303\, RT_1T_2}{(T_2 - T_1)} \log x$$

$$11.231 \text{ kcal mol}^{-1} = 2.303 \times 2 \times 10^{-3} \text{kcal mol}^{-1}\text{K}^{-1} \frac{\times 270 \times 300 \log x}{(300 - 270)}$$

$$\log x = 0.9031 = 0.3010 \times 3$$
$$= 3\log 2$$
$$= \log 2^3$$
$$x = 2^3 = 8$$

23. (9) Both are first order reactions (unit of k being s^{-1})

$$k_1 = k_2$$

$$10^{13} e^{-\frac{152.3 \text{kJ mol}^{-1}}{RT}} = 10^{14}\, e^{-157.7 \text{kJ mol}^{-1}/RT}$$

$$10 = e^{157.7 - 152.3/RT}$$

$$\log_e 10 = \log_e e^{5.4 \text{ kJ}/RT}$$

$$2.303 \log_{10} 10 = \frac{5.4 \text{ kJ}}{RT}$$

$$= \frac{5.4 \text{kJ mol}^{-1}}{8.314 \times 10^{-3} \text{ kJ mol}^{-1} \text{ K}^{-1} \text{ T}}$$

$$\therefore \quad T = \frac{5.4}{8.314 \times 10^{-3} \times 2.303}$$

$$T = 282 \text{ K} = 9°\text{C}$$

24. (9) $k_{obs} = k_3 \left(\frac{k_1}{k_2}\right)^{\frac{1}{2}}$

$$Ae^{-E_{obs}/RT} = Ae^{-E_3/RT} \left(\frac{Ae^{-E_1/RT}}{Ae^{-E_2/RT}}\right)^{\frac{1}{2}}$$

$$e^{-E_{obs}/RT} = e^{-E_3/RT} \left(e^{\frac{E_2 - E_1}{2RT}}\right)$$

$$\log_e e^{-E_{obs}/RT} = \log_e e^{-E_3/Rt} + \log_e e^{\frac{E_2 - E_1}{2RT}}$$

$$\frac{-E_{obs}}{RT} = \frac{-E_3}{RT} + \frac{E_2 - E_1}{2RT}$$

$$\therefore \quad E_{obs} = E_3 + \left(\frac{E_1 - E_2}{2}\right) = 7 + \frac{10 - 6}{2} = 9 \text{kJ mol}^{-1}$$

DPP-5 Collision Theory and Reaction Mechanism

1. (d) By slow step II,

$$\left(\frac{dx}{dt}\right) = k[X][B]$$

By reversible fast step I,

$$K_C = \frac{[X]}{[A]^2}$$

$$\therefore \quad [X] = K_C[A]^2$$

$$\therefore \quad \frac{dx}{dt} = K_C k[A]^2[B]$$
$$= k'[A]^2[B]$$

2. (c) In I, slow step is

$$NO_2 + O_3 \xrightarrow{k} NO_3 + O_2$$

Thus, $\quad \frac{-d\,[NO_2]}{dt} = k\,[NO_2]\,[O_3]$

In II slow step is

$$NO_2 + O \xrightarrow{k_1} NO_3$$

$$\therefore \quad \frac{-d[NO_2]}{dt} = k_1\,[NO_2]\,[O]$$

In fast reversible step,

$$K_C = \frac{[O_2]\,[O]}{[O_3]}$$

$$\therefore \quad [O] = K_C \frac{[O_3]}{[O_2]}$$

$$\therefore \quad \frac{-d\,[NO_2]}{dt} = k_1 K_C \frac{[NO_2]\,[O_3]}{[O_2]}$$
$$= k[NO_2]\,[O_3]$$

3. (b) Stage I with minimum value of $\Delta G°$ is the reaction intermediate. Stage II with a peak represents activated state.

4. (c) Slow step I is the rate-determining step.

$$\therefore \quad \left(\frac{dx}{dt}\right) = k_1\,[NO_2]^2$$

5. (b) By slow step II,

$$\frac{dx}{dt} = k_3[NO_2F_2]$$

By fast step I,

$$k_1\,[NO_2]\,[F_2] = k_2[NO_2F_2]$$

$$\therefore \quad [NO_2F_2] = \frac{k_1}{k_2}\,[NO_2]\,[F_2]$$

$$\therefore \quad \frac{dx}{dt} = \frac{k_1k_3}{k_2}\,[NO_2][F_2]$$

Net reaction is

$$2NO_2(g) + F_2(g) \longrightarrow 2NO_2F(g)$$

$$\left(\frac{dx}{dt}\right)_{net} = k[NO_2]^2[F_2]$$

Thus, rate law is based on the slow step.

6. (b)

(a) Slow step is the rate-determining step, thus by

rate law $= k[A][B]$

Net reaction, $\quad A + 2B \longrightarrow C + D$

rate law $= k\,[A]\,[B]^2$

Thus, this mechanism is not consistent with the stoichiometry. but is consistent with rate law.

(b) By slow step,

Rate law $= k_3[A][B_2]$

By fast step, $\quad \frac{[B_2]}{[B]^2} = \frac{k_1}{k_2}$

$$\therefore \quad [B_2] = \frac{k_1}{k_2}\,[B]^2$$

$$\therefore \quad \text{Rate law} = \frac{k_1k_3}{k_2}\,[A][B]^2$$

Thus, this mechanism is consistent with the stoichiometry of the net reaction but not with the rate law.

Thus, (b) is correct.

7. (b) $\left(\frac{dx}{dt}\right)_{II} = k_1 [N_2O_2][H_2]$

If step II is fast and reversible, then

$$K_C = \frac{[N_2O_2]}{[NO]^2}$$

$\therefore$ $[N_2O_2] = K_C [NO]^2$

$\therefore$ $\left(\frac{dx}{dt}\right) = k_1 K_C [NO]^2 [H_2] = k[NO]^2[H_2]$

8. (a) $(CH_3)_3CCl + H_2O \longrightarrow$

H_2O is a weak nucleophile and cannot attack tertiary carbon of maximum electron density directly. Thus, rate-determining slow step is

$$(CH_3)_3CCl \xrightarrow{\text{Slow}} (CH_3)_3C^+ + Cl^-$$

and rate law is

$$\left(\frac{dx}{dt}\right) = k[(CH_3)_3CCl]$$

and $[H_2O]$ is not involved.

It is S_N1 reaction.

9. (a) Slow step II is the rate-determining step.

$$\left(\frac{dx}{dt}\right) = k_2[N_2O_2][H_2] \quad \ldots(i)$$

By step I in equilibrium,

$$K_C = \frac{k_1}{k_{-1}} = \frac{[N_2O_2]}{[NO]^2}$$

$\therefore$ $[N_2O_2] = \frac{k_1}{k_{-1}}[NO]^2 \quad \ldots(ii)$

Put the value of $[N_2O_2]$ in Eq. (i), we get

$\therefore$ $\left(\frac{dx}{dt}\right) = \frac{k_1 k_2}{k_{-1}}[NO]^2[H_2]$

10. (b) Slow step is the rate-determining step.

From (I), rate $= k[Cl_2][H_2S]$

which is the rate law as given.

From (II), rate $= k[Cl_2][HS^-]$

and $K_{eq} = \frac{[H^+][HS^-]}{[H_2S]}$

$\therefore$ $[HS^-] = \frac{K_{eq}[H_2S]}{[H^+]}$

$\therefore$ Rate $= kK_{eq}\frac{[H_2S][Cl_2]}{[H^+]}$

which is not the rate law.

11. (b) From slow step,

$$\frac{d[NOBr]}{dt} = k[NOBr_2][NO]$$

From equilibrium step,

$$K_C = \frac{[NOBr_2]}{[NO][Br_2]}$$

$\therefore$ $[NOBr_2] = K_C[NO][Br_2]$

$\therefore$ $\frac{d[NOBr]}{dt} = kK_C[NO]^2[Br_2] = k'[NO]^2[Br_2]$

Thus, order of reaction = 2 w.r.t. NO

and 1 w.r.t. Br_2.

12. (d) I. $\frac{d[C]}{dt} = k[A]^2[B]$

It is second order in A and first order in B thus, total order = 3.

II. $K = \frac{[A_2]}{[A]^2}$

$\therefore$ $[A_2] = K[A]^2$

$$\frac{d[C]}{dt} = k[A_2][B] = kK[A]^2[B]$$

Thus, it is second order in A and first order in B.

Total order = 3

III. $k_1[A]^2 = k_2[A_2]$ by steady state approximation

$$\frac{d[C]}{dt} = k_3[A_2][B] = \frac{k_3k_1}{k_2}[A]^2[B] = k[A]^2[B]$$

Thus, it is second order in A and first order in B.

Thus total order = 3

13. (c) $A_2 \rightleftharpoons 2A \quad K_C = 4 = \frac{[A]^2}{[A_2]}$

$$[A]^2 = 4[A_2]$$

$$[A] = 2[A_2]^{\frac{1}{2}}$$

$$A + B \rightleftharpoons C, K_C' = 16 = \frac{[C]}{[A][B]}$$

$$[C] = 16[A][B] = 16 \times 2[A_2]^{\frac{1}{2}}[B]$$

$$\frac{d[D]}{dt} = k[A_2][C]$$

$$= 1 \times 10^{-2}[A_2]\,32\,[A_2]^{\frac{1}{2}}[B]$$

$$= 0.32\,[A_2]^{3/2}[B]$$

14. (a) Slow step (II) is the rate-determining step.

$$\left(\frac{dx}{dt}\right) = k_3[NO_3][NO]$$

By fast step I in equilibrium,

$$k_1[NO][O_2] = k_2[NO_3]$$

$\therefore$ $[NO_3] = \frac{k_1}{k_2}[NO][O_2]$

$\therefore$ $\left(\frac{dx}{dt}\right) = \frac{k_1k_3}{k_2}[NO]^2[O_2]$

Explanations for Q. Nos. (15-16)

Step I is fast and reversible

$$K_{eq} = \frac{[O_2][O]}{[O_3]}$$

$$[O] = K_{eq}\frac{[O_3]}{[O_2]}$$

Step II is slow and rate-determining step.

$\therefore$ $\left(\frac{dx}{dt}\right) = k_1[O_3][O]$

$\therefore$ $\left(\frac{dx}{dt}\right) = k_1K_{eq}\frac{[O_3]^2}{[O_2]}$

$\therefore$ $\frac{-d[O_3]}{dt} = k_1K_{eq}\frac{[O_3]^2}{[O_2]}$

$$= \frac{k[O_3]^2}{[O_2]}$$

where, $k_1K_{eq} = k = 2.0 \times 10^{-3} \times 5 = 1.0 \times 10^{-2}$

	$2O_3(g)$ $\longrightarrow$	$3O_2(g)$
Initial	1.0	0
At 50% reaction	2(1 − 0.5)	3 × 0.5
	1.0	1.5

$$-\frac{d[O_3]}{dt} = \frac{k[O_3]^2}{[O_2]}$$

$$= \frac{1.0 \times 10^{-2}(1.0)^2}{1.5}$$

$$= 6.67 \times 10^{-3}\ \text{mol L}^{-1}\text{s}^{-1}$$

15. (b) **16.** (a)

Explanations for Q. Nos. (17-18)

$$NO_2 + NO_3 \xrightarrow{k_1} NO_2 + O_2 + NO$$

$$\therefore \quad \frac{d[O_2]}{dt} = k_1[NO_2][NO_3]$$

Also, $$N_2O_5 \overset{K_{eq}}{\rightleftharpoons} NO_2 + NO_3$$

$$\therefore \quad K_{eq} = \frac{[NO_2][NO_3]}{[N_2O_5]}$$

$$\therefore \quad [NO_2][NO_3] = K_{eq}[N_2O_5]$$

$$\therefore \quad \frac{d[O_2]}{dt} = k_1 K_{eq}[N_2O_5]$$

Half-life period is independent of $[N_2O_5]$.

Hence, it is first order reaction.

$$t_{50} \propto \left(\frac{1}{[N_2O_5]}\right)^{n-1}$$

$$n - 1 = 0$$

$$\therefore \quad n = 1$$

Thus, $\frac{d[N_2O_5]}{dt} = k[N_2O_5]$

17. (a) **18.** (b)

Explanations for Q. Nos. (19-20)

Molecularity of each step is two thus, true.

Thus, (a) is correct.

The limiting step of a complex reaction is always the slowest, thus (b) is correct.

It involves more than one elementary step, thus, it is a complex reaction. Thus, (d) is also correct.

HOCl appears in step I and disappears in step II, Thus it is an intermediate.

HOI appears in step II and disappears in step III. Thus, HOI is also an intermediate.

Thus, (c) is incorrect.

Slow step II is the rate-determining step.

Thus, $$\left(\frac{dx}{dt}\right) = k_2[I^-][HOCl]$$

By step I, $$k_1[OCl^-] = k_{-1}[HOCl][OH^-]$$

$$\therefore \quad [HOCl] = \frac{k_1[OCl^-]}{k_{-1}[OH^-]}$$

$$\therefore \quad \left(\frac{dx}{dt}\right) = \frac{k_2[I^-]k_1[OCl^-]}{k_{-1}[OH^-]} = \frac{k_1k_2}{k_{-1}}\frac{[I^-][OCl^-]}{[OH^-]}$$

Note H_2O *as reactant is not taken in equilibrium and rate expression.*

19. (c) **20.** (a)

21. (4) Rate determining step is $A_2 + C \longrightarrow D + 2A$

$$\frac{d[D]}{dt} = k[A_2][C]$$

Also, $$k_1 = \frac{[A_2]}{[A]^2}$$

$$\therefore \quad [A_2] = k_1[A]^2$$

and $$k_2 = \frac{[C]}{[A][B]}$$

$$\therefore \quad [C] = k_2[A][B]$$

Put the value of $[A_2]$ and $[C]$ into rate law

$$\frac{d[D]}{dt} = kk_1[A]^2 k_2[A][B] = kk_1k_2[A]^3[B]$$

The reaction is third order in A and first order in B.

Total order = 3 + 1 = 4

22. (5) By I, $\left(\frac{dx}{dt}\right)_I = k_1[A][B] = 2 \times 10^{-5}[A][B]$

In mechanism II, $k_2' = 1 \times 10^{10} M^{-1} s^{-1}$

Thus, $E + B \longrightarrow C + D$ is fast and rate-determining step is thus,

$$A \longrightarrow A^* \longrightarrow E$$

$$\therefore \quad \left(\frac{dx}{dt}\right)_{II} = k_2[A] = 1 \times 10^{-4}[A]$$

For $\left(\frac{dx}{dt}\right)_I = \left(\frac{dx}{dt}\right)_{II}$

$$2 \times 10^{-5}[A][B] = 1 \times 10^{-4}[A]$$

$$\therefore \quad [B] = \frac{1 \times 10^{-4}}{2 \times 10^{-5}} = 5.0\text{ M}$$

Revisal Problems (JEE Main)

1. (b) A steady state is obtained when

$$\left(\frac{dx}{dt}\right)_f = \left(\frac{dx}{dt}\right)_b$$

$$k_1[A][B] = k_2[C][D]$$

But $k_1 = k_2$, hence $[A][B] = [C][D]$

(a) $0.1 \times 0.2 \neq 0.3 \times 0.4$, incorrect

(b) $0.4 \times 0.25 = 0.2 \times 0.5$

$0.1 = 0.1$, correct

2. (d) $P + Q \longrightarrow R + S$

$$\left(\frac{dx}{dt}\right) = k[P]^a[Q]^b$$

Time taken for 75% reaction if P is twice the time taken for 50% reaction of P.

Thus, w.r.t. P, order = a = 1.

Concentration of Q decreases linearly with time. Thus, w.r.t. Q rate = $k[Q]^0 = k$ = constant

Thus, w.r.t. Q, order = zero

Total order = 1 + 0 = 1

3. (c) For zero order reaction,

$$k = \frac{x}{t}$$

When $x = \frac{a}{2}, t = t_{50}$

$$\therefore \quad k = \frac{a}{2t_{50}} = \frac{2.0\text{ mol L}^{-1}}{2 \times 1\text{ h}}$$

$$k = 1.0\text{ mol L}^{-1}\text{h}^{-1}$$

Also, $$k = \frac{x_2 - x_1}{t_2 - t_1}$$

$$\therefore \quad \Delta t = \frac{\Delta x}{k} = \frac{0.50 - 0.25}{1} = 0.25\text{ h}$$

4. (a) Radioactive substance decays 20% in 10 min.

If $$a = 100$$

$$(a - x) = 80$$

$$\therefore \quad k = \frac{2.303}{10}\log\frac{100}{80} = 0.0223\text{ min}^{-1}$$

Initial atoms = 5×10^{20}

Final atoms = 5×10^{17}

$\therefore$ Time = t min

$$k = \frac{2.303}{t}\log\frac{5 \times 10^{20}}{5 \times 10^{17}}$$

$$\frac{2.303}{10}\log\left(\frac{100}{80}\right) = \frac{2.303}{t}\log 10^3$$

$$\frac{2.303}{10}\log 1.25 = \frac{2.303 \times 3 \times \log 10}{t}$$

$$\frac{0.0969}{10} = \frac{3}{t}$$

$$\therefore \quad t = \frac{10 \times 3}{0.0969} = 310 \text{ min}$$

Alternate Method

$$k = 0.0223 \text{ min}^{-1}$$

$$\therefore \quad t_{50} = \frac{0.693}{k} = \frac{0.693}{0.0223} \text{ min} = 31.0 \text{ min}$$

Initial atoms = 5×10^{20} atoms

Final atoms = 5×10^{17} atoms

$$\text{Reaction completed} = 100 - \frac{5 \times 10^{17}}{5 \times 10^{20}} \times 100 = 99.9\%$$

We know for first order reaction, time of 99.9% reaction = $10 \times t_{50}$

$$\therefore \quad t = 10 \times 31 = 310 \text{ min}$$

5. (b) Virus inactivation follows first order kinetics based on unit of $k(\text{s}^{-1})$.

$$t_{50} = \frac{0.693}{k} = \frac{0.693}{3.3 \times 10^{-4}\ \text{s}^{-1}} = 2100 \text{ s}$$

$$\therefore \quad t_{75} = 2 \times t_{50} = 4200 \text{ s} = \frac{4200}{60} = 70 \text{ min}$$

6. (a) $2N_2O_5(g) \longrightarrow 4NO_2(g) + O_2(g)$

$t = 0$	p_i	0	0
$t = 30$ min	$(p_i - x)$	$2x$	$\frac{x}{2}$
At the end	$(p_i - p_i)$	$2p_i$	$\frac{p_i}{2}$

Given at the end of the reaction,

$$\text{total pressure} = 2p_i + \frac{p_i}{2} = 500 \text{ mm}$$

$$\therefore \quad 2.5\, p_i = 500 \text{ mm}$$

$$\therefore \quad p_i = 200 \text{ mm}$$

At the end of 30 min

$$\text{pressure} = p_i - x + 2x + \frac{x}{2}$$

$$350 = p_i + 1.5x$$

$$\therefore \quad 350 = 200 + 1.5x$$

$$1.5x = 150$$

$$x = 100 \text{ mm}$$

$$\therefore \quad k = \frac{2.303}{t}\log\frac{p_i}{p_i - x} = \frac{2.303}{30}\log\frac{200}{200 - 100}$$

$$= \frac{2.303}{30}\log 2 = 2.31 \times 10^{-2} \text{ min}^{-1}$$

7. (c) Initial concentration = a

Final concentration = $\frac{3a}{4}$, Amount reacted = $a/4$

Time taken = $t_{1/4}$

$$k = \frac{2.303}{t}\log\left(\frac{a}{a - x}\right) = \frac{2.303}{t_{1/4}}\log\frac{a}{3a/4}$$

$$t_{1/4} = \frac{2.303}{k}\log\left(\frac{4}{3}\right) = \frac{2.303 \times 0.125}{k} = \frac{0.29}{k}$$

8. (a) $CH_3COCH_3(g) \longrightarrow C_2H_6(g) + CO(g)$

Initial pressure	p_i	0	0
Pressure at time t	$(pi - x)$	x	x

$\therefore$ Given $p_i = 2.0$ atm

$$(p_i - x) = 0.3 \text{ atm}$$

$$k = \frac{2.303 \log 2}{t_{50}} = \frac{2.303 \times 0.3010}{810 \text{ s}}$$

$$\therefore \text{Also,} \quad k = \frac{2.303}{t}\log\left(\frac{a}{a - x}\right) = \frac{2.303}{t}\log\frac{p_i}{(p_i - x)}$$

$$\frac{2.303 \times 0.3010}{810 \text{ s}} = \frac{2.303}{t}\log\frac{p_i}{(p_i - x)}$$

$$t = \frac{810}{0.3010}\log\frac{2}{0.3} = 2217 \text{ s}$$

$$= 36.95 \text{ min} \approx 37 \text{ min}$$

9. (d) For first order kinetics,

$$C = C_0\left(\frac{1}{2}\right)^y$$

$$\text{where, } y = \left(\frac{\text{Total time}}{t_{50}}\right)$$

$$t_{50} = \frac{2.303 \log 2}{k} = \frac{0.693}{60 \text{ s}^{-1}}$$

$$C_0 = 100$$

$$C = \left(\frac{100}{16}\right)$$

$$\therefore \quad \left(\frac{1}{16}\right) = \left(\frac{1}{2}\right)^y$$

$$\left(\frac{1}{2}\right)^4 = \left(\frac{1}{2}\right)^y$$

$$\therefore \quad y = 4$$

$$\therefore \quad \text{Total time} = t_{50} \times y$$

$$= \frac{0.693 \times 4}{60} \text{s} = 0.0462 \text{ s}$$

10. (a) For nth order reaction,

$$t_{50} \propto \left(\frac{1}{a}\right)^{n-1}$$

$$\frac{t_2}{t_1} = \left(\frac{a_1}{a_2}\right)^{n-1}$$

$$\frac{t_2}{t_1} = 2, \left(\frac{a_1}{a_2}\right) = \frac{1}{2}$$

$$(2) = \left(\frac{1}{2}\right)^{n-1}$$

$$\left(\frac{1}{2}\right)^{-1} = \left(\frac{1}{2}\right)^{n-1}$$

$$\therefore \quad n - 1 = -1$$

$$n = 0$$

11. (b) Rate = $k[A]$

Thus, it is first order reaction.

When $[A] = 0.1$ M from graph

$$\text{Rate} = 1 \times 10^{-3} = k(0.1)$$

$$\therefore \quad k = 1 \times 10^{-2} \text{ s}^{-1}$$

When $[A] = 0.3$ M from graph

$$3 \times 10^{-3} = k(0.3)$$

$$\therefore \quad k = 1 \times 10^{-2} \text{ s}^{-1}$$

12. (d) $\frac{d[C]}{dt} = k[A]^a[B]^b$

I. $1.2 \times 10^{-3} = k(0.1)^a(0.1)^b$ ...(i)

II. $1.2 \times 10^{-3} = k(0.1)^a(0.2)^b$ ...(ii)

III. $2.4 \times 10^{-3} = k(0.2)^a(0.1)^b$ ...(iii)

Divide Eq. (i) by Eq. (ii), we get

$$1 = (2)^b$$

$$\therefore \quad (2)^0 = (2)^b$$

$$\therefore \quad b = 0$$

Divide Eq. (ii) by Eq. (iii), we get

$$(2) = (2)^a$$

$$\therefore \quad a = 1$$

Thus, order w.r.t. $A = 1$

and w.r.t. $B = 0$

$$\therefore \quad \frac{d[C]}{dt} = k[A][B]^0 = k[A]$$

13. (a) $k = Ae^{-E_a/RT}$

$$k = Ae^{-E_a/R \times 500}$$

$$k' = Ae^{-(E_a - 20)/R \times 400}$$

Rates are equal,hence

$$k = k'$$

$$\therefore \quad Ae^{-E_a/R \times 500} = Ae^{-(E_a - 20)R \times 400}$$

$$\therefore \quad \frac{E_a}{500} = \frac{E_a - 20}{400}$$

$$4E_a = 5E_a - 100$$

$$\therefore \quad E_a = 100 \text{ kJ mol}^{-1}$$

14. (a) Rate (300 K) $= k_1[A]^n$

Rate (310 K) $= k_2[A]^n = 2$ Rate (300 K)

Thus, $\frac{k_2}{k_1} = 2$

$$E_a = \frac{2.303\, RT_1T_2}{(T_2 - T_1)}\log\frac{k_2}{k_1}$$

$$= \frac{2.303 \times 2 \times 10^{-3} \text{ kcal mol}^{-1}\text{K}^{-1} \times 300 \text{ K} \times 310 \text{ K}}{(310 - 300) \text{ K}}\log 2$$

$$= 12.89 \text{ kcal mol}^{-1}$$

15. (b) $K_C = \frac{k_1}{k_2} = 10^4$

$$\therefore \quad k_1 = k_2 \times 10^4$$

Also, $k_2 = 10^{12}e^{-E_a/RT}$

$$= 10^{12}\, e^{-42/8.3143 \times 10^{-3} \times 298}$$

$$= 10^{12} \times 4.34 \times 10^{-8}\text{s}^{-1} = 4.34 \times 10^4 \text{ s}^{-1}$$

$$\therefore \quad k_1 = k_2 \times 10^4$$

$$= 4.34 \times 10^4 \times 10^4 = 4.34 \times 10^8 \text{ s}^{-1}$$

$$k_1 + k_2 = 4.34 \times 10^4 + 4.34 \times 10^8 = 4.34 \times 10^8\text{s}^{-1}$$

16. (c) $\frac{-d[A]}{dt} = k[A]$

It follows first order kinetics.

$C_0 = [A]_0 = 10$ M, initial concentration

$C = [A] = (a - x)$, after one natural life time $\left(= \frac{1}{k}\right)$

For first order kinetics,

$$C = C_0e^{-kt} = 10\,e^{-k/k} = 10\,e^{-1} = \frac{10}{e}$$

$$\therefore \quad C = \frac{10}{e}$$

17. (b) Time required for 99.9% reaction is 10 times of half-life period.

$$T_{99.9} = 10T_{50}$$

But $T_{50} = \frac{0.693}{k}$

$$\therefore \quad T_{av} = \frac{1}{k}$$

and $T_{50} = 0.693T_{av}$

$$\therefore \quad T_{99.9} = 10 \times 0.693\, T_{av} = 6.93\, T_{av}$$

18. (c) It is a gaseous phase reaction.

	$A(g) \longrightarrow$	$B(g) +$	$C(g)$
Initial pressure	p_i	0	0
At time t	$(p_i - x)$	x	x

Thus, total pressure at time $t = p_i - x + x + x = (p_i + x)$

$$\therefore \quad x = (p_t - p_i)$$

$$\therefore \quad (p_i - x) = (2p_i - p_t)$$

$$\therefore \quad = (200 - 120) = 80 \text{ mm}$$

$$k = \frac{2.303}{t}\log\frac{p_i}{(p_i - x)}$$

$$= \frac{2.303}{10 \text{ min}}\log\frac{100}{80} = \left(\frac{2.303}{10}\log\frac{5}{4}\right)\text{min}^{-1}$$

19. (d) In both acids, first order kinetics is followed.

Rate constant $= k_1$ with $HA = 2 \times 10^{-3}\text{min}^{-1}$

Rate constant $= k_2$ with $HB = 1 \times 10^{-3}\text{min}^{-1}$

Relative strength $= \frac{k_1}{k_2} = \frac{2 \times 10^{-3}}{1 \times 10^{-3}} = 2$

20. (b) $\frac{-d[A]}{dt} = k[A]$

Thus, it follows first order kinetics.

$$T_{50} = \frac{2.303 \log 2}{k}$$

$$k = \frac{2.303}{t}\log\left(\frac{a}{a - x}\right)$$

$$a = 2 \text{ M}$$

$$x = \frac{2 \times 10}{100} = 0.2 \text{ M after 10 min}$$

$$\therefore \quad k = \frac{2.303}{10}\log\frac{2}{1.8} = 2.303 \times 0.004575$$

$$\therefore \quad T_{50} = \frac{2.303\log 2}{k} = \frac{2.303 \times 0.3010}{2.303 \times 0.004575}$$

$$= 65.79 \approx 66.0 \text{ min}$$

21. (c) $\frac{-d[N_2O]}{dt} = k[N_2O]$

It follows first order kinetics.

$$t_{50} = \frac{2.303 \log 2}{k}$$

$$\therefore \quad k = \frac{2.303 \log 2}{t_{50}} = \frac{2.303 \log 2}{3.58 \times 10^3 \text{min}}$$

	$N_2O(g) \longrightarrow$	$N_2(g) +$	$\frac{1}{2}O_2(g)$
Initial pressure	p_i	0	0
At time t,	$(p_i - x)$	x	$\frac{x}{2}$

Total pressure at time $t = p_i - x + x + \frac{x}{2}$

$$= \left(p_i + \frac{x}{2}\right) = \left(2.10 + \frac{x}{2}\right)$$

$$k = \frac{2.303}{t}\log\frac{p_i}{(p_i - x)}$$

$$\frac{2.303\log 2}{3.58 \times 10^3} = \frac{2.303}{3.58 \times 10^3}\log\frac{2.10}{(2.10 - x)}$$

$$\log \frac{2.10}{(2.10 - x)} = \log 2$$

$$\frac{2.10}{(2.10 - x)} = 2$$

$\therefore$ $$2.10 - x = \frac{2.10}{2} = 1.05$$

$$x = 1.05 \text{ atm}$$

Total pressure $= 2.10 + \frac{1.05}{2} = 2.625$ atm

Alternate Method

$p_{N_2O} = 2.10$ atm

At half change, $p_{N_2O} = 1.05$ atm

$\therefore$ Total pressure $= p_i + \frac{x}{2} = 2.10 + \frac{1.05}{2} = 2.625$ atm

22. (a) $k = \frac{2.303}{t} \log \left(\frac{a}{a-x}\right)$

$$\frac{kt}{2.303} = \log a - \log(a - x)$$

$$\frac{kt_1}{2.303} = \log a - \log(a - x_1) \quad \text{at time } t_1$$

$$\frac{kt_2}{2.303} = \log a - \log(a - x_2) \quad \text{at time } t_2$$

$\therefore$ $$\frac{k}{2.303}(t_2 - t_1) = \log\left(\frac{a - x_1}{a - x_2}\right)$$

Rate, $r_1 = k(a - x_1)$

$r_2 = k(a - x_2)$

$\therefore$ $$\frac{(a - x_1)}{(a - x_2)} = \frac{r_1}{r_2} \Rightarrow \frac{0.04}{0.03} = \frac{4}{3}$$

$$\frac{k(20 - 10)}{2.303} = \log\frac{4}{3}$$

$\therefore$ $$k = \frac{2.303}{10}\log\frac{4}{3} = \frac{2.303}{10}\log 1.33$$

$$= \frac{2.303 \times 0.125}{10} = 0.02878 \text{ min}^{-1}$$

$$T_{50} = \frac{0.693}{0.02878} = 24.1 \text{ min}$$

23. (a) Rate law, $\frac{dx}{dt} = k[A][B]^2$

1.0×10^{-2} M s^{-1} $= k(1\text{ M})(1\text{ M})^2$

$k = 1.0 \times 10^{-2}$ M^{-2} s^{-1}

When $[A] = 0.5$ M, $[B] = 0.5$ M

$$\frac{dx}{dt} = 1.0 \times 10^{-2} \text{ M}^{-2}\text{s}^{-1}(0.5)(0.5)^2$$

$$= 1.25 \times 10^{-3} \text{ M s}^{-1}$$

24. (b) $X \underset{k_2}{\overset{k_1}{\rightleftharpoons}} Y$

	X	Y
at $t = 0$	$[X]_0$	0
at time t	$[X]$	$[X]_0 - [X]$

Thus, net rate $\frac{-d[X]}{dt} = k_1[X] - k_2([X]_0 - [X])$

$$= k_1[X] - k_2[X]_0 + k_2[X]$$
$$= (k_1 + k_2)[X] - k_2[X_0]$$

25. (a) $\text{KE} = \frac{1}{2}mu^2 = \frac{3}{2}RT$

Thus, u (rate) $\propto T$

Hence, rate increases with increase in temperature.

Distribution of molecules (N) with velocity (u) at two temperatures is shown as

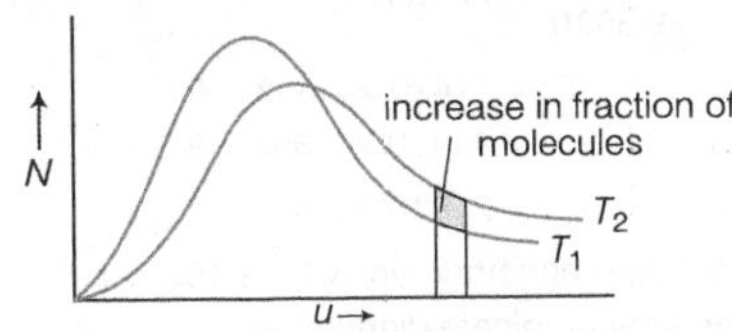

$T_2 > T_1$

When temperature is increased, fraction of molecules with energy excess of E_a is increased.

Thus, both Statements I and II are correct and Statement II is the correct explanation of Statement I.

26. (a) There are collisions among molecules which increase the rate of reaction. These collisions are only effective if they have proper orientations. In the absence of proper orientations rate becomes slow. Thus, Statements I and II both are correct and Statement II is the correct explanation of Statement I.

27. (a) Diamond $\longrightarrow$ graphite $\Delta G < 0$

This reaction is thermodynamically feasible due to ($\Delta G < 0$). But E_a is very high.

Thus, rate becomes slow. Thus, Statements I and II are correct and Statement II is the correct explanation for Statement I.

Revisal Problems (JEE Advanced)

1. (a)

	$A(g)$	$\longrightarrow$	$B(g)$	$+$	$C(g)$
Initial pressure	p_i		0		0
At time, t (20 min)	$(p_i - x)$		x		x

Total pressure at time $t = p_A + p_B + p_C = p_i - x + x + x$

$p_t = (p_i + x)$

Given, $p_t = 250$ mm, $p_i = 200$ mm

$\therefore$ $x = 50$ mm

$\therefore$ $(p_i - x) = 200 - 50 = 150$ mm

$\therefore$ $$k = \frac{2.303}{t}\log\frac{p_i}{(p_i - x)}$$

$$= \frac{2.303}{20}\log\frac{200}{150} = 0.0143 \text{ min}^{-1}$$

t_{50} (first order reaction) $= \frac{0.693}{k}$

$$= \frac{0.693}{0.01439 \text{ min}^{-1}} = 48 \text{min}$$

2. (b) $\left(\frac{dx}{dt}\right)_1 = k[M]^n$

$$8\left(\frac{dx}{dt}\right)_1 = k[2M]^n$$

$\therefore$ $8 = (2)^n$

$(2)^3 = (2)^n$

$\therefore$ $n = 3$

3. (c) By rate law, Rate $= k[P]^x[Q]^y[R]^z$

x = order w.r.t. P

y = order w.r.t. Q

z = order w.r.t. R

(1) $8.0 \times 10^{-5} = k(0.2)^x(0.5)^y(0.4)^z$

(2) $3.2 \times 10^{-4} = k(0.4)^x(0.5)^y(0.4)^z$

(3) $1.28 \times 10^{-3} = k(0.4)^x(2.0)^y(0.4)^z$

(4) $4.0 \times 10^{-5} = k(0.1)^x(0.25)^y(1.6)^z$

From (1) and (2), $\dfrac{3.2\times10^{-4}}{8.0\times10^{-5}} = (2)^x$

$$(4) = (2)^x \quad \Rightarrow \quad x = 2$$

From (2) and (3), $\dfrac{1.28\times10^{-3}}{3.2\times10^{-4}} = (4)^y$

$$(4) = (4)^y \quad \Rightarrow \quad y = 1$$

From (1) and (4), $\dfrac{8\times10^{-5}}{4\times10^{-5}} = (2)^x\,(2)^y\left(\dfrac{1}{4}\right)^z = (4)\,(2)\left(\dfrac{1}{4}\right)^z$

$$\left(\frac{1}{4}\right) = \left(\frac{1}{4}\right)^z \quad \Rightarrow \quad z = 1$$

Thus, order w.r.t. $P = 2$

w.r.t. $Q = 1$

w.r.t. $R = 1$

4. (a) By Arrhenius equation

$$\log_{10} k = -\frac{E_a}{2.303RT} + \log A$$

$\therefore \quad \dfrac{E_a}{2.303R} = 11420\text{K}$

$\therefore \quad E_a = 11420 \times 2.303 \times R\,K$

$$= 11420 \times 2.303 \times 8.314 \times 10^{-3}\text{kJ mol K}^{-1}\text{K}$$

$$= 218.67 \text{ kJ mol}^{-1}$$

Given reaction is endothermic, $\Delta H = +103.1\text{kJ}$.

Thus, reaction-coordinate diagram is

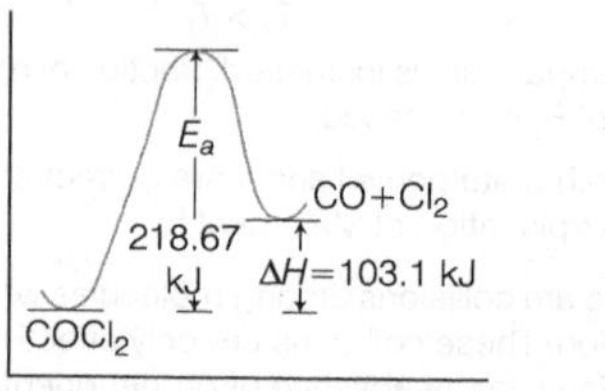

5. (b) By Arrhenius equation,

$$k = Ae^{-E_a/RT}$$

when $T \longrightarrow \infty$

$$k = Ae^{-E_a/\infty} = Ae^0$$

$$k = A$$

$\therefore$ A (pre-exponential factor) $= k$

$$= 1.2 \times 10^3 \text{mol}^{-1}\text{L s}^{-1}$$

6. (a) $T_{50} = \dfrac{0.693}{k}$

$\therefore \quad k = \dfrac{0.693}{t_{50}} = \dfrac{0.693}{100}\text{day}^{-1}$

$$\frac{1}{k} = \frac{100}{0.693} = 144.3\text{days} = \text{Average life}$$

Thus, given time $t = \dfrac{1}{k}$ (Average life)

$$C = Coe^{-kt} = Coe^{-k/k} = \frac{Co}{e}$$

Thus, concentration is reduced to $\left(\dfrac{1}{e}\right)$ of the initial value.

7. (a) $-\dfrac{dP}{dt} = k[A]$ for first order kinetics

$$-\frac{dp}{dt} = k(P)$$

$\therefore \quad -\dfrac{dp}{P} = k \cdot dt$

$$-\int \frac{dp}{P} = k\int dt$$

$$-\log_e p = kt + C \text{ (constant)}$$

At $t = 0$, $p = p_0$ (Initial pressure)

$\therefore \quad C = -\log_e p_0$

$$-\log_e p_A = kt - \log p_0$$

$$\log_e p_A - \log p_0 = -kt$$

(p_A = partial pressure of A after a given time)

$$p_A = p_0 e^{-kt}$$

p_B = partial pressure of $B = n[p_0 - p_0 e^{-kt}]$

$$= n\,p_0[1 - e^{-kt}]$$

Thus, total pressure $= p_A + p_B$

$$= p_0 e^{-kt} + n\,p_0[1 - e^{-kt}]$$

$$= p_0 e^{-kt} + n\,p_0 - n\,p_0 e^{-kt}]$$

$$= p_0[n + (1-n)_e{}^{-kt}]$$

8. (d) For a second order reaction

$$k = \frac{1}{t}\left[\frac{1}{a-x} - \frac{1}{a}\right]$$

$$kt = \frac{1}{a-x} - \frac{1}{a}$$

$$\left(\frac{1}{a-x}\right) = kt + \frac{1}{a} \qquad \ldots\text{(i)}$$

$$y = mx + c \qquad \ldots\text{(ii)}$$

Thus, (i) represents a straight line.

Hence, $\dfrac{1}{(x)}\left(= \dfrac{1}{(a-x)}\right)$ when plotted against time will give a straight line.

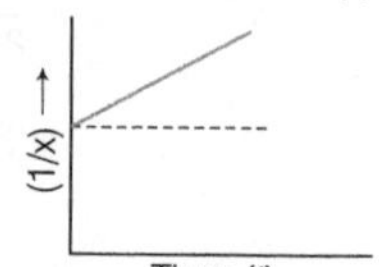

9. (c) $NH_4^+ + OH^- \xrightarrow{k_1} NH_4OH$

$$\frac{d\,[NH_4OH]}{dt} = k_1\,[NH_4^+]\,[OH^-] = 4.0 \times 10^{10}\,[NH_4^+]\,[OH^-] \qquad \ldots\text{(i)}$$

$$NH_3 + H_2O \xrightarrow{k_2} NH_4OH$$

(protonation of NH_3 by H_2O)

$$\frac{d[NH_4OH]}{dt} = k_2[NH_3]\,[H_2O] = k_2[NH_3] \qquad \ldots\text{(ii)}$$

$[(H_2O) = \text{constant}]$

From Eqs. (i) and (iii), we get

$$4.0 \times 10^{10}[NH_4^+]\,[OH^-] = k_2[NH_3]$$

Also, $\quad NH_3(aq) \rightleftharpoons NH_4^+ + OH^-$

$$K_b = \frac{[NH_4^+]\,[OH^-]}{[NH_3]} = 1.8 \times 10^{-5}$$

$\therefore \quad 4.0 \times 10^{10}\dfrac{[NH_4^+]\,[OH^-]}{[NH_3]} = k_2$

$$4.0 \times 10^{10} \times 1.8 \times 10^{-5} = k_2$$

$\therefore \quad k_2 = 7.2 \times 10^5\,\text{s}^{-1}$ (being first order)

10. (d) This represents a second order reaction

$$k = \frac{1}{t}\left[\frac{1}{(a-x)} - \frac{1}{a}\right]$$

Given, $[A]_0 = a$ = Initial concentration

$[A] = (a - x)$ = Concentration after time t

$\therefore \quad k = \dfrac{1}{t}\left[\dfrac{1}{[A]} - \dfrac{1}{[A]_0}\right]$

$\therefore \quad kt = \dfrac{1}{[A]} - \dfrac{1}{[A]_0}$

Thus, (a) is correct.

$$\frac{1}{[A]} = kt + \frac{1}{[A_0]}$$

$$\frac{1}{[A]} = \frac{kt[A]_0 + 1}{[A]_0}$$

$$\therefore \quad [A] = \frac{[A]_0}{kt[A]_0 + 1}$$

Thus, (b) is correct.

when $t = t_{50}$, $\quad [A] = \frac{[A]_0}{2}$

$$\therefore \quad k = \frac{1}{t_{50}}\left[\frac{2}{[A]_0} - \frac{1}{[A]_0}\right] = \frac{1}{t_{50}[A]_0}$$

$$\therefore \quad t_{50} = \frac{1}{k[A]_0}$$

Thus, (c) is also correct.

11. (a) $k = \frac{2.303}{t(a-b)} \log \frac{b(a-x)}{a(b-x)}$

If $\quad a >> b, (a-b) \approx a$

and $\quad (a-x) \approx a$

Thus, $\quad k = \frac{2.303}{t_a} \log \frac{ba}{a(b-x)}$

$$k_a = \frac{2.303}{t} \log\left(\frac{b}{b-x}\right)$$

$$k' = \frac{2.303}{t} \log\left(\frac{b}{b-x}\right)$$

This represents unimolecular reaction.

12. (a) $\left(\frac{dx}{dt}\right)_{\text{initial}} = k[A][B]$

$$[A] = 0.1\text{ M}, [B]_0 = 6.0\text{ M} >>> [A]_0$$

Thus, concentration of B almost remains constant and reaction is first order w.r.t. A only.

$$\text{rate} = k[A][B] = 5.0 \times 10^{-3}[A] \times 6 = 3.0 \times 10^{-2}[A]$$

For first order, $\quad k = \frac{2.303}{t} \log\left(\frac{a}{a-x}\right)$

$$t = 100\text{ s}$$

$$\therefore \quad \log\left(\frac{a}{a-x}\right) = \frac{kt}{2.303} = \frac{3.0 \times 10^{-2} \times 100}{2.303}$$

$$\log\left(\frac{a}{a-x}\right) = \frac{3.0}{2.303} = 1.3026 = \log 20$$

$$\left(\frac{a}{a-x}\right) = 20$$

$$\therefore \quad (a-x) = \frac{a}{20} = \frac{0.1\text{M}}{20} = 0.005\text{ M}$$

13. (a, c, d) $\log_{10} k\ (\text{min}^{-1}) = 5 - \frac{2000\text{k}}{T}$

$$\log_{10} k = \log_{10} A - \frac{E_a}{2.303\,RT}$$

(a) Unit of k is min^{-1}, hence it is first order reaction thus (a) is correct.

(b) $\log_{10} A = 5$

$$\therefore \quad A = 10^5\ \text{min}^{-1}$$

Pre-exponential factor A is 10^5min^{-1} hence, statement (b) is incorrect.

(c) On comparison

$$\frac{E_a}{2.303R} = 2000\text{K}$$

$$E_a = 2.303 \times R \times 2000\text{K}$$

$$= 2.303 \times 2 \times 10^{-3}\ \text{kcal mol}^{-1}\ \text{K}^{-1} \times 2000\ \text{K}$$

$$= 9.212\ \text{kcal mol}^{-1}$$

Thus, (c) is correct.

(d) Given equation represents a straight line

$$y = c - mx$$

Hence, variation of $\log_{10} k$ with $\frac{1}{T}$ is linear.

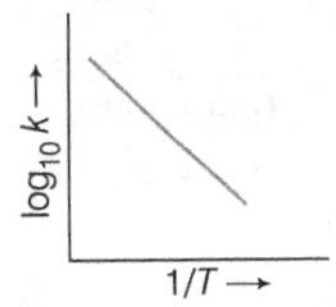

Thus, (d) is correct.

14. (a,b,d) For zero order reaction

$$\left(\frac{dx}{dt}\right) = k$$

Thus, rate remains constant.

Thus, (a) is correct.

Also, $\quad dx = k\,dt$

$\therefore \quad x = kt$

Thus, $\quad x \propto \text{time } t$

Thus, (b) is correct.

In (c), rate increases with time.

Thus, (c) is incorrect.

$$x = kt$$

$$\log x = \log t + \log k$$

$$y = mx + c$$

Hence $\log(x)$ varies as $\log(t)$ increases.

Thus, (d) is also correct.

15. (a, b, c, d)

(a) For zero order reaction

$$k = \frac{x}{t}$$

If reaction takes place 50%, $x = \frac{a}{2}, t = t_{50}$

$$\therefore \quad k = \frac{a}{2t_{50}}$$

$$\therefore \quad t_{50} = \frac{a}{2k}$$

$$t_{50} \propto a$$

Thus, (a) is correct.

(b) Average life $= \frac{1}{k}$ for first order reaction

$$k = \frac{0.693}{t_{50}}$$

$$t_{50} = \frac{0.693}{k}$$

$$\therefore \quad \frac{t_{50}}{0.693} = \frac{1}{k} = \text{average life}$$

$$1.44 \times t_{50} = \text{average life}$$

Thus, (b) is correct.

(c) For second order reaction

$$k = \frac{1}{t}\left[\frac{1}{a-x} - \frac{1}{a}\right]$$

At $t = t_{50}$, $x = \frac{a}{2}$, $(a-x) = \frac{a}{2}$

$$\therefore \quad k = \frac{1}{t_{50}}\left[\frac{2}{a} - \frac{1}{a}\right] = \frac{1}{t_{50}\,a}$$

At $t = t_{75}$, $x = \frac{3a}{4}$, $(a-x) = \frac{a}{4}$

$$k = \frac{1}{t_{75}}\left[\frac{4}{a} - \frac{1}{a}\right]$$

$$k = \frac{3}{t_{75}a}$$

$$\therefore \quad \frac{1}{t_{50}a} = \frac{3}{t_{75}a}$$

$$\therefore \quad t_{75} = 3t_{50}$$

Thus, (c) is correct.

(d) $k = 0.0693 \text{ min}^{-1}$

Unit of k indicates first order reaction

If $a = 100$, $\quad x = 99.9, t = t_{99.9}$

$(a - x) = 0.1$

$$\therefore \quad k = \frac{2.303}{t_{99.9}} \log \frac{100}{0.1}$$

$$t_{99.9} = \frac{2.303 \log 1000}{k} = \frac{2.303 \times 3}{k}$$

$$t_{99.9} = \frac{2.303 \times 3}{0.0693} = 100 \text{ min}$$

Thus, (d) is also correct.

16. (a, b, c, d) $3BrO^- \longrightarrow BrO_3^- + 2Br^-$

$$-\frac{\Delta[BrO^-]}{3\Delta t} = \frac{\Delta[BrO_3^-]}{\Delta t} = \frac{1}{2}\frac{\Delta[Br^-]}{\Delta t}$$

Also,
$$-\frac{\Delta[BrO^-]}{\Delta t} = k_1[BrO^-]$$

$$\frac{\Delta[BrO_3^-]}{\Delta t} = k_2[BrO^-]$$

$$\frac{\Delta[Br^-]}{\Delta t} = k_3[BrO^-]$$

$$\therefore \quad \frac{k_1}{3}[BrO^-] = k_2[BrO^-] = \frac{k_3}{2}[BrO^-]$$

$$\therefore \quad k_2 = \frac{k_1}{3} = \frac{0.056}{3} = 0.019 \text{ L mol}^{-1}\text{s}^{-1}$$

Thus, (a) is correct.

$$k_3 = \frac{2}{3}k_1 = \frac{2}{3} \times 0.056 = 0.037 \text{ L mol}^{-1}\text{s}^{-1}$$

Thus, (b) is correct.

BrO^- has been oxidised to BrO_3^- and has been reduced to Br^-.

Thus, it is a redox as well as disproportionation reaction.

Thus, (c) and (d) are correct.

17. (d)

(i) $\left(\frac{dx}{dt}\right) = k[A]^x$

$$\log\left(\frac{dx}{dt}\right) = n\log[A] + \log k$$

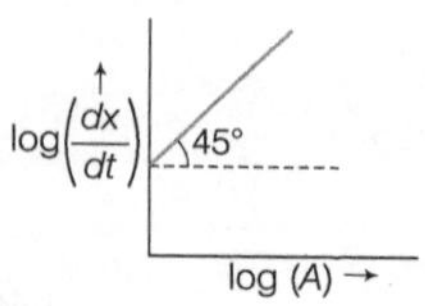

Slope = $\tan 45° = n$

$\therefore \quad n = 1$

Half-life period $\left(\frac{1}{[A]}\right)^{n-1}$

$n - 1 = 0$ (Independent of $[A]$)

$n = 1$ (first order)

Thus, **(i) → (q)**

(ii) Slope = $0 = n$

$\therefore$ Order = 0

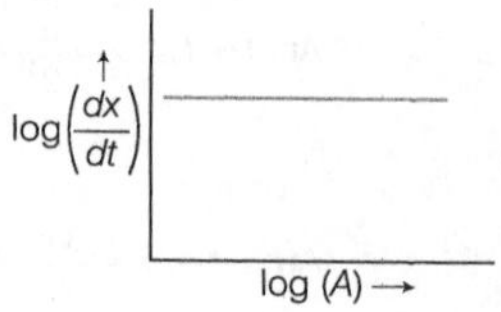

$$T_{50} \propto \left(\frac{1}{[A]}\right)^{n-1}$$

$$T_{50} \propto \left(\frac{1}{[A]}\right)^{0-1}$$

$$T_{50} \propto [A]$$

Thus, **(ii) → (p)**

(iii) $[A] = [(H^+)] = 10^{-pH}$

$$R = k[(H^+)]^n - k(10^{-pH})^n$$

$$100\,R = k(10^{-(pH-1)})^n$$

$$100 = (10)^n$$

$$(10)^2 = (10)^n$$

$$\therefore \quad n = 2$$

$$T_{50} \propto \left(\frac{1}{[A]}\right)^{n-1}$$

$$\propto \frac{1}{[A]}$$

Thus, **(iii) → (s)**

(iv)
$$R = k[A]^n$$

$$8R = k[2A]^n$$

$$8 = (2)^n$$

$$(2)^3 = (2)^n$$

$$\therefore \quad n = 3$$

$$T_{50} \propto \left(\frac{1}{a}\right)^{n-1}$$

$$\propto \left(\frac{1}{a}\right)^2 \text{ for third order reaction}$$

$$\frac{T_2}{T_1} = \left(\frac{a_2}{a_1}\right)^2$$

$$\left(\frac{25}{100}\right) = \left(\frac{1}{2}\right)^2$$

Thus, **(iv) → (r)**

Explanations for Q. Nos. (18-19)

$$\frac{d[O_2]}{dt} = k[O][O_3]$$

$$0.15 \text{ mol L}^{-1}\text{s}^{-1} = 1.5 \times 10^7 \text{ L mol}^{-1}\text{s}^{-1}[O][O_3]$$

$$pV = nRT$$

For ozone, $\frac{n}{V} = \text{mol L}^{-1} = \frac{p}{RT} = \frac{0.76}{760 \times 0.0821 \times 298}$

$$= 4.087 \times 10^{-5} \text{ mol L}^{-1}$$

$$\therefore \quad [O] = \frac{0.15 \text{ mol L}^{-1}\text{s}^{-1}}{1.5 \times 10^7 \text{ L mol}^{-1}\text{ s}^{-1}(4.087 \times 10^{-5}\text{mol L}^{-1})}$$

$$= 2.447 \times 10^{-4} \text{ mol L}^{-1}$$

$$-\frac{d[O]}{dt} = \frac{1}{2}\frac{d[O_2]}{dt}$$

$$\therefore \quad \frac{d[O_2]}{dt} = 2\left(-\frac{d[O]}{dt}\right)$$

$$= 2 \times 0.15 = 0.30 \text{ mol L}^{-1}\text{s}^{-1}$$

18. (a) **19.** (c)

Explanations for Q. Nos. (20-21)

For reverse reaction

Rate = $1.22 \times 10^{-4} \text{ s}^{-1} [Fe(dipy)_3^{2+}]$

It follows first order kinetics with rate constant

$$k = 1.22 \times 10^{-4}$$

$$\text{Half-life} = \frac{0.693}{k}$$

$$= \frac{0.693}{1.22 \times 10^{-4} \text{ s}^{-1}}$$

$$= 5680.3 \text{ s} = \frac{5680.3}{3600} \text{ h}$$

$$= 1.578 \text{ h}$$

Stability constant is obtained when rate of forward reaction = rate of reverse reaction

$$2.44 \times 10^{13}\ L^3mol^{-3}s^{-1}[Fe^{2+}]\ [dipy]^3 = 1.22 \times 10^{-4}\ [Fe(dipy)_3]^{2+}$$

$$\therefore \quad \frac{2.44 \times 10^{13}}{1.22 \times 10^{-4}} = \frac{[Fe(dipy)_3]^{2+}}{[Fe^{2+}]\ [dipy]^3}$$

$$\therefore \quad K = 2 \times 10^{17}$$

$$\therefore \quad pk = -\log k = -\log 2 \times 10^{17} = -17.30$$

20. (a) **21.** (b)

Explanations for Q. Nos. (22-23)

$$A \longrightarrow \text{Product}$$

$$t_{50} = t_1$$

$$-\frac{d[A]}{dt} = k_1[A]$$

It follows first order kinetics

$$t_1 = \frac{0.693}{k_1}$$

$$B \longrightarrow \text{Product}$$

$$t_{50} = f_2$$

$$-\frac{d[B]}{dt} = k_2[B]^2$$

It follows second order kinetics.

$$t_2 = \frac{1}{k_2 a}, \quad a = \text{initial concentration}$$

Rate at the start

$$R_1 = k_1 a$$

$$R_2 = k_2 a^2$$

$$\frac{R_1}{R_2} = \frac{k_1 a}{k_2 a^2} = \frac{k_1}{k_2 a}$$

Also, $$t_1 = t_2$$

$$\therefore \quad \frac{0.693}{k_1} = \frac{1}{k_2 a}$$

$$\therefore \quad \frac{k_1}{k_2 a} = 0.693$$

Also, $$0.693 = 2.303 \log 2$$

$$t_1 = \frac{0.693}{k_1}$$

$$\therefore \quad 0.693 = k_1 t_1$$

$$\therefore \quad k_1 t_1 = 2.303 \log 2 = 0.693$$

After one half-life $= [A] = [B] = \frac{a}{2}$

$$\therefore \quad R_1 = k_1\left(\frac{a}{2}\right)$$

$$R_2 = k_2\left(\frac{a}{2}\right)^2 = \frac{k_2 a^2}{4}$$

$$\frac{R_1}{R_2} = \frac{2k_1}{k_2 a} = 2 \times 0.693 = 1.386 = 2 \times k_1 t_1$$

d

22. (d) **23.** (c)

Explanations for Q. Nos. (24-25)

By Arrhenius equation

$$\log_{10} k = \log_{10} A - \frac{E_a}{2.303\ RT}$$

$$\log_{10} k = 14.34 - \frac{1.25 \times 10^4\ k}{T}$$

On comparison,

$$\frac{E_a}{2.303 R} = 1.25 \times 10^4$$

$$\therefore \quad E_a = 2.303 R \times 1.25 \times 10^4$$

$$= 2.303 \times 8.3143 \times 10^{-3}\ kJ\ mol^{-1}K^{-1} \times 1.25 \times 10^4$$

$$= 239.35\ kJ\ mol^{-1}$$

Unit of k indicates that reaction of first order. Rate constant at 670 K is

$$\therefore \quad \log_{10} k(s^{-1}) = 14.34 - \frac{1.25 \times 10^4}{670}$$

$$= 14.34 - 18.66 = -4.32$$

$$\therefore \quad k = 4.78 \times 10^{-5} s^{-1}$$

24. (a) **25.** (b)

Explanations for Q. Nos. (26-27)

In $CCl_4(l)$, N_2O_5 and N_2O_4 are soluble. Hence, volume of O_2 is only formed.

Initial $N_2O_5 = [N_2O_5]_0 = 80 g = \frac{80}{108} mol$

Final $N_2O_5 = [N_2O_5]_t = 2.5\ g = \frac{2.5}{108} mol$

$$\therefore \quad k = \frac{2.303}{t} \log \frac{(80/108)}{(2.5/108)}$$

$$6.2 \times 10^{-4}\ s^{-1} = \frac{2.303}{t} \log\left(\frac{80}{2.5}\right)$$

$$t = \frac{2.303}{6.2 \times 10^{-4}} \times 1.505 = 5591\ s$$

$$= \frac{5591\ s}{3600\ s\ h^{-1}} = 1.553\ h$$

$$N_2O_5 \text{ reacted} = \left(\frac{80 - 2.5}{108}\right) mol = 0.7176\ mol$$

$$O_2 \text{ formed} = \frac{0.7176}{2} mol = 0.3588\ mol$$

$$= 0.3588 \times 22.4\ L \text{ at STP} = 8.037\ L$$

26. (a) **27.** (b)

28. (0)

Time (min)	$[R]/M$	Δt	$\Delta[R]$	$\frac{\Delta[R]}{\Delta t}$
0	1.0			
0.05	0.75	0.05	0.25	$\frac{0.025}{0.05} = 5$
0.12	0.40	0.12	0.60	$\frac{0.60}{0.12} = 5$
0.18	0.10	0.18	0.90	$\frac{0.90}{0.18} = 5$

$$\left|\frac{\Delta[R]}{\Delta t}\right| = k[R]^n = 5 M\ min^{-1}$$

Rate remains constant throughout thus, order = zero

29. (8) $$\left(\frac{dx}{dt}\right) = k[NO]^2\ [O_2]$$

$$\left(\frac{dx}{dt}\right)_1 = k\left(\frac{n(NO)}{V}\right)^2 \left(\frac{n(O_2)}{V}\right)$$

$$\left(\frac{dx}{dt}\right)_2 = k\left(\frac{n(NO)^2}{V/2}\right)^2 \left(\frac{n(O_2)}{V/2}\right)$$

$$= 8k\left(\frac{n(NO)^2}{V}\right)^2 \left(\frac{n(O_2)}{V}\right) = 8\left(\frac{dx}{dt}\right)_1$$

Thus, rate increases 8 times of its initial value.

30. (6) $\left(\frac{dx}{dt}\right)_{\text{initial}} = k\,p_A\,p_B^2 = k(0.60)\,(0.80)^2$

$$\begin{array}{ccccc} A(g) & + & 2B(g) & \longrightarrow C(g) + & D(g) \\ 0.60 & & 0.80 & 0 & 0 \\ (0.60-x) & & (0.80-2x) & x & x \end{array}$$

$$p_C = X = 0.20 \text{ atm}$$

$\therefore$ $$p_A = 0.60 - 0.20 = 0.40 \text{ atm}$$

$$p_B = 0.80 - 0.40 = 0.40 \text{ atm}$$

$\therefore$ $$\left(\frac{dx}{dt}\right)_{\text{final}} = k\,p_A\,p_B^2 = k(0.40)\,(0.40)^2$$

$\therefore$ $$\frac{\left(\frac{dx}{dt}\right)_{\text{initial}}}{\left(\frac{dx}{dt}\right)_{\text{final}}} = \frac{k(0.60)\,(0.80)^2}{k(0.40)\,(0.40)^2} = 6$$

Thus, $y = 6$

31. (4) Rate constant $= k = 4 \times 10^3 \text{L mol}^{-1}\text{s}^{-1}$. Unit of k indicates that rates of formation of HCO_3 is a second order reaction.

$$\frac{d[HCO_3^-]}{dt} = k[CO_2]\,(OH^-)$$

$$= 4.0 \times 10^3 (10^{-6})\,(0.1)$$

$$= 4.0 \times 10^{-4} \text{ mol L}^{-1}\text{ s}^{-1}$$

$$= x \times 10^{-4} \text{ mol L}^{-1}\text{s}^{-1}$$

Thus, $x = 4$

32. (4) $A \longrightarrow$ Product

$$t_{50} = 54\text{min}$$

$$y = \frac{54}{54} = 1$$

C_t (Concentration of A after 54 min) $= C_0\left(\frac{1}{2}\right)^y$

$$= C_0\left(\frac{1}{2}\right)^1 = C_0\left(\frac{1}{2}\right)$$

C_t' (concentration of B after 54 min) $= C_0\left(\frac{1}{2}\right)^y$

$$y = \frac{54}{18} = 3$$

$\therefore$ $$C_t' = C_0\left(\frac{1}{2}\right)^3 = \frac{1}{8}(C_6)$$

$\therefore$ $$\frac{C_t}{C_t'} = \frac{C_0}{2}\Big/\frac{C_0}{8} = 4$$

33. (2) $$\frac{d[O]}{dt} = -k[O]\,[OH]$$

$$\left(\frac{d[O]}{dt}\right)_1 = -k[1.0 \times 10^{-4}]\,[1.0 \times 10^{-4}]$$

$$\left(\frac{[O]}{dt}\right)_2 = -k[1.0 \times 10^{-4}]\,(2.0 \times 10^{-4})$$

$\therefore$ $$\frac{\left(\frac{d[O]}{dt}\right)_2}{\left(\frac{d[O]}{dt}\right)_1} = 2$$

34. (2) t_1 = time to decrease concentration from 4×10^{-4} mol dm^{-3} to 2×10^{-4} mol dm^{-3}.

t_2 = time to decrease concentration from 2×10^{-4} mol dm^{-3} to 1×10^{-4} mol dm^{-3}

In both case, reaction takes place 50%

$$k \text{ (zero order)} = \frac{x}{t}$$

For 50% reaction, $x = \frac{a}{2}$

$\therefore$ $$t_{50} = \frac{a}{2k}$$

$$t_1 = \frac{4 \times 10^{-4}}{2 \times 5 \times 10^{-8}} = 4 \times 10^3\text{s}$$

$$t_2 = \frac{2 \times 10^{-4}}{2 \times 5 \times 10^{-8}} = 2 \times 10^3\text{s}$$

$\therefore$ $$\frac{t_1}{t_2} = \frac{4 \times 10^3}{2 \times 10^3} = 2$$

35. (5) One curie $= 3.7 \times 10^{10}$dps (disintegration of atoms per second)

$$\left(\frac{dx}{dt}\right) = 5 \text{ curie} = 5 \times 3.7 \times 10^{10} \text{ dps}$$

Also, $$\left(\frac{dx}{dt}\right) = k\frac{w}{A}N_0$$

(w = weight, A = atomic mass)

$$k = \frac{0.693}{t_{50}} = \frac{0.693}{1620}\text{ yr}^{-1} = \frac{0.693}{1620 \times 3.154 \times 10^7}\text{ s}^{-1}$$

$$\therefore 5 \times 3.7 \times 10^{10} = \frac{0.693 \times w \times 6.02 \times 10^{23}}{1620 \times 3.154 \times 10^7 \times 226}$$

$\therefore$ $$w = \frac{5 \times 3.7 \times 10^{10} \times 1620 \times 3.154 \times 10^7 \times 226}{0.693 \times 6.02 \times 10^{23}}$$

$$= 5.12 \approx 5.0\text{ g}$$

36. (6) **By set I**

4 units of reactant change to 2 units in 2 min

Thus, half-life is 2 min.

For zero order reaction, $t_{50} \propto \left(\frac{1}{a}\right)^{n-1}$

$$n = 0, t_{50} \propto (a)$$

By set II Units of reactant = 8

Thus, half-life = 4 min

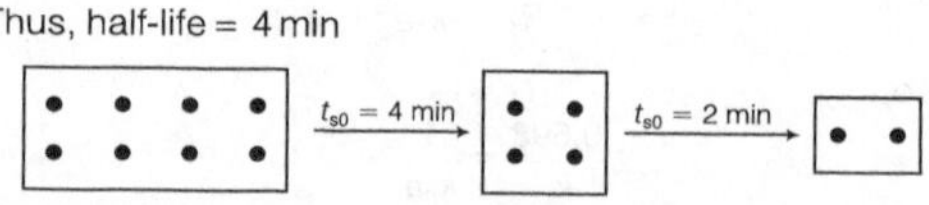

Then in set II, 8 units of reactant change to 2 units in time $= 4 + 2 = 6$ min.

37. (9) Decomposition follows first order kinetics.

$$k = \frac{2.303}{t}\log\left(\frac{a}{a-x}\right)$$

Case I $a = a$, $(a - x) = \frac{a}{8}$

Then $t = t_{1/8}$

$\therefore$ $$t_{1/8} = \frac{2.303}{k}\log\left(\frac{a}{a/8}\right)$$

$$= \frac{2.303}{R}\log 8 = \frac{2.303}{k}\log 2^3$$

$$= 3\left(\frac{2.303}{k}\log 2\right)$$

$$= 3\left(\frac{2.303}{k} \times 0.3010\right) = \frac{2.303}{k} \times 0.9$$

Case II $a = a$, $(a - x) = \frac{a}{10}$

then $t = t_{1/10}$

$$\therefore\ t_{1/10} = \frac{2.303}{k}\log\frac{a}{(a/10)} = \frac{2.303}{k}\log 10 = \frac{2.303}{k}$$

$\therefore$ $$\frac{t_{1/8}}{t_{1/10}} = 0.9$$

$\therefore$ $$\frac{10\,t_{1/8}}{t_{1/10}} = 0.9 \times 10 = 9$$

JEE Main & AIEEE Archive

1. (b) The probability that more than three molecules can collide and react simultaneously is very small. Thus, order of the reaction greater than three is very rare.

2. (d) Rate law equation is given by

$$\frac{dc}{dt} = k\,[A]^x\,[B]^y$$

$\therefore$ $\frac{dc}{dt}$ for C at various concentrations are

$$1.2 \times 10^{-3} = k(0.1)^x(0.1)^y \quad \ldots(i)$$

$$1.2 \times 10^{-3} = k(0.1)^x(0.2)^y \quad \ldots(ii)$$

$$2.4 \times 10^{-3} = k(0.2)^x(0.1)^y \quad \ldots(iii)$$

Solving for x [divide Eq. (iii) by Eq. (i)]

$$\frac{2.4 \times 10^{-3}}{1.2 \times 10^{-3}} = \frac{k\,(0.2)^x(0.1)^y}{k\,(0.1)^x(0.1)^y}$$

$$2 = (2)^x$$

$\Rightarrow$ $x = 1$

Solving for y [divide Eq. (iii) by Eq. (i)]

$$\frac{2.4 \times 10^{-3}}{1.2 \times 10^{-3}} = \frac{k\,(0.2)^x(0.1)^y}{k\,(0.1)^x(0.2)^y}$$

$$2 = (2)^x\left(\frac{0.1}{0.2}\right)^y$$

Put $x = 1$ in above equation, we get

$$2 = (2)^1\left(\frac{0.1}{0.2}\right)^y$$

$$1 = (0.5)^y$$

$$(1)^0 = (0.5)^y$$

$\Rightarrow$ $y = 0$

Thus, $\frac{dc}{dt} = k\,[A]^1\,[B]^0$

$\Rightarrow$ $\frac{dc}{dt} = k\,[A]^1$

3. (a) From Arrhenius equation,

$$\log\frac{k_2}{k_1} = \frac{-E_a}{2.303\,R}\left(\frac{1}{T_2} - \frac{1}{T_1}\right)$$

Given, $\frac{k_2}{k_1} = 2$; $T_2 = 310$ K

$T_1 = 300$ K

On putting values,

$\Rightarrow$ $\log 2 = \frac{-E_a}{2.303 \times 8.314}\left(\frac{1}{310} - \frac{1}{300}\right)$

$\Rightarrow$ $E_a = 53603.93$ J/mol

$= 53.6$ kJ/mol

4. (b) Rate constant for the first order reaction is given by

$$k = \frac{2.303}{t}\log\left(\frac{a}{a-x}\right)$$

$a = 0.1$ M

$(a - x) = 0.025$ M

$t = 40$ min

$\therefore$ $k = \frac{2.303}{40}\log\frac{0.1}{0.025\text{ M}}$

$= 0.0347\ \text{min}^{-1}$

$$\text{Rate} = \left(\frac{dx}{dt}\right) = k[A]^1 = 0.0347 \times 0.1$$

$= 3.47 \times 10^{-4}$ M min^{-1}

5. (c) For every 10°C rise of temperature, rate is doubled. Thus, temperature coefficient of the reaction = 2

When temperature is increased by 50°, rate becomes

$$= 2^{(50/10)} = 2^5 \text{ times} = 32 \text{ times}$$

6. (b) $A \xrightarrow{k_1} B,\ A \xrightarrow{k_2} C,$

By Arrhenius equation,

$$k_1 = A'e^{-E_{a_1}/RT}$$

and $k_2 = A'e^{-E_{a_2}/RT}$ [A' is Arrhenius constant]

$\because$ $E_{a_2} = 2E_{a_1}$

$\therefore$ $k_2 = A'e^{-2E_{a_1}/RT}$

$$\frac{k_1}{k_2} = \frac{A'e^{-E_{a_1}/RT}}{A'e^{-2E_{a_1}/RT}} = e^{E_{a_1}/RT}$$

$\therefore$ $k_1 = k_2 e^{E_{a_1}/RT}$

7. (c) Half- life for a zero order reaction is given by

$$t_{1/2} = \frac{[A_0]}{2k}$$

where $[A_0]$ = initial concentration of reactant

$\Rightarrow$ $k_0 = \frac{[A]_0}{2t_{1/2}} = \frac{2.0\ \text{mol L}^{-1}}{2 \times 1\text{h}} = 1.0\ \text{mol L}^{-1}\ \text{h}^{-1}$

Rate constant for a zero order reaction is given by

$$k = \frac{1}{t}[(A_0) - (A)]$$

$\Rightarrow$ $t = \frac{1}{k}[(A_0) - (A)]$

$$t = \frac{(0.50) - (0.25)\ \text{mol L}^{-1}}{1\ \text{mol L}^{-1}\ \text{h}^{-1}}$$

$\Rightarrow$ $t = 0.25$ h

8. (d) Slowest step is the rate determining step. Thus, in case (I), rate law is given as rate $= k[Cl_2][H_2S]$

While for the reaction given in case (II), rate law is given as rate $= k[H_2S][Cl_2][H^+]^{-1}$.

Hence, only mechanism (I) is consistent with the given rate law.

9. (c) Half-life = 6.93 min, $k_1 = \frac{0.693}{6.93} = 0.1\ \text{min}^{-1}$

Rate constant for the first order reaction is given by

$$k = \frac{2.303}{t}\log\frac{[A_0]}{[A]}$$

where, $[A_0]$ = initial concentration

$[A]$ = final concentration

$$0.1 = \frac{2.303}{t} \times \log\frac{100}{1}$$

$$0.1 = \frac{2.303}{t}\log 10^2$$

$$t = \frac{2.303 \times 2}{0.1} \times 2\log 10 = 46.06 \text{ min}$$

10. (b) $\frac{1}{2}A \longrightarrow 2B$

Remember for $aA \longrightarrow bB$

$$-\frac{1}{a}\cdot\frac{d[A]}{dt} = \frac{1}{b}\cdot\frac{d[B]}{dt} = \text{Rate of reaction}$$

For the given reaction

$$-\frac{2d[A]}{dt} = \frac{1}{2}\cdot\frac{d[B]}{dt} = \text{Rate of reaction}$$

$$\text{Rate of disappearance of } A = -\frac{d[A]}{dt} = \frac{1}{2 \times 2}\cdot\frac{d[B]}{dt} = \frac{1}{4}\cdot\frac{d[B]}{dt}$$

11. (a) $2A + B \longrightarrow$ Products

$[B]$ is doubled, half-life didn't change.

Half-life is independent of change in concentration of reactant i.e. first order

First order w.r.t. B

When $[A]$ is doubled, rate increased by two times.

$\Rightarrow$ First order w.r.t. A

Hence, net order of reaction = 1 + 1 = 2

Unit for the rate constant = $\text{conc}^{(1-n)}t^{-1}$

$$= (\text{mol L}^{-1})^{-1} \cdot \text{s}^{-1}$$

$$= \text{Lmol}^{-1}\ \text{s}^{-1}$$

12. (d) $A_2 + B_2 \rightleftharpoons 2AB$

$$E_a \text{ (forward)} = 180 \text{ kJ mol}^{-1}$$

$$E_a \text{ (backward)} = 200 \text{ kJ mol}^{-1}$$

In the presence of catalyst,

$$E_a \text{(forward)} = 180 - 100 = 80 \text{ kJ mol}^{-1}$$

$$E_a \text{ (backward)} = 200 - 100 = 100 \text{ kJ mol}^{-1}$$

$$\Delta H = E_a \text{ (forward)} - E_a \text{ (backward)}$$

$$= 80 - 100$$

$$= -20 \text{ kJ mol}^{-1}$$

For the value of enthalpy change of the reaction, negative sign can be neglected.

13. (d) Activity $\propto N$

$$\frac{N}{N_0} = \left(\frac{1}{2}\right)^n$$

or $\quad \frac{1}{10} = \left(\frac{1}{2}\right)^n$ or $10 = 2^n$

Taking log on both sides

$$\log 10 = n \log 2$$

$$n = \frac{1}{0.301} = 3.32$$

$$\text{Time} = n \times \text{half-life} = 3.32 \times 30$$

$$= 99.6 \text{ days}$$

$$\approx 100 \text{ days}$$

14. (d) Rate $= k\ [NOBr_2]\ [NO]$ ...(i)

But $NOBr_2$ is in equilibrium.

$$\therefore \quad K_{eq} = \frac{[NOBr_2]}{[NO]\ [Br_2]}$$

$$[NOBr_2] = K_{eq}\ [NO]\ [Br_2]$$

Putting the $[NOBr_2]$ in Eq. (i), we get

$$\text{Rate} = k \cdot K_{eq}\ [NO]\ [Br_2]\ [NO]$$

Hence, $\quad \text{Rate} = k \cdot K_{eq}\ [NO]^2\ [Br_2]$

$$\text{Rate} = k'\ [NO]^2\ [Br_2]$$

where $\quad k' = k \cdot K_{eq}$

Thus, order w.r.t. NO is 2.

15. (b) According to Arrhenius equation

$$k = Ae^{-E/RT}$$

where, E is the activation energy below which the colliding molecules will not react.

16. (c) If order $= n$ then on making concentration m-times, rate becomes m^n times

$$n = 2, m = 2$$

Thus, new rate $= 2^2 = 4$ times

17. (d) There are two different reactants (say A and B).

$$A + B \longrightarrow \text{product}$$

Thus, it is a bimolecular reaction.

If $\frac{dx}{dt} = k\ [A]\ [B]$

It is second order reaction.

If $\left(\frac{dx}{dt}\right) = k\ [A] = k\ [B]$

It is first order reaction.

Molecularity is independent of rate but is the sum of the reacting substances. Thus, it cannot be unimolecular reaction.

18. (c)

	$A \longrightarrow$	Product
Initially	a	0
After time t	$(a - x)$	x
After $t_{1/4}$	$\left(a - \frac{a}{4}\right)$	$\frac{a}{4}$

For first order kinetics,

$$k = \frac{2.303}{t} \log\left(\frac{a}{a - x}\right)$$

$$\therefore \quad k = \frac{2.303}{t_{1/4}} \log \frac{a}{\frac{3a}{4}}$$

$$t_{1/4} = \frac{2.303 \log \frac{4}{3}}{k} = \frac{0.29}{k}$$

19. (d) $X \rightarrow Y$ is an endothermic reaction, $\Delta H = +$ ve

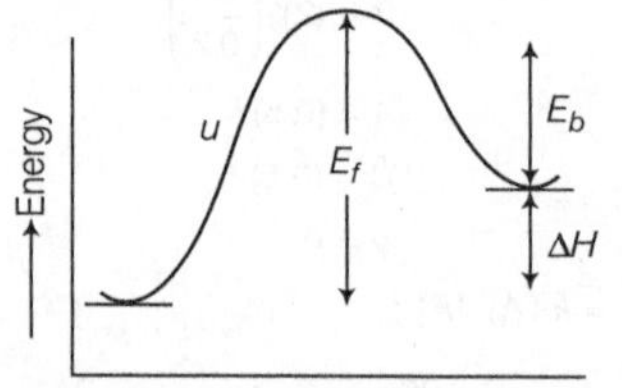

E_b = energy of activation of backward reaction

E_f = energy of activation of forward reaction

ΔH = heat of reaction

Thus, $\quad E_f = E_b + \Delta H, \quad E_f > E_b$

20. (d) $2A + B \longrightarrow C$

$$\text{Rate} = k\ [A]\ [B]$$

It represents second order reaction.

Thus, unit of k is L $\text{mol}^{-1}\ \text{s}^{-1}$.

Hence, (a) is false.

$t_{1/2}$ is dependent of concentration but not constant.

Hence, (b) is false.

And $\quad -\frac{1}{2}\frac{d\ [A]}{dt} = \frac{d\ [C]}{dt}$

Thus, (c) is also false.

unit of $k = \text{conc}^{1-x}\ \text{time}^{-1}$

By rate law $n = 2$

Thus, unit is $\text{conc}^{-1}\ \text{time}^{-1}$

(L $\text{mol}^{-1}\ \text{s}^{-1}$ if concentration in mol L^{-1} and time in second)

21. (a) Order = 1

Concentration changes from 0.8 M to 0.4 M in (50%).

15 min, thus half-life = 15 min = T_{50}

A change from 0.1 M to 0.025 is 75% and for first order reaction

$$T_{75} = 2 \times T_{50} = 2 \times 15 = 30 \text{ min}$$

Alternate Method

$$T_{50} = 15 \text{ min}$$

$$k = \frac{2.303 \log 2}{T_{50}} = \frac{2.303 \log 2}{15}$$

$$a = 0.1 \text{ M}$$

$$(a - x) = 0.025 \text{ M}$$

For first order

$$k = \frac{2.303}{t} \log \left(\frac{a}{a - x}\right)$$

$$\frac{2.303 \log 2}{15} = \frac{2.303}{t} \log \frac{0.1}{0.025} = \frac{2.303}{t} \log 4$$

$$\therefore \quad \frac{2.303 \log 2}{15} = \frac{2 \times 2.303 \log 2}{t}$$

$$\therefore \quad t = 30 \text{ min}$$

22. (c) By Arrhenius equation,

$k = Ae^{-E_a/RT}$, k = Rate constant

A = Pre-exponential, frequency factor

E_a = Activation energy, R = Gas constant

T = Temperature

23. (d) Rate, $r = k\,[A]^n [B]^m$

On doubling $[A]$ and halving $[B]$, the rate becomes

$$r' = k\,[2A]^n\,[1/2\,B]^m$$

$$\Rightarrow \quad \frac{r'}{r} = \frac{k[2A]^n[1/2\,B]^m}{k[A]^n[B]^m} = \frac{(2)^n}{(2)^m} = (2)^n \cdot (2)^{-m} = (2)^{n-m}$$

24. (c) $\left(\frac{dx}{dt}\right) = k\,[\text{NO}]^2\,[O_2] = k\left(\frac{n_{\text{NO}}}{V}\right)^2 \left(\frac{n_{O_2}}{V}\right)$

$$\left(\frac{dx}{dt}\right) = \frac{k}{V^3}(n_{\text{NO}})^2\,(n_{O_2})$$

$$\left(\frac{dx}{dt}\right)' = \frac{k\,(n_{\text{NO}})^2\,(n_{O_2})}{\left(\frac{V}{2}\right)^3} = 8\left(\frac{dx}{dt}\right)$$

25. (b) $H_2 + I_2 \longrightarrow 2HI$

$$\text{Rate of reaction} = \frac{-d\,[H_2]}{dt} = \frac{-d\,[I_2]}{dt} = \frac{1}{2}\frac{d\,[\text{HI}]}{dt}$$

$$\text{or} \quad \frac{-2d\,[H_2]}{dt} = \frac{-2d\,[I_2]}{dt} = \frac{d\,[\text{HI}]}{dt}$$

26. (b) Rate of the reaction depend upon the concentration of A and B both, hence it is a second order reaction.

27. (b)

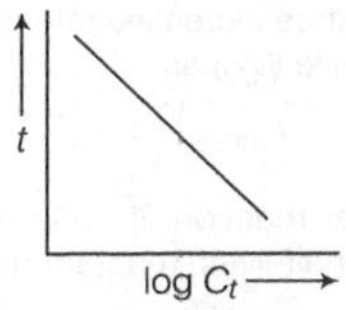

$$kt = -\log_e C_t + \log_e C_0$$

$$t = -\frac{1}{k}\log_e C_t + \log_e C_0$$

$y = mx + c$, straight line, negative slope.

28. (d) Unit of rate constant,

$$k = \frac{1}{\text{Time}} \cdot \frac{1}{(\text{conc})^{n-1}}$$

For k_1, $n = 0$,

$$k = \frac{1}{t} \cdot \frac{1}{(\text{conc})^{0-1}}$$

Hence, conc $t^{-1} \equiv \text{M} \cdot \text{s}^{-1}$

For k_2, $n = 1$,

$$k = \frac{1}{t} \cdot \frac{1}{(\text{conc.})^{1-1}} = \frac{1}{t} = \text{s}^{-1}$$

JEE Advanced & IIT JEE Archive

1. (b) For the elementary reaction, $M \rightarrow N$

Rate law can be written as Rate $\propto [M]^n$

$$\text{Rate} = k\,[M]^n \quad \ldots\text{(i)}$$

When we double the concentration of $[M]$, rate becomes 8 times, hence new rate law can be written as

$$8 \times \text{Rate} = k\,[2M]^n \quad \ldots\text{(ii)}$$

$$\frac{\text{Rate}}{8 \times \text{Rate}} = \frac{k\,[M]^n}{k\,[2M]^n}, \quad \frac{1}{8} = \frac{1}{[2]^n}$$

$$\Rightarrow \quad [2]^n = 8 = [2]^3$$

$$n = 3$$

2. (d) **Plan** Time of 75% reaction is twice the time taken for 50% reaction if it is first order reaction w.r.t P. From graph, $[Q]$ decreases linearly with time, thus it is zeroth order reaction w.r.t. Q

$$\frac{dx}{dt} = k[P]^a[Q]^b$$

Order w.r.t. $P = a = 1$

Order w.r.t. $Q = b = 0$

Thus, overall order of the reaction = 1 + 0 = 1

3. (9) For a first order process,

$$kt = \ln\frac{[A]_0}{[A]}$$

where, $[A]_0$ = initial concentration

$[A]$ = concentration of reactant remaining at time t.

$$\Rightarrow \quad kt_{1/8} = \ln\frac{[A]_0}{[A]_0/8} = \ln 8 \quad \ldots\text{(i)}$$

$$\text{and} \quad kt_{1/10} = \ln\frac{[A]_0}{[A]_0/10} = \ln 10 \quad \ldots\text{(ii)}$$

$$\text{Therefore,} \quad \frac{t_{1/8}}{t_{1/10}} = \frac{\ln 8}{\ln 10}$$

$$= \log 8$$

$$= 3 \log 2$$

$$= 3 \times 0.3$$

$$= 0.9$$

$$\Rightarrow \quad \frac{t_{1/8}}{t_{1/10}} \times 10 = 0.9 \times 10$$

$$= 9$$

4. (a,b,d)

(a) For a first order reaction, the concentration of reactant remaining after time t is given by $[A] = [A]_0 e^{-kt}$

Therefore, concentration of reactant decreases exponentially with time.

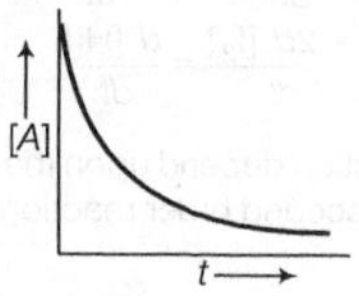

(b) Rise in temperature increases rate constant (k) and therefore decreases half-life ($t_{1/2}$) as

$$t_{1/2} = \frac{\ln 2}{k}$$

(d) For a first order reaction, if 100 moles of reactant is taken initially, after n half-lives, reactant remaining is given by

$$\% A = 100\left(\frac{1}{2}\right)^n = 100\left(\frac{1}{2}\right)^8 = 0.3906$$

$\Rightarrow$ A reacted $= 100 - 0.3906 = 99.6\%$

(c) Half-life of first order reaction is independent of initial concentration.

5. (a) According to Arrhenius equation, rate constant increases exponentially with temperature :

$$k = Ae^{-E_a/RT}$$

6. (0)

Time (min)	$[R]/M$	Δt	$\Delta[R]$	$\frac{\Delta[R]}{\Delta t}$
0	1.0			
0.05	0.75	0.05	0.25	$\frac{0.025}{0.05} = 5$
0.12	0.40	0.12	0.60	$\frac{0.60}{0.12} = 5$
0.18	0.10	0.18	0.90	$\frac{0.90}{0.18} = 5$

$$\left|\frac{\Delta |R|}{\Delta t}\right| = k[R]^n = 5 \text{ M min}^{-1}$$

Rate remains constant throughout thus, order = zero

7. (d) The logarithmic form of Arrhenius equation is

$$\log k = \log A - \frac{E_a}{2.303\,RT} \quad \ldots(i)$$

Given, $$\log k = 6 - \frac{2000}{T} \quad \ldots(ii)$$

Comparing the above Eqs. (i) and (ii),

$$\log A = 6$$

$$\Rightarrow \quad A = 10^6$$

and $$\frac{E_a}{2.303\,R} = 2000$$

$$\Rightarrow \quad E_a = 2000 \times 2.303 \times 8.314 \text{ J}$$

$$= 38.3 \text{ kJ}$$

8. (a) For first order reaction,

$$t_{1/2} = \frac{\ln 2}{k_1} = 40 \text{ s} \quad \ldots(i)$$

For zero order reaction,

$$t_{1/2} = \frac{[A]_0}{2k_0} = 20 \text{ s} \quad \ldots(ii)$$

On dividing Eq. (ii) by Eq. (i), we get

$$\frac{1}{2} = \frac{[A]_0}{2k_0} \times \frac{k_1}{\ln 2}$$

$$\Rightarrow \quad \frac{k_1}{k_0} = \frac{\ln 2}{[A]_0} = \frac{0.693 \text{ s}^{-1}}{1.386 \text{ mol}^{-3}\text{dm}^{-3}\text{s}^{-1}}$$

$$= 0.5 \text{ mol}^{-1}\text{dm}^{-1}$$

9. (d)

$$(\text{Rate}) = k[G]^m [H]^n \quad \ldots(i)$$

$$8\,(\text{Rate})_1 = (\text{Rate})_2 = k[2G]^m [2H]^n \quad \ldots(ii)$$

$$2\,(\text{Rate})_1 = (\text{Rate})_3 = k[2G]^m [H]^n \quad \ldots(iii)$$

From Eqs. (i) and (iii),

$$(2)^m = 2$$

$$\Rightarrow \quad m = 1$$

From Eqs. (ii) and (iii),

$$(2)^n = 4 = 2^2$$

$$\Rightarrow \quad n = 2$$

Thus, total order = 1 + 2 = 3

10. (d) Order of a reaction can take any real value, i.e. negative, integer, fraction, etc.

Last 3 Years' Questions
JEE Main & Advanced
(2016-2018)

JEE Main

2016

1. 18 g of glucose ($C_6H_{12}O_6$) is added to 178.2 g water. The vapour pressure of water (in torr) for this aqueous solution is **[Solutions]**

(a) 76.0 (b) 752.4
(c) 759.0 (d) 7.6

2. Decomposition of H_2O_2 follows a first order reaction. In 50 min, the concentration of H_2O_2 decreases from 0.5 to 0.125 M in one such decomposition. When the concentration of H_2O_2 reaches 0.05 M, the rate of formation of O_2 will be **[Chemical Kinetics]**

(a) 6.93×10^{-4} mol min^{-1} (b) 2.66 L min^{-1} at STP
(c) 1.34×10^{-2} mol min^{-1} (d) 6.93×10^{-2} mol min^{-1}

2017

3. A metal crystallises in a face centred cubic structure. If the edge length of its unit cell is 'a', the closest approach between two atoms in metallic crystal will be **[Solid State]**

(a) $2a$ (b) $2\sqrt{2}\,a$
(c) $\sqrt{2}\,a$ (d) $\dfrac{a}{\sqrt{2}}$

4. Two reactions R_1 and R_2 have identical pre- exponential factors. Activation energy of R_1 exceeds that of R_2 by 10 kJ mol^{-1}. If k_1 and k_2 are rate constants for reactions R_1 and R_2, respectively at 300 K, then $\ln\left(\dfrac{k_2}{k_1}\right)$ is equal to ($R = 8.314$ J mol^{-1}K^{-1}) **[Chemical Kinetics]**

(a) 8 (b) 12 (c) 6 (d) 4

5. The freezing point of benzene decreases by 0.45°C when 0.2 g of acetic acid is added to 20 g of benzene. If acetic acid associates to form a dimer in benzene, percentage association of acetic acid in benzene will be (K_f for benzene = 5.12 K kg mol^{-1}) **[Solutions]**

(a) 64.6 % (b) 80.4 %
(c) 74.6 % (d) 94.6 %

6. Given, $E^\circ_{Cl_2/Cl^-} = 1.36$ V,

$E^\circ_{Cr^{3+}/Cr} = -0.74$ V

$E^\circ_{Cr_2O_7^{2-}/Cr^{3+}} = 1.33$ V, $E^\circ_{MnO_4^-/Mn^{2+}} = 1.51$ V

Among the following, the strongest reducing agent is **[Electrochemistry]**

(a) Cr (b) Mn^{2+}
(c) Cr^{3+} (d) Cl^-

2018

7. Which type of 'defect' has the presence of cations in the interstitial sites? **[Solid State]**

(a) Schottky defect (b) Vacancy defect
(c) Frenkel defect (d) Metal deficiency defect

8. For 1 molal aqueous solution of the following compounds, which one will show the highest freezing point? **[Solutions]**

(a) $[Co(H_2O)_6]Cl_3$
(b) $[Co(H_2O)_5Cl]Cl_2 \cdot H_2O$
(c) $[Co(H_2O)_4Cl_2]Cl \cdot 2H_2O$
(d) $[Co(H_2O)_3Cl_3] \cdot 3H_2O$

9. At 518°C, the rate of decomposition of a sample of gaseous acetaldehyde, initially at a pressure of 363 Torr, was 1.00 Torr s^{-1} when 5% had reacted and 0.5 Torr s^{-1} when 33% had reacted. The order of the reaction is **[Chemical Kinetics]**

(a) 2
(b) 3
(c) 1
(d) 0

10. How long (approximate) should water be electrolysed by passing through 100 amperes current so that the oxygen released can completely burn 27.66 g of diborane?
(Atomic weight of $B = 10.8\,\mu$) **[Electrochemistry]**

(a) 6.4 hours
(b) 0.8 hours
(c) 3.2 hours
(d) 1.6 hours

JEE Advanced

2016

1. According to the Arrhenius equation, **[Chemical Kinetics, Type 2]**

(a) a high activation energy usually implies a fast reaction
(b) rate constant increases with increase in temperature. This is due to a greater number of collisions whose energy exceeds the activation energy
(c) higher the magnitude of activation energy, stronger is the temperature dependence of the rate constant
(d) the pre-exponential factor is a measure of the rate at which collisions occur, irrespective of their energy

2. The mole fraction of a solute in a solution is 0.1. At 298 K, molarity of this solution is the same as its molality. Density of this solution at 298 K is 2.0 g cm^{-3}. The ratio of the molecular weights of the solute and solvent, $\left(\frac{m_{solute}}{m_{solvent}}\right)$ is ... **[Solution, Type 3]**

3. For the following electrochemical cell at 298 K,

$$Pt(s)\,|\,H_2(g, 1\,bar)\,|\,H^+(aq, 1\,M)\,||\,M^{4+}(aq), M^{2+}(aq)\,|\,Pt(s), \quad E_{cell} = 0.092\text{ V}$$

when $\frac{[M^{2+}(aq)]}{[M^{4+}(aq)]} = 10^x$

Given : $E^{\circ}_{M^{4+}/M^{2+}} = 0.151\text{ V};\ 2.303\frac{RT}{F} = 0.059\text{ V}$

The value of x is **[Electrochemistry, Type 1]**

(a) – 2
(b) – 1
(c) 1
(d) 2

4. The correct statement(s) for cubic close packed (ccp) three dimensional structure is (are) **[Solid State, Type 2]**

(a) The number of the nearest neighbours of an atom present in the topmost layer is 12
(b) The packing efficiency of atom is 74%
(c) The number of octahedral and tetrahedral voids per atom are 1 and 2, respectively
(d) The unit cell edge length is $2\sqrt{2}$ times the radius of the atom

5. Mixture(s) showing positive deviation from Raoult's law at 35°C is (are) **[Solutions, Type 2]**

(a) carbon tetrachloride + methanol
(b) carbon disulphide + acetone
(c) benzene + toluene
(d) phenol + aniline

Note *Type-1 Only one correct option, Type-2 More then one correct option, Type-3 Single digit Answer, Type-4 Numerical value based*

2017

6. For a solution formed by mixing liquids L and M, the vapour pressure of L plotted against the mole fraction of M in solution is shown in the following figure. Here x_L and x_M represent mole fractions of L and M, respectively, in the solution. The correct statement(s) applicable to this system is (are) **[Solutions, Type 2]**

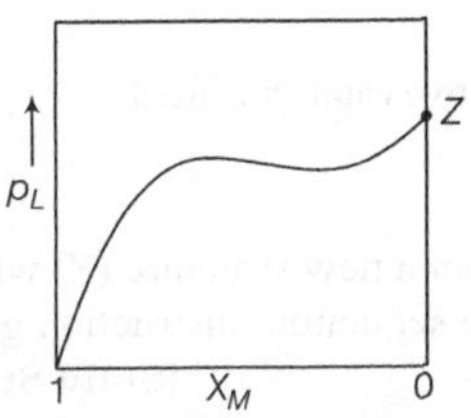

(a) The point Z represents vapour pressure of pure liquid M and Raoult's law is obeyed from $x_L = 0$ to $x_L = 1$

(b) Attractive intermolecular interactions between L - L in pure liquid L and M - M in pure liquid M are stronger than those between L - M when mixed in solution

(c) The point Z represents vapour pressure of pure liquid M and Raoult's law is obeyed when $x_L \to 0$

(d) The point Z represents vapour pressure of pure liquid L and Raoult's law is obeyed when $x_L \to 1$

7. A crystalline solid of a pure substance has a face-centred cubic structure with a cell edge of 400 pm. If the density of the substance in the crystal is 8 g cm^{-3}, then the number of atoms present in 256 g of the crystal is $N \times 10^{24}$. The value of N is **[Solid State, Type 3]**

8. The conductance of a 0.0015 M aqueous solution of a weak monobasic acid was determined by using a conductivity cell consisting of platinised Pt electrodes. The distance between the electrodes is 120 cm with an area of cross section of 1 cm^2. The conductance of this solution was found to be 5×10^{-7} S. The pH of the solution is 4. The value of limiting molar conductivity (Λ°_m) of this weak monobasic acid in aqueous solution is $Z \times 10^2$ S cm^{-1} mol^{-1}. The value of Z is **[Electrochemistry, Type 3]**

9. Pure water freezes at 273 K and 1 bar. The addition of 34.5 g of ethanol to 500 g of water changes the freezing point of the solution. Use the freezing point depression constant of water as 2 K kg mol^{-1}. The figures shown below represent plots of vapour pressure (V.P.) *versus* temperature (T). [Molecular weight of ethanol is 46 g mol^{-1}]

Among the following, the option representing change in the freezing point is **[Solution, Type 1]**

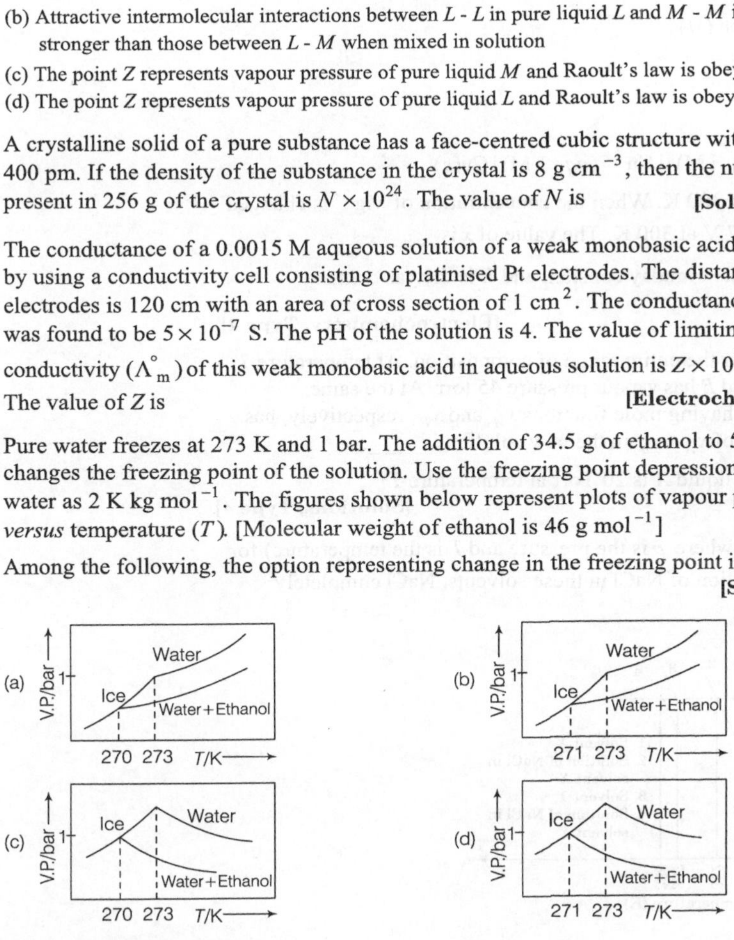

10. For the following cell,

$Zn(s)|\ ZnSO_4(aq)\ ||\ CuSO_4(aq)|\ Cu(s)$

when the concentration of Zn^{2+} is 10 times the concentration of Cu^{2+}, the expression for ΔG (in J mol^{-1}) is [F is Faraday constant; R is gas constant; T is temperature; E° (cell) = 1.1 V] **[Electrochemistry, Type 1]**

(a) 2.303 RT + 1.1 F

(b) 1.1 F

(c) 2.303 RT − 2.2 F

(d) −2.2 F

11. In a bimolecular reaction, the steric factor P was experimentally determined to be 4.5. the correct option(s) among the following is(are) **[Chemical Kinetics, Type 2]**

(a) The activation energy of the reaction is unaffected by the value of the steric factor

(b) Experimentally determined value of frequency factor is higher than that predicted by Arrhenius equation

(c) The value of frequency factor predicted by Arrhenius equation is higher than that determined experimentally

(d) Since $P = 4.5$, the reaction will not proceed unless an effective catalyst is used

2018

12. Consider an ionic solid MX with NaCl structure. Construct a new structure (Z) whose unit cell is constructed from the unit cell of MX following the sequential instruction given below. Neglect the charge balance. **[Solid State, Type 4]**

(a) Remove all the anions (X) except the central one

(b) Replace all the face centered cations (M) by anions (X)

(c) Remove all the corner cations (M)

(d) Replace the central anion (X) with cation (M)

The value of $\left(\frac{\text{Number of anions}}{\text{Number of cations}}\right)$ in Z is ____

13. For the electrochemical cell,

$$\mathrm{Mg}(s)|\mathrm{Mg}^{2+}(aq,\ 1\,\mathrm{M})\,||\,\mathrm{Cu}^{2+}(aq,\ 1\,\mathrm{M})\,|\,\mathrm{Cu}(s)$$

The standard emf of the cell is 2.70 V at 300 K. When the concentration of Mg^{2+} is changed to x M, the cell potential changes to 2.67 V at 300 K. The value of x is ____.

(Given, $\frac{F}{R} = 11500\,\mathrm{K\,V^{-1}}$, where F is the Faraday constant and R is the gas contant, In (10) = 2.30) **[Electrochemistry, Type 4]**

14. Liquids A and B form ideal solution over the entire range of composition. At temperature T, equimolar binary solution of liquids A and B has vapour pressure 45 torr. At the same temperature, a new solution of A and B having mole fractions x_A and x_B, respectively, has vapour pressure of 22.5 torr. The value of x_A / x_B in the new solution is ____.

(Given that the vapour pressure of pure liquid A is 20 Torr at temperature T) **[Solutions, Type 4]**

15. The plot given below shows $p-T$ curves (where p is the pressure and T is the temperature) for two solvents X and Y and isomolal solution of NaCl in these solvents. NaCl completely dissociates in both the solvents.

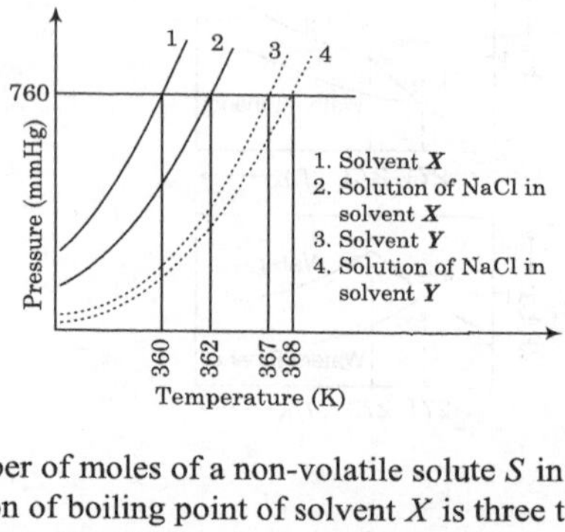

On addition of equal number of moles of a non-volatile solute S in equal amount (in kg) of these solvents, the elevation of boiling point of solvent X is three times that of solvent Y. Solute S is known to undergo dimerisation in these solvents. If the degree of dimerisation is 0.7 in solvent Y, the degree of dimerisation in solvent X is ____. **[Solution, Type 4]**

16. For a first order reaction $A(g) \longrightarrow 2B(g) + C(g)$ at constant volume and 300 K, the total pressure at the beginning $(t = 0)$ and at time t are p_0 and p_t, respectively.
Initially, only A is present with concentration $[A]_0$, and $t_{1/3}$ is the time required for the partial pressure of A to reach 1/3 rd of its initial value. The correct option(s) is (are)
(Assume that all these gases behave as ideal gases) **[Chemical Kinetics, Type 2]**

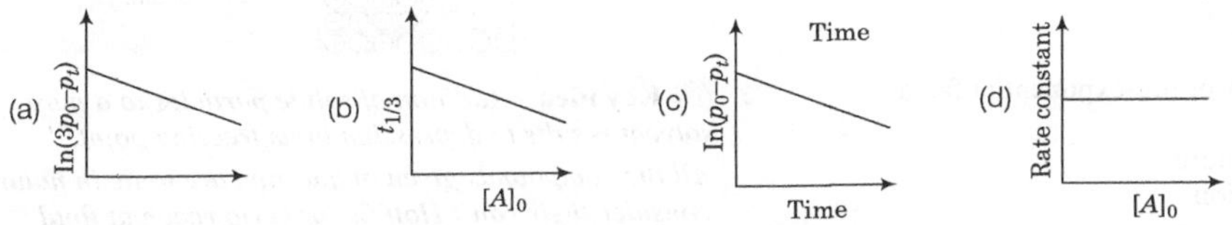

17. Consider the following reversible reaction,

$$A(g) + B(g) \rightleftharpoons AB(g)$$

The activation energy of the backward reaction exceeds that of the forward reaction by $2RT$ (in J mol^{-1}). If the pre-exponential factor of the forward reaction is 4 times that of the reverse reaction, the absolute value of $\Delta G^{\ominus}$ (in J mol^{-1}) for the reaction at 300 K is ....
(Given ; $\ln(2) = 0.7$ $RT = 2500$ J mol^{-1} at 300 K and G is the Gibbs energy)
[Chemical Kinetics, Type 4]

18. Consider an electrochemical cell : $A(s)\,|\,A^{n+}(aq, 2\text{M})\,||\,B^{2n+}(aq, 1\text{M})\,|\,B(s)$. The value of $\Delta H^{\ominus}$ for the cell reaction is twice of $\Delta G^{\ominus}$ at 300 K. If the emf of the cell is zero, the $\Delta S^{\ominus}$ (in J K^{-1} mol^{-1}) of the cell reaction per mole of B formed at 300 K is .... .
(Given : $\ln(2) = 0.7$, R (universal gas constant) = 8.3 J K^{-1} mol^{-1}. H, S and G are enthalpy, entropy and Gibbs energy, respectively.) **[Electrochemistry, Type 4]**

Answer *with* Explanations

JEE Main

1. *(b)* **Key concept** Vapour pressure of water $(p^\circ) = 760$ torr

$$\text{Number of moles of glucose} = \frac{\text{Mass (g)}}{\text{Molecular mass (g/mol)}}$$

$$= \frac{18 \text{ g}}{180 \text{ gmol}^{-1}} = 0.1\text{mol}$$

Molar mass of water = 18 g/mol

Mass of water (given) = 178.2g

$$\text{Number of moles of water} = \frac{\text{Mass of water}}{\text{Molar mass of water}}$$

$$= \frac{178.2\text{g}}{18 \text{ g/mol}} = 9.9 \text{ mol}$$

Total number of moles = (0.1 + 9.9) moles = 10 moles

Now, mole fraction of glucose in solution = Change in pressure with respect to initial pressure

i.e. $\dfrac{\Delta p}{p^\circ} = \dfrac{0.1}{10}$

or $\Delta p = 0.01p^\circ = 0.01 \times 760 = 7.6$ torr

$\therefore$ Vapour pressure of solution

$= (760 - 7.6)$ torr $= 752.4$ torr

2. *(a)* For first order reaction, $k = \dfrac{2.303}{t} \log \dfrac{a}{a-x}$

Given, $t = 50$ min, $a = 0.5$ M,

$a - x = 0.125$ M

$$\therefore \quad k = \frac{2.303}{50} \log \frac{0.5}{0.125} = 0.0277 \text{ min}^{-1}$$

Now, as per reaction

$$2H_2O_2 \longrightarrow 2H_2O + O_2$$

$$-\frac{1}{2}\frac{d\,[H_2O_2]}{dt} = \frac{1}{2}\frac{d\,[H_2O]}{dt} = \frac{d\,[O_2]}{dt}$$

$$\text{Rate of reaction, } -\frac{d\,[H_2O_2]}{dt} = k[H_2O_2]$$

$$\therefore \quad \frac{d\,[O_2]}{dt} = -\frac{1}{2}\frac{d\,[H_2O_2]}{dt} = \frac{1}{2}k[H_2O_2] \quad \ldots\text{(i)}$$

When the concentration of H_2O_2 reaches 0.05 M,

$$\frac{d\,[O_2]}{dt} = \frac{1}{2} \times 0.0277 \times 0.05 \quad \text{[from Eq. (i)]}$$

or $$\frac{d\,[O_2]}{dt} = 6.93 \times 10^{-4} \text{ mol min}^{-1}$$

Alternative Method In 50 min, the concentration of H_2O_2 decreases from 0.5 to 0.125 M or in one half-life, concentration of H_2O_2 decreases from 0.5 to 0.25 M. In two half-lives, concentration of H_2O_2 decreases from 0.5 to 0.125 M or $2t_{1/2} = 50$ min, $t_{1/2} = 25$ min

$$\therefore \quad k = \left(\frac{0.693}{25}\right) \text{min}^{-1}$$

or $$\frac{d\,[O_2]}{dt} = -\frac{1}{2}\frac{d\,[H_2O_2]}{dt} = \frac{k\,[H_2O_2]}{2}$$

$$= 6.93 \times 10^{-4} \text{mol min}^{-1}$$

3. *(d)* For fcc arrangement, $4r = \sqrt{2}a$

where, r = radius and a = edge length

$\therefore$ Closest distance $= 2r = \dfrac{\sqrt{2}\,a}{2} = \dfrac{a}{\sqrt{2}}$

4. *(d)* According to Arrhenius equation

$$k = Ae^{-E_a/RT}$$

where, A = collision number or pre-exponential factor.
R = gas constant
T = absolute temperature
E_a = energy of activation

For reaction R_1, $k_1 = Ae^{-E_{a_1}/RT}$ ...(i)

For reaction R_2, $k_2 = Ae^{-E_{a_2}/RT}$ ...(ii)

On dividing Eq. (ii) by Eq. (i), we get

$$\frac{k_2}{k_1} = e^{-\frac{(E_{a_2} - E_{a_1})}{RT}} \quad ...(iii)$$

[$\because$ Pre-exponential factor 'A' is same for both reactions]
Taking ln on both the sides of Eq. (iii), we get

$$\ln\left(\frac{k_2}{k_1}\right) = \frac{E_{a_1} - E_{a_2}}{RT}$$

Given, $E_{a_1} = E_{a_2} + 10\,\text{kJ mol}^{-1}$
$= E_{a_2} + 10{,}000\,\text{J mol}^{-1}$

$$\therefore \quad \ln\frac{k_2}{k_1} = \frac{10{,}000\ \text{J mol}^{-1}}{8.314\ \text{J mol}^{-1}\text{K}^{-1} \times 300\ \text{K}} = 4$$

5. *(d)* Let the degree of association of acetic acid (CH_3COOH) in benzene is α, then

$$2CH_3COOH \rightleftharpoons (CH_3COOH)_2$$

	$2CH_3COOH$	$(CH_3COOH)_2$
Initial moles	1	0
Moles at equilibrium	$1-\alpha$	$\alpha/2$

$\therefore$ Total moles $= 1 - \alpha + \dfrac{\alpha}{2} = 1 - \dfrac{\alpha}{2}$

or $\quad i = 1 - \dfrac{\alpha}{2}$

Now, depression in freezing point (ΔT_f) is given as

$$\Delta T_f = iK_f m \quad ...(i)$$

where, K_f = molal depression constant or cryoscopic constant.
m = molality

$$\text{Molality} = \frac{\text{number of moles of solute}}{\text{weight of solvent (in kg)}} = \frac{0.2}{60} \times \frac{1000}{20}$$

Putting the values in Eq. (i)

$$\therefore \quad 0.45 = \left[1 - \frac{\alpha}{2}\right](5.12)\left[\frac{0.2}{60} \times \frac{1000}{20}\right]$$

$$1 - \frac{\alpha}{2} = \frac{0.45 \times 60 \times 20}{5.12 \times 0.2 \times 1000}$$

$$\Rightarrow \quad 1 - \frac{\alpha}{2} = 0.527 \Rightarrow \frac{\alpha}{2} = 1 - 0.527$$

$$\therefore \quad \alpha = 0.946$$

Thus, percentage of association = 94.6%

6. *(a)* The substances which have lower reduction potentials are stronger reducing agents. Therefore, Cr ($E^\circ_{Cr^{3+}/Cr} = -0.74$ V) is the strongest reducing agent among all the other given options.

7. *(c)* It is the "Frenkel defect" in which cations leave their original site and occupy interstitial site as shown below.

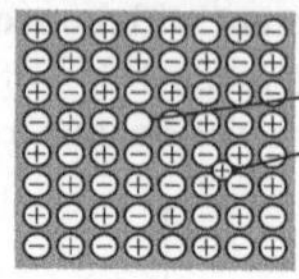

8. *(d)* **Key idea** *"Addition of solute particles to a pure solvent results to depression in its freezing point."*
All the compounds given in question are ionic in nature so, consider their van't Hoff factor (i) to reach at final conclusion.

The solution with maximum freezing point must have minimum number of solute particles. This generalisation can be done with the help of van't Hoff factor (i) i.e.
Number of solute particles $\propto$ van't Hoff factor (i)
Thus, we can say directly
Solution with maximum freezing point will be the one in which solute with minimum van't Hoff factor is present
Now, for $Co(H_2O)_6]Cl_3 \rightleftharpoons [Co(H_2O)_6]^{3+} + 3Cl^-$
van't Hoff factor (i) is 4. Similarly for,

$$[Co(H_2O)_5Cl]Cl_2 \cdot H_2O \rightleftharpoons [Co(H_2O)_5Cl]^{2+} + 2Cl^- \quad \text{'}i\text{' is 3}$$

$$[Co(H_2O)_4Cl_2]Cl \cdot 2H_2O \rightleftharpoons [Co(H_2O)_4Cl_2]^+ + Cl^- \quad \text{'}i\text{' is 2}$$

and for $[Co(H_2O)_3Cl_3] \cdot 3H_2O$, '$i$' is 1 as it does not show ionisation. Hence, $[Co(H_2O)_3Cl_3] \cdot 3H_2O$ have minimum number of particles in the solution.
So, freezing point of its solution will be maximum.

9. *(a)* For the reaction,

$$CH_3CHO(g) \xrightarrow{\text{Decomposes}} CH_4 + CO$$

Let order of reaction with respect to CH_3CHO is m.
Its given, $r_1 = 1$ torr/sec. when CH_3CHO is 5% reacted i.e. 95% unreacted. Similarly, $r_2 = 0.5$ torr/sec when CH_3CHO is 33% reacted i.e., 67% unreacted.
Use the formula, $r \propto (a - x)^m$
where $(a - x)$ = amount unreacted

so, $\quad \dfrac{r_1}{r_2} = \dfrac{(a - x_1)^m}{(a - x_2)^m}$ or $\dfrac{r_1}{r_2} = \left[\dfrac{a - x_1}{a - x_2}\right]^m$

Now putting the given values

$$\frac{1}{0.5} = \left(\frac{0.95}{0.67}\right)^m$$

$$\Rightarrow \quad 2 = (1.41)^m \text{ or } m = 2$$

10. *(c)* Given that, $i = 100$ amp. also, 27.66 g of diborane (B_2H_6)
Molecular mass of $B_2H_6 = 10.8 \times 2 + 6 = 27.6$
Number of moles of B_2H_6 in 27.66 g

$$= \frac{\text{Given mass}}{\text{Molar mass}} = \frac{27.66}{27.6} \approx 1$$

Now consider the equation

$$B_2H_6 + 3O_2 \longrightarrow B_2O_3 + 3H_2O$$

From the equation we can interpret that 3 moles of oxygen is required to burn 1 mole (i.e. 27.6 g) B_2H_6 completely.

Also consider the electrolysis reaction of water i.e.

$$H_2O \rightleftharpoons 2H^+ + O^{--}$$

$$2H^+ \xrightarrow[\text{Cathode}]{+2e^-} 2H \longrightarrow H_2 \uparrow$$

$$O^{--} \xrightarrow[-2e^-]{\text{Anode}} O \xrightarrow[\text{atoms}]{\text{2 such}} O_2 \uparrow$$

From the above equation it can be easily interpreted that in electrolysis of water for the production of 1 mole of oxygen from 1 mole of H_2O at anode 4 moles electrons are required.

Likewise for the production of 3 moles of O_2 $12(3 \times 4)$ moles of electrons will be needed.

So, the total amount of charge required to produce 3 moles of oxygen will be $12 \times F$ or 12×96500

We know $Q = it$

So, $12 \times 96500 = 100 \times t$ in seconds

or $\dfrac{12 \times 96500}{100 \times 3600} = t$ in hours $= 3.2$ hours

JEE Advanced

1. *(b, c, d)* Rate constant, $k = Ae^{-E_a/RT}$ where,

E_a = activation energy and

A = pre exponential factor

(a) If E_a is high, it means lower value of k hence, slow reaction. Thus, incorrect.

(b) On increasing temperature, molecules are raised to higher energy (greater than E_a), hence number of collisions increases. Thus, correct.

(c) $\log k = \log A - \dfrac{E_a}{RT} \Rightarrow \dfrac{d(\log k)}{dT} = \dfrac{E_a}{RT^2}$

Thus, when E_a is high, stronger is the temperature dependence of the rate constant. Thus, correct.

(d) Pre-exponential factor (A) is a measure of rate at which collisions occur. Thus, correct.

2. *(9)* Moles of solute, $n_1 = \dfrac{w_1}{m_1}$; Moles of solvent, $n_2 = \dfrac{w_2}{m_2}$

χ_1 (solute) = 0.1 and χ_2 (solvent) = 0.9

$$\therefore \quad \frac{\chi_1}{\chi_2} = \frac{n_1}{n_2} = \frac{w_1}{m_1} \cdot \frac{m_2}{w_2} = \frac{1}{9}$$

$$\text{Molarity} = \frac{\text{Solute (moles)}}{\text{Volume (L)}} = \frac{w_1 \times 1000 \times 2}{m_1 (w_1 + w_2)}$$

$$\textbf{Note } \text{Volume} = \frac{\text{Total mass of solution}}{\text{Density}} = \left(\frac{w_1 + w_2}{2}\right) \text{mL}$$

$$\text{Molality} = \frac{\text{Solute (moles)}}{\text{Solvent (kg)}} = \frac{w_1 \times 1000}{m_1 \times w_2}$$

Given, molarity = molality

$$\text{hence,} \quad \frac{2000 w_1}{m_1 (w_1 + w_2)} = \frac{1000\, w_1}{m_1\, w_2}$$

$$\therefore \quad \frac{w_2}{w_1 + w_2} = \frac{1}{2}$$

$$\Rightarrow \quad w_1 = w_2 = 1$$

$$\therefore \quad \frac{w_1\, m_2}{m_1\, w_2} = \frac{1}{9}$$

$$\Rightarrow \quad \frac{m_1 \text{(solute)}}{m_2 \text{(solvent)}} = 9$$

3. *(d)* Oxidation at anode

$$H_2(g) \longrightarrow 2H^+(aq) + 2e^-;\ E^\circ_{SHE} = 0.00 \text{ V}$$

Reduction at cathode

$$M^{4+}(aq) + 2e^- \longrightarrow M^{2+}(aq);$$

$$E^\circ_{M^{4+}/M^{2+}} = 0.151 \text{ V}$$

Net: $M^{4+}(aq) + H_2(g) \longrightarrow M^{2+}(aq) + 2H^+(aq);$

$$K = \frac{[M^{2+}][H^+]^2}{[M^{4+}]\, p_{H_2}} \quad (E^\circ_{cell} = 0.151 \text{ V})$$

$$= \frac{[M^{2+}]}{[M^{4+}]}$$

$$E_{cell} = E^\circ_{cell} - \frac{0.059}{2} \log K$$

$$0.092 = 0.151 - \frac{0.059}{2} \log \frac{[M^{2+}]}{[M^{4+}]}$$

$$0.059 = \frac{0.059}{2} \log 10^x$$

$$\therefore \quad \log 10^x = 2$$

$$\therefore \quad x = 2$$

4. *(b,c,d)*

(a) Nearest neighbour in the topmost layer of ccp structure is 9 thus, incorrect.

(b) Packing efficiency is 74% thus, correct.

(c) Tetrahedral voids = 2

Octahedral voids = 1 per atom thus, correct.

(d) Edge length, $a = \dfrac{4}{\sqrt{2}} r = 2\sqrt{2} r$

thus, correct explanation

Edge length $= a$, Radius $= r$

$$AC^2 = AB^2 + BC^2\ (4r)^2 = a^2 + a^2 = 2a^2$$

$$4r = \sqrt{2}a \Rightarrow r = \frac{\sqrt{2}}{4} a = \frac{a}{2\sqrt{2}}$$

$$\therefore \quad a = 2\sqrt{2}\, r$$

In ccp structure, the number of spheres is 4.

Hence, volume of 4 spheres $= 4\left(\dfrac{4}{3}\pi r^3\right)$

Total volume of unit cell $= a^3 = (2\sqrt{2}r)^3$ % of packing efficiency

$$= \frac{\text{Volume of 4 spheres}}{\text{Volume of unit cell}}$$

$$= \frac{4\left(\frac{4}{3}\pi r^3\right)}{[2(\sqrt{2}r)]^3} \times 100 = 74.05\% \simeq 74\%$$

5. *(a, b)* When intermolecular attraction between two components A and B in the mixture is same as between A and A or B and B, hence it is a case of ideal solution.

When intermolecular attraction between A and B in a mixture is smaller than that between A and A or B and B, then mixture is more vaporised, bp is lowered. It is a case of positive deviation from Raoult's law.

When intermolecular attraction between A and B is higher than that between A and A or B and B, then mixture is less vaporised, bp is increased. It is a case of negative deviation.

(a) Methanol molecules (CH_3OH) are hydrogen bonded. In a mixture of CCl_4 and CH_3OH, extent of H-bonding is decreased. Mixture is more vaporised thus, positive deviation from Raoult's law.

(b) Acetone molecules have higher intermolecular attraction due to dipole-dipole interaction. With CS_2, this interaction is decreased thus, positive deviation.

(c) Mixture of benzene and toluene forms ideal solution.

(d) Phenol and aniline have higher interaction due to intermolecular H-bonding. Hence, negative deviation.

6. *(a,c)* The graph shown indicates that there is positive deviation because the observed vapour pressure of L is greater than the ideal pressure

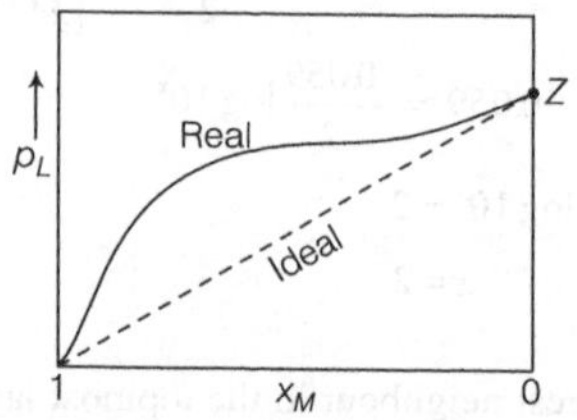

Since, deviation is positive, the intermolecular force between L and M is smaller than the same in pure L and pure M.

Also as $x_L \to 1, x_M \to 0$, the real curve approaching ideal curve where Raoult's law will be obeyed.

7. *(2)* Density $(\rho) = 8 = \dfrac{4 \times M}{N_A (4 \times 10^{-8} \text{cm})^3}$

$$\Rightarrow \quad M = 128 \times 10^{-24} N_A$$

$$\Rightarrow \quad \text{No of atoms} = \frac{256}{M} \times N_A$$

$$= \frac{256}{128 \times 10^{-24} N_A} \times N_A = 2 \times 10^{24}$$

8. *(6)* $\text{pH} = C\alpha = 10^{-4} \Rightarrow \alpha = \dfrac{10^{-4}}{0.0015}$

Also, conductance $(G) = \kappa\left(\dfrac{A}{l}\right)$

$$\Rightarrow \quad \kappa = G\left(\frac{l}{A}\right) = 5 \times 10^{-7} \times \frac{120}{1} = 6 \times 10^{-5}$$

$$\Rightarrow \quad \Lambda^c = \frac{\kappa \times 1000}{C} = \frac{6 \times 10^{-5} \times 1000}{0.0015}$$

$$\Rightarrow \Lambda^\infty = \frac{\Lambda^c}{\alpha} = \frac{6 \times 10^{-5} \times 1000}{0.0015} \times \frac{0.0015}{10^{-4}}$$

$$= 600 = 6 \times 10^2 \text{ S cm}^{-1}\text{mol}^{-1}$$

9. *(b)* $-\Delta T_f = ik_f m_2$

$$= 1 \times 2 \times \frac{34.5}{46 \times 500} \times 1000 = 3$$

Vapour pressure curves shown in *(b)* is in agreement with the calculated value of $-\Delta T_f$. *(a)* is wrong, vapour pressure decreases on cooling.

10. *(a)* The redox reaction is :

$$Zn(s) + Cu^{2+} \longrightarrow Zn^{2+} + Cu$$

The Nernst equation is

$$E = E^\circ - \frac{2.303RT}{2F} \log 10$$

$$= 1.1 - \frac{2.303RT}{2F}$$

Also, $\Delta G = -nEF = -2F\left(1.1 - \dfrac{2.303RT}{2F}\right)$

$$= -2.2F + 2.303RT$$

$$= 2.303RT - 2.2F$$

11. *(a, c)* If steric factor is considered, the corrected Arrhenius equation will be

$k = pAe^{\frac{-E_a}{RT}}$ where A = frequency factor by Arrhenius.

$\because p > 1, pA > A$ hence, (a) is correct.

Activation energy is not related to steric factor.

12. *(3)* The unit cell of initial structure of ionic solid MX looks like

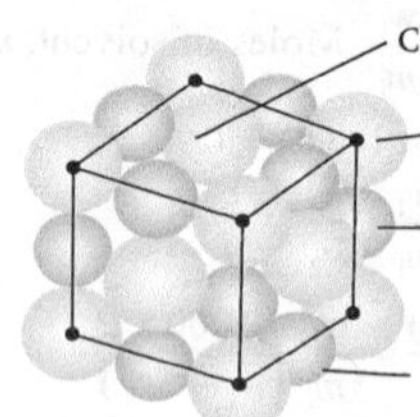

In NaCl type of solids cations (Na^+) occupy the octahedral voids while anions (Cl^-) occupy the face centre positions.

However, as per the demand of problem the position of cations and anions are swapped.

We also know that (for 1 unit cell)

(A) Total number of atoms at FCC = 4

(B) Total number of octahedral voids = 4

(as no. of atoms at FCC = No. of octahedral voids)

Now taking the conditions one by one

(i) If we remove all the anions except the central one than number of left anions.

$$= 4 - 3 = 1$$

(ii) If we replace all the face centred cations by anions than effective number of cations will be $= 4 - 3 = 1$
Likewise effective number of anions will be $= 1 + 3 = 4$

(iii) If we remove all the corner cations then effective number of cations will be $1 - 1 = 0$

(iv) If we replace central anion with cation then effective number of cations will be $0 + 1 = 1$
Likewise effective number of anions will be $4 - 1 = 3$

Thus, as the final outcome, total number of cations present in Z after fulfilling all the four sequential instructions = 1

Likewise, total number of anions = 3

Hence, the value of $\frac{\text{Number of anions}}{\text{Number of cations}} = \frac{3}{1} = 3$

13. *(10)* Equation of cell reaction according to the cell notation given, is

Given, $E^\circ_{cell} = 2.70\text{ V}, T = 300\text{ K}$

with $[Mg^{2+}(aq)] = 1\text{ M}$ and $[Cu^{2+}(aq)] = 1\text{ M}$

and $n = 2$

Further, $E_{cell} = 2.67\text{ V}$

with $[Cu^{2+}(aq)] = 1\text{M}$ and $[Mg^{2+}(aq)] = x\text{M}$

and $\frac{F}{R} = 11500\text{KV}^{-1}$

where F = Faraday constant, R = gas constant

From the formula,

$$E_{cell} = E^\circ_{cell} - \frac{RT}{nF}\ln\frac{[Mg^{2+}(aq)]}{[Cu^{2+}(aq)]}$$

After putting the given values

$$2.67 = 2.70 - \frac{RT}{2F}\ln\frac{x}{1}$$

or $$2.67 = 2.70 - \frac{R \times 300}{2F} \times \ln x$$

$$-0.03 = \frac{-R \times 300}{2F} \times \ln x$$

or $$\ln x = \frac{0.03 \times 2}{300} \times \frac{F}{R}$$

$$= \frac{0.03 \times 2 \times 11500}{300} = 2.30$$

So, $\ln x = 2.30$ or $x = 10$ (as given $\ln(10) = 2.30$)

14. *(19)* **Key Idea** *Use the formula*

$$p_{Total} = p^\circ_A \times \chi_A + p^\circ_B \times \chi_B$$

and for equimolar solutions $\chi_A = \chi_B = 1/2$

Given, $p_{Total} = 45$ torr for equimolar solution

$p^\circ_A = 20$ torr

So, $$45 = p^\circ_A \times \frac{1}{2} + p^\circ_B \times \frac{1}{2} = \frac{1}{2}(p^\circ_A + p^\circ_B)$$

or $p^\circ_A + p^\circ_B = 90$ torr ...(i)

But we know $p^\circ_A = 20$ torr

so, $p^\circ_B = 90 - 20 = 70$ torr (From Eq. (i))

Now, for the new solution from the same formula

$p_{Total} = 22.5$ torr

$$22.5 = 20\chi_A + 70(1 - \chi_A) \text{ (As } \chi_A + \chi_B = 1)$$

or $$22.5 = 70 - 50\chi_A$$

So, $$\chi_A = \frac{70 - 22.5}{50} = 0.95$$

Thus $\chi_B = 1 - 0.95 = 0.05$ (as $\chi_A + \chi_B = 1$)

Hence, the ratio

$$\frac{\chi_A}{\chi_B} = \frac{0.95}{0.05} = 19$$

15. *(0.05)* From the graph we can note

ΔT_b for solution X i.e.,

$$\Delta T_{b(X)} = 362 - 360 = 2$$

Likewise, ΔT_b for solution Y i.e.,

$$\Delta T_{b(Y)} = 368 - 367 = 1$$

Now by using the formula

$$\Delta T_b = i \times \text{molality of solution} \times K_b$$

For solution X

$$2 = i \times m_{NaCl} \times K_{b(X)} \quad \text{...(i)}$$

Similarly for solution y

$$1 = i \times m_{NaCl} \times K_{b(Y)} \quad \text{...(ii)}$$

from Eq. (i) and (ii) above

$$\frac{K_{b(X)}}{K_{b(Y)}} = \frac{2}{1} \text{ or } 2 \text{ or } K_{b(X)} = 2K_{b(Y)}$$

For solute S

$2S \longrightarrow S_2$ (given due to dimerisation)

	$2S$	S_2
Initial	α	0
Final	$(1-\alpha)$	$\alpha/2$

So, here $$i = \left(1 - \frac{\alpha}{2}\right)$$

$$\Delta T_{b[X](s)} = \left(1 - \frac{\alpha_1}{2}\right)K_{b(X)}$$

$$\Delta T_{b[Y](s)} = \left(1 - \frac{\alpha_2}{2}\right)K_{b(Y)}$$

Given, $\Delta T_{b(X)(s)} = 3\Delta T_{b(Y)(s)}$

$$\left(1 - \frac{\alpha_1}{2}\right)K_{b(X)} = 3 \times \left(1 - \frac{\alpha_2}{2}\right) \times K_{b(Y)}$$

or $$2\left(1 - \frac{\alpha_1}{2}\right) = 3\left(1 - \frac{\alpha_2}{2}\right) \quad [\because K_{b(X)} = 2K_{b(Y)}]$$

or $$2\left(1 - \frac{\alpha_1}{2}\right) = 3\left(1 - \frac{0.7}{2}\right) \quad \text{(as given, } \alpha_2 = 0.7)$$

or $4 - 2\alpha_1 = 6 - 2.1$ or $2\alpha_1 = 0.1$

so, $\alpha_1 = 0.05$

16. *(a, d)* Given for the reaction (at T= 300 K and constant volume=V)

	$A(g)$	$\longrightarrow$	$2B(g)$	$+ C(g)$
at $t = 0$	p_0		–	–
at $t = t$	$p_0 - x$		$2x$	x
at $t = t_{1/3}$	$\left[p_0 - \frac{2p_0}{3}\right] = \frac{p_0}{3}$		$\frac{4p_0}{3}$	$\frac{2p_0}{3}$

We can calculate,

$$p_t = p_0 - x + 2x + x = p_0 + 2x$$

or $$2x = p_t - p_0 \text{ or } x = \frac{p_t - p_0}{2}$$

Now for first order reaction,

$$t = \frac{1}{k}\ln\frac{p_0}{(p_0 - x)}$$

Putting the value of x in the equation,

$$t = \frac{1}{k}\ln\frac{p_0}{p_0 - \left(\frac{p_t - p_0}{2}\right)}$$

$$= \frac{1}{k} \ln \frac{2p_0}{2p_0 - p_t + p_0}$$

or $$kt = \ln \frac{2p_0}{(3p_0 - p_t)}$$

or $$kt = \ln 2p_0 - \ln (3p_0 - p_t)$$

or $\ln (3p_0 - p_t) = -kt + \ln 2p_0$

It indicates graph between $\ln (3p_0 - p_t)$ *vs* '*t*' will be a straight line with negative slope , so option (a) is correct

$$t_{1/3} = \frac{1}{k} \ln \frac{p_0}{p_0/3} = \frac{1}{k} \ln 3$$

It indicates $t_{1/3}$ is independent of initial concentration so, option (b) is incorrect.

Likewise, rate constant also does not show its dependence over initial concentration. Thus, graph between rate constant and $[A]_0$ will be a straight line parallel to *X*-axis .

$$(a - x) \text{at time } t = \frac{yp_i - p_t}{(y - 1)}$$

17. ***(+ 8500 J/mol)***

For the reaction,

$$A(g) + B(g) \rightleftharpoons AB(g)$$

Given $$E_{a_b} = E_{a_f} + 2RT$$

or $$E_{a_b} - E_{a_f} = 2RT$$

Further $A_f = 4A_b$ or $\frac{A_f}{A_b} = 4$

Now, rate constant for forward reaction,

$$k_f = A_f e^{-E_{a_f}/RT}$$

Likewise, rate constant for backward reaction,

$$k_b = A_b e^{-E_{a_b}/RT}$$

At equilibrium,

Rate of forward reaction = Rate of backward reaction

i.e., $$k_f = k_b$$

or $$\frac{k_f}{k_b} = k_{eq}$$

so $$k_{eq} = \frac{A_f e^{-E_{a_f}/RT}}{A_b e^{-E_{a_b}/RT}}$$

$$= \frac{A_f}{A_b} e^{-(E_{a_f} - E_{a_b})/RT}$$

After putting the given values

$$k_{eq} = 4e^2 \quad (\text{as } E_{a_b} - E_{a_f} = 2RT \text{ and } \frac{A_f}{A_b} = 4)$$

Now, $$\Delta G^\circ = -RT \ln K_{eq}$$

$$= -2500 \ln(4e^2)$$
$$= -2500 (\ln 4 + \ln e^2)$$
$$= -2500 (1.4 + 2)$$
$$= -2500 \times 3.4$$
$$= -8500 \text{ J/mol}$$

Absolute value = 8500 J/mol

18. ***(– 11.62 JK^{-1} mol^{-1})***

Given, $A(s) | A^{n+}(aq, 2\text{ M}) \| B^{2n+}(aq, 1\text{ M}) | B(s)$

So, reactions at respective electrode will be

Anode $A(s) \longrightarrow A^{n+} + ne^- \times 2$

Cathode $B^{2n+} + 2ne^- \longrightarrow B(s)$

Overall reaction

$$2A(s) + B^{2n+}(aq) \longrightarrow 2A^{n+}(aq) + B(s)$$

Further,

$\Delta H^\circ = 2\Delta G^\circ$ and $E_{cell} = 0$ is also given

Now by using the Nernst equation

$$E_{cell} = E^\circ_{cell} - \frac{RT}{nF} \ln \frac{[\text{Product}]}{[\text{Reactant}]}$$

After putting the values

$$0 = E^\circ_{cell} - \frac{RT}{2nF} \ln \frac{[A^{n+}]^2}{[B^{2n+}]}$$

or $$E^\circ = \frac{RT}{2nF} \ln \frac{[2]^2}{[1]} = \frac{RT}{2nF} \ln 4 \quad \ldots \text{(i)}$$

Further from the formula,

$$\Delta G^\circ = -nFE^\circ$$
$$\Delta G^\circ = -2nFE^\circ$$

Now putting the value of E° from eq. (i)

$$\Delta G^\circ = -2nF \times \frac{RT}{2nF} \ln 4 \quad \ldots \text{(ii)}$$
$$\Delta G^\circ = -RT \ln 4$$

Finally, using the formula

$$\Delta G^\circ = \Delta H^\circ - T\Delta S^\circ$$
$$\Delta G^\circ = 2\Delta G^\circ - T\Delta S^\circ \; (\text{as } \Delta H^\circ = 2\Delta G^\circ, \text{ given})$$
$$\Delta G^\circ = T\Delta S^\circ$$

or $$\Delta S^\circ = \frac{\Delta G^\circ}{T} = \frac{-RT \ln 4}{T}$$

(from eq. (ii), $\Delta G^\circ = -RT \ln 4$)

$$= -R \ln 4 = -8.3 \times 2 \times 0.7$$

(as all values given)

$$= -11.62 \text{ J/K-mol}$$